W9-ATC-009

CALCULUS

CALCULUS

for the Managerial, Life, and Social Sciences

Ernest F. Haeussler, Jr.

Richard S. Paul

Department of Mathematics
The Pennsylvania State University

RESTON PUBLISHING COMPANY, INC.
A Prentice-Hall Company
Reston, Virginia

Library of Congress Cataloging in Publication Data

Haeussler, Ernest F
Calculus, for the managerial, life, and social sciences.

Includes index.
1. Calculus. I. Paul, Richard S., joint author.
II.Title.
QA303.H15 515 79-17419
ISBN 0-8359-0628-0

10 9 8 7 6 5 4

Printed in the United States of America.

Contents

Preface

This text is designed to provide a foundation in those topics of calculus that are relevant to students in the managerial, life, and social sciences. It begins with functions and progresses through both single variable and multivariable calculus. Technical proofs, conditions, etc., are sufficiently described but not overdone. An algebra review is given in an appendix and provides a convenient reference for students.

An abundance and variety of applications that are meaningful to the intended audience appear in the exposition, examples, and exercises. These cover such diverse areas as business, economics, biology, medicine, sociology, psychology, ecology, statistics, and archeology. An index of applications is provided on the endpapers. Many of these real-world situations are drawn from literature and are documented by references. Such documentation is indicated by a boldface number enclosed in brackets, such as **[1]**, and the source appears in the reference list at the back of the book. In many of the applications drawn from source material, the background and context are given in order to stimulate interest. As a result of blending applied problems throughout the text, students continually see ways in which the mathematics they are learning can be used. The text is virtually self-contained in the sense that it assumes no prior exposure to the concepts on which the applications are based.

Interspersed throughout the text are many warnings to the student that point out commonly made errors. These warnings are indicated under the format "PITFALL."

More than 2500 exercises are included. In each exercise set, the problems are given in increasing order of difficulty. However, there are always a sufficient number that the student can do without difficulty so that confidence may be gained and the basic concepts studied may be fixed firmly in mind. Answers to odd-numbered problems appear at the end of the book. For many of the differentiations problems in Chapter 3, the answers appear in both unsimplified and simplified forms. This allows students to readily check their procedures and work.

Each chapter has a review section that contains: a listing of important terms and symbols together with page numbers that indicate where these terms and symbols first appear; a programmed review; and numerous review problems.

As an aid in planning a course outline, we point out that the following sections are optional and may be omitted: 7-1, 7-5, 8-4, and 8-6. Certain sections of Chapter 1 may also be omitted at the discretion of the instructor. Should time not permit, any of the following sections may be deleted: 2-5, 5-7, 6-7, 7-3, 7-4, 7-7, 8-8, 8-9, and 8-10.

Available from the publisher is an extensive instructor's manual that contains answers to all problems and detailed solutions to a great many of them, including all applied problems. Included also are problems suitable for examinations.

We express our appreciation to the following colleagues who contributed comments and suggestions that were valuable to us in the development of the manuscript: Ron Atkinson of the *University of Tennessee at Nashville*, Wilson Banks of *Illinois State University*, Anne Grams of the *University of Tennessee at Nashville*, Robert Hathway of *Illinois State University*, Burt C. Horne, Jr. of *Virginia Polytechnic Institute and State University*, Jon Laible of *Eastern Illinois University*, Ralph McWilliams of *Florida State University*, Glenda K. Owens of *Central State University*, and Franklin E. Schroeck, Jr., of *Florida Atlantic University*.

We are indebted to Jerry Covert of the Department of Biology of *The Pennsylvania State University* for his review and comments relative to biological applications.

In addition, we especially wish to acknowledge and sincerely thank those colleagues who reviewed the entire manuscript and offered many helpful suggestions for its improvement: John Bishir of *North Carolina State University at Raleigh*, Micheal Dyer of the *University of Oregon*, Burt C. Horne, Jr., of *Virginia Polytechnic Institute and State University*, Franklin E. Schroeck, Jr., of *Florida Atlantic University*, and Donald R. Sherbert of the *University of Illinois at Urbana–Champaign*.

Ernest F. Haeussler, Jr.
Richard S. Paul

CALCULUS

CHAPTER 1

Functions

1-1 FUNCTIONS

In 1694 Gottfried Wilhelm Leibniz, one of the developers of calculus, introduced the word *function* into the mathematical vocabulary. The concept of a "function" is no doubt one of the most basic in all of mathematics.

To introduce it, let us consider the equation

$$y = x + 2.$$

Replacing x by various numbers, we get corresponding values of y. For example,

$$\text{if } x = 0, \quad \text{then } y = 0 + 2 = 2;$$
$$\text{if } x = 1, \quad \text{then } y = 1 + 2 = 3.$$

Notice that for *each* value of x that "goes into" the equation, only *one* value of y "comes out."

Think of the equation $y = x + 2$ as defining a rule: Add 2 to x. This rule assigns to each *input number* x exactly one *output number* y:

$$\underset{\text{input number}}{x} \longrightarrow \underset{\text{output number}}{y} \quad (= x + 2).$$

We call this rule a *function* in the following sense:

DEFINITION. *A* ***function*** *is a rule that assigns to each input number exactly one output number. The set of all input numbers to which the rule applies is called the* **domain** *of the function. The set of all output numbers is called the* **range**.

In general the inputs or outputs need not be numbers. For example, a table of U.S. cities and their populations assigns to each city (which is not a number) its population (exactly one output). Nonetheless, for now we shall use the word "function" in the restricted sense of having a domain and range consisting of only numbers.

For the function given by $y = x + 2$, if the input x is any real number, then $x + 2$ is defined. Unless otherwise stated, the domain of a function consists of all real numbers for which the function is defined. Thus the domain of the above function is all real numbers. To the input number 0 is assigned the output number 2:

$$x \to y \; (= x + 2),$$
$$0 \to 2 \; (= 0 + 2).$$

Thus 2 is in the range.

A variable that represents input numbers for a function is called an **independent variable.** One that represents output numbers is a **dependent variable** because its value *depends* on the value of the independent variable. We say that the dependent variable is a *function of* the independent variable. Thus, in the equation $y = x + 2$ the independent variable is x, the dependent variable is y, and y is a function of x.

Not all equations define y as a function of x, as Example 1 shows.

EXAMPLE 1

Let $y^2 = x$.

a. Suppose x is an input number, say $x = 9$. Then $y^2 = 9$ and so $y = \pm 3$. Thus, with the input number 9 there are assigned not one but *two* output numbers, $+3$ and -3. Hence y **is not** a function of x.

b. Now, suppose y is an input number:

$$y \to x\,(= y^2).$$

It determines exactly one output number, x. For example, if $y = 3$, then $x = y^2 = 3^2 = 9$. Thus x is a function of y. Since y can be any real number, the domain is all real numbers. The independent variable is y, and the dependent variable is x.

In some cases the domain of a function is restricted for physical or economic reasons, as Example 2 will show.

EXAMPLE 2

Suppose that consumers will buy (that is, demand) q units of a certain product per week at a price of p dollars per unit, where $p = 100/q$. This equation is called a *demand equation* for the product. If q is an input number, then to each value of q there is assigned exactly one output number p:

$$q \to \frac{100}{q} = p.$$

For example,

$$20 \to \frac{100}{20} = 5.$$

Thus price p is a function of quantity demanded q. Here q is the independent variable and p is the dependent variable. Since q cannot be 0 (division by 0) and cannot be negative (q represents quantity), the domain is all values of q such that $q > 0$. This function is called a **demand function.**

Usually, letters such as f, g, h, F, G, etc. are used to name functions. Suppose we let f represent the function defined by $y = x + 2$. Then the notation

$f(x)$, which is read "f of x,"

means the *output number* corresponding to the input number x:

$$\underbrace{f(\overset{\text{input}}{\overset{\downarrow}{x}})}_{\text{output}}.$$

Thus $f(x)$ is the same as y. But since $y = x + 2$ we may write

$$f(x) = x + 2. \qquad (1)$$

To find $f(3)$, the output corresponding to the input 3, replace x in Eq. (1) by 3:

$$f(3) = 3 + 2 = 5.$$

Similarly,

$$f(8) = 8 + 2 = 10,$$
$$f(-4) = -4 + 2 = -2,$$
$$f(0) = 0 + 2 = 2.$$

Sometimes output numbers, such as $f(3)$, $f(8)$, etc., are called *functional values*. They are in the range of f.

PITFALL. $f(x)$ ***does not*** *mean "f times x."*

Quite often, functions are defined by "functional notation." For example, $g(x) = x^3 + x^2$ defines the function g which assigns to an input number x the output number $x^3 + x^2$. This correspondence is represented by the following notation:

$$g: x \to x^3 + x^2.$$

Some functional values are

$$g(0) = 0^3 + 0^2 = 0,$$
$$g(2) = 2^3 + 2^2 = 12,$$
$$g(-1) = (-1)^3 + (-1)^2 = -1 + 1 = 0,$$
$$g(t) = t^3 + t^2,$$
$$g(x + 1) = (x + 1)^3 + (x + 1)^2.$$

Note that $g(x + 1)$ was found by replacing each x in $x^3 + x^2$ by the input $x + 1$. That is, g adds the cube and the square of an input number.

When referring to the function g defined by $g(x) = x^3 + x^2$, we shall use some literary freedom and speak of "the function $g(x) = x^3 + x^2$." Similarly, we speak of the function $y = x + 2$.

Special names are given to functions having particular forms, as the next examples indicate.

EXAMPLE 3

a. The function $f(x) = 2x + 3$ is called a **linear function**. It has the form $f(x) = ax + b$ where a and b are constants and $a \neq 0$. Its domain is all real numbers. Some functional values are

$$f(0) = 2(0) + 3 = 3,$$
$$f(t + 7) = 2(t + 7) + 3 = 2t + 14 + 3 = 2t + 17.$$

b. The equation $g(x) = 2$ defines a *constant function*. For any input the output is 2. The domain of g is all real numbers and the range is 2. For example,

$$g(.10) = 2; \quad g(-420) = 2; \quad g(0) = 2.$$

In general, a **constant function** g is one of the form $g(x) = c$ where c is a fixed real number.

c. $y = h(x) = -3x^2 + x - 5$ is a **quadratic function**, that is, one of the form $h(x) = ax^2 + bx + c$ where a, b, and c are constants and $a \neq 0$. The domain of h is all real numbers, and y is a function of x. Examples of functional values are

$$h(2) = -3(2)^2 + 2 - 5 = -15,$$
$$h\left(\frac{1}{t}\right) = -3\left(\frac{1}{t}\right)^2 + \left(\frac{1}{t}\right) - 5 = -\frac{3}{t^2} + \frac{1}{t} - 5,$$
$$h(r^2) = -3(r^2)^2 + (r^2) - 5 = -3r^4 + r^2 - 5,$$
$$h(x + t) - h(x) = \left[-3(x + t)^2 + (x + t) - 5\right] - [-3x^2 + x - 5]$$
$$= -6xt - 3t^2 + t.$$

PITFALL. *Do not be confused by notation. If $f(x) = x^2$, then $f(x + h) = (x + h)^2$.* ***Do not*** *write the function and add h:*

$$f(x + h) \neq x^2 + h.$$

Also, ***do not*** *use the distributive law on $f(x + h)$. It is not a multiplication:*

$$f(x + h) \neq f(x) + f(h).$$

EXAMPLE 4

A function of the form

$$f(x) = c_0x^n + c_1x^{n-1} + \ldots + c_{n-1}x + c_n,$$

where n is a positive integer and $c_0, c_1, \ldots, c_n$ are constants with $c_0 \neq 0$, is called a **polynomial function** (in x). The term with the greatest power of x is c_0x^n. The exponent n is called the **degree** of the function, and c_0 is the **leading coefficient**. Thus $f(x) = 3x^2 - 8x + 9$ is a polynomial function of degree 2 and has a leading coefficient of 3. Likewise, $g(x) = 3x^3 - 7x^8$ has degree 8 and leading coefficient -7. Constant functions are also considered polynomial functions. A nonzero constant function, such as $f(x) = 5$, is said to have degree 0. The domain of a polynomial function is all real numbers.

All of the functions in Example 3 are polynomial functions. However, those in Example 5 are not.

EXAMPLE 5

a. $f(t) = \frac{2t}{t^2 - 1}$.

Here the input number is t. The domain of f is all real numbers except ± 1 (to avoid division by 0). Some functional values are

$$f\left(\frac{1}{2}\right) = \frac{2(\frac{1}{2})}{(\frac{1}{2})^2 - 1} = -\frac{4}{3}, \qquad f(-t) = \frac{2(-t)}{(-t)^2 - 1} = -\frac{2t}{t^2 - 1}.$$

b. $g(x) = \sqrt{x}$.

We want input numbers to give rise only to real output numbers. For $\sqrt{x}$ to be a real number, x cannot be negative. Thus the domain of g is all $x \geqslant 0$. Examples of functional values are

$$g(0) = \sqrt{0} = 0, \qquad g(4) = \sqrt{4} = 2,$$

$$\frac{g(x + h) - g(x)}{h} = \frac{\sqrt{x + h} - \sqrt{x}}{h}.$$

c. $h(q) = \sqrt{q - 3}$.

Here we must have $q - 3 \geqslant 0$. Thus the domain of h is all $q \geqslant 3$. Some functional values are

$$h(3) = \sqrt{3 - 3} = 0, \qquad h(5) = \sqrt{5 - 3} = \sqrt{2},$$

$$h(q^2) = \sqrt{q^2 - 3}.$$

d. $f(x) = x^{2/3}$.

This function can be written $f(x) = (\sqrt[3]{x})^2$. The domain is all real numbers. Some functional values are

$$f(8) = (\sqrt[3]{8})^2 = 2^2 = 4,$$

$$f(-8) = (\sqrt[3]{-8})^2 = (-2)^2 = 4,$$

$$f(2000) = (\sqrt[3]{2000})^2 = (\sqrt[3]{1000}\,\sqrt[3]{2})^2 = (10\sqrt[3]{2})^2$$
$$= (10)^2\sqrt[3]{2^2} = 100\sqrt[3]{4}.$$

EXAMPLE 6

The function

$$F(s) = \begin{cases} 1, & \text{if } -1 \leqslant s < 1, \\ 0, & \text{if } 1 \leqslant s \leqslant 2, \\ s - 3, & \text{if } 2 < s \leqslant 3 \end{cases}$$

is defined in three parts. Here s represents input numbers and the domain of F is all s such that $-1 \leqslant s \leqslant 3$. The value of an input number determines which part to use.

Find $F(0)$: *Since $-1 \leqslant 0 < 1$, we have $F(0) = 1$.*
Find $F(2)$: *Since $1 \leqslant 2 \leqslant 2$, we have $F(2) = 0$.*
Find $F(9/4)$: *Since $2 < 9/4 \leqslant 3$, we substitute $s = 9/4$ in $s - 3$.*

$$F\left(\frac{9}{4}\right) = \frac{9}{4} - 3 = -\frac{3}{4}.$$

The symbol $|x|$ is read "**absolute value of x**" and is defined:

$$|x| = \begin{cases} x, & \text{if } x > 0, \\ 0, & \text{if } x = 0, \\ -x, & \text{if } x < 0. \end{cases}$$

For example, since $2 > 0$, $|2| = 2$. Since $-4 < 0$, $|-4| = -(-4) = 4$. Of course, $|0| = 0$. Notice that $|x| \geqslant 0$ for all x.

EXAMPLE 7

$f(x) = |x|$ is called the *absolute value function*. Its domain is all real numbers. Some functional values are

$$f(16) = |16| = 16,$$
$$f\left(-\frac{4}{3}\right) = \left|-\frac{4}{3}\right| = -\left(-\frac{4}{3}\right) = \frac{4}{3},$$
$$f(0) = |0| = 0,$$
$$f(2x + 3) = |2x + 3|,$$
$$f(x^2 + 1) = |x^2 + 1| = x^2 + 1, \text{ since } x^2 + 1 > 0 \text{ for all } x.$$

EXAMPLE 8

Suppose two black guinea pigs are bred and produce exactly five offspring. Under certain conditions it can be shown that the probability P that exactly r of the offspring will be brown and the others black is a function of r, say

$P = P(r)$, where

$$P(r) = \frac{5!\left(\frac{1}{4}\right)^r\left(\frac{3}{4}\right)^{5-r}}{r!\,(5-r)!}, \qquad r = 0, 1, 2, \ldots, 5.$$

Here the domain of P is all integers from 0 to 5 inclusive. The symbol $r!$, where r is a positive integer, is read "r factorial." It represents the product $r(r-1)(r-2)\cdots(1)$. Thus $5! = 5\cdot4\cdot3\cdot2\cdot1 = 120$. We define $0!$ to be 1. Find the probability that exactly three guinea pigs will be brown.

We want to find $P(3)$.

$$P(3) = \frac{5!\left(\frac{1}{4}\right)^3\left(\frac{3}{4}\right)^2}{3!2!}$$

$$= \frac{120\left(\frac{1}{64}\right)\left(\frac{9}{16}\right)}{6(2)} = \frac{45}{512}.$$

For some equations containing two variables, either variable may be considered a function of the other. For example, if $y = 2x$, then y is a function of x and x is the independent variable. Letting f denote this function, we have $y = f(x) = 2x$. However, since $x = y/2$, then x is a function of y where y is the independent variable. If g denotes this function, then $x = g(y) = y/2$.

We have seen that a function is essentially a *correspondence* whereby to each input number in the domain there is assigned exactly one output number in the range. The correspondence given by $f(x) = x^2$ is shown by the arrows in Fig. 1-1.

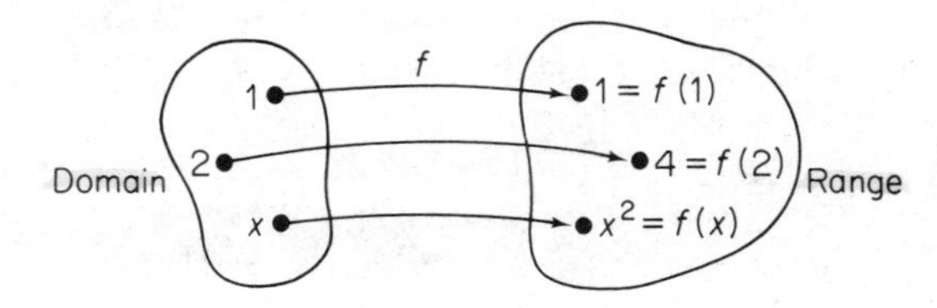

FIG. 1-1

EXAMPLE 9

The table in Fig. 1-2 is a *supply schedule*. It gives a correspondence between the price p of a certain product and the quantity q that producers will supply per week at that price.

If p is the independent variable, then q is a function of p, say $q = f(p)$, and

$$f(500) = 11, \quad f(600) = 14, \quad f(700) = 17, \quad \text{and} \quad f(800) = 20.$$

Supply schedule

p Price per unit in dollars	q Quantity supplied per week
500	11
600	14
700	17
800	20

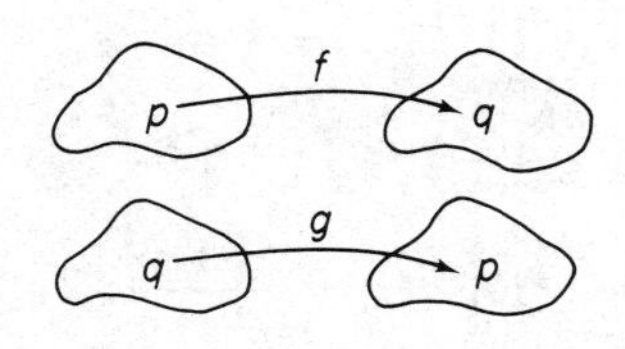

FIG. 1-2

If q is the independent variable, then p is a function of q, say $p = g(q)$, and

$$g(11) = 500, \quad g(14) = 600, \quad g(17) = 700, \quad \text{and} \quad g(20) = 800.$$

We speak of f and g as **supply functions**. Notice from the supply schedule that as price per unit increases, the producers are willing to supply more units per week.

Exercise 1-1

For each function in Problems **1–26**, *determine the independent variable and domain. Also find the indicated functional expressions.*

1. $f(x) = 4x$; $f(0)$, $f(3)$, $f(-\frac{1}{4})$, $f(t)$.
2. $F(x) = x + 8$; $F(4)$, $F(-\frac{1}{2})$, $F(12)$, $F(-3)$, $F(x_1)$.
3. $g(x) = 1 - 2x$; $g(0)$, $g(-u)$, $g(7)$, $g(-2x)$, $g(x + h)$.
4. $h(x) = -\frac{9}{2}$; $h(0)$, $h[(13)^2]$, $h(t)$, $h(x + 1)$.
5. $G(t) = 1.02$; $G(5)$, $G(-107.3)$, $G(x^2)$, $G(2 + x)$.
6. $F(s) = 2(4 - s)$; $F(0)$, $F\left(\frac{1}{s}\right)$, $F(1 + \frac{1}{3})$, $F\left(\frac{s}{2}\right)$.
7. $q = h(p) = \dfrac{3(4p + 1)}{2}$; $h(1)$, $h\left(\frac{p}{2}\right)$, $h\left(\frac{1}{p}\right)$.
8. $g(x) = 3x^2$; $g(-4)$, $g(-3u)$, $g(x^3)$, $g(2/x)$.
9. $f(p) = p^2 + 2p + 1$; $f(0)$, $f(2)$, $f(x_1)$, $f(w)$, $f(p + h)$.
10. $x = H(y) = 2y^2 - 3y + 1$; $H(1)$, $H(-\frac{1}{2})$, $H(z)$, $H(z + 1)$.
11. $y = G(t) = (t + 4)^2$; $G(0)$, $G(2)$, $G(2 + h)$, $\dfrac{G(2 + h) - G(2)}{h}$.
12. $F(x) = |x - 3|$; $F(10)$, $F(3)$, $F(-3)$, $F(4t + 2)$.
13. $f(q) = |2q - 7|$; $f(6)$, $f(2)$, $f(7/2)$, $f(x^2 + 5)$.

14. $s = h(t) = \sqrt{6t}$; $h(0)$, $h(6)$, $h(\frac{2}{3})$, $h(6t)$.

15. $H(x) = \sqrt{4 + x}$; $H(0)$, $H(-4)$, $H(-3)$, $H(x + 1) - H(x)$.

16. $y = F(t) = \dfrac{t}{t - 3}$; $F(0)$, $F(4)$, $F(-1)$, $F(t + 2)$.

17. $g(s) = \dfrac{4}{s^2 - 9}$; $g(1)$, $g(-1)$, $g(-2w)$, $g(s - 1)$, $g(s) - 1$.

18. $h(z) = \left(\dfrac{z + 1}{z - 1}\right)^2$; $h(0)$, $h(1)$, $h(-\frac{1}{2})$, $h(z - 1)$.

19. $f(x) = \begin{cases} 4, & \text{if } x \geqslant 0 \\ 3, & \text{if } x < 0 \end{cases}$; $f(3)$, $f(-4)$, $f(\frac{17}{3})$, $f(-7.3)$.

20. $H(x) = \begin{cases} 1, & \text{if } x > 1 \\ x + 1, & \text{if } -1 \leqslant x \leqslant 1 \\ 1, & \text{if } x < -1 \end{cases}$; $H(7)$, $H(-7)$, $H(.5)$, $H(-\frac{1}{2})$.

21. $h(r) = \begin{cases} 3r - 1, & \text{if } r > 2 \\ r^2 - 4r + 7, & \text{if } r < -2 \end{cases}$; $h(3)$, $h(-3)$, $h(5)$, $h(-5)$.

22. $y = g(x) = \dfrac{1}{x - 2} + \dfrac{1}{x + 3}$; $g(-2)$, $g(3)$, $g(0)$.

23. $f(x) = x^{4/3}$; $f(0)$, $f(64)$, $f(\frac{1}{8})$, $f(-16)$.

24. $g(x) = x^{2/5}$; $g(1)$, $g(32)$, $g(-64)$, $g(t^{10})$.

25. $y = f(x) = \dfrac{1}{\sqrt{x}}$; $f(1)$, $f(\sqrt{16})$.

26. $F(t) = \dfrac{2}{3(2t^2 - 3t - 5)}$; $F(1)$, $F(-2)$, $F(t + 1)$, $F(2t)$.

27. If $z = 4x^2$, can z be considered a function of x? Can x be considered a function of z?

28. If $2p = 3q - 2$, can p be considered a function of q? Can q be considered a function of p?

In Problems **29–32**, *find* $\dfrac{f(x + h) - f(x)}{h}$.

on test or quiz

29. $f(x) = 3x - 4$.

30. $f(x) = \dfrac{x}{2}$.

31. $f(x) = x^2 + 2x$.

32. $f(x) = 2x^2 - 3x - 5$.

33. If q units of a certain product are sold (q is nonnegative), the profit P is given by the equation $P = 1.25q$. Is P a function of q? What is the dependent variable; the independent variable?

34. If a principal of P dollars is invested at a simple annual interest rate of r for t years, express the total accumulated amount of the principal and interest as a function of t. Is your result a linear function of t?

35. In manufacturing a component for a machine, the initial cost of a die is \$850 and all other additional costs are \$3 per unit produced. Express the total cost C (in dollars) as a function of the number q of units produced.

36. In Example 8 find the probability that all five offspring will be brown.

37. Under certain conditions, if two brown-eyed parents have exactly three children, the probability P that there will be exactly r blue-eyed children is given by the function $P = P(r)$, where

$$P(r) = \frac{3!\left(\frac{1}{4}\right)^r\left(\frac{3}{4}\right)^{3-r}}{r!\,(3-r)!}, \qquad r = 0, 1, 2, 3.$$

Find the probability that exactly two of the children will be blue-eyed.

38. Table 1-1 is called a *demand schedule*. It gives a correspondence between the price p of a product and the quantity q that consumers will demand (that is, purchase) at that price. (a) If $p = f(q)$, list the numbers in the domain of f. Find $f(2900)$ and $f(3000)$. (b) If $q = g(p)$, list the numbers in the domain of g. Find $g(10)$ and $g(17)$.

TABLE 1-1 Demand Schedule

p PRICE PER UNIT IN DOLLARS	q QUANTITY DEMANDED PER WEEK
10	3000
12	2900
17	2300
20	2000

39. Bacteria are growing in a culture. The time t (in hours) for the number of bacteria to double in number (generation time) is a function of the temperature T (in °C) of the culture. If this function is given by (adapted from **[23]**)

$$t = f(T) = \begin{cases} \frac{1}{24}T + \frac{11}{4}, & \text{if } 30 \leqslant T \leqslant 36, \\ \frac{4}{3}T - \frac{175}{4}, & \text{if } 36 < T \leqslant 39, \end{cases}$$

(a) determine the domain of f, and (b) find $f(30)$, $f(36)$, and $f(39)$.

40. In a paired-associate learning experiment **[26]**, the probability of a correct response as a function of the number n of trials has the form

$$P(n) = 1 - \frac{1}{2}(1-c)^{n-1}, \qquad n \geqslant 1,$$

where the estimated value of c is .344. Find $P(1)$ and $P(2)$ by using this value of c.

41. A psychophysical experiment (adapted from **[1]**) was conducted to analyze human response to electrical shocks. The subjects received a shock of a certain intensity. They were told to assign a magnitude of 10 to this particular shock, called the standard stimulus. Then other shocks (stimuli) of various intensities were given. For each one the response R was to be a number that indicated the perceived magnitude of the shock relative to that of the standard stimulus. It was found that R was a function of the intensity I of the shock (I in microamperes) and was estimated by

$$R = f(I) = \frac{I^{4/3}}{2500}, \qquad 500 \leqslant I \leqslant 3500.$$

Evaluate (a) $f(1000)$ and (b) $f(2000)$. (c) Suppose I_0 and $2I_0$ are in the domain of f. Express $f(2I_0)$ in terms of $f(I_0)$. What effect does the doubling of intensity have on response?

42. In examining the records of a group of individuals hospitalized for a particular illness, it was found that the total proportion who had been discharged at the end of t days of hospitalization is given by $f(t)$, where

$$f(t) = 1 - \left(\frac{300}{300 + t}\right)^3.$$

Evaluate (a) $f(0)$, (b) $f(100)$, and (c) $f(300)$. (d) At the end of how many days was .999 of the group discharged?

1-2 COMBINATIONS OF FUNCTIONS

If f and g are functions we can combine them to create new functions. For example, suppose

$$f(x) = x^2 \quad \text{and} \quad g(x) = x + 1.$$

Adding $f(x)$ and $g(x)$ in the obvious way gives

$$f(x) + g(x) = x^2 + (x + 1).$$

This sum defines a new function—let us call it H:

$$H: x \to f(x) + g(x) = x^2 + (x + 1).$$

Thus,

$$H(x) = f(x) + g(x) = x^2 + x + 1.^{\dagger}$$

†We assume that x is in the domains of both f and g.

Similarly we can create other functions:

$$f(x) - g(x) = x^2 - (x + 1),$$
$$f(x) \cdot g(x) = x^2(x + 1),$$
$$\frac{f(x)}{g(x)} = \frac{x^2}{x + 1}, \quad \text{if } g(x) \neq 0.$$

EXAMPLE 1

Let $f(x) = 3x - 1$ *and* $g(x) = x^2 + 3x + 3$.

a. $f(x) + g(x) = (3x - 1) + (x^2 + 3x + 3)$

$$= x^2 + 6x + 2 = H(x).$$

Thus, $f(2) + g(2) = H(2) = 2^2 + 6(2) + 2 = 18$.

b. $f(x) - g(x) = (3x - 1) - (x^2 + 3x + 3) = -4 - x^2$.

c. $f(x) \cdot g(x) = (3x - 1)(x^2 + 3x + 3) = 3x^3 + 8x^2 + 6x - 3$.

d. $\dfrac{f(x)}{g(x)} = \dfrac{3x - 1}{x^2 + 3x + 3}$.

EXAMPLE 2

Let $f(x) = \sqrt{x}$, $g(x) = 6x + 1$, *and* $H(x) = f(x) + g(x)$.

a. $H(4x) = f(4x) + g(4x)$

$$= \sqrt{4x} + 6(4x) + 1 = 2\sqrt{x} + 24x + 1.$$

b. $H(x + h) = f(x + h) + g(x + h) = \sqrt{x + h} + 6(x + h) + 1$.

It is possible to combine functions in yet another way. In Fig. 1-3 we see that x is in the domain of g. Applying g to x, we get the number

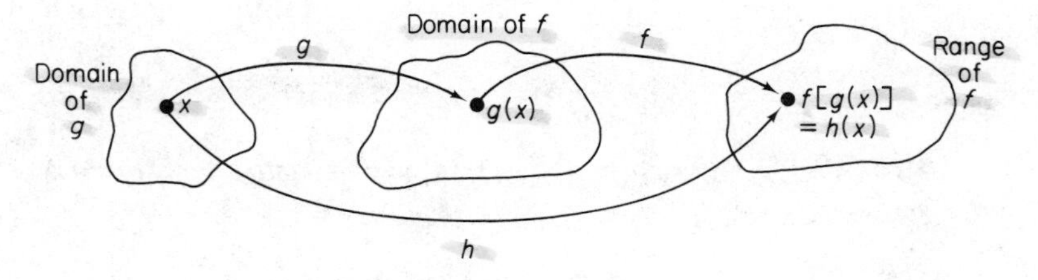

FIG. 1-3

$g(x)$, which we shall assume is the domain of f. By applying f to $g(x)$ we get $f[g(x)]$, which is in the range of f. This procedure of applying g and then f defines a so-called "composite" function h. This function assigns to the input number x the output number $f[g(x)]$. Thus $h(x) = f[g(x)]$.

DEFINITION. *If f and g are functions, then the* ***composition of f with g*** *is the function h defined by*

$$h(x) = f[g(x)],$$

where the domain of h is the set of all x in the domain of g such that $g(x)$ is in the domain of f.

EXAMPLE 3

Let $f(x) = \sqrt{x}$ and $g(x) = x + 1$. Find the following compositions.

a. $f[g(x)]$.

f takes the square root of an input number. But the input number here is $g(x)$ or $x + 1$. Thus,

$$f[g(x)] = f[x + 1] = \sqrt{x + 1}.$$

The domain of g is all real numbers x and the domain of f is all nonnegative reals. Hence the domain of the composition is all x for which $g(x) = x + 1$ is nonnegative. That is, the domain is all $x \geqslant -1$.

b. $g[f(x)]$.

g adds 1 to any input number. But the input number is $f(x)$ or $\sqrt{x}$. Thus g adds 1 to $\sqrt{x}$.

$$g[f(x)] = g[\sqrt{x}] = \sqrt{x} + 1.$$

The domain of f is all $x \geqslant 0$ and the domain of g is all reals. Hence the domain of the composition is all $x \geqslant 0$ for which $f(x) = \sqrt{x}$ is real, namely all $x \geqslant 0$.

From Example 3 we see that $f[g(x)] \neq g[f(x)]$.

PITFALL. *Do not confuse $f[g(x)]$ with the product $f(x) \cdot g(x)$. If $f(x) = \sqrt{x}$ and $g(x) = x + 1$, then*

$$f[g(x)] = \sqrt{x + 1},$$

$$\text{but} \quad f(x) \cdot g(x) = \sqrt{x}\,(x + 1).$$

EXAMPLE 4

If $F(p) = p^2 + 4p - 3$ and $G(p) = 2p + 1$, find $F[G(p)]$ and $G[F(p)]$.

a. $F[G(p)] = F[2p + 1] = (2p + 1)^2 + 4(2p + 1) - 3 = 4p^2 + 12p + 2.$

b. $G[F(p)] = 2(p^2 + 4p - 3) + 1 = 2p^2 + 8p - 5.$

EXAMPLE 5

If $f(x) = 5x^2 - 2$ and $g(x) = \sqrt{10 - x^2}$, find $f[g(-1)]$ and $g[f(2)]$.

a. $f[g(-1)] = f\left[\sqrt{10 - (-1)^2}\right] = f[3] = 5(3)^2 - 2 = 43.$

b. $g[f(2)] = g[5(2)^2 - 2] = g[18] = \sqrt{10 - (18)^2}$, which is not a real number. Thus you must be careful with a composition; it may not be defined.

EXAMPLE 6

If $f(s) = \dfrac{1}{s - 1}$ and $g(t) = t^2 + t + 1$, find $f[g(t)]$ and $g[f(s)]$.

a. $$f[g(t)] = f[t^2 + t + 1] = \frac{1}{(t^2 + t + 1) - 1} = \frac{1}{t^2 + t} = \frac{1}{t(t + 1)}.$$

b. $$g[f(s)] = g\left[\frac{1}{s - 1}\right] = \left(\frac{1}{s - 1}\right)^2 + \frac{1}{s - 1} + 1 = \frac{1 + (s - 1) + (s - 1)^2}{(s - 1)^2} = \frac{s^2 - s + 1}{(s - 1)^2}.$$

EXAMPLE 7

The function $y = (x^2 + 2x + 3)^3$ can be considered a composition. If we let

$$f(x) = x^3 \quad \text{and} \quad g(x) = x^2 + 2x + 3,$$

then

$$f[g(x)] = f[x^2 + 2x + 3] = (x^2 + 2x + 3)^3.$$

Thus $y = f[g(x)]$.

Exercise 1-2

In each of Problems **1–6**, *find*

a. $f(x) + g(x)$, b. $f(x) - g(x)$, c. $f(x) \cdot g(x)$,

d. $\dfrac{f(x)}{g(x)}$, e. $f[g(x)]$, f. $g[f(x)]$.

1. $f(x) = x + 5$, $g(x) = x + 4$.

2. $f(x) = 3$, $g(x) = -1$.

3. $f(x) = 3x + 4$, $g(x) = x^2 - 1$.

4. $f(x) = x^2$, $g(x) = x^3 + 1$.

5. $f(x) = x^2 + 3x - 4$, $g(x) = 2x^2 - 7$.

6. $f(x) = 1/x$, $g(x) = 4x + 5$.

Problems **7–9** *refer to the following functions*:

$H(x) = f(x) + g(x)$, $F(x) = f(x) - g(x)$,

$G(x) = f(x) \cdot g(x)$, $Q(x) = \dfrac{f(x)}{g(x)}$.

7. If $f(x) = 6x - 5$ and $g(x) = 4x + 1$, find

a. $H(3)$, b. $H(2x)$, c. $F(-1)$, d. $F(-w)$,

e. $G(\frac{1}{2})$, f. $G(x + 1)$, g. $Q(0)$, h. $Q(t^2)$.

8. $f(x) = 4x$ and $g(x) = x^2 + 6x - 1$, find

a. $H(1)$, b. $H(2x + 1)$, c. $F(-3)$, d. $F(x/2)$,

e. $G(-.5)$, f. $G(1/x)$, g. $Q(0)$, h. $Q(-z^2)$.

9. If $f(x) = 2x^2 + 3$ and $g(x) = 1 - 3x$, find

a. $H(-2)$, b. $H(x + h)$, c. $F(-2)$, d. $F(x + h)$,

e. $G(0)$, f. $G(\sqrt{x})$, g. $Q(1)$, h. $Q(2p/3)$.

10. If $f(x) = 4x$ and $g(x) = x^2 + 6x - 1$, find $f[g(1)]$ and $g[f(1)]$.

11. If $f(x) = 2x^2 + 3$ and $g(x) = 1 - 3x$, find $f[g(2)]$ and $g[f(2)]$.

12. If $f(p) = \dfrac{4}{p}$ and $g(p) = \dfrac{p - 2}{3}$, find $f[g(p)]$ and $g[f(p)]$.

13. If $F(t) = t^2 + 3t + 1$ and $G(t) = \dfrac{2}{t - 1}$, find $F[G(t)]$ and $G[F(t)]$.

14. If $F(s) = \sqrt{s}$ and $G(t) = 3t^2 + 4t + 2$, find $F[G(t)]$ and $G[F(s)]$.

15. If $f(w) = \dfrac{1}{w^2 + 1}$ and $g(v) = \sqrt{v + 2}$, find $f[g(v)]$ and $g[f(w)]$.

16. If $f(x) = x^2 + 3$, find $f[f(x)]$.

17. Let $f(x) = 3x + 8$ and $g(x) = 2$. (a) Find $f[g(x)]$ and give the domain and range of this composition. (b) Find $g[f(x)]$ and give the domain and range of this composition.

In Problems **18–24**, *find functions f and g such that* $h(x) = f[g(x)]$.

18. $h(x) = (x^2 + 2)^2$

19. $h(x) = \sqrt{x - 2}$.

20. $h(x) = \sqrt[5]{\dfrac{x + 1}{3}}$.

21. $h(x) = \left(\dfrac{4x - 5}{x^2 + 1}\right)^{2/3}$.

22. $h(x) = (x^2 - 1)^2 + 2(x^2 - 1)$.

23. $h(x) = (3x^3 - 2x)^3 - (3x^3 - 2x)^2 + 7$.

24. $h(x) = \dfrac{x + 1}{(x + 1)^2 + 2}$.

25. Studies have been conducted **[28]** concerning the statistical relations among a person's status, education, and income. Let S denote a numerical value of status based on annual income I. For a certain population, suppose

$$S = f(I) = .45(I - 1000)^{.53}.$$

Furthermore, suppose a person's income I is a function of the number E of years of education, where

$$I = g(E) = 7202 + .29E^{3.68}.$$

Find $f[g(E)]$. What does this function describe?

1-3 GRAPHS OF FUNCTIONS

Functions can be represented in a coordinate plane. If f is a function and x is the independent variable, then the **graph** of f is all points $(x, f(x))$, where x is in the domain of f.

For example, let us graph

$$f(x) = x^2 + 2x - 3.$$

We begin by substituting values for x.

If $x = 1$, then $f(x) = 1^2 + 2(1) - 3 = 0$. Thus the point $(1, 0)$ is on the graph. Similarly, if $x = -2$, then $f(x) = (-2)^2 + 2(-2) - 3 = -3$, and so $(-2, -3)$ is also on the graph. By choosing other values for x we get more points [see table in Fig. 1-4(a)]. In Fig. 1-4(b) we have plotted the points given in the table.

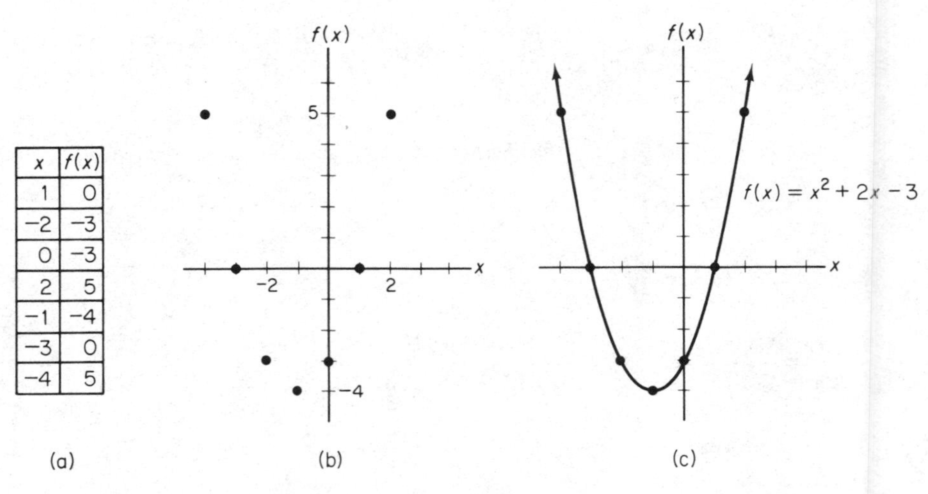

x	$f(x)$
1	0
−2	−3
0	−3
2	5
−1	−4
−3	0
−4	5

FIG. 1-4

But the graph has infinitely many points, so it seems impossible to determine the graph precisely. However, we are concerned only with the graph's general shape. For this reason we locate only enough points so that we may guess its general behavior. Then we join these points by a smooth curve wherever conditions permit. We start with the point having the least x-coordinate, namely $(-4, 5)$, and progress through the points having increasingly larger x-coordinates. We finish with the point having the greatest x-coordinate, namely $(2, 5)$. See Fig. 1-4(c). Of course, the more points we plot, the better is our graph. Here we assume that the graph extends indefinitely upward, which is indicated by the arrows.

In a later chapter you will see that calculus is a *great* aid in graphing because it helps to determine the "shape" of a graph. It provides powerful techniques for determining whether or not a curve "wiggles" between points.

EXAMPLE 1

a. *Graph* $f(x) = \sqrt{x}$.

The domain of f is all $x \geqslant 0$. See Fig. 1-5.

b. *Graph* $p = G(q) = |q|$ *(absolute value function).*

We use the independent variable q to label the horizontal axis. The vertical axis can be labeled either $G(q)$ or p. See Fig. 1-6.

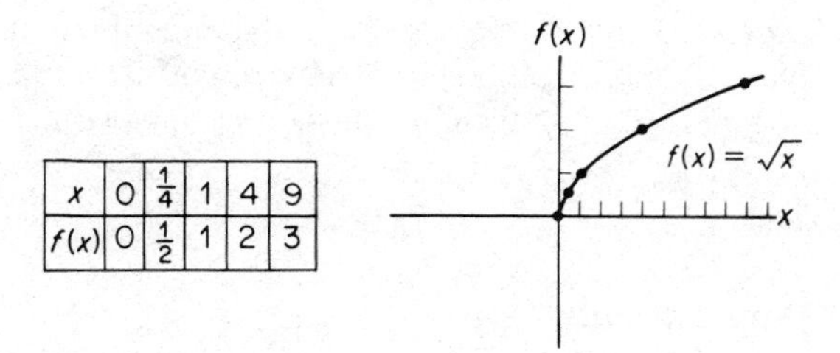

x	0	$\frac{1}{4}$	1	4	9
$f(x)$	0	$\frac{1}{2}$	1	2	3

FIG. 1-5

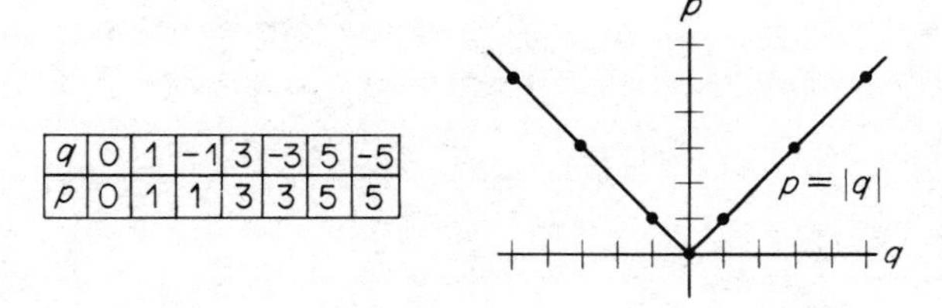

q	0	1	−1	3	−3	5	−5
p	0	1	1	3	3	5	5

FIG. 1-6

c. *Graph* $s = f(t) = \dfrac{100}{t}$.

Using t for the horizontal axis and s for the vertical, we get Fig. 1-7. Notice that since 0 is not in the domain of f, the graph has no point on the s-axis.

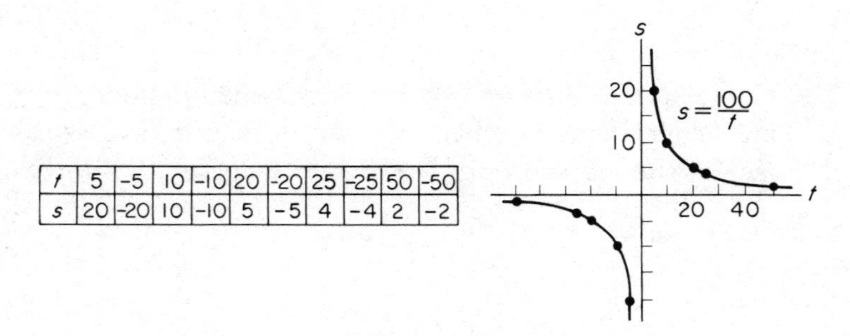

t	5	-5	10	-10	20	-20	25	-25	50	-50
s	20	-20	10	-10	5	-5	4	-4	2	-2

FIG. 1-7

EXAMPLE 2

In a discussion of contemporary waters of shallow seas, Odum **[36]** *states that in such waters the total organic matter y (in milligrams per liter) is a function of species diversity x (in number of species per thousand individuals). If* $y = f(x) = 100/x$, *graph f for* $x > 0$.

Observe that the function is identical in form to that of Example 1(c). Here, however, $x > 0$. That is, the independent variable is restricted to positive values. The graph of this function would be identical to the portion of the graph in Fig. 1-7 which is in the first quadrant.

EXAMPLE 3

Sketch the graph of

$$f(x) = \begin{cases} x, & \text{if } 0 \leqslant x < 3, \\ x - 1, & \text{if } 3 \leqslant x \leqslant 5, \\ 4, & \text{if } 5 < x \leqslant 7. \end{cases}$$

The domain of f is $0 \leqslant x \leqslant 7$. The graph is given in Fig. 1-8, where the *hollow dot* means that the point is **not** included in the graph.

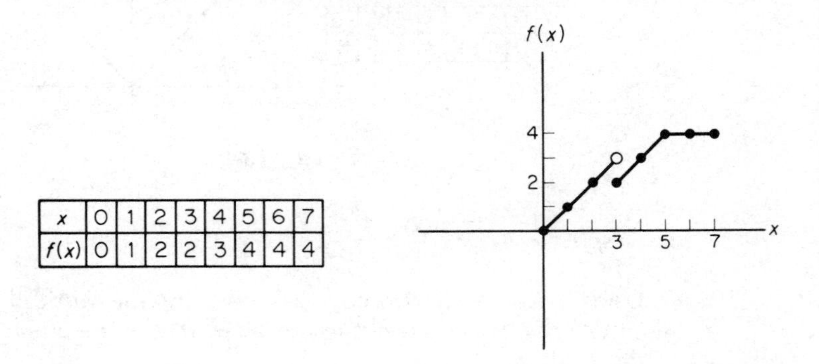

x	0	1	2	3	4	5	6	7
$f(x)$	0	1	2	2	3	4	4	4

FIG. 1-8

Figure 1-9 shows the graph of some function $y = f(x)$. Corresponding to the input number x on the horizontal axis is the output number $f(x)$ on the vertical axis. For example, $f(4) = 3$. From the graph it seems

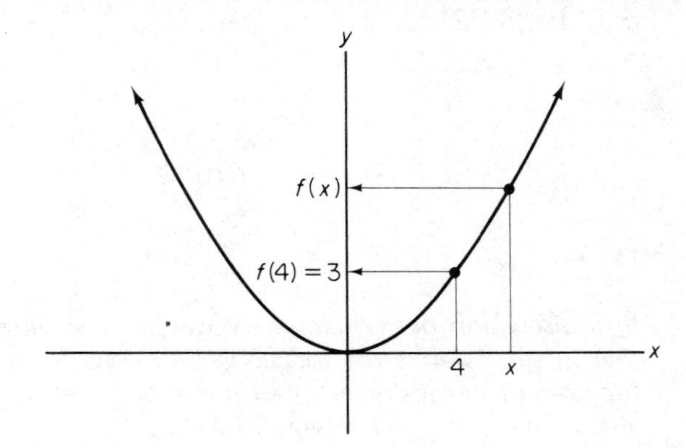

FIG. 1-9

reasonable to assume that there is an output number for any value of x and that the domain of f is all real numbers. Notice that the y-coordinates of all points on the graph are nonnegative, and for any $y \geqslant 0$ there is an x such that $y = f(x)$. Thus the range of f is all $y \geqslant 0$. This shows that we may make an "educated" guess about the domain and range of a function by looking at its graph. For example, from Fig. 1-6 we conclude that the domain of $p = G(q) = |q|$ is all real numbers and the range is all nonnegative numbers ($p \geqslant 0$). From Fig. 1-7 it is clear that both the domain and range of $f(t) = 100/t$ are all real numbers except 0.

EXAMPLE 4

Figure 1-10 shows the graph of a function F. To the right of 4 assume that the graph repeats itself indefinitely. Thus the domain of F is all $t \geqslant 0$. The range is $-1 \leqslant s \leqslant 1$. Some functional values are

$$F(0) = 0, \qquad F(1) = 1, \qquad F(2) = 0, \qquad F(3) = -1.$$

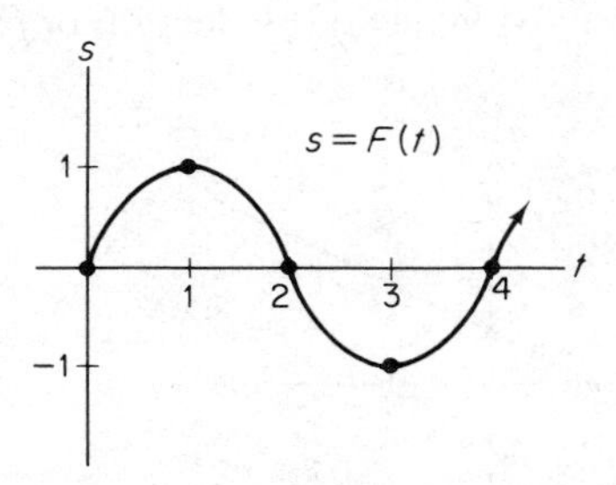

FIG. 1-10

In the leftmost diagram in Fig. 1-11, notice that with the given x there are associated *two* values of y—namely y_1 and y_2. Thus the curve *is not* the graph of a function of x.

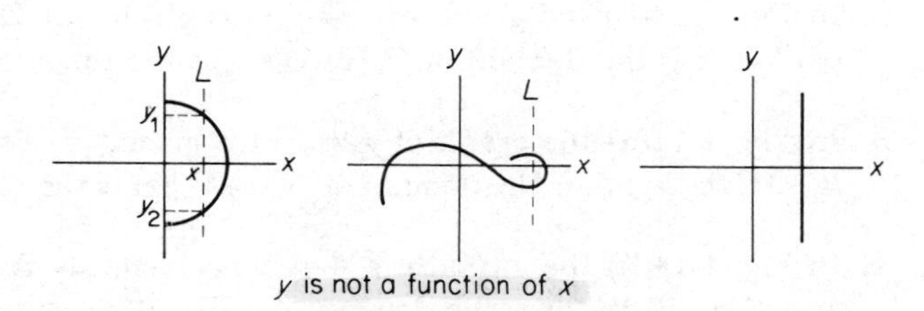

FIG. 1-11

In general, if a *vertical* line L can be drawn which meets a curve in at least two points, then the curve *is not* the graph of a function of x. When no such vertical line can be drawn, the curve *is* the graph of a function of x. Thus the curves in Fig. 1-11 do not represent functions of x, but those in Fig. 1-12 do.

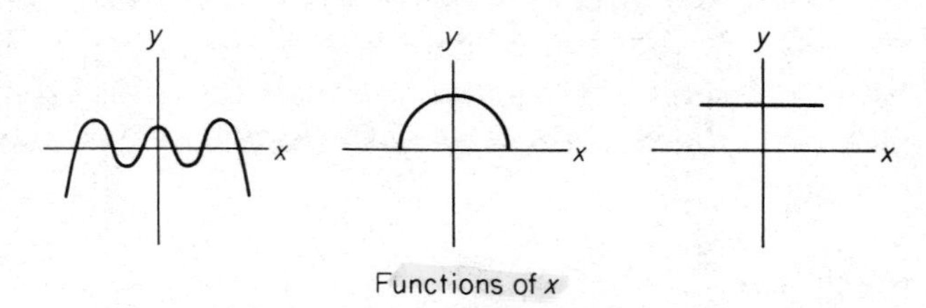

Functions of x

FIG. 1-12

Exercise 1-3

1. In Fig. 1-13(a) the graph of $y = f(x)$ is given. (a) Estimate $f(0)$, $f(2)$, $f(4)$, and $f(-2)$. (b) What is the domain of f? (c) What is the range of f?

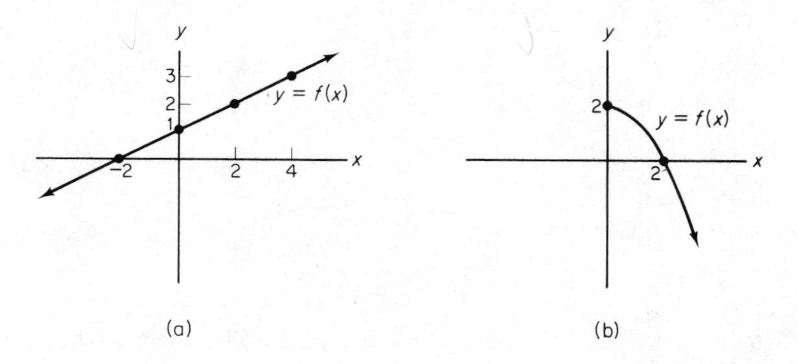

(a) (b)

FIG. 1-13

2. In Fig. 1-13(b) the graph of $y = f(x)$ is given. (a) Estimate $f(0)$ and $f(2)$. (b) What is the domain of f? (c) What is the range of f?

3. In Fig. 1-14(a) the graph of $y = f(x)$ is given. (a) Estimate $f(0)$, $f(1)$, and $f(-1)$. (b) What is the domain of f? (c) What is the range of f?

4. In Fig. 1-14(b) the graph of $y = f(x)$ is given. (a) Estimate $f(0)$, $f(2)$, $f(3)$, and $f(4)$. (b) What is the domain of f? (c) What is the range of f?

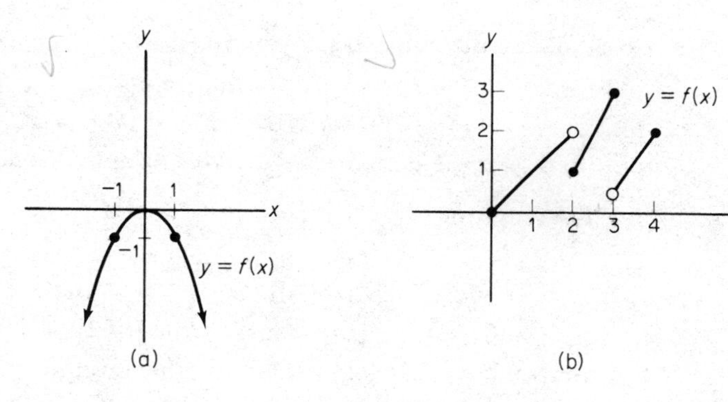

FIG. 1-14

In Problems **5–20**, *sketch the graph of each function and give the domain and range.*

5. $f(x) = 2x + 2$.

6. $g(x) = 4 - x$.

7. $h(x) = (x + 1)^2$.

8. $f(x) = 5 - 2x^2$.

9. $y = g(x) = 2$.

10. $G(s) = -8$.

11. $y = h(x) = x^2 - 4x + 1$.

12. $y = f(x) = x^2 + 2x - 8$.

13. $f(t) = -t^3$.

14. $p = h(q) = q(2 - q)$.

15. $s = F(r) = \sqrt{r - 5}$.

16. $F(r) = -\dfrac{1}{r}$.

17. $f(x) = |2x - 1|$.

18. $v = H(u) = |u - 3|$.

19. $F(t) = \dfrac{16}{t^2}$.

20. $y = f(x) = \dfrac{2}{x - 4}$.

21. Graph $c = \begin{cases} p, & \text{if } 0 \leqslant p < 2, \\ 2, & \text{if } p \geqslant 2, \end{cases}$ and label the horizontal axis with p.

22. Graph $f(x) = \begin{cases} 2x + 1, & \text{if } -1 \leqslant x < 2, \\ 9 - x^2, & \text{if } x \geqslant 2. \end{cases}$

23. Graph $g(x) = \begin{cases} x + 6, & \text{if } x \geqslant 3, \\ x^2, & \text{if } x < 3. \end{cases}$

24. Graph $f(x) = \begin{cases} x + 1, & \text{if } 0 < x \leqslant 3, \\ 4, & \text{if } 3 < x \leqslant 5, \\ x - 1, & \text{if } x > 5. \end{cases}$

25. Unicellular protein coagulates (denatures) at temperatures T above 60°C. This is indicated by noticing a decrease in solubility of the protein in a certain solution for $T > 60$. Suppose y is the percentage of protein that remains soluble at time t. If y is estimated by

$$y = f(T) = \begin{cases} 100, & \text{if } 20 \leqslant T \leqslant 60, \\ 160 - T, & \text{if } 60 < T \leqslant 90, \end{cases}$$

graph f and give the domain and range. (Adapted from [40].)

26. Sketch the graph of

$$y = f(x) = \begin{cases} -100x + 600, & \text{if } 0 \leqslant x < 5, \\ -100x + 1100, & \text{if } 5 \leqslant x < 10, \\ -100x + 1600, & \text{if } 10 \leqslant x < 15. \end{cases}$$

A function such as this might describe the inventory y of a company at time x.

27. In a psychological experiment on visual information, a subject briefly viewed an array of letters and was then asked to recall as many letters from the array as possible. The procedure was repeated several times. Suppose y is the average number of letters recalled from arrays with x letters. The graph of the results is approximately the graph of (adapted from [30])

$$y = f(x) = \begin{cases} x, & \text{if } 0 \leqslant x \leqslant 4, \\ \frac{1}{2}x + 2, & \text{if } 4 < x \leqslant 5, \\ 4.5, & \text{if } 5 < x \leqslant 12. \end{cases}$$

Graph this function.

28. Table 1-2 is called a *demand schedule*. It indicates the quantities of Brand X that consumers will demand (that is, purchase) each week at alternative prices per unit (in dollars). Plot each quantity-price pair by choosing the vertical axis for the possible prices. Connect the points with a smooth curve. In this way we approximate points in between the given data. The result is called a *demand curve*. From the graph determine the relationship between the price of Brand X and the amount that will be demanded. (That is, as price decreases, what happens to the quantity demanded?) Is price per unit a function of quantity demanded?

TABLE 1-2 Demand Schedule

q QUANTITY DEMANDED	p PRICE PER UNIT
5	20
10	10
20	5
25	4

29. Given the supply schedule (see Example 9 on page 8) in Table 1-3, plot each quantity-price pair by choosing the horizontal axis for the possible quantities. Approximate the points in between the data by connecting the data points with a smooth curve. Thus you get a *supply curve*. From the graph determine the relationship between price and supply. (That is, as price increases, what happens to the quantity supplied?) Is price per unit a function of quantity supplied?

TABLE 1-3 Supply Schedule

p PRICE PER UNIT IN DOLLARS	q QUANTITY SUPPLIED PER WEEK
10	30
20	100
30	150
40	190
50	210

1-4 LINEAR FUNCTIONS

In Fig. 1-15 the points (x_1, y_1) and (x_2, y_2) lie on the straight line. Using

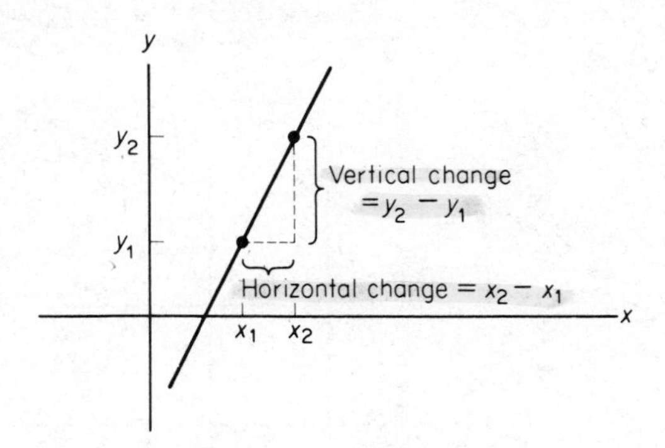

FIG. 1-15

these points we define the *slope* of the line.

DEFINITION. *Suppose that (x_1, y_1) and (x_2, y_2) are two different points on a line. The* **slope** *m of the line is the number (if it exists)*

$$m = \frac{y_2 - y_1}{x_2 - x_1}\left(= \frac{\text{vertical change}}{\text{horizontal change}}\right).$$

For example, a slope of 3 means that for each one-unit increase in the x-coordinate of a point, the y-coordinate *increases* by 3. See Fig. 1-16(a). On the other hand, a slope of -3 means that for each one-unit

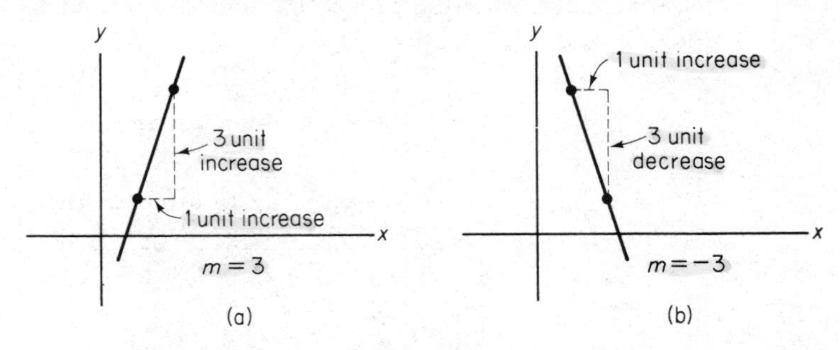

FIG. 1-16

increase in the x-coordinate of a point, the y-coordinate *decreases* by 3. See Fig. 1-16(b).

EXAMPLE 1

To find the slope of the *horizontal* line through (2, 2) and (3, 2) [see Fig. 1-17], we let $(x_1, y_1) = (2, 2)$ and $(x_2, y_2) = (3, 2)$. Then

$$m = \frac{y_2 - y_1}{x_2 - x_1} = \frac{2 - 2}{3 - 2} = \frac{0}{1} = 0.$$

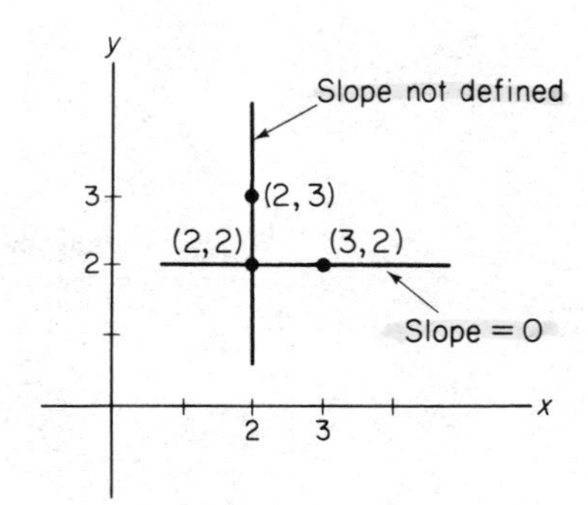

FIG. 1-17

The slope of the *vertical* line through (2, 2) and (2, 3) is [see Fig. 1-17]

$$m = \frac{3 - 2}{2 - 2} = \frac{1}{0}, \text{ which is not defined.}$$

In fact, **the slope of every horizontal line is 0** and **the slope of every vertical line is not defined.**

Basically, slope indicates the "steepness" of a line. Figure 1-18

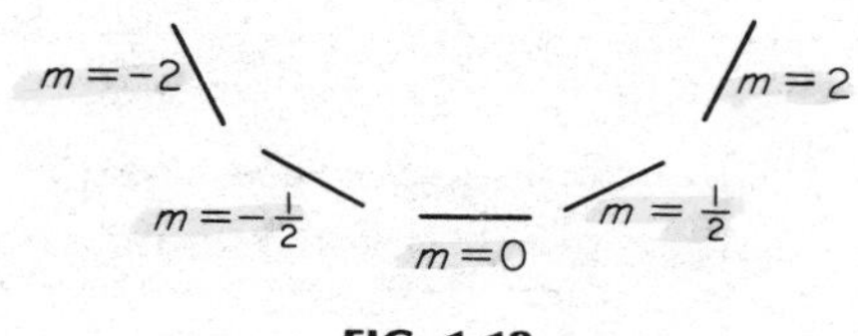

FIG. 1-18

shows some lines together with their slopes. Notice that

1. A horizontal line has slope 0.
2. A line rising from left to right has positive slope.
3. A line falling from left to right has negative slope.
4. The closer the slope is to 0, the more nearly horizontal is the line.
5. The greater the absolute value of the slope, the more nearly vertical is the line.

Suppose that line L has slope m, that it passes through (x_1, y_1), and that (x, y) is *any* other point on L (see Fig. 1-19). We can find a

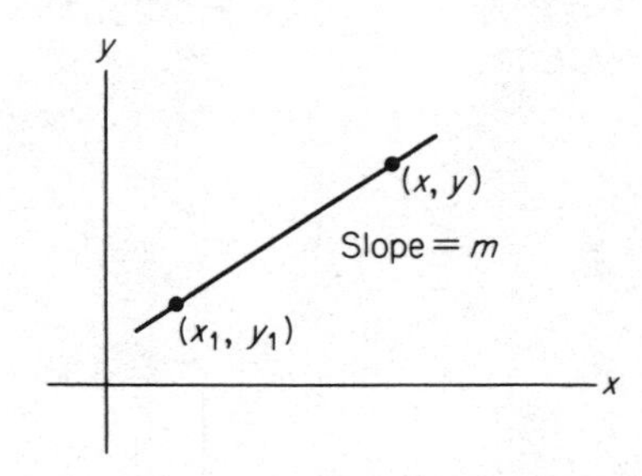

FIG. 1-19

relationship between x and y. By the slope formula,

$$\frac{y - y_1}{x - x_1} = m,$$

$$y - y_1 = m(x - x_1). \tag{1}$$

That is, every point on L satisfies Eq. (1). It is also true that any point satisfying Eq. (1) must lie on L. Thus we say that

$$y - y_1 = m(x - x_1)$$

is the **point-slope form** of an equation of the line through (x_1, y_1) and having slope m.

EXAMPLE 2

Determine an equation of the line that has slope 2 *and passes through* (1, −3).

Here $m = 2$ and $(x_1, y_1) = (1, -3)$. Using a point-slope form,

$$y - (-3) = 2(x - 1),$$
$$y + 3 = 2x - 2,$$
$$y = 2x - 5.$$

A point $(0, b)$ where a graph intersects the y-axis is called a ***y*-intercept** (Fig. 1-20). If the slope and y-intercept of a line L are

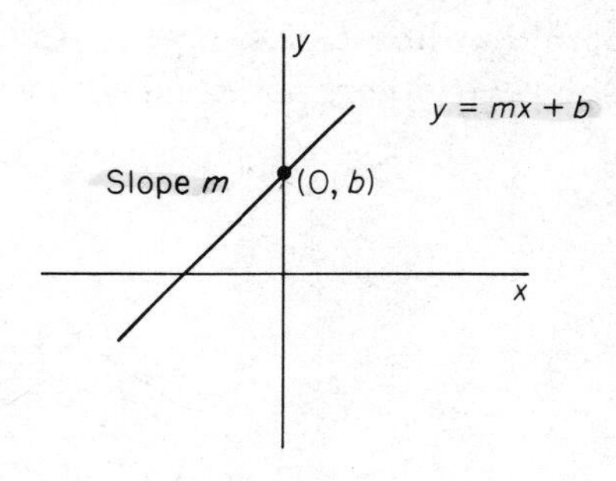

FIG. 1-20

known, an equation for L is (using a point-slope form)

$$y - b = m(x - 0).$$

Solving for y, we have

$$y = mx + b$$

which is the **slope-intercept form** of an equation of the line with slope m and y-intercept $(0, b)$.

EXAMPLE 3

a. An equation of the line with slope 3 and y-intercept $(0, -4)$ is

$$y = mx + b,$$
$$y = 3x + (-4),$$
$$y = 3x - 4.$$

b. The equation $y = 5(x - 3)$ can be written $y = 5x - 15$, which has the form $y = mx + b$ where $m = 5$ and $b = -15$. Thus its graph is a line with slope 5 and y-intercept $(0, -15)$.

In Sec. 1-1 (Example 3a) a *linear function* was described. More formally we have the following definition.

DEFINITION. *A function f is a* **linear function** *if and only if $f(x)$ can be written in the form $f(x) = ax + b$, where a and b are constants and $a \neq 0$.*

Suppose $f(x) = ax + b$ is a linear function and we let $y = f(x)$. Then $y = ax + b$, which is an equation of a straight line with slope a and y-intercept $(0, b)$. Thus **the graph of a linear function is a straight line**. We say that $f(x) = ax + b$ has slope a.

EXAMPLE 4

a. *Graph $f(x) = 2x - 1$.*

Here f is a linear function (with slope 2), and so its graph is a straight line. Since two points determine a straight line, we need only plot two points and then draw a line through them. See Fig. 1-21(a). Note that one of the

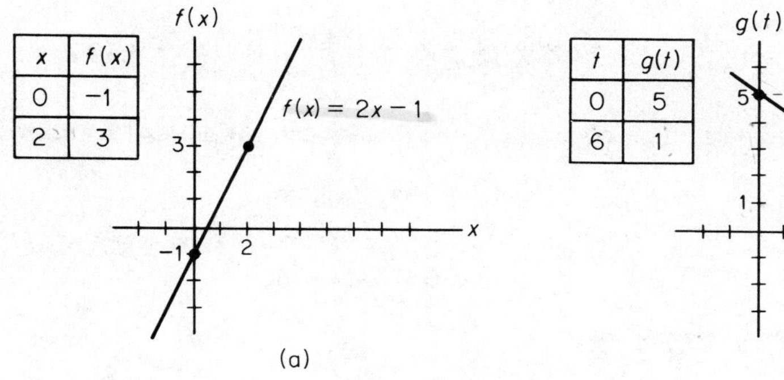

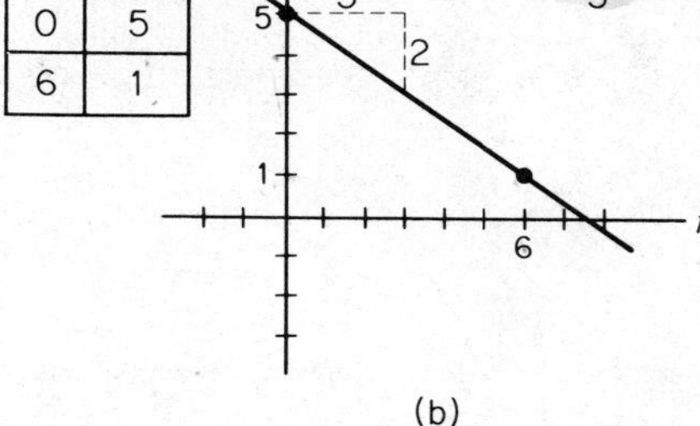

FIG. 1-21

points plotted is the vertical-axis intercept that occurs when $x = 0$.

b. *Graph* $g(t) = \dfrac{15 - 2t}{3}$.

Notice that

$$g(t) = \frac{15 - 2t}{3} = \frac{15}{3} - \frac{2t}{3} = -\frac{2}{3}t + 5.$$

Thus $g(t)$ is a linear function. See Fig. 1-21(b). Observe that since the slope is $-2/3$, then as t increases by 3 units, $g(t)$ decreases by 2.

EXAMPLE 5

Suppose f is a linear function with slope 2 *and such that* $f(4) = 8$. *Find* $f(x)$.

Since f is linear it has the form $f(x) = ax + b$. The slope is 2 and so $a = 2$:

$$f(x) = 2x + b. \tag{2}$$

Now we determine b. Since $f(4) = 8$, in Eq. (2) we replace x by 4 and solve for b.

$$\begin{aligned} f(4) &= 2(4) + b, \\ 8 &= 8 + b, \\ 0 &= b. \end{aligned}$$

Hence $f(x) = 2x$.

EXAMPLE 6

If $y = f(x)$ *is a linear function such that* $f(-2) = 6$ *and* $f(1) = -3$, *find* $f(x)$.

The points $(-2, 6)$ and $(1, -3)$ lie on the graph of f which is a straight line. If $(x_1, y_1) = (-2, 6)$ and $(x_2, y_2) = (1, -3)$, then the slope of the line is

$$m = \frac{y_2 - y_1}{x_2 - x_1} = \frac{-3 - 6}{1 - (-2)} = \frac{-9}{3} = -3.$$

We can find an equation of the line by using a point-slope form.

$$\begin{aligned} y - y_1 &= m(x - x_1), \\ y - 6 &= -3[x - (-2)], \\ y - 6 &= -3x - 6. \end{aligned}$$

Solving for y gives

$$y = -3x.$$

Thus $f(x) = -3x$.

In many studies data are collected and plotted on a coordinate system. An analysis of the results may indicate a functional relationship between the variables involved. For example, the data points may be approximated by points on a straight line. This would indicate a linear functional relationship, such as the one in Example 7 below.

EXAMPLE 7

In testing an experimental diet for hens, it was determined that the average live weight w (in grams) of a hen was statistically a linear function of the number of days d after the diet was begun, where $0 \leqslant d \leqslant 50$. Suppose the average weight of a hen beginning the diet was 40 g and 25 days later it was 675 g. (a) Determine w as a linear function of d. (b) Find the average weight of a hen when $d = 10$.

a. Since w is a linear function of d, its graph is a straight line. When $d = 0$ (the beginning of the diet) then $w = 40$. Thus (0, 40) lies on the graph. See Fig. 1-22. Similarly, (25, 675) lies on the graph. If $(d_1, w_1) = (0, 40)$ and

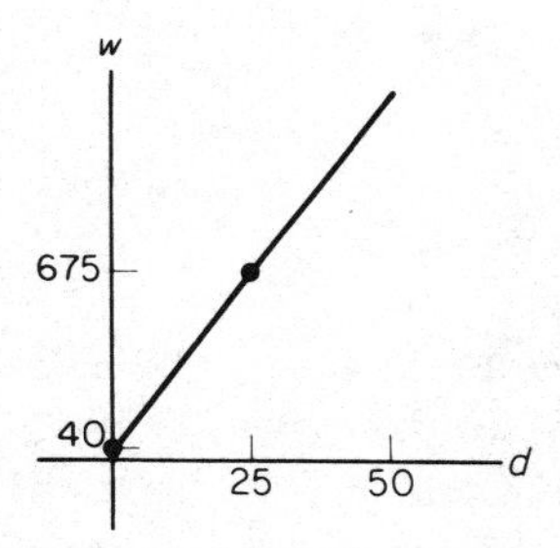

FIG. 1-22

$(d_2, w_2) = (25, 675)$, then the slope of the line is

$$m = \frac{w_2 - w_1}{d_2 - d_1} = \frac{675 - 40}{25 - 0} = \frac{635}{25} = \frac{127}{5}.$$

Using a point-slope form, we have

$$w - w_1 = m(d - d_1),$$

$$w - 40 = \frac{127}{5}(d - 0),$$

$$w - 40 = \frac{127}{5}d.$$

Solving for w gives

$$w = \frac{127}{5}d + 40,$$

which expresses w as a linear function of d.

b. When $d = 10$, then $w = \frac{127}{5}(10) + 40 = 254 + 40 = 294$. Thus the average weight of a hen 10 days after the beginning of the diet is 294 g.

Exercise 1-4

In Problems **1–6**, *find the slope and vertical-axis intercept of the linear function and sketch the graph.*

1. $y = f(x) = -4x$.

2. $y = f(x) = x + 1$.

3. $g(t) = 2t - 4$.

4. $g(t) = 2(4 - t)$.

5. $h(q) = \frac{7 - q}{2}$.

6. $h(q) = .5q + .25$.

In Problems **7–14**, *determine* $f(x)$ *if* f *is a linear function that has the given properties.*

7. slope $= 5$, $f(3) = 1$.

8. $f(0) = 4$, $f(2) = -6$.

9. $f(2) = 3$, $f(-1) = 12$.

10. slope $= -6$, $f(\frac{1}{2}) = -2$.

11. slope $= -1/2$, $f(-1/2) = 4$.

12. $f(1) = 1$, $f(2) = 2$.

13. $f(-1) = -2$, $f(-3) = -4$.

14. slope $= .01$, $f(.1) = .01$.

15. For sheep maintained at high environmental temperatures, respiratory rate r (per minute) increases as wool length l (in centimeters) decreases. Suppose sheep with a wool length of 2 cm have a (average) respiratory rate of 160, and those with a wool length of 4 cm have a respiratory rate of 125. (Adapted from **[13]**.) If r is a linear function of l, (a) determine this function and (b) determine the respiratory rate of sheep with a wool length of 1 cm.

16. The result of Sternberg's psychological experiment **[30]** on information retrieval is that a person's reaction time R, in milliseconds, is statistically a linear function of memory set size N as follows:

$$R = 38N + 397.$$

Sketch the graph for $1 \leqslant N \leqslant 5$. What is the slope?

17. In a certain learning experiment involving repetition and memory **[17]**, the proportion p of items recalled was estimated to be a linear function of effective study time t (in seconds), where $5 \leqslant t \leqslant 9$. For an effective study time of 5 seconds, the proportion of items recalled was .32. For each one-second increase in study time, the proportion recalled increased by .059. (a) Determine the linear function. (b) What proportion of items were recalled with 9 seconds of effective study time?

18. In testing an experimental diet for pigs, it was determined that the (average) live weight w (in kilograms) of a pig was statistically a linear function of the number of days d after the diet was initiated, where $0 \leqslant d \leqslant 100$. If the weight of a pig beginning the diet was 20 kg and thereafter the pig gained 6.6 kg every 10 days, determine w as a function of d and find the weight of a pig 50 days after the beginning of the diet.

1-5 QUADRATIC FUNCTIONS

In Sec. 1-1 (Example 3c) a quadratic function was described. More formally we have the following definition.

DEFINITION. *A function f is a* ***quadratic function*** *if and only if $f(x)$ can be written in the form $f(x) = ax^2 + bx + c$ where a, b, and c are constants and $a \neq 0$.*

For example, $f(x) = x^2 - 3x + 2$ and $F(t) = -3t^2$ are quadratic functions. However, $g(x) = \dfrac{1}{x^2 + 1}$ is *not* a quadratic function since it cannot be written in the form $g(x) = ax^2 + bx + c$.

The graph of the quadratic function $y = f(x) = ax^2 + bx + c$ is called a **parabola** and has a shape such as the curves in Fig. 1-23. If $a > 0$, the parabola extends upward indefinitely and we say that the parabola *opens upward* [Fig. 1-23(a)]. If $a < 0$, the parabola *opens downward* [Fig. 1-23(b)].

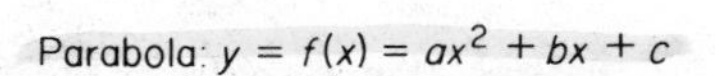

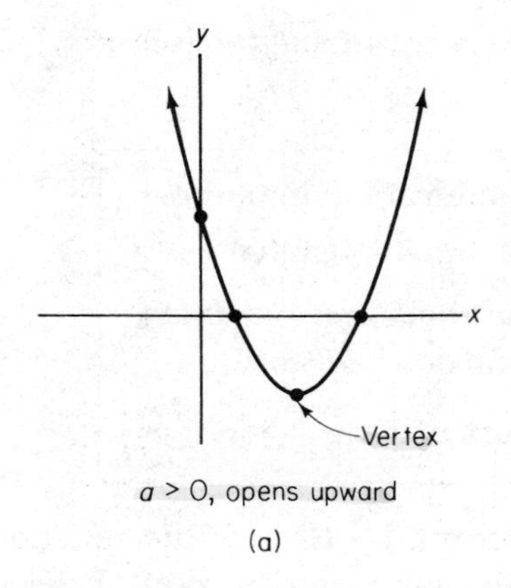

$a > 0$, opens upward

(a)

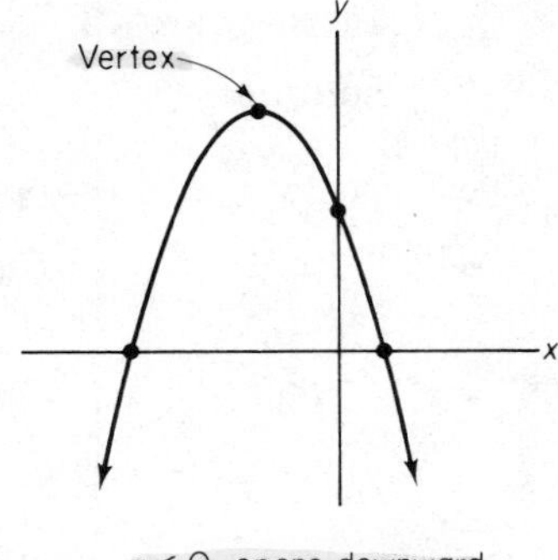

$a < 0$, opens downward

(b)

FIG. 1-23

Figure 1-23 shows points labeled **vertex**. If $a > 0$, the vertex is the "lowest" point on the parabola. This means that at this point $f(x)$ has a minimum value. By performing algebraic manipulations on $ax^2 + bx + c$ we can determine not only this minimum value but also where it occurs.

$$f(x) = ax^2 + bx + c$$
$$= (ax^2 + bx) + c.$$

Adding and subtracting $\frac{b^2}{4a}$ gives

$$f(x) = \left(ax^2 + bx + \frac{b^2}{4a}\right) + c - \frac{b^2}{4a}$$
$$= a\left(x^2 + \frac{b}{a}x + \frac{b^2}{4a^2}\right) + c - \frac{b^2}{4a}.$$
$$f(x) = a\left(x + \frac{b}{2a}\right)^2 + c - \frac{b^2}{4a}.$$

Since $\left(x + \frac{b}{2a}\right)^2 \geqslant 0$ and $a > 0$, it follows that $f(x)$ has a minimum value when $x + \frac{b}{2a} = 0$, that is, when $x = -\frac{b}{2a}$. The minimum value is $c - \frac{b^2}{4a}$. Thus the vertex is the point $\left(-\frac{b}{2a}, c - \frac{b^2}{4a}\right)$. Since the y-coordinate of this point is $f\ -\frac{b}{2a}$, we have

$$\text{vertex} = \left(-\frac{b}{2a}, f\left(-\frac{b}{2a}\right)\right).$$

This is also the vertex of a parabola that opens downward ($a < 0$), but in this case $f\left(-\frac{b}{2a}\right)$ is the *maximum* value of $f(x)$ [see Fig.1-23(b)]. In summary:

> The graph of the quadratic function $y = f(x) = ax^2 + bx + c$ is a parabola.
>
> 1. If $a > 0$, the parabola opens upward. If $a < 0$, it opens downward.
> 2. The vertex occurs at $\left(-\frac{b}{2a}, f\left(-\frac{b}{2a}\right)\right)$.

We can quickly sketch the graph of a quadratic function by first locating the vertex and a few other points on the graph. Frequently it is convenient to choose these other points to be those where the parabola intersects the x- and y-axes. These are called x- and *y-intercepts*, respectively. A y-intercept $(0, y)$ is obtained by setting $x = 0$ in $y = ax^2 + bx + c$ and solving for y. The x-intercepts $(x, 0)$ are obtained by setting $y = 0$ and solving for x. Once the intercepts and vertex are found, it is then relatively easy to pass the appropriate parabola through these points. In the event that the x-intercepts are very close to the vertex, or that no x-intercepts exist, we find a point on each side of the vertex so that we can give a reasonable sketch of the parabola.

EXAMPLE 1

Graph the following quadratic functions.

a. $y = f(x) = 12 - 4x - x^2$.

Here $a = -1$, $b = -4$, and $c = 12$. Since $a < 0$, the parabola opens downward. If the vertex is (x, y), then

$$x = -\frac{b}{2a} = -\frac{-4}{2(-1)} = -2,$$

and $y = f(-2) = 12 - 4(-2) - (-2)^2 = 16$. Thus the vertex (highest point) is $(-2, 16)$.

If $x = 0$, then $y = 12 - 4(0) - 0^2 = 12$. Hence the y-intercept is $(0, 12)$. If $y = 0$, then

$$0 = 12 - 4x - x^2,$$
$$0 = (6 + x)(2 - x).$$

Thus $x = -6$ or $x = 2$, and the x-intercepts are $(-6, 0)$ and $(2, 0)$.

Now we plot the vertex and intercepts [see Fig. 1-24(a)]. Through these points we draw a parabola opening downward. See Fig. 1-24(b).

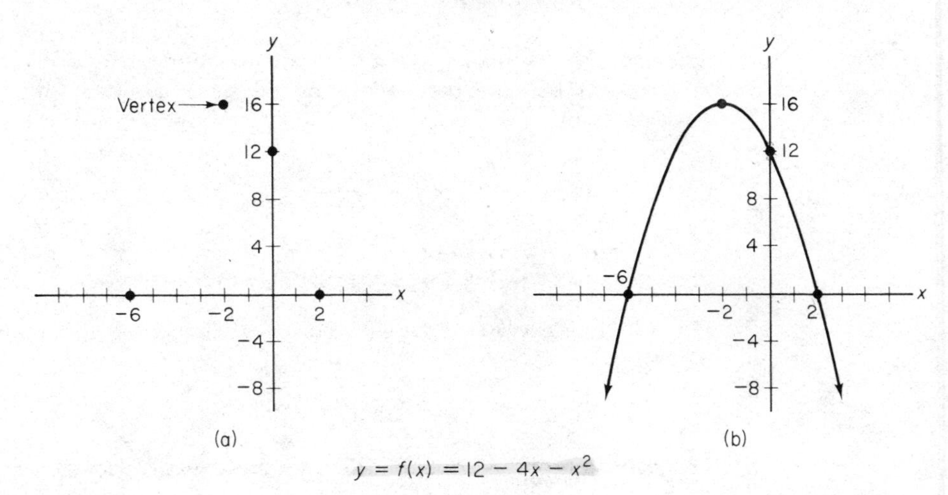

FIG. 1-24

b. $p - 2q^2 = 0$.

Since $p = 2q^2 + 0q + 0$, p is a quadratic function of q where $a = 2$, $b = 0$, and $c = 0$. The parabola opens upward since $a > 0$. If the vertex is (q, p), then

$$q = -\frac{b}{2a} = -\frac{0}{2(2)} = 0,$$

and $p = 2(0)^2 = 0$. Thus the vertex is $(0, 0)$.

A parabola opening upward with vertex at $(0, 0)$ cannot have any other intercepts. Hence, to draw a reasonable graph we plot a point on each side of the vertex and pass a parabola through the three points. See Fig. 1-25(a).

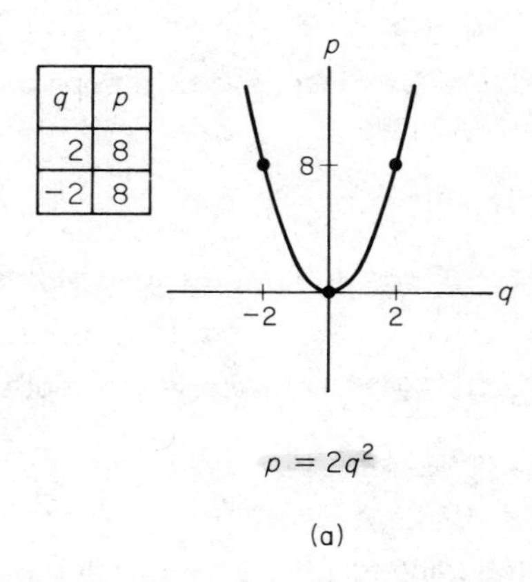

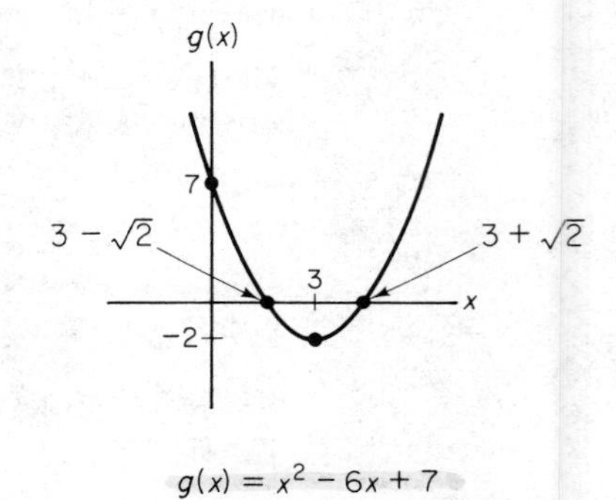

FIG. 1-25

c. $g(x) = x(x - 6) + 7$.

Since $g(x) = x^2 - 6x + 7$, g is a quadratic function where $a = 1$, $b = -6$, and $c = 7$. The parabola opens upward since $a > 0$. If the vertex is $(x, g(x))$, then

$$x = -\frac{b}{2a} = -\frac{-6}{2(1)} = 3,$$

and $g(3) = 3^2 - 6(3) + 7 = -2$. Thus the vertex is $(3, -2)$.

If $x = 0$, then $g(x) = 7$. Thus the vertical-axis intercept is $(0, 7)$. If $g(x) = 0$, then

$$0 = x^2 - 6x + 7.$$

The right side of this equation does not factor easily. Hence we shall use the quadratic formula to solve for x.

$$x = \frac{-b \pm \sqrt{b^2 - 4ac}}{2a} = \frac{-(-6) \pm \sqrt{(-6)^2 - 4(1)(7)}}{2(1)}$$

$$= \frac{6 \pm \sqrt{8}}{2} = \frac{6 \pm \sqrt{4 \cdot 2}}{2} = \frac{6 \pm 2\sqrt{2}}{2}$$

$$= \frac{6}{2} \pm \frac{2\sqrt{2}}{2} = 3 \pm \sqrt{2}\,.$$

Thus the x-intercepts are $(3 + \sqrt{2}\,, 0)$ and $(3 - \sqrt{2}\,, 0)$.

After plotting the vertex and intercepts, we draw a parabola opening upward. See Fig. 1-25(b).

EXAMPLE 2

Graph $y = f(x) = 2x^2 + 2x + 3$ and find the range of f.

This function is quadratic with $a = 2$, $b = 2$, and $c = 3$. Since $a > 0$, the graph is a parabola opening upward. If the vertex is (x, y), then

$$x = -\frac{b}{2a} = -\frac{2}{2(2)} = -\frac{1}{2},$$

and $y = 2(-\frac{1}{2})^2 + 2(-\frac{1}{2}) + 3 = \frac{5}{2}$. Thus the vertex is $(-1/2, 5/2)$.

If $x = 0$, then $y = 3$, and the y-intercept is $(0, 3)$. A parabola opening upward with its vertex above the x–axis has no x-intercepts.

In Fig. 1-26 we plotted the y-intercept, the vertex, and an additional point to

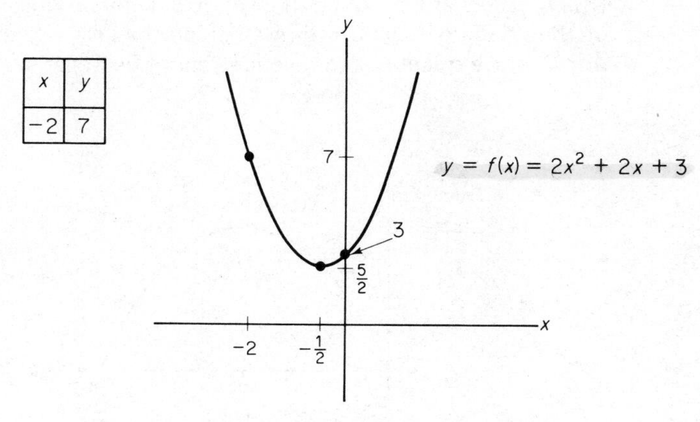

FIG. 1-26

the left of the vertex. Passing a parabola through these points gives the desired graph. From Fig. 1-26 we see that the range of f is all $y \geqslant 5/2$.

EXAMPLE 3

Suppose the demand function for a manufacturer's product is $p = f(q) = 1000 - 2q$, where p is the price (in dollars) per unit when q units are demanded (per week) by consumers. Find the level of production that will maximize the manufacturer's total revenue, and determine this revenue.

Total revenue r is given by

$$\text{Total revenue} = (\text{price})(\text{quantity}),$$

$$r = pq,$$

$$r = (1000 - 2q)q,$$

$$r = 1000q - 2q^2.$$

Note that r is a quadratic function of q, with $a = -2$, $b = 1000$, and $c = 0$. Since $a < 0$ (parabola opens downward), then r is maximum when

$$q = -\frac{b}{2a} = -\frac{1000}{2(-2)} = 250.$$

The maximum value of r is

$$r = 1000(250) - 2(250)^2,$$

$$r = 250{,}000 - 125{,}000 = 125{,}000.$$

Thus the maximum revenue that the manufacturer can receive is \$125,000 which occurs at a production level of 250 units. Figure 1-27 shows the graph of the revenue function. Only that portion for which $q \geqslant 0$ and $r \geqslant 0$ is drawn, since quantity and revenue cannot be negative.

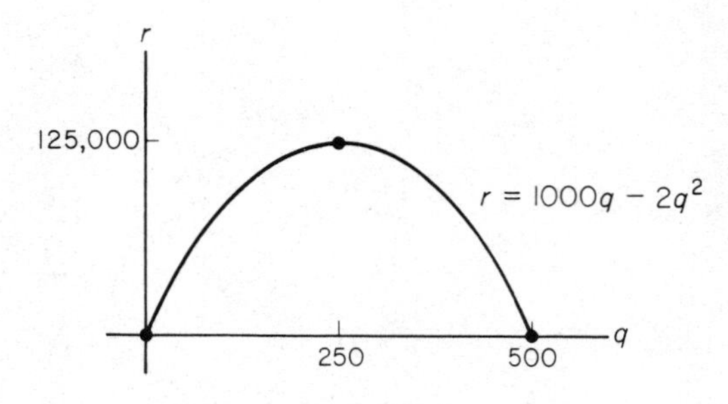

FIG. 1-27

Exercise 1-5

In Problems **1–8**, *state whether the function is linear, quadratic, or neither.*

1. $f(x) = 26 - 3x$.

2. $g(x) = (7 - x)^2$.

3. $g(x) = 4x^2$.

4. $h(s) = 6(4s + 1)$.

5. $h(q) = \dfrac{1}{2q - 4}$.

6. $f(t) = 2t(3 - t) + 4t$.

7. $f(s) = \dfrac{s^2 - 4}{2}$.

8. $g(t) = (t^2 - 1)^2$.

In Problems **9–12**, *do not include a graph.*

9. For the parabola $y = f(x) = -4x^2 + 8x + 7$, (a) find the vertex. (b) Does the vertex correspond to the highest point, or the lowest point, on the graph?

10. Repeat Problem 9 if $y = f(x) = 8x^2 + 4x - 1$.

11. For the parabola $y = f(x) = x^2 + 2x - 8$, find (a) the y-intercept, (b) the x-intercepts, and (c) the vertex.

12. Repeat Problem 11 if $y = f(x) = 3 + x - 2x^2$.

In Problems **13–26**, *graph the functions. For those functions that are linear, give the slope and the vertical-axis intercept. For those that are quadratic, give the intercepts and vertex, and state the range.*

13. $y = f(x) = x^2 - 6x + 5$.

14. $y = f(x) = -3x^2$.

15. $y = f(x) = 4x - 3$.

16. $y = f(x) = x^2 - 1$.

17. $y = g(x) = -2x^2 - 6x$.

18. $y = g(x) = \frac{1}{2}(x + 1) - 4$.

19. $s = h(t) = (t + 1)^2$.

20. $s = h(t) = 2t^2 + 3t - 2$.

21. $y = H(x) = 3\left(5 - \dfrac{x}{2}\right)$.

22. $y = H(x) = 1 - x - x^2$.

23. $y = f(x) = 2x(4 - x) - 9$.

24. $y = f(x) = -4(1 + 2x)$.

25. $t = f(s) = s^2 - 8s + 13$.

26. $t = f(s) = s^2 + 6s + 11$.

27. Determine the minimum value of $f(x) = 100x^2 - 20x + 25$.

28. Determine the maximum value of $g(t) = 4t - (50 + .1t^2)$.

29. The demand function for a manufacturer's product is $p = f(q) = 1200 - 3q$, where p is the price (in dollars) per unit when q units are demanded (per week). Find the level of production that maximizes the manufacturer's total revenue and determine this revenue.

30. A sociologist is hired by a city to study various programs that aid the education of preschool-age children. The sociologist estimates that n years after the beginning of a particular program, $f(n)$ thousand preschoolers will be enrolled, where

$$f(n) = \frac{10}{9}n(12 - n), \qquad 0 \leqslant n \leqslant 12.$$

Estimate the maximum number of preschoolers that will be enrolled in the program. Sketch the graph of f.

31. A group of biologists studied the nutritional effects on rats that were fed a diet containing 10 percent protein (adapted from **[4]**). The protein consisted of yeast and corn flour. By varying the percentage P of yeast in the protein mix, the group estimated that the average weight gain (in grams) of a rat over a period of time was

$$f(P) = -\tfrac{1}{50}P^2 + 2P + 20, \qquad 0 \leqslant P \leqslant 100.$$

Find the maximum weight gain. Sketch the graph of f.

1-6 REVIEW

IMPORTANT WORDS AND SYMBOLS IN CHAPTER 1

function *(p. 2)*
domain *(p. 2)*
range *(p. 2)*
independent variable *(p. 2)*
dependent variable *(p. 2)*
demand equation *(p. 3)*
demand function *(p. 3)*
$f(x)$ *(p. 3)*
linear function *(p. 5)*
constant function *(p. 5)*
quadratic function *(p. 5)*
polynomial function *(p. 6)*
absolute value *(p. 7)*
$|x|$ *(p. 7)*
$r!$ *(p. 8)*
supply function *(p. 9)*
$f(x) + g(x)$ *(p. 12)*
$f(x) - g(x)$ *(p. 13)*
$f(x) \cdot g(x)$ *(p. 13)*
$f(x)/g(x)$ *(p. 13)*
composition of functions *(p. 14)*
$f[g(x)]$ *(p. 14)*
graph of function *(p. 17)*
equation of line:
  slope-intercept form *(p. 28)*
  point-slope form *(p. 28)*

REVIEW SECTION

1. If f is a function, the set of all input numbers is called the __(a)__ of f. The set of all output numbers is the __(b)__ of f.

Ans. (a) domain; (b) range.

2. If $f(x) = -x^2 - 1$, then $f(-1) =$ __(a)__ . If $g(x) = 3$, then $g(1) =$ __(b)__ . If $h(x) = |6x - 15|$, then $h(2) =$ __(c)__ .

Ans. (a) -2; (b) 3; (c) 3.

3. If $h(u) = 2u$, then $h(t + 1) = 2($______$)$.

Ans. $t + 1$.

4. If $f(x) = x$, then the domain of f is ______.

Ans. all real numbers.

5. True or false:

(a) 12 is in the domain of $f(x) = 5x + 3$. ______

(b) 6 is in the domain of $g(x) = \sqrt{25 - x^2}$. ______

(c) 0 is in the domain of $h(z) = \dfrac{z}{z^2 - 9}$. ______

(d) -3 is in the domain of $F(t) = \dfrac{t}{t^2 - 9}$. ______

Ans. (a) true; (b) false; (c) true; (d) false.

6. The domain of the function

$$f(x) = \frac{3(x - 1)(x + 6)}{(x - 4)(x + 2)}$$

consists of all real numbers except ______ and ______.

Ans. 4 and -2.

7. A variable representing input numbers of a function is called (a dependent) (an independent) variable.

Ans. an independent.

8. If $h(x) = f(x) + g(x)$ where $f(x) = x + 2$ and $g(x) = 5x$, then $h(1) =$ ______.

Ans. 8.

9. If $f(x) = x^2$ and $g(x) = x + 1$, then $g[f(x)] =$ __(a)__ and $f[g(x)] =$ __(b)__ .

Ans. (a) $x^2 + 1$; (b) $(x + 1)^2$.

10. Which of the graphs in Fig. 1-28 represent functions of x? ________.

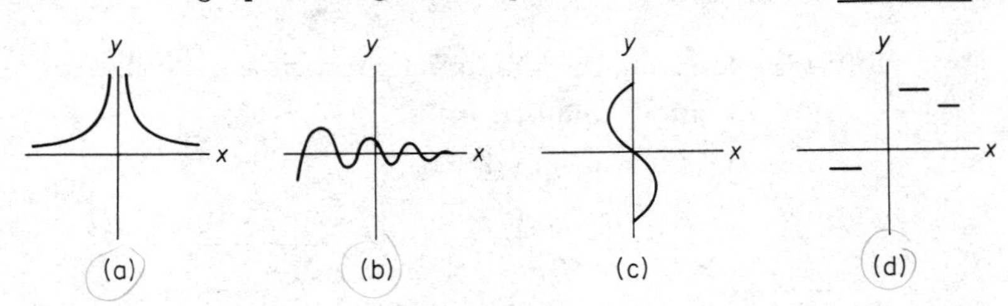

FIG. 1-28

Ans. (a), (b), (d).

11. The domain of the function whose graph is in Fig. 1-29 is __(a)__ and its range is __(b)__.

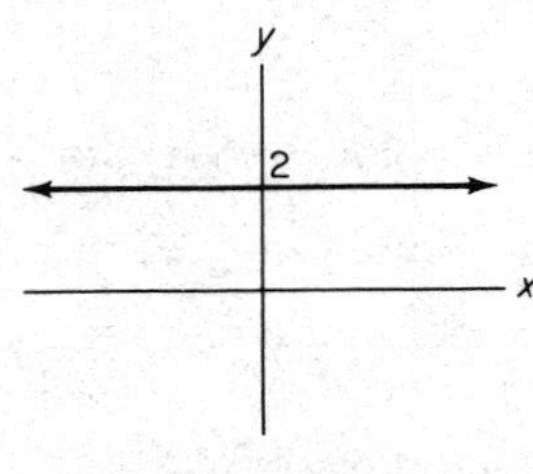

FIG. 1-29

Ans. (a) all real numbers; (b) 2.

12. The graph of a linear function is a ________.

Ans. straight line.

13. The graph of $f(x) = 7x^2 + 3x - 5$ is called a ________.

Ans. parabola.

14. The y-intercept of the graph of $y = f(x) = 3x^2 - 5x + 2$ is the point ________.

Ans. (0, 2).

15. The parabola $g(x) = x^2 - 1$ opens (upward) (downward).

Ans. upward.

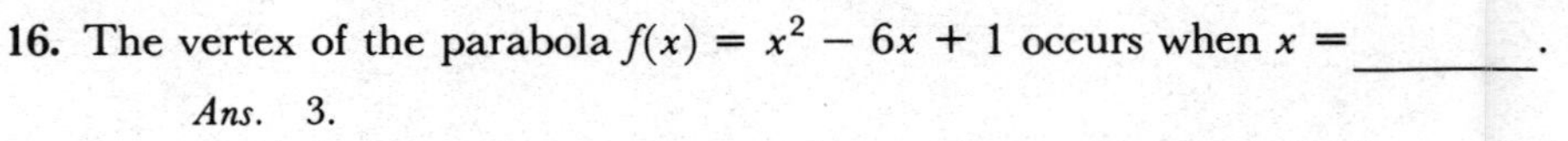

16. The vertex of the parabola $f(x) = x^2 - 6x + 1$ occurs when $x =$ ________.

Ans. 3.

17. True or false: A quadratic function is a polynomial function. ________.

Ans. true.

REVIEW PROBLEMS

In Problems **1–8**, *find the given functional expressions and the domain of the function.*

1. $f(x) = 3x^2 - 4x + 7;\quad f(0),\quad f(-3),\quad f(5),\quad f(x^2).$

2. $g(t) = \dfrac{t-3}{t+4};\quad g(3),\quad g(-1),\quad g(2),\quad g(2+h) - g(2).$

3. $H(u) = 6u^2;\quad H(\frac{1}{2}),\quad H(-\sqrt{3}),\quad H(\sqrt[4]{u}),\quad H(u+h) - H(u).$

4. $f(x) = \sqrt{2x-1};\quad f(\frac{1}{2}),\quad f(5),\quad f(x+3),\quad f(s/2).$

5. $g(z) = 2z^{-2/3};\quad g(1),\quad g(8),\quad g(-1/8),\quad g(80).$

6. $f(t) = t^{1/2} - t^{3/4};\quad f(0),\quad f(1),\quad f(16),\quad f(32).$

7. $f(x) = \begin{cases} 4, & \text{if } x < 2 \\ 8 - x^2, & \text{if } x > 2 \end{cases};\quad f(4),\quad f(-2),\quad f(0),\quad f(10).$

8. $h(q) = \begin{cases} q, & \text{if } -1 \leqslant q < 0 \\ 3 - q, & \text{if } \quad 0 \leqslant q < 3; \\ 2q^2, & \text{if } \quad 3 \leqslant q \leqslant 5 \end{cases}\quad h(0),\quad h(4),\quad h(-1/2),\quad h(1/2).$

In Problems **9** *and* **10** *find* (a) $f(x) + g(x)$, (b) $f(x) - g(x)$, (c) $f(x) \cdot g(x)$, (d) $f(x)/g(x)$, (e) $f[g(x)]$, *and* (f) $g[f(x)]$.

9. $f(x) = 4 - 3x;\quad g(x) = 2x - 8.$

10. $f(x) = x^2 + 7x - 3;\quad g(x) = 2x + 1.$

11. If $f(s) = s^2 + 5s - 3$ and $g(r) = \sqrt{r + 13}$, find $f[g(10)]$ and $g[f(-2)]$.

12. If $F(x) = \dfrac{1}{2x}$ and $G(x) = \dfrac{3x+1}{2}$, find $F[G(4x)]$ and $G[F(x^2)]$.

In Problems **13–26**, *graph each function and give its domain and range. For those that are linear, also give the slope and the vertical-axis intercept. For those that are quadratic, give all intercepts and the vertex.*

13. $y = f(x) = 4 - 2x.$

14. $y = f(x) = |x| + 1.$

15. $s = g(t) = \dfrac{2}{t-4}.$

16. $s = g(t) = 8 - 2t - t^2.$

17. $y = f(x) = 9 - x^2.$

18. $y = f(x) = 3x - 7.$

19. $y = h(t) = t^2 - 4t - 5.$

20. $y = h(t) = 1 + 3t.$

21. $p = g(t) = 3t.$

22. $p = g(t) = \sqrt{4t}.$

23. $y = F(x) = -(x^2 + 2x + 3).$

24. $y = F(x) = (2x - 1)^2.$

25. $y = f(x) = \begin{cases} 1 - x, & \text{if } x \leqslant 0, \\ 1, & \text{if } x > 0. \end{cases}$ **26.** $y = f(x) = \frac{x}{3} - 2.$

27. Suppose f is a linear function such that $f(1) = 5$ and $f(x)$ decreases by four units for every three-unit increase in x. Find $f(x)$.

28. If f is a linear function such that $f(-1) = 8$ and $f(2) = 5$, find $f(x)$.

29. In psychology the term *semantic memory* refers to our knowledge of the meaning and relationships of words as well as the means by which we store and retrieve such information **[30]**. In a network model of semantic memory, there is a hierarchy of levels at which information is stored. In an experiment by Collins and Quillian based on a network model, data were obtained on the reaction time to respond to simple questions about nouns. The graph of the results shows that, on the average, reaction time R (in milliseconds) is a linear function of the level L at which a characterizing property of the noun is stored. At level 0 the reaction time is 1310; at level 2 the reaction time is 1460. (a) Find the linear function. (b) Find the reaction time at level 1. (c) Find the slope and its significance.

30. Celsius temperature C is a linear function of Fahrenheit temperature F. Use the facts that 32°F is the same as 0°C and 212°F is the same as 100°C to find this function. Also find C when F = 50.

31. The demand function for a manufacturer's product is $p = f(q) = 200 - 2q$, where p is the price (in dollars) per unit when q units are demanded (per week). Find the level of production that maximizes the manufacturer's total revenue and determine this revenue.

32. Graph the "post office function"

$$c = f(x) = \begin{cases} 15, & \text{if } 0 < x \leqslant 1, \\ 28, & \text{if } 1 < x \leqslant 2, \\ 41, & \text{if } 2 < x \leqslant 3, \\ & \text{etc.,} \end{cases}$$

for $0 < x \leqslant 3$. Here c is the cost (in cents) of mailing a first-class letter of weight x (in ounces) in January 1980.

CHAPTER 2

Limits and Continuity

2-1 LIMITS

Our study of calculus will begin in the next chapter. However, since the notion of a *limit* lies at the foundation of calculus, we must develop not only some understanding of that concept, but also insight. We shall first give you a "feeling" for limits by some examples.

Suppose we examine the function

$$f(x) = 2x - 1,$$

when x is "near" 2 but not equal to 2. Some values of $f(x)$ for x less than 2 and then greater than 2 are given in Table 2-1. It is apparent that as x takes on values closer to 2, regardless of whether x approaches 2 *from the left* ($x < 2$) or *from the right* ($x > 2$), the corresponding values of $f(x)$ become closer to one number, 3. This is also clear from the graph of f in Fig. 2-1. To express our conclusion we say that 3 is the **limit** of $f(x)$ as x

TABLE 2-1

$x < 2$	$x > 2$
$f(1.7) = 2.4$	$f(2.3) = 3.6$
$f(1.8) = 2.6$	$f(2.2) = 3.4$
$f(1.9) = 2.8$	$f(2.1) = 3.2$
$f(1.99) = 2.98$	$f(2.01) = 3.02$
$f(1.999) = 2.998$	$f(2.001) = 3.002$

approaches 2. Symbolically we write

$$\lim_{x \to 2} (2x - 1) = 3.$$

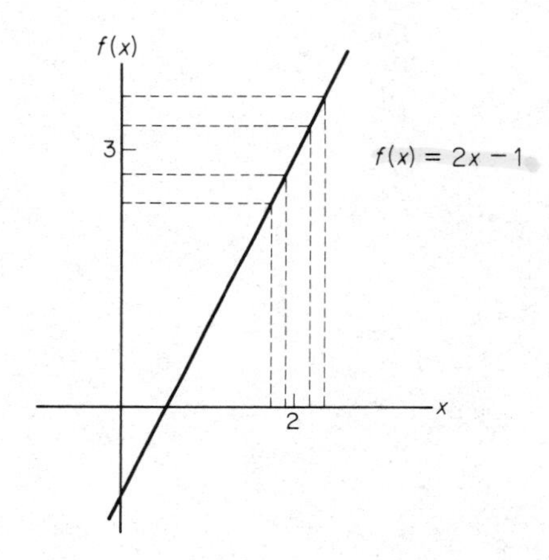

FIG. 2-1

Actually we can make the number $f(x)$ as close to 3 as we wish by taking x sufficiently close to 2.

You may think that you can find the limit of a function as x approaches some number a by just evaluating the function when x is a. For the function above, this is true: $f(2) = 2(2) - 1 = 3$, which was also the limit. But this method of substitution does not always work. For example, consider the function

$$g(x) = \begin{cases} 2x - 1, & \text{if } x \neq 2, \\ 1, & \text{if } x = 2. \end{cases}$$

Notice that $g(2) = 1$. Let us find the limit of $g(x)$ as x approaches 2, that is, as $x \to 2$. From the graph of g in Fig. 2-2, you can see that as x gets

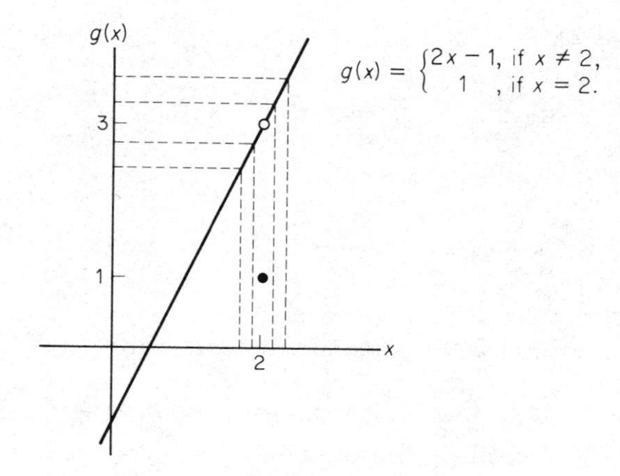

FIG. 2-2

closer to 2 (but *not equal* to 2), then $g(x)$ gets closer to 3. Thus,

$$\lim_{x \to 2} g(x) = 3,$$

which is *not* the same as $g(2)$.

Our results can be generalized to any function f. To say "The limit of $f(x)$, as x approaches a, is L," written

$$\lim_{x \to a} f(x) = L,$$

means that $f(x)$ will be as close to the number L as we please for all x sufficiently close to the number a but not equal to a. Again, we are not concerned with what happens to $f(x)$ when x *equals* a, but only with what happens to it when x is *close to* a. We emphasize that a limit is independent of the way in which $x \to a$. The limit must be the same whether x approaches a from the left or from the right (for $x < a$ or $x > a$, respectively).

We shall now state some properties of limits which may seem reasonable to you.

I. **If $f(x) = c$ is a constant function, then $\lim_{x \to a} f(x) = \lim_{x \to a} c = c$.**

II. **$\lim_{x \to a} x^n = a^n$, for any positive integer n.**

EXAMPLE 1

a. $\lim_{x\to 2} 7 = 7;\quad \lim_{x\to -5} 7 = 7.$

b. $\lim_{x\to 6} x^2 = 6^2 = 36.$

c. $\lim_{t\to -2} t^4 = (-2)^4 = 16.$

Some other properties of limits are

If $\lim_{x\to a} f(x) = L_1$ and $\lim_{x\to a} g(x) = L_2$, where L_1 and L_2 are real numbers, then

III. $\lim_{x\to a} [f(x) \pm g(x)] = \lim_{x\to a} f(x) \pm \lim_{x\to a} g(x) = L_1 \pm L_2.$

This property can be extended to the limit of a finite number of sums and differences.

IV. $\lim_{x\to a} [f(x) \cdot g(x)] = \lim_{x\to a} f(x) \cdot \lim_{x\to a} g(x) = L_1 \cdot L_2.$

V. $\lim_{x\to a} [c\, f(x)] = c \cdot \lim_{x\to a} f(x) = cL_1$, **where c is a constant.**

EXAMPLE 2

a. $\lim_{x\to 2} (x^2 + x) = \lim_{x\to 2} x^2 + \lim_{x\to 2} x = 2^2 + 2 = 6.$

b. $\lim_{s\to 3} (s^3 - s) = \lim_{s\to 3} s^3 - \lim_{s\to 3} s = 3^3 - 3 = 24.$

c. $\lim_{x\to -1} (x^3 - x + 1) = \lim_{x\to -1} x^3 - \lim_{x\to -1} x + \lim_{x\to -1} 1$

$= (-1)^3 - (-1) + 1 = 1.$

d. $\lim_{x\to 2} [(x + 1)(x - 3)] = \lim_{x\to 2} (x + 1) \cdot \lim_{x\to 2} (x - 3)$

$= [\lim_{x\to 2} x + \lim_{x\to 2} 1] \cdot [\lim_{x\to 2} x - \lim_{x\to 2} 3]$

$= [2 + 1] \cdot [2 - 3] = 3[-1] = -3.$

e. $\lim_{x\to -2} 3x^3 = 3 \lim_{x\to -2} x^3 = 3(-2)^3 = -24.$

EXAMPLE 3

Let $f(x) = c_0x^n + c_1x^{n-1} + \ldots + c_{n-1}x + c_n$ define a polynomial function f. Then

$$\begin{aligned}\lim_{x\to a} f(x) &= \lim_{x\to a} (c_0x^n + c_1x^{n-1} + \ldots + c_{n-1}x + c_n)\\ &= c_0 \cdot \lim_{x\to a} x^n + c_1 \cdot \lim_{x\to a} x^{n-1} + \ldots + c_{n-1} \cdot \lim_{x\to a} x + \lim_{x\to a} c_n\\ &= c_0a^n + c_1a^{n-1} + \ldots + c_{n-1}a + c_n = f(a).\end{aligned}$$

Thus, **if f is a polynomial function, then**

$$\lim_{x\to a} f(x) = f(a).$$

The result of Example 3 allows us to find many limits by just substituting a for x. For example,

$$\lim_{x\to -3} (x^3 + 4x^2 - 7) = (-3)^3 + 4(-3)^2 - 7 = 2,$$

$$\lim_{h\to 3} [2(h-1)] = 2(3-1) = 4.$$

Our final two properties will concern limits involving quotients and roots.

If $\lim_{x\to a} f(x) = L_1$ and $\lim_{x\to a} g(x) = L_2$, where L_1 and L_2 are real numbers, then

VI. $$\lim_{x\to a} \frac{f(x)}{g(x)} = \frac{\lim_{x\to a} f(x)}{\lim_{x\to a} g(x)} = \frac{L_1}{L_2}, \text{ if } L_2 \neq 0.$$

VII. $$\lim_{x\to a} \sqrt[n]{f(x)} = \sqrt[n]{\lim_{x\to a} f(x)} = \sqrt[n]{L_1}, \text{ if } \sqrt[n]{L_1} \text{ is defined}^{\dagger}.$$

EXAMPLE 4

a. $$\lim_{x\to 2} \frac{x^2+1}{4x-1} = \frac{\lim_{x\to 2}(x^2+1)}{\lim_{x\to 2}(4x-1)} = \frac{5}{7}.$$

b. $$\lim_{x\to 1} \frac{2x^2+x-3}{x^3+4} = \frac{\lim_{x\to 1}(2x^2+x-3)}{\lim_{x\to 1}(x^3+4)} = \frac{2+1-3}{1+4} = \frac{0}{5} = 0.$$

†Strictly speaking, $\sqrt[n]{f(x)}$ must be defined on an open interval containing a.

c. $\lim_{t\to 4} \sqrt{t^2+1} = \sqrt{\lim_{t\to 4}(t^2+1)} = \sqrt{17}$.

d. $\lim_{x\to 3} \sqrt[3]{x^2+7} = \sqrt[3]{\lim_{x\to 3}(x^2+7)} = \sqrt[3]{16} = \sqrt[3]{8\cdot 2} = 2\sqrt[3]{2}$.

EXAMPLE 5

a. *Find* $\lim_{h\to 0} \dfrac{(2+h)^2-4}{h}$.

As $h \to 0$, both numerator and denominator approach zero. Thus we cannot use property VI. However, since what happens when h equals zero is of no concern, we can assume $h \neq 0$ and write

$$\frac{(2+h)^2-4}{h} = \frac{4+4h+h^2-4}{h} = \frac{4h+h^2}{h} = \frac{h(4+h)}{h} = 4+h.$$

The algebraic manipulation on the original function $\dfrac{(2+h)^2-4}{h}$ yielded a *new* function $4+h$ which agrees with the original except when $h=0$. Thus

$$\lim_{h\to 0} \frac{(2+h)^2-4}{h} = \lim_{h\to 0}(4+h) = 4.$$

Notice that although the original function is not defined at zero, it *does* have a limit as $h \to 0$.

b. *Find* $\lim_{x\to -1} \dfrac{x^2-1}{x+1}$.

Since both numerator and denominator approach 0 as $x \to -1$, we try to express $(x^2-1)/(x+1)$ in a different form for $x \neq -1$. Here we shall first factor the numerator.

$$\lim_{x\to -1} \frac{x^2-1}{x+1} = \lim_{x\to -1} \frac{(x-1)(x+1)}{x+1} = \lim_{x\to -1}(x-1) = -2.$$

In Example 5 the method of finding a limit by substitution does not work. In (a), replacing h by 0 gives $0/0$ which has no meaning. Similarly, in (b), replacing x by -1 gives $0/0$. When the meaningless form $0/0$ arises, algebraic manipulation (as in Example 5) may result in a form for which the limit *can* be determined. In fact, most important limits cannot be evaluated by substitution.

Exercise 2-1

In Problems **1–36**, *find the limits.*

1. $\lim_{x\to 3} 14$.

2. $\lim_{x\to 0} 4x$.

3. $\lim_{x\to 6} (x - 17)$.

4. $\lim_{s\to 1} 2$.

5. $\lim_{t\to -2} (t^2 + 1)$.

6. $\lim_{t\to 1/2} (3t - 5)$.

7. $\lim_{x\to 0.3} (3 - 2x^2)$.

8. $\lim_{x\to -3} (x^3 - 4)$.

9. $\lim_{h\to 6} (h^2 - 5h - 6)$.

10. $\lim_{x\to -2} (x^2 - 2x + 1)$.

11. $\lim_{x\to -1} (x^3 - 3x^2 - 2x + 1)$.

12. $\lim_{r\to 9} \dfrac{4r - 3}{11}$.

13. $\lim_{t\to -3} \dfrac{t - 2}{t + 5}$.

14. $\lim_{x\to -6} \dfrac{x^2 + 6}{x - 6}$.

15. $\lim_{h\to 0} \dfrac{h}{h^2 - 7h + 1}$.

16. $\lim_{h\to 0} \dfrac{h^2 - 2h - 4}{h^3 - 1}$.

17. $\lim_{p\to 4} \sqrt{p^2 + p + 5}$.

18. $\lim_{y\to 9} \sqrt{y + 3}$.

19. $\lim_{x\to -2} \sqrt{\dfrac{4x - 1}{x + 1}}$.

20. $\lim_{x\to -1} \sqrt[3]{x^2}$.

21. $\lim_{x\to 2} \dfrac{(x + 3)\sqrt{x^2 - 1}}{(x - 4)(x + 1)}$.

22. $\lim_{t\to 3} \sqrt{\dfrac{2t + 3}{3t - 5}}$.

23. $\lim_{t\to 1} \dfrac{t^2}{\sqrt[3]{(t^2 - 2)^2}}$.

24. $\lim_{t\to 2} \dfrac{(t + 3)(t + 7)}{(t - 1)(t + 4)}$.

25. $\lim_{x\to -1} \dfrac{x^2 + 2x + 1}{x + 1}$.

26. $\lim_{t\to 1} \dfrac{t^2 - 1}{t - 1}$.

27. $\lim_{x\to 3} \dfrac{x - 3}{x^2 - 9}$.

28. $\lim_{x\to 0} \dfrac{x^2 - 2x}{x}$.

29. $\lim_{x\to 4} \dfrac{x^2 - 9x + 20}{x^2 - 3x - 4}$.

30. $\lim_{x\to 2} \dfrac{x^2 - 2x}{x - 2}$.

31. $\lim_{x\to 1/2} \dfrac{2x^2 + 5x - 3}{4x^2 - 2x}$.

32. $\lim_{x\to -4} \dfrac{x^2 + 2x - 8}{x^2 + 5x + 4}$.

33. $\lim_{x\to 2} \dfrac{3x^2 - x - 10}{x^2 + 5x - 14}$.

34. $\lim_{x\to 0} \dfrac{(x+2)^2 - 4}{x}$.

35. $\lim_{h\to 0} \dfrac{(2+h)^2 - 2^2}{h}$.

36. $\lim_{x\to a} \dfrac{x^4 - a^4}{x^2 - a^2}$.

37. Find $\lim_{h\to 0} \dfrac{(x+h)^2 - x^2}{h}$ by treating x as a constant.

38. Find $\lim_{h\to 0} \dfrac{2(x+h)^2 + 5(x+h) - 2x^2 - 5x}{h}$ by treating x as a constant.

39. If $f(x) = x + 5$, show that $\lim_{h\to 0} \dfrac{f(x+h) - f(x)}{h} = 1$. (See Problem 37.)

40. If $f(x) = x^2$, show that $\lim_{h\to 0} \dfrac{f(x+h) - f(x)}{h} = 2x$. (See Problem 37.)

41. The maximum theoretical efficiency E of a power plant is given in **[43]** by

$$E = \frac{T_h - T_c}{T_h},$$

where T_c and T_h are the respective absolute temperatures of the hotter and colder reservoirs. Find (a) $\lim_{T_c\to 0} E$ and (b) $\lim_{T_c\to T_h} E$.

2-2 LIMITS, CONTINUED

Figure 2-3 shows the graph of a function f. Notice that $f(x)$ is not defined when $x = 0$. As x approaches 0 *from the right*, $f(x)$ approaches 1.

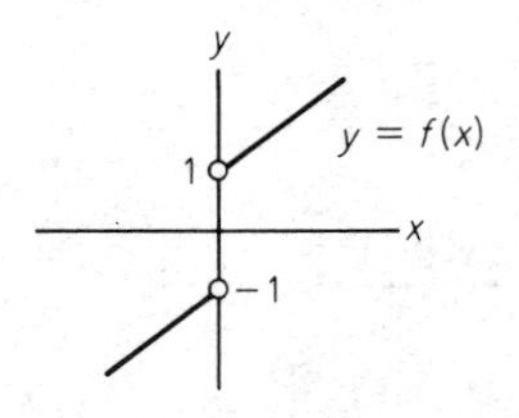

FIG. 2-3

We write this as

$$\lim_{x\to 0^+} f(x) = 1.$$

On the other hand, as x approaches 0 *from the left*, $f(x)$ approaches -1

and we write

$$\lim_{x\to 0^-} f(x) = -1.$$

Limits like these are called **one-sided limits**. From the last section we know that the limit of a function as $x \to a$ is independent of the way x approaches a. Thus the limit will exist if and only if both one-sided limits exist and are equal. We therefore conclude that

$$\lim_{x\to 0} f(x) \text{ does not exist.}$$

As another example of a one-sided limit, consider $f(x) = \sqrt{x-3}$ as x approaches 3 (see Fig. 2-4). Since f is defined only when $x \geqslant 3$, we

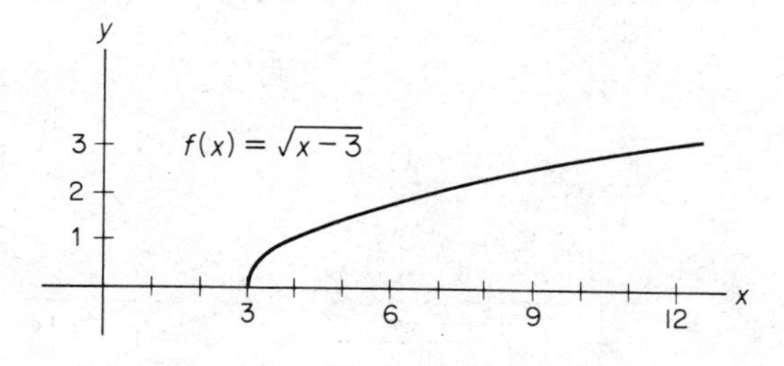

FIG. 2-4

speak of the limit as x approaches 3 from the right. From the diagram it is clear that

$$\lim_{x\to 3^+} \sqrt{x-3} = 0.$$

Now let us look at $y = f(x) = 1/x^2$ near $x = 0$. Figure 2-5 shows a table of values of $f(x)$ for x near 0, together with the graph of the function. Notice that as $x \to 0$, both from the left and from the right,

x	$f(x)$
± 1	1
± 0.5	4
± 0.1	100
± 0.01	10,000
± 0.001	1,000,000

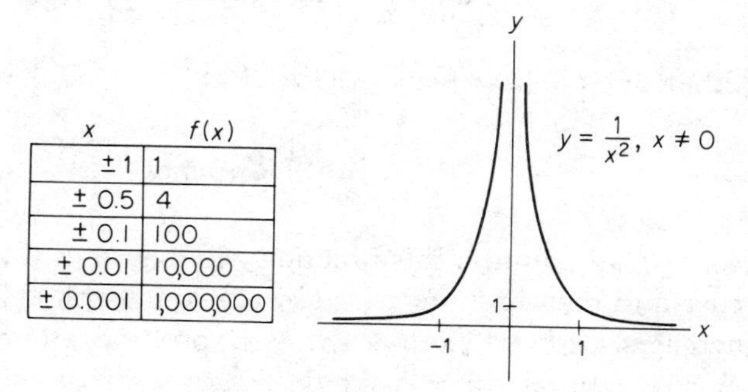

FIG. 2-5

$f(x)$ increases without bound. Hence no limit exists at 0. We say that as $x \to 0$, $f(x)$ becomes positively infinite and symbolically we write

$$\lim_{x\to 0} \frac{1}{x^2} = \infty.$$

PITFALL. *The use of the "equals" sign in this situation does not mean that the limit exists. On the contrary, the symbolism here* (∞) *is a way of saying specifically that there is no limit and it indicates* ***why*** *there is no limit.*

Consider the graph of $y = f(x) = 1/x$ for $x \neq 0$ (see Fig. 2-6). As x approaches 0 from the right, $1/x$ becomes positively infinite; as x

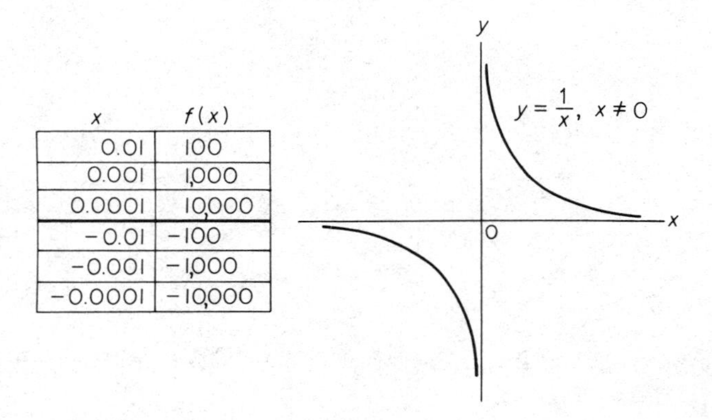

x	$f(x)$
0.01	100
0.001	1,000
0.0001	10,000
−0.01	−100
−0.001	−1,000
−0.0001	−10,000

FIG. 2-6

approaches 0 from the left, $1/x$ becomes negatively infinite. Symbolically we write

$$\lim_{x\to 0^+} \frac{1}{x} = \infty \quad \text{and} \quad \lim_{x\to 0^-} \frac{1}{x} = -\infty.$$

Either one of these facts implies that

$$\lim_{x\to 0} \frac{1}{x} \text{ does not exist.}$$

Now let us examine this function as x becomes infinite first in a positive sense and then in a negative sense. From Table 2-2 you can see that as x increases without bound through positive values, the values of $f(x)$ approach 0. Likewise, as x decreases without bound through negative

TABLE 2-2

x	$f(x)$
1,000	.001
10,000	.0001
100,000	.00001
1,000,000	.000001

x	$f(x)$
−1,000	−.001
−10,000	−.0001
−100,000	−.00001
−1,000,000	−.000001

values, the values of $f(x)$ also approach 0. Symbolically we write

$$\lim_{x\to\infty} \frac{1}{x} = 0 \quad \text{and} \quad \lim_{x\to-\infty} \frac{1}{x} = 0.$$

EXAMPLE 1

Find the limit (if it exists).

a. $\lim_{x\to-1^+} \frac{2}{x+1}$.

As x approaches -1 from the right, $x + 1$ approaches 0 but is always positive. Since we are dividing 2 by positive numbers close to 0, the results, $2/(x + 1)$, are positive numbers that are arbitrarily large. Thus,

$$\lim_{x\to-1^+} \frac{2}{x+1} = \infty,$$

and no limit exists.

b. $\lim_{x\to 2} \frac{x+2}{x^2-4}$.

As $x \to 2$ the numerator approaches 4 and the denominator approaches 0. Thus we are dividing numbers near 4 by numbers near 0. The results are numbers arbitrarily large in magnitude. At this stage we can write

$$\lim_{x\to 2} \frac{x+2}{x^2-4} \text{ does not exist.}$$

However, let us see if we can use the symbol ∞ or $-\infty$ to be more specific about "does not exist." Notice that

$$\lim_{x\to 2} \frac{x+2}{x^2-4} = \lim_{x\to 2} \frac{x+2}{(x+2)(x-2)} = \lim_{x\to 2} \frac{1}{x-2}.$$

Since $\lim_{x\to 2^+} \frac{1}{x-2} = \infty$ and $\lim_{x\to 2^-} \frac{1}{x-2} = -\infty$,

then $\lim_{x\to 2} \frac{x+2}{x^2-4}$ is neither ∞ nor $-\infty$.

c. $\lim_{t\to 2} \frac{t-2}{t^2-4}$.

As $t \to 2$ both numerator and denominator approach 0. Thus we first simplify the fraction, as we did in Sec. 2-1.

$$\lim_{t\to 2} \frac{t-2}{t^2-4} = \lim_{t\to 2} \frac{t-2}{(t+2)(t-2)} = \lim_{t\to 2} \frac{1}{t+2} = \frac{1}{4}.$$

d. $\lim_{x\to\infty} \frac{4}{(x-5)^3}$.

As x becomes very large, so does $x - 5$. Since the cube of a large number is also large, $(x-5)^3 \to \infty$. Dividing 4 by very large numbers results in numbers near 0. Thus,

$$\lim_{x\to\infty} \frac{4}{(x-5)^3} = 0.$$

In our next discussion we shall need to know a certain limit, namely $\lim_{x\to\infty} 1/x^p$ where $p > 0$. As x becomes very large, so does x^p. Dividing 1 by very large numbers results in numbers near 0. Thus,

$$\boxed{\lim_{x\to\infty} \frac{1}{x^p} = 0 \quad \text{for } p > 0.}$$

Let us now turn to the limit of a quotient of two polynomials where the variable becomes infinite. For example, consider

$$\lim_{x\to\infty} \frac{8x^2 + 2x + 3}{2x^3 + 3x - 1}.$$

It is clear that as $x \to \infty$, both numerator and denominator become infinite. However, the form of the quotient can be changed so that we can draw a conclusion as to whether or not a limit exists. Since $x \to \infty$, we are concerned only with those values of x which are very large. Thus we can assume $x \neq 0$. A frequently used "gimmick" in a case like this is to divide both the numerator and denominator by the largest power of x which occurs in either the numerator or denominator. In our example it is x^3. Thus,

$$\lim_{x\to\infty} \frac{8x^2+2x+3}{2x^3+3x-1} = \lim_{x\to\infty} \frac{\dfrac{8x^2+2x+3}{x^3}}{\dfrac{2x^3+3x-1}{x^3}}$$

$$= \lim_{x\to\infty} \frac{\dfrac{8}{x}+\dfrac{2}{x^2}+\dfrac{3}{x^3}}{2+\dfrac{3}{x^2}-\dfrac{1}{x^3}}$$

$$= \frac{8\cdot\lim\limits_{x\to\infty}\dfrac{1}{x}+2\cdot\lim\limits_{x\to\infty}\dfrac{1}{x^2}+3\cdot\lim\limits_{x\to\infty}\dfrac{1}{x^3}}{\lim\limits_{x\to\infty}2+3\cdot\lim\limits_{x\to\infty}\dfrac{1}{x^2}-\lim\limits_{x\to\infty}\dfrac{1}{x^3}}.$$

Since $\lim\limits_{x\to\infty}\dfrac{1}{x^p} = 0$ for $p > 0$, then

$$\lim_{x\to\infty}\frac{8x^2+2x+3}{2x^3+3x-1} = \frac{8(0)+2(0)+3(0)}{2+3(0)-0} = \frac{0}{2} = 0.$$

EXAMPLE 2

a. $$\lim_{x\to\infty}\frac{2x+5}{3x+2} = \lim_{x\to\infty}\frac{\dfrac{2x+5}{x}}{\dfrac{3x+2}{x}} = \lim_{x\to\infty}\frac{2+\dfrac{5}{x}}{3+\dfrac{2}{x}} = \frac{2+0}{3+0} = \frac{2}{3}.$$

b. $$\lim_{x\to-\infty}\frac{x^2-5x}{x^4+2x^2+1} = \lim_{x\to-\infty}\frac{\dfrac{x^2-5x}{x^4}}{\dfrac{x^4+2x^2+1}{x^4}}$$

$$= \lim_{x\to-\infty}\frac{\dfrac{1}{x^2}-\dfrac{5}{x^3}}{1+\dfrac{2}{x^2}+\dfrac{1}{x^4}}$$

$$= \frac{0-0}{1+0+0} = \frac{0}{1} = 0.$$

c. *Find* $\lim\limits_{x\to-\infty}\dfrac{10x^2}{x}$.

Here it is not necessary to divide both numerator and denominator by x^2. Instead, we'll first simplify the fraction.

$$\lim_{x\to-\infty} \frac{10x^2}{x} = \lim_{x\to-\infty} 10x.$$

As $x \to -\infty$, the factor 10 remains the same and the factor x becomes negatively infinite. As a result, the product becomes negatively infinite. Thus,

$$\lim_{x\to-\infty} \frac{10x^2}{x} = \lim_{x\to-\infty} 10x = -\infty,$$

and no limit exists.

PITFALL. *To find* $\lim_{x\to 0} \frac{2x+5}{3x+2}$, *we have*

$$\lim_{x\to 0} \frac{2x+5}{3x+2} = \frac{0+5}{0+2} = \frac{5}{2}.$$

Here we do not first divide numerator and denominator by x since x does not approach ∞ *or* $-\infty$.

EXAMPLE 3

In examining the records of a group of individuals hospitalized for a particular illness, it is found that the total proportion that is discharged at the end of t days of hospitalization is given by f(t), where

$$f(t) = 1 - \left(\frac{300}{300+t}\right)^3.$$

Find $\lim_{t\to\infty} f(t)$ *and interpret the answer.*

As $t \to \infty$, then $300 + t$ becomes infinite and so $\frac{300}{300+t}$ approaches 0. Since the cube of a number near 0 is also near 0, we have

$$\begin{aligned}\lim_{t\to\infty} f(t) &= \lim_{t\to\infty}\left[1 - \left(\frac{300}{300+t}\right)^3\right]\\ &= \lim_{t\to\infty} 1 - \lim_{t\to\infty}\left(\frac{300}{300+t}\right)^3 = 1 - 0 = 1.\end{aligned}$$

This indicates that as the number of days of hospitalization increases without bound, practically all the individuals will be discharged. See Fig. 2-7.

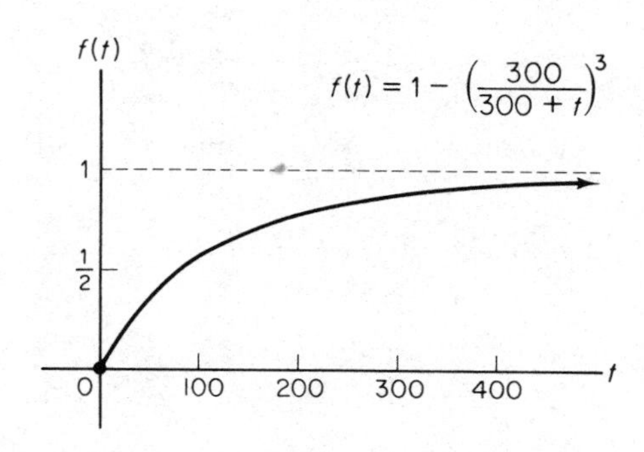

FIG. 2-7

Exercise 2-2

1. For the function f given in Fig. 2-8(a), find the following limits if they exist. Where appropriate, use the symbols ∞ or $-\infty$.

(a) $\lim_{x\to 1^-} f(x)$, (b) $\lim_{x\to 1^+} f(x)$, (c) $\lim_{x\to 1} f(x)$,
(d) $\lim_{x\to \infty} f(x)$, (e) $\lim_{x\to -2^-} f(x)$, (f) $\lim_{x\to -2^+} f(x)$,
(g) $\lim_{x\to -2} f(x)$, (h) $\lim_{x\to -\infty} f(x)$, (i) $\lim_{x\to -1^+} f(x)$,
(j) $\lim_{x\to -1^-} f(x)$, (k) $\lim_{x\to -1} f(x)$.

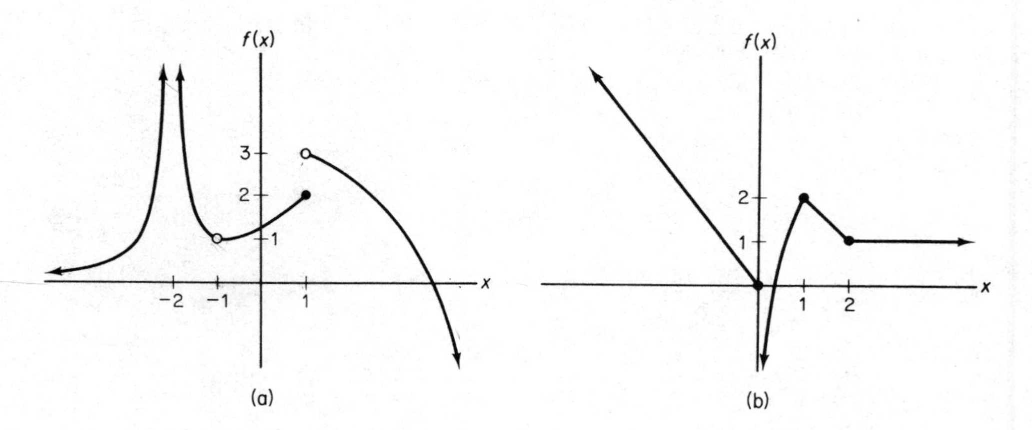

FIG. 2-8

2. For the function f given in Fig. 2-8(b), find the following limits if they exist. Where appropriate, use the symbols ∞ or $-\infty$.

(a) $\lim_{x\to 0^-} f(x)$, (b) $\lim_{x\to 0^+} f(x)$, (c) $\lim_{x\to 0} f(x)$,

(d) $\lim_{x\to-\infty} f(x)$, (e) $\lim_{x\to 1} f(x)$, (f) $\lim_{x\to 2^-} f(x)$,

(g) $\lim_{x\to 2^+} f(x)$, (h) $\lim_{x\to\infty} f(x)$.

In each of Problems **3–46**, *find the limit if it exists. Where appropriate, use the symbols* ∞ *or* $-\infty$.

3. $\lim_{x\to 3^+} (x-2)$.

4. $\lim_{x\to -1^-} (1-x^2)$.

5. $\lim_{x\to-\infty} 5x$.

6. $\lim_{x\to\infty} 3$.

7. $\lim_{x\to 0^-} \dfrac{6x}{x^4}$.

8. $\lim_{x\to 0} \dfrac{5}{x-1}$.

9. $\lim_{x\to-\infty} x^2$.

10. $\lim_{t\to\infty} (t-1)^3$.

11. $\lim_{h\to 0^+} \sqrt{h}$.

12. $\lim_{h\to 5^-} \sqrt{5-h}$.

13. $\lim_{x\to 5} \dfrac{3}{x-5}$.

14. $\lim_{x\to 0^-} 2^{1/2}$.

15. $\lim_{x\to 1^+} (4\sqrt{x-1})$.

16. $\lim_{x\to 2^+} (x\sqrt{x^2-4})$.

17. $\lim_{x\to\infty} \dfrac{7}{2x+1}$.

18. $\lim_{x\to-\infty} \dfrac{1}{(4x-1)^3}$.

19. $\lim_{x\to\infty} \dfrac{x+2}{x+3}$.

20. $\lim_{x\to\infty} \dfrac{2x-4}{3-2x}$.

21. $\lim_{x\to-\infty} \dfrac{x^2-1}{x^3+4x-3}$.

22. $\lim_{r\to\infty} \dfrac{r^3}{r^2+1}$.

23. $\lim_{t\to\infty} \dfrac{5t^2+2t+1}{4t+7}$.

24. $\lim_{x\to-\infty} \dfrac{2x}{3x^6-x+4}$.

25. $\lim_{x\to\infty} \dfrac{3-4x-2x^3}{5x^3-8x+1}$.

26. $\lim_{x\to\infty} \dfrac{7-2x-x^4}{9-3x^4+2x^2}$.

27. $\lim_{x\to 3^-} \dfrac{x+3}{x^2-9}$.

28. $\lim_{x\to -2^+} \dfrac{2x}{4-x^2}$.

29. $\lim_{w\to\infty} \dfrac{2w^2-3w+4}{5w^2+7w-1}$.

30. $\lim_{x\to\infty} \dfrac{4-3x^3}{x^3-1}$.

31. $\lim_{x\to -5} \dfrac{2x^2+9x-5}{x^2+5x}$.

32. $\lim_{t\to 2} \dfrac{t^2+2t-8}{2t^2-5t+2}$.

33. $\lim_{x \to 1} \dfrac{x^2 - 3x + 1}{x^2 + 1}$.

34. $\lim_{x \to -1} \dfrac{3x^3 - x^2}{2x + 1}$.

35. $\lim_{x \to 1^+} \left[1 + \dfrac{1}{x - 1} \right]$.

36. $\lim_{x \to -\infty} \dfrac{x^3 + 2x^2 + 1}{x^3 - 4}$.

37. $\lim_{x \to 0^+} \dfrac{2}{x + x^2}$.

38. $\lim_{x \to \infty} \left(x + \dfrac{1}{x} \right)$.

39. $\lim_{x \to 1} x(x - 1)^{-1}$.

40. $\lim_{x \to 1/2} \dfrac{1}{2x - 1}$.

41. $\lim_{x \to 0^+} \left(-\dfrac{3}{x} \right)$.

42. $\lim_{x \to 0} \left(-\dfrac{3}{x} \right)$.

43. $\lim_{x \to 0} |x|$.

44. $\lim_{x \to 0} \left| \dfrac{1}{x} \right|$.

45. $\lim_{x \to -\infty} \dfrac{x + 1}{x}$.

46. $\lim_{x \to \infty} \left[\dfrac{2}{x} - \dfrac{x^2}{x^2 - 1} \right]$.

In Problems **47–50**, *sketch the graphs of the functions and find the indicated limits if they exist. Where appropriate, use the symbols* ∞ *or* $-\infty$.

47. $f(x) = \begin{cases} 2, \text{ if } x \leq 2 \\ 1, \text{ if } x > 2 \end{cases}$; (a) $\lim_{x \to 2^+} f(x)$, (b) $\lim_{x \to 2^-} f(x)$, (c) $\lim_{x \to 2} f(x)$, (d) $\lim_{x \to \infty} f(x)$, (e) $\lim_{x \to -\infty} f(x)$.

48. $f(x) = \begin{cases} x, \text{ if } x \leq 1 \\ 2, \text{ if } x > 1 \end{cases}$; (a) $\lim_{x \to 1^+} f(x)$, (b) $\lim_{x \to 1^-} f(x)$, (c) $\lim_{x \to 1} f(x)$, (d) $\lim_{x \to \infty} f(x)$, (e) $\lim_{x \to -\infty} f(x)$.

49. $g(x) = \begin{cases} x, \text{ if } x < 0 \\ -x, \text{ if } x > 0 \end{cases}$; (a) $\lim_{x \to 0^+} g(x)$, (b) $\lim_{x \to 0^-} g(x)$, (c) $\lim_{x \to 0} g(x)$, (d) $\lim_{x \to \infty} g(x)$, (e) $\lim_{x \to -\infty} g(x)$.

50. $g(x) = \begin{cases} x^2, \text{ if } x < 0 \\ x, \text{ if } x > 0 \end{cases}$; (a) $\lim_{x \to 0^+} g(x)$, (b) $\lim_{x \to 0^-} g(x)$, (c) $\lim_{x \to 0} g(x)$, (d) $\lim_{x \to \infty} g(x)$, (e) $\lim_{x \to -\infty} g(x)$.

51. For a particular host-parasite relationship it was determined that when host density (number of hosts per unit of area) is x, then the number of hosts parasitized over a period of time is y, where

$$y = \frac{900x}{10 + 45x}.$$

If the host density were to increase without bound, what value would y approach?

52. The population N of a certain small city t years from now is predicted to be

$$N = 20{,}000 + \frac{10{,}000}{(t+2)^2}.$$

Determine the population in the long run; that is, find $\lim_{t\to\infty} N$.

53. If c is the total cost in dollars to produce q units of a product, then the average cost per unit $\bar{c}$ for an output of q units is given by $\bar{c} = c/q$. Thus, if the total cost equation is $c = 5000 + 6q$, then $\bar{c} = (5000/q) + 6$. For example, the total cost of an output of 5 units is \$5030, and the average cost per unit at this level of production is \$1006. By finding $\lim_{q\to\infty} \bar{c}$, show that the average cost approaches a level of stability if the producer continually increases output. What is the limiting value of the average cost? Sketch the graph of the average cost function.

54. Repeat Problem 53 if $c = 12{,}000 + 7q$.

In Problems **55–58**, *find* $\lim_{h\to 0} \frac{f(x+h) - f(x)}{h}$ *by treating x as a constant.*

55. $f(x) = 2x + 3$.

56. $f(x) = 4 - x$.

57. $f(x) = x^2 + x + 1$.

58. $f(x) = x^2 - 3$.

59. *Calculator Problem.* Evaluate $f(x) = x^{2x}$ when $x = 1, .5, .2, .1, .01, .001$, and $.0001$. From your results, draw a conclusion about $\lim_{x\to 0^+} x^{2x}$.

2-3 CONTINUITY

Let us consider the functions

$$f(x) = x \quad \text{and} \quad g(x) = \begin{cases} x, & \text{if } x \neq 1, \\ 2, & \text{if } x = 1. \end{cases}$$

Their graphs appear in Fig. 2-9 and Fig. 2-10, respectively. The significant difference between the graphs is that there is a "break" in the graph of g when $x = 1$, while there is *no* "break" at all in the graph of f. Stated another way, if you were to trace both graphs with a pencil, you would have to lift the pencil on the graph of g when $x = 1$, but you

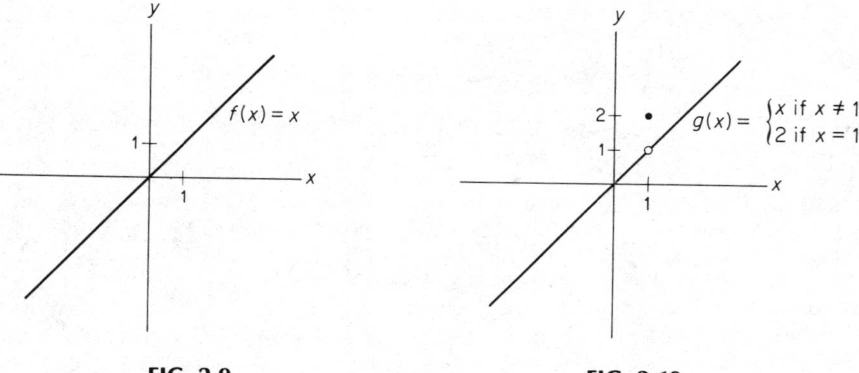

FIG. 2-9

FIG. 2-10

would not have to lift it on the graph of f. These situations can be expressed by limits. As x approaches 1,

$$\lim_{x\to 1} f(x) = 1 = f(1),$$

while

$$\lim_{x\to 1} g(x) = 1 \neq g(1) = 2.$$

The limit of f as $x \to 1$ is the same as $f(1)$, but the limit of g as $x \to 1$ is *not* the same as $g(1)$. For these reasons we say that f is *continuous* at $x = 1$ and g is *discontinuous* at $x = 1$.

DEFINITION. *A function f is* ***continuous*** *at $x = a$ if and only if the following three conditions are met:*

(1) *$f(x)$ is defined at $x = a$,*

(2) *$\lim_{x\to a} f(x)$ exists,*

(3) *$\lim_{x\to a} f(x) = f(a)$.*

DEFINITION. *A function is* ***discontinuous*** *at $x = a$ if and only if it is not continuous at $x = a$.*

Continuous functions have many useful properties that discontinuous functions do not have. These properties are important not only from a purely mathematical point of view, but also from the point of view of practical situations. We shall say more about this in Sec. 2-5.

EXAMPLE 1

a. *Show that $f(x) = 5$ is continuous at $x = 7$.*

We must verify that three conditions are met. First, f is indeed defined at $x = 7$. Second,

$$\lim_{x\to 7} f(x) = \lim_{x\to 7} 5 = 5.$$

Thus f has a limit as $x \to 7$. Third,

$$\lim_{x\to 7} f(x) = 5 = f(7).$$

Therefore $f(x) = 5$ is continuous at $x = 7$.

b. *Show that $g(x) = x^2 - 3$ is continuous at $x = -4$.*

The function g is defined at $x = -4$; $g(-4) = 13$. Also,

$$\lim_{x\to -4} g(x) = \lim_{x\to -4} (x^2 - 3) = 13 = g(-4).$$

Therefore $g(x) = x^2 - 3$ is continuous at $x = -4$.

c. *Show that $f(x) = \dfrac{7}{x^2}$ is continuous at $x = -2$.*

$$f(-2) = \frac{7}{4},$$

$$\lim_{x\to -2} \frac{7}{x^2} = \frac{7}{4}.$$

Since the limit as $x \to -2$ equals $f(-2)$, the function is continuous at $x = -2$.

We say that a function is **continuous everywhere** if it is continuous at each real number. Such functions are simply called **continuous**. A continuous function has a graph that is connected everywhere. In Sec. 2-1 (Example 3) we showed that for *any* polynomial function f, we have $\lim_{x\to a} f(x) = f(a)$. Thus **all polynomial functions are continuous everywhere.**

EXAMPLE 2

The functions $f(x) = 7$ and $g(x) = x^3 - 9x + 3$ are polynomial functions. Therefore they are continuous. For example, they are continuous at $x = 3$.

If a function is not defined at a, it is automatically discontinuous there. If it *is* defined at a, then it is discontinuous at a if

(1) it has no limit as $x \to a$,

or (2) as $x \to a$ it has a limit that is different from $f(a)$.

In Fig. 2-11 we can find points of discontinuity by inspection.

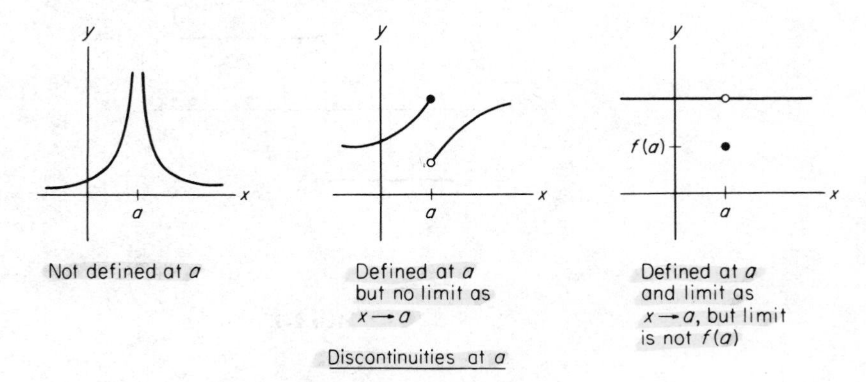

FIG. 2-11

EXAMPLE 3

a. Let $f(x) = 1/x$ (see Fig. 2-12). Since f is not defined at $x = 0$, it is discontinuous there. Moreover, $\lim_{x\to 0^+} f(x) = \infty$ and $\lim_{x\to 0^-} f(x) = -\infty$. A function is said to have an **infinite discontinuity** at $x = a$ when at least one of the one-sided limits is either ∞ or $-\infty$ as $x \to a$. Hence f has an *infinite discontinuity* at $x = 0$.

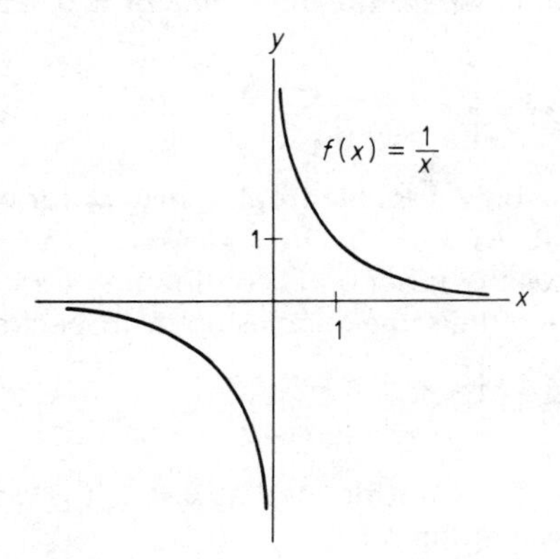

FIG. 2-12

b. Let $f(x) = \begin{cases} 1, \text{ if } x > 0 \\ 0, \text{ if } x = 0 \\ -1, \text{ if } x < 0 \end{cases}$ (see Fig. 2-13). Although f is defined at $x = 0$, $\lim_{x \to 0} f(x)$ does not exist. Thus f is discontinuous at $x = 0$.

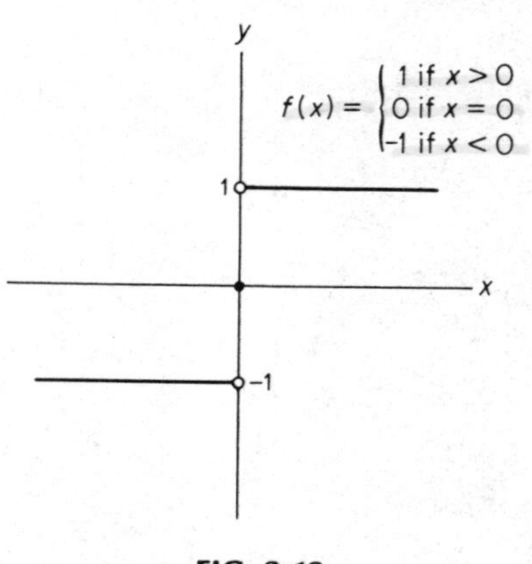

FIG. 2-13

EXAMPLE 4

Find any points of discontinuity for each of the following.

a. $f(x) = \dfrac{x^4 - 3x^3 + 2x - 1}{x^2 - 4}$.

The denominator is zero when $x = \pm 2$. Hence f is not defined at ± 2 and is therefore discontinuous at these points. Otherwise the function is "well-behaved." In fact, **any quotient of polynomials is discontinuous at points where the denominator is 0, and is continuous elsewhere.**

b. $g(x) = \begin{cases} x + 6, \text{ if } x \geqslant 3, \\ x^2, \text{ if } x < 3. \end{cases}$

The only possible trouble may occur when $x = 3$. We know $g(3) = 3 + 6 = 9$. As $x \to 3^+$, then $g(x) \to 3 + 6 = 9$. As $x \to 3^-$, then $g(x) \to 3^2 = 9$. Thus the function is continuous at $x = 3$ as well as at all other x. We can reach the same conclusion by inspecting the graph of g (Fig. 2-14).

c. $f(x) = \begin{cases} x + 2, \text{ if } x > 2, \\ x^2, \text{ if } x < 2. \end{cases}$

Since f is not defined at $x = 2$, it is discontinuous there. It is continuous for all other x.

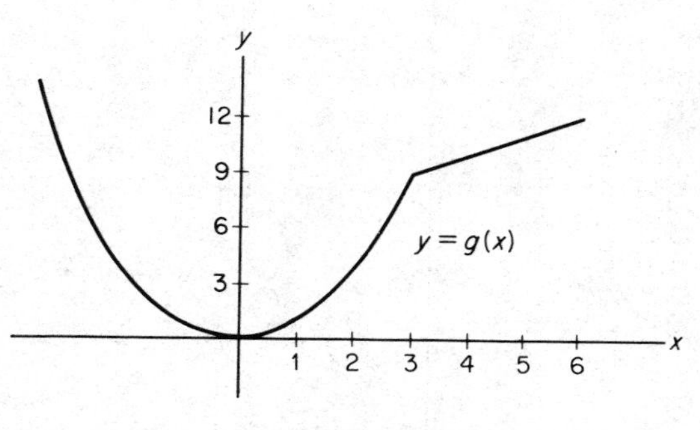

FIG. 2-14

EXAMPLE 5

Table 2-3 lists redemption values of a fifty-dollar savings bond for the first six successive periods after the date of issue. If y is the redemption value x years after the date of issue, then y is a function of x: $y = f(x)$. It is clear from the graph of this function in Fig. 2-15 that f has discontinuities when $x = \frac{1}{2}$, 1, $1\frac{1}{2}$, 2, and $2\frac{1}{2}$ and is constant for values of x between successive discontinuities. Such a function is called a *step function* because of the appearance of its graph.

TABLE 2-3

NUMBER OF YEARS AFTER ISSUE DATE (greater than)	—	(not more than)	REDEMPTION VALUE
0	—	$\frac{1}{2}$	\$37.50
$\frac{1}{2}$	—	1	38.10
1	—	$1\frac{1}{2}$	39.02
$1\frac{1}{2}$	—	2	39.90
2	—	$2\frac{1}{2}$	40.80
$2\frac{1}{2}$	—	3	41.76

Exercise 2-3

In Problems **1–6**, *use the definition of continuity to show that the given function is continuous at the indicated point.*

1. $f(x) = x^3 - 5x, \quad x = 2.$

2. $f(x) = \dfrac{x-3}{9x}, \quad x = -3.$

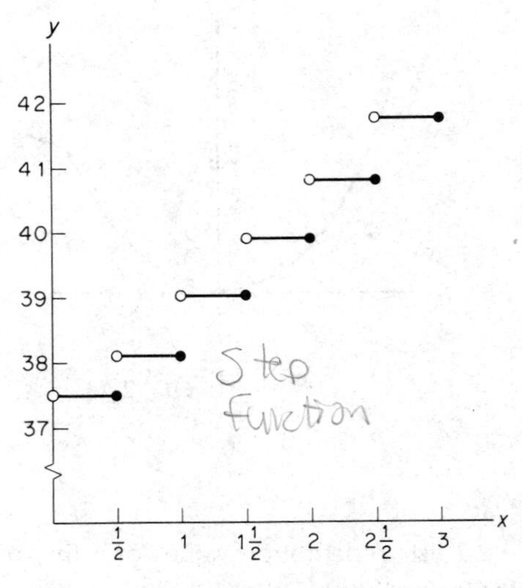

FIG. 2-15

3. $g(x) = \sqrt{2 - 3x}, \quad x = 0.$

4. $f(x) = \frac{1}{8}, \quad x = 2.$

5. $h(x) = \dfrac{x - 4}{x + 4}, \quad x = 4.$

6. $f(x) = \sqrt[3]{x}, \quad x = -1.$

In Problems **7–12**, *determine whether the function is continuous at the given values of* x.

7. $f(x) = \dfrac{x + 4}{x - 2}; \quad -2, 0.$

8. $f(x) = \dfrac{x^2 - 4x + 4}{6}; \quad 2, -2.$

9. $g(x) = \dfrac{x - 3}{x^2 - 9}; \quad 3, -3.$

10. $h(x) = \dfrac{3}{x^2 + 4}; \quad 2, -2.$

11. $F(x) = \begin{cases} x + 2, & \text{if } x \geq 2 \\ x^2, & \text{if } x < 2 \end{cases}; \quad 2, 0.$

12. $f(x) = \begin{cases} \dfrac{1}{x}, & \text{if } x \neq 0 \\ 0, & \text{if } x = 0 \end{cases}; \quad 0, -1.$

In Problems **13–16**, *state why the functions are continuous everywhere.*

13. $f(x) = 2x^2 - 3.$

14. $f(x) = \dfrac{x + 2}{5}.$

15. $f(x) = \dfrac{x - 1}{x^2 + 4}.$

16. $f(x) = x(1 - x).$

In Problems **17–34**, *find all points of discontinuity.*

17. $f(x) = 3x^2 - 3.$

18. $h(x) = x - 2.$

19. $f(x) = \dfrac{3}{x-4}$.

20. $f(x) = \dfrac{x^2 + 3x - 4}{x + 4}$.

21. $g(x) = \dfrac{(x^2 - 1)^2}{5}$.

22. $f(x) = \begin{cases} 5, & \text{if } x \geq 3, \\ 2x - 1, & \text{if } x < 3. \end{cases}$

23. $f(x) = \dfrac{x^2 + 6x + 9}{x^2 + 2x - 15}$.

24. $g(x) = \dfrac{x - 3}{x^2 + x}$.

25. $h(x) = \dfrac{x - 7}{x^3 - x}$.

26. $f(x) = \dfrac{x}{x}$.

27. $p(x) = \dfrac{x}{x^2 + 1}$.

28. $f(x) = \dfrac{x^4}{x^4 - 1}$.

29. $f(x) = \begin{cases} x^2, & \text{if } x > 2, \\ x - 1, & \text{if } x < 2. \end{cases}$

30. $f(x) = \begin{cases} \dfrac{1}{x}, & \text{if } x \neq 3, \\ 5, & \text{if } x = 3. \end{cases}$

31. $f(x) = \begin{cases} \dfrac{1}{x - 3}, & \text{if } x \geq 4, \\ 5 - x, & \text{if } x < 4. \end{cases}$

32. $f(x) = \begin{cases} 10x - 3, & \text{if } x \geq 1, \\ \dfrac{1}{x + 1}, & \text{if } x < 1. \end{cases}$

33. $f(x) = \begin{cases} \dfrac{-3}{x - 2}, & \text{if } x > 0, \\ 4 - x, & \text{if } x \leq 0. \end{cases}$

34. $f(x) = \dfrac{5x + 2}{3} - \dfrac{7}{x}$.

35. Bacteria are growing in a culture. The time t (in hours) for the number of bacteria to double in number (generation time) is a function of the temperature T (in °C) of the culture and is given by

$$t = f(T) = \begin{cases} \frac{1}{24}T + \frac{11}{4}, & \text{if } 30 \leq T \leq 36, \\ \frac{4}{3}T - \frac{175}{4}, & \text{if } 36 < T \leq 39. \end{cases}$$

Show that f is continuous at $T = 36$.

36. The *greatest integer function*, $f(x) = [x]$, is sometimes called the square-bracket function and is defined to be the greatest integer less than or equal to x, where x is any real number. For example, $[3] = 3$, $[1.999] = 1$, $[\frac{1}{4}] = 0$, and $[-4.5] = -5$. Sketch the graph of this function for $-3.5 \leq x \leq 3.5$. Use your graph to determine the values of x at which discontinuities occur.

37. Suppose the long distance rate for a telephone call from Hazleton, Pa. to Washington, D.C. is \$1.85 for the first three minutes and \$0.30 for each additional minute or fraction thereof. If $y = f(t)$ is a function that indicates the total charge y for a call of t minutes' duration, sketch the graph of f for $0 < t \leq 6$. Use your graph to determine the values of t at which discontinuities occur.

2-4 CONTINUITY APPLIED TO INEQUALITIES

In this section you will see how continuity can be applied to solve inequalities such as $x^2 + 3x - 4 > 0$. But first we must take a moment to go over some concepts that will be involved in our method.

Frequently we shall use the term *interval* to describe certain sets of numbers. For example, the set of all numbers x for which $a \leqslant x \leqslant b$ is called a **closed interval**. It includes the *endpoints* a and b, and is denoted by $[a, b]$. The set of all x for which $a < x < b$ is called an **open interval** and is denoted by (a, b). The endpoints are not part of this set. In Fig. 2-16 a square bracket indicates that an endpoint *is* included, while a parenthesis indicates that an endpoint *is not* included in the interval.

FIG. 2-16

Extending these concepts, we have the intervals shown in Fig. 2-17, where the symbols ∞ and $-\infty$ are not numbers but merely a convenience for indicating that an interval extends indefinitely in some direction.

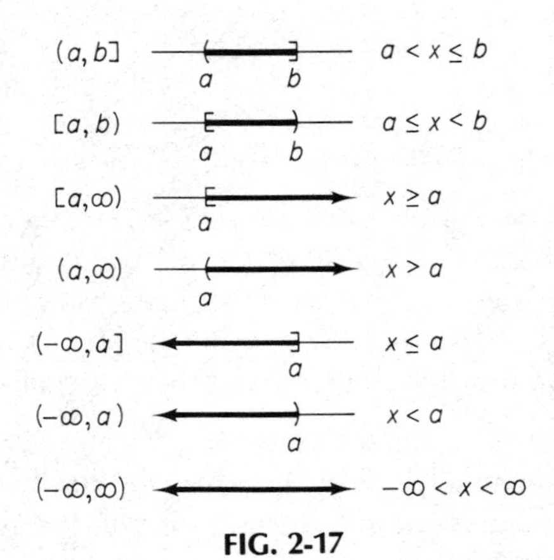

FIG. 2-17

If a function is continuous at each point in an interval, we say that it is *continuous on the interval*.

Now, we wish to draw your attention to the relationship between the x-intercepts of the graph of a function g (that is, the points where the graph meets the x-axis) and the roots of the equation $g(x) = 0$. If the graph of g has an x-intercept $(r, 0)$, then $g(r) = 0$ and so r is a root of the equation $g(x) = 0$. Hence, from the graph of $y = g(x)$ in Fig. 2-18,

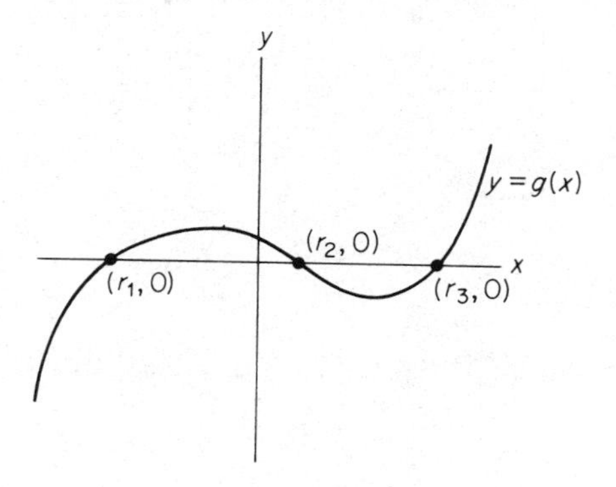

FIG. 2-18

we conclude that r_1, r_2, and r_3 are roots of $g(x) = 0$. On the other hand, if r is any real root of the equation $g(x) = 0$, then $g(r) = 0$ and hence $(r, 0)$ lies on the graph of g. This means that all real roots of the equation $g(x) = 0$ can be represented by the points where the graph of g meets the x-axis. Note also that in Fig. 2-18 these points determine four open intervals on the x-axis:

$$(-\infty, r_1), (r_1, r_2), (r_2, r_3), \text{ and } (r_3, \infty).$$

Now we can solve inequalities. Returning to $x^2 + 3x - 4 > 0$, we shall let $f(x) = x^2 + 3x - 4 = (x + 4)(x - 1)$. Since f is a polynomial function it is continuous everywhere. The roots of $f(x) = 0$ are -4 and 1; hence the graph of f has x-intercepts $(-4, 0)$ and $(1, 0)$ [see Fig. 2-19]. The roots, or to be more precise the intercepts, determine three intervals on the x-axis:

$$(-\infty, -4), \quad (-4, 1), \quad \text{and} \quad (1, \infty).$$

Consider the interval $(-\infty, -4)$. Since f is continuous on this interval, we claim that either $f(x) > 0$ or $f(x) < 0$ *throughout* the interval. Suppose $f(x)$ did indeed change sign there. Then by the continuity of f there would be a point where the graph intersects the x-axis, for

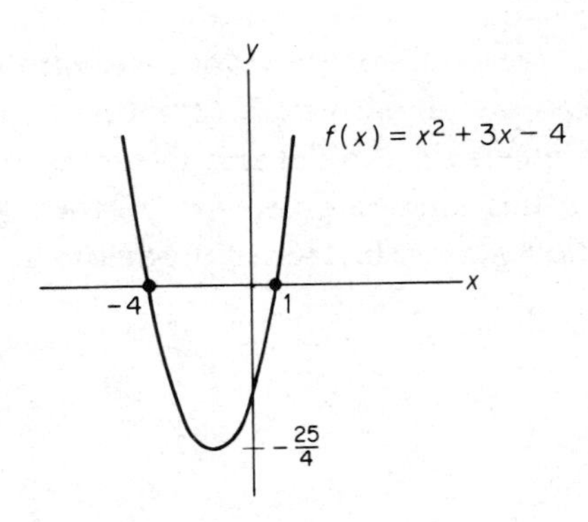

FIG. 2-19

example at $(x_0, 0)$ [refer to Fig. 2-20]. But then x_0 would be a root of the

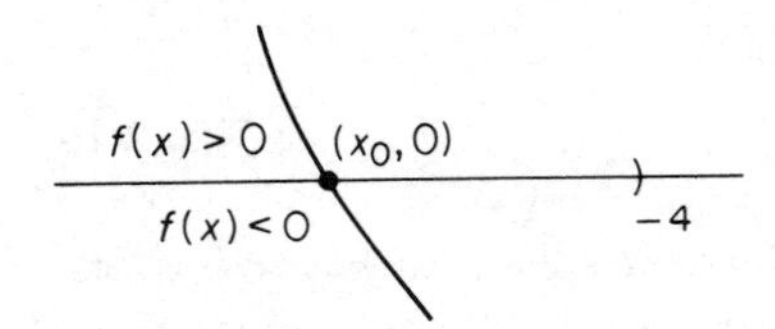

FIG. 2-20

equation $f(x) = 0$. This cannot be, since no root of $x^2 + 3x - 4 = 0$ is less than -4. Hence $f(x)$ must be strictly positive or strictly negative on $(-\infty, -4)$ as well as on the other intervals.

To determine the sign of $f(x)$ on any of these intervals, it is sufficient to determine its sign at any point in the interval. For instance, -5 is in $(-\infty, -4)$ and $f(-5) = 6 > 0$. Thus $f(x) > 0$ on $(-\infty, -4)$. Since 0 is in $(-4, 1)$ and $f(0) = -4 < 0$, then $f(x) < 0$ on $(-4, 1)$. Similarly, 3 is in $(1, \infty)$ and $f(3) = 14 > 0$; thus, $f(x) > 0$ on $(1, \infty)$ [see Fig. 2-21]. Therefore, $x^2 + 3x - 4 > 0$ for $x < -4$ and for $x > 1$, so we have solved the inequality. These results are obvious from the graph in Fig. 2-19.

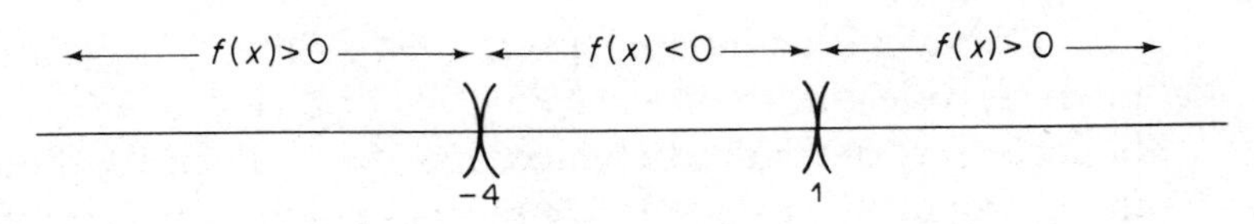

FIG. 2-21

EXAMPLE 1

Solve $x(x-1)(x+4) \leqslant 0.$

If $f(x) = x(x-1)(x+4)$, then f is continuous everywhere. The roots of $f(x) = 0$ are 0, 1, and -4 which are shown in Fig. 2-22.

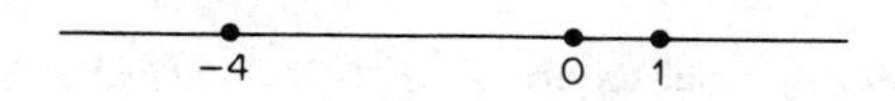

FIG. 2-22

These roots determine four intervals:

$$(-\infty, -4),\ (-4, 0),\ (0, 1), \text{ and } (1, \infty).$$

Since -5 is in $(-\infty, -4)$, the sign of $f(x)$ on $(-\infty, -4)$ is the same as that of $f(-5)$. Since

$$f(x) = x(x-1)(x+4),$$

then

$$f(-5) = -5(-5-1)(-5+4) = (-)(-)(-) = (-),$$

so $f(x) < 0$ on $(-\infty, -4)$. For the other intervals we find that

$$f(-2) = (-)(-)(+) = (+), \text{ and so } f(x) > 0 \text{ on } (-4, 0),$$

$$f\left(\tfrac{1}{2}\right) = (+)(-)(+) = (-), \text{ and so } f(x) < 0 \text{ on } (0, 1),$$

and $$f(2) = (+)(+)(+) = (+), \text{ and so } f(x) > 0 \text{ on } (1, \infty).$$

Figure 2-23 gives a summary of the signs. Thus $x(x-1)(x+4) \leqslant 0$ for

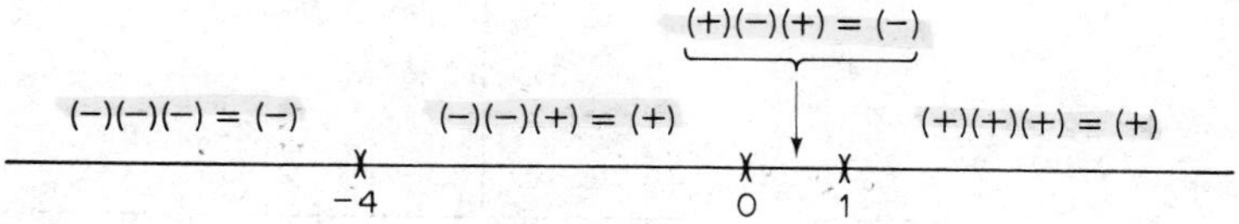

(+)(−)(+) = (−)

(−)(−)(−) = (−) (−)(−)(+) = (+) (+)(+)(+) = (+)

−4 0 1

FIG. 2-23

$x \leqslant -4$ and $0 \leqslant x \leqslant 1$. Note that -4, 0, and 1 are included in the solution because these roots satisfy the equality ($=$) part of the inequality ($\leqslant$).

EXAMPLE 2

Solve $\dfrac{x^2 - 6x + 5}{x} \leqslant 0.$

Let $f(x) = \dfrac{x^2 - 6x + 5}{x} = \dfrac{(x-1)(x-5)}{x}$. For a quotient we solve the inequality by considering the intervals determined by the roots of $f(x) = 0$, namely 1 and 5, and the points where f is discontinuous. The function is discontinuous at $x = 0$ and continuous otherwise. In Fig. 2-24 we have

FIG. 2-24

placed a hollow dot at 0 to indicate that f is not defined there. We thus consider the intervals

$$(-\infty, 0), (0, 1), (1, 5), \text{and } (5, \infty).$$

Determining the sign of $f(x)$ at a point in each interval, we find that

$$f(-1) = \frac{(-)(-)}{(-)} = (-), \text{ so } f(x) < 0 \text{ on } (-\infty, 0);$$

$$f\left(\frac{1}{2}\right) = \frac{(-)(-)}{(+)} = (+), \text{ so } f(x) > 0 \text{ on } (0, 1);$$

$$f(2) = \frac{(+)(-)}{(+)} = (-), \text{ so } f(x) < 0 \text{ on } (1, 5);$$

and

$$f(6) = \frac{(+)(+)}{(+)} = (+), \text{ so } f(x) > 0 \text{ on } (5, \infty).$$

Therefore, $f(x) \leqslant 0$ for $x < 0$ and $1 \leqslant x \leqslant 5$ (see Fig. 2-25). Why are 1 and 5 included but 0 excluded?

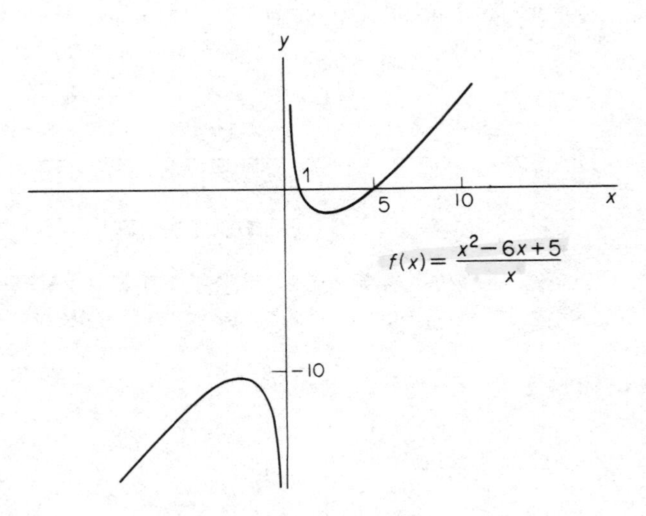

FIG. 2-25

EXAMPLE 3

Solve the following inequalities.

a. $x^2 + 1 > 0$.

The equation $x^2 + 1 = 0$ has no real roots. Thus the graph of $f(x) = x^2 + 1$ has no x-intercepts. Since f is continuous everywhere, $f(x)$ is always positive or is always negative. But x^2 is always positive or zero, so $x^2 + 1$ is always positive. Thus the solution of $x^2 + 1 > 0$ is $-\infty < x < \infty$.

b. $x^2 + 1 < 0$.

From part (a), $x^2 + 1$ is always positive, so the inequality $x^2 + 1 < 0$ has no solution.

Exercise 2-4

By the technique discussed in this section, solve the following inequalities.

1. $x^2 - 3x - 4 > 0$.

2. $x^2 - 8x + 15 > 0$.

3. $x^2 - 5x + 6 \leqslant 0$.

4. $14 - 5x - x^2 \leqslant 0$.

5. $2x^2 + 11x + 14 < 0$.

6. $x^2 - 4 < 0$.

7. $x^2(x^2 - 1) < 0$.

8. $2x^2 - x - 2 \leqslant 0$.

9. $(x + 2)(x - 3)(x + 6) \leqslant 0$.

10. $(x - 5)(x - 2)(x + 3) \geqslant 0$.

11. $-x(x - 5)(x + 4) > 0$.

12. $(x + 2)^2 > 0$.

13. $x^3 + 4x \geqslant 0$.

14. $(x + 2)^2(x^2 - 1) < 0$.

15. $x^3 + 2x^2 - 3x > 0$.

16. $x^3 - 4x^2 + 4x > 0$.

17. $\dfrac{x}{x^2 - 1} < 0$.

18. $\dfrac{x^2 - 1}{x} < 0$.

19. $\dfrac{4}{x - 1} \geqslant 0$.

20. $\dfrac{3}{x^2 - 5x + 6} > 0$.

21. $\dfrac{x^2 - x - 6}{x^2 + 4x - 5} \geqslant 0$.

22. $\dfrac{x^2 + 2x - 8}{x^2 + 3x + 2} \geqslant 0$.

23. $\dfrac{3}{x^2 + 6x + 8} \leqslant 0$.

24. $\dfrac{2x + 1}{x^2} \leqslant 0$.

25. $x^2 + 2x \geqslant 2$.

26. $x^4 - 16 \geqslant 0$.

27. Suppose that consumers will purchase q units of a product when the price of *each* unit is $20 - .1q$ dollars. How many units must be sold in order that sales revenue will be no less than \$750?

28. For a particular host-parasite relationship it is found that when the host density (number of hosts per unit of area) is x, then the number of hosts that are parasitized is y, where

$$y = 10\left(1 - \frac{1}{1 + 2x}\right), \quad x \geqslant 0.$$

Find the minimum host density for which at least eight hosts are parasitized.

2-5 WHY CONTINUOUS FUNCTIONS?

Often it is helpful to describe a situation by a continuous function. For example, the demand schedule in Table 2-4 indicates the number of units of a particular product that consumers will demand per week at

TABLE 2-4 Demand Schedule

p (PRICE/UNIT IN DOLLARS)	q (QUANTITY PER WEEK)
20	0
10	5
5	15
4	20
2	45
1	95

various prices. This information can be given graphically as in Fig. 2-26(a) by plotting each quantity-price pair as a point. Clearly this graph does not represent a continuous function. Furthermore, it gives us no information as to the price at which, say, 35 units would be demanded. However, if we connect the points in Fig. 2-26(a) by a smooth curve [see Fig. 2-26(b)], we get a so-called *demand curve*. From it we could guess that at about \$2.50 per unit, 35 units would be demanded.

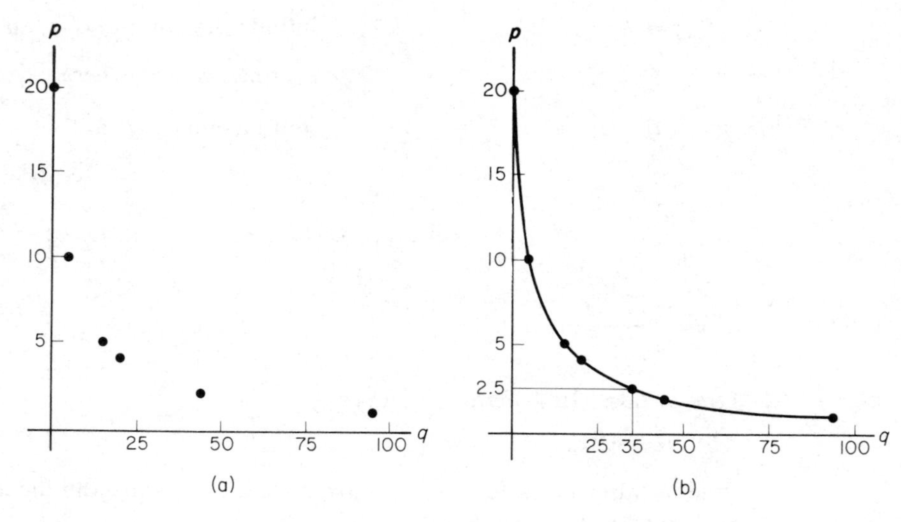

FIG. 2-26

Frequently, it is possible and useful to describe a graph, as in Fig. 2-26(b), by means of an explicit equation which defines a continuous function f. Such a function not only gives us a demand equation, $p = f(q)$, which allows us to anticipate corresponding prices and quantities demanded, but it also permits a convenient mathematical analysis of the nature and basic properties of demand. Of course some care must be used in working with equations such as $p = f(q)$. Mathematically, f may be defined when $q = \sqrt{37}$, but from a practical standpoint a demand of $\sqrt{37}$ units could be meaningless to our particular situation. For example, if a unit is an egg, then a demand of $\sqrt{37}$ eggs makes no sense.

In general it will be our desire to view practical situations in terms of continuous functions whenever possible so that we may be better able to analyze their nature.

2-6 REVIEW

IMPORTANT TERMS AND SYMBOLS IN CHAPTER 2

$\lim_{x \to a} f(x) = L$ *(p. 47)*

$\lim_{x \to a} f(x) = \infty$ *(p. 54)*

$\lim_{x \to \infty} f(x) = L$ *(p. 55)*

one-sided limits *(p. 53)*

continuous *(p. 63)*

discontinuous *(p. 63)*

$\lim_{x \to -\infty} f(x) = L$ *(p. 55)*

$\lim_{x \to a^+} f(x)$ *(p. 52)*

$\lim_{x \to a^-} f(x)$ *(p. 53)*

infinite discontinuity *(p. 65)*

continuous everywhere *(p. 64)*

step function *(p. 67)*

REVIEW SECTION

1. $\lim_{x \to a} x =$ ________.

Ans. a.

2. True or false: In general, $\lim_{x \to a} f(x) = f(a)$. ________

Ans. false.

3. True or false: For a function to have a limit at a point, the function must be defined at that point. ________

Ans. false.

4. If $\lim_{x \to a} f(x) = \infty$, then $\lim_{x \to a} \dfrac{1}{f(x)} =$ ________.

Ans. 0.

5. If $\lim_{x \to a} f(x) = 10$, then $\lim_{x \to a} [6f(x)] =$ ________.

Ans. 60.

6. $\lim_{x \to \infty} \dfrac{1}{x} =$ (a)________, $\lim_{x \to \infty} \dfrac{1}{x^2} =$ (b)________, and $\lim_{x \to \infty} \dfrac{3x^2}{6x} =$ (c)________.

Ans. (a) 0; (b) 0; (c) ∞.

7. $\lim_{h \to 0} (x + h) =$ ________.

Ans. x.

8. To solve $x(x - 2) > 0$ we consider how many intervals? ________

Ans. 3.

9. If $\lim_{x \to a^+} f(x) = L_1$ and $\lim_{x \to a^-} f(x) = L_2$ and $L_1 \neq L_2$, then f (is) (is not) continuous at $x = a$. ________

Ans. is not.

10. The function $f(x) = (4 - 2x)/(2 + x)$ is discontinuous at $x =$ ________.

Ans. -2.

11. If f is continuous at $x = a$, then $\lim_{x \to a} f(x) =$ ________.

Ans. $f(a)$.

REVIEW PROBLEMS

In Problems **1–20**, *find the limits if they exist. Where appropriate, use the symbols* ∞ *or* $-\infty$.

1. $\lim_{x \to -1} (2x^2 + 6x - 1)$.

2. $\lim_{x \to 0} \frac{2x^2 - 3x + 1}{2x^2 - 2}$.

3. $\lim_{x \to 3} \frac{x^2 - 9}{x^2 - 3x}$.

4. $\lim_{x \to -2} \frac{x + 1}{x^2 - 2}$.

5. $\lim_{h \to 0} (x + h)$.

6. $\lim_{x \to 2} \frac{x^2 - 4}{x^2 - 3x + 2}$.

7. $\lim_{x \to \infty} \frac{2}{x + 1}$.

8. $\lim_{x \to \infty} \frac{x^2 + 1}{x^2}$.

9. $\lim_{x \to \infty} \frac{3x - 2}{5x + 3}$.

10. $\lim_{x \to -\infty} \frac{1}{x^4}$.

11. $\lim_{t \to 3} \frac{2t - 3}{t - 3}$.

12. $\lim_{x \to -\infty} \frac{x^6}{x^5}$.

13. $\lim_{x \to -\infty} \frac{x + 3}{1 - x}$.

14. $\lim_{x \to 4} \sqrt{4}$.

15. $\lim_{y \to 5^+} \sqrt{y - 5}$.

16. $\lim_{x \to 1} f(x)$ if

$$f(x) = \begin{cases} x^2, & \text{if } 0 \leqslant x < 1, \\ x, & \text{if } x > 1. \end{cases}$$

17. $\lim_{x \to \infty} \frac{x^2 - 1}{(3x + 2)^2}$.

18. $\lim_{x \to 1} \frac{x^2 + x - 2}{x - 1}$.

19. $\lim_{x \to 3^-} \frac{x + 3}{x^2 - 9}$.

20. $\lim_{x \to 2} \frac{2 - x}{x - 2}$.

21. For a particular predator-prey relationship, it was determined that the number y of prey consumed by an individual predator over a period of time was a function of prey density x (the number of prey per unit of area). Suppose

$$y = f(x) = \frac{10x}{1 + .1x}.$$

If the prey density were to increase without bound, what value would y approach?

22. For a particular host-parasite relationship, it was determined that when the host density (number of hosts per unit of area) is x, then the number of hosts parasitized over a certain period of time is y, where

$$y = 11\left[1 - \frac{1}{1 + 2x}\right].$$

If the host density were to increase without bound, what value would y approach?

23. Using the definition of continuity, show that $f(x) = x + 5$ is continuous at $x = 7$.

24. Using the definition of continuity, show that $f(x) = (x - 3)/(x^2 + 4)$ is continuous at $x = 3$.

25. State whether $f(x) = x/4$ is continuous everywhere. Give a reason for your answer.

26. State whether $f(x) = x^2 - 2$ is continuous everywhere. Give a reason for your answer.

In Problems **27–30**, *determine the points of discontinuity for each function.*

27. $f(x) = \dfrac{x^2}{x + 3}.$

28. $f(x) = \dfrac{0}{x^3}.$

29. $f(x) = \begin{cases} x + 4, & \text{if } x > -2, \\ 3x - 1, & \text{if } x \leqslant -2. \end{cases}$

30. $f(x) = \begin{cases} \dfrac{x}{x + 1}, & \text{if } x > 1, \\ \dfrac{3}{x + 4}, & \text{if } x < 1. \end{cases}$

In Problems **31–38**, *solve the given inequalities.*

31. $x^2 + 4x - 12 > 0.$

32. $2x^2 - 6x + 4 \leqslant 0.$

33. $x^3 \geqslant 2x^2.$

34. $x^3 + 8x^2 + 15x \geqslant 0.$

35. $\dfrac{x + 5}{x^2 - 1} < 0.$

36. $\dfrac{x(x + 5)(x + 8)}{3} < 0.$

37. $\dfrac{x^2 + 3x}{x^2 + 2x - 8} \geqslant 0.$

38. $\dfrac{x^2 - 4}{x^2 + 2x + 1} \geqslant 0.$

CHAPTER 3

Differentiation

Now we begin our study of calculus. The ideas involved in calculus are completely different from those of algebra and geometry. Moreover, the power and importance of these ideas and their applications will be evident to you later in the text. The objective of this chapter is not only to understand what the so-called "derivative" of a function is and means, but also to learn techniques of finding derivatives by properly applying rules.

3-1 THE DERIVATIVE

One of the main problems with which calculus deals is finding the slope of the *tangent line* at a point on a curve. In geometry you probably thought of a tangent line, or *tangent*, to a circle as a line which meets the circle at exactly one point (Fig. 3-1). But this idea of a tangent is not very useful for other kinds of curves.

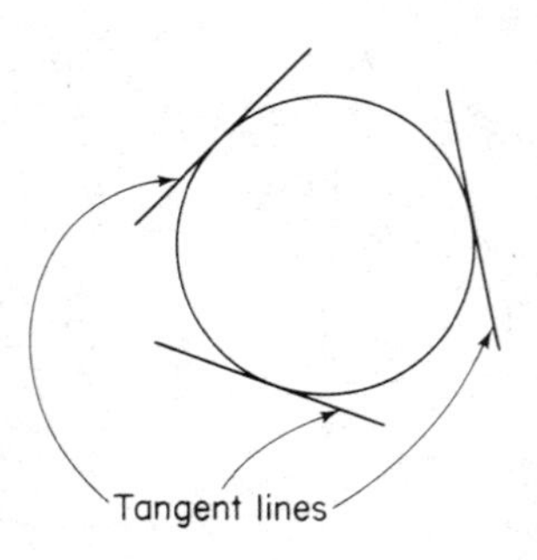

FIG. 3-1

For example, in Fig. 3-2(a) the lines L_1 and L_2 intersect the curve at exactly one point. Intuitively we would not think of L_2 as the tangent at this point, but it seems natural that L_1 is. Also, in Fig. 3-2(b) we

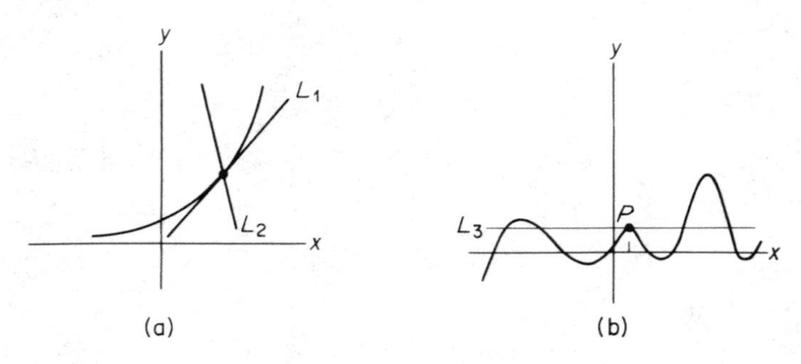

FIG. 3-2

would consider L_3 to be the tangent at point P even though it intersects the curve at other points. From these examples you can see that we must drop the idea that a tangent intersects a curve at only one point. To develop a suitable definition of tangent line, we use the limit concept.

Look at the graph of the function $y = f(x)$ in Fig. 3-3. Here $P(x_1, y_1)$ and $Q(x_2, y_2)$ are two different points on the curve. The line PQ passing through them is called a *secant line*. By the slope formula, the slope of PQ is

$$m_{PQ} = \frac{y_2 - y_1}{x_2 - x_1}.$$

If Q moves along the curve and approaches P, the secant line has a limiting position as shown in Fig. 3-4. As Q approaches P from the right, the positions of the secant lines are PQ', PQ'', etc. As Q approaches P from the left, they are PQ_1, PQ_2, etc. *In both cases, the* ***same***

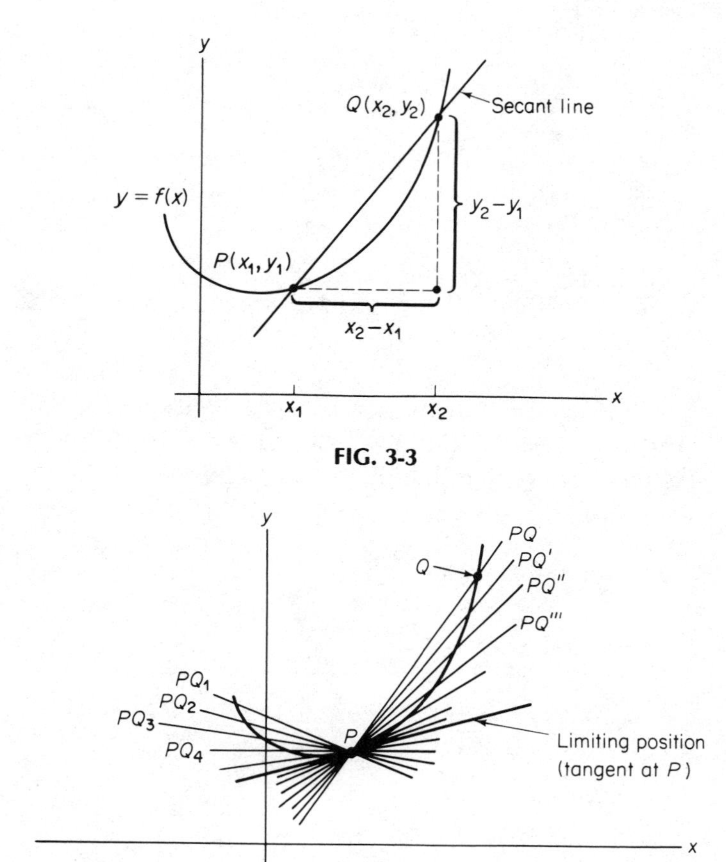

FIG. 3-3

FIG. 3-4

limiting position is obtained. This common limiting position of secant lines is called the **tangent line** to the curve at P. This definition seems reasonable and avoids the difficulties mentioned at the beginning of this section.

Not every curve has a tangent at each of its points. For example, the curve $y = |x|$ does not have a tangent at (0, 0). In Fig. 3-5, at (0, 0) a secant line from the right must always be the line $y = x$, and from the left it is the line $y = -x$. Since there is no common limiting position, there is no tangent.

Now that we have a suitable definition of a tangent to a curve at a point, we can define the *slope of a curve* at a point.

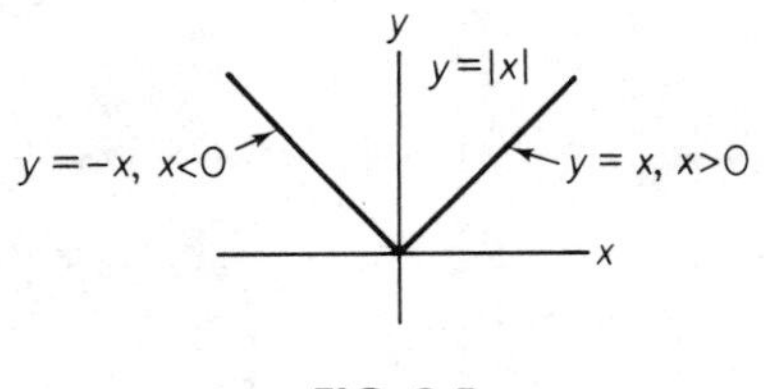

FIG. 3-5

DEFINITION. *The **slope of a curve at a point P** is the slope of the tangent line at P.*

Since the tangent is a limiting position of secant lines, the slope of the tangent is the limiting value of the slopes of the secant lines PQ as Q approaches P. We shall now find an expression for the slope of the curve $y = f(x)$ at the point $(x_1, f(x_1))$. In Fig. 3-6 the slope of the secant line PQ is

$$m_{PQ} = \frac{f(x_2) - f(x_1)}{x_2 - x_1}.$$

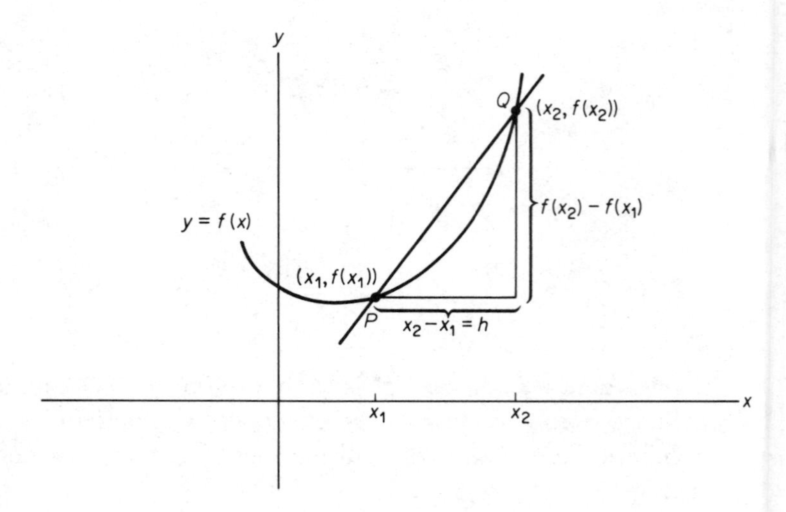

FIG. 3-6

If the difference $x_2 - x_1$ is h, then we can write x_2 as $x_1 + h$. Here $h \neq 0$, for if $h = 0$ then $x_2 = x_1$ and no secant line would exist. Thus,

$$m_{PQ} = \frac{f(x_1 + h) - f(x_1)}{(x_1 + h) - x_1} = \frac{f(x_1 + h) - f(x_1)}{h}.$$

As Q moves along the curve towards P, then $x_2 \to x_1$. This means that h is getting closer and closer to zero. The limiting value of the slopes of the secant lines at the point $(x_1, f(x_1))$—which is the slope of the tangent line at $(x_1, f(x_1))$—is the limit:

$$\lim_{h \to 0} \frac{f(x_1 + h) - f(x_1)}{h}. \tag{1}$$

EXAMPLE 1

The slope of the curve $y = f(x) = x^2$ at the point (1, 1) is

$$\begin{aligned}\lim_{h \to 0} \frac{f(1 + h) - f(1)}{h} &= \lim_{h \to 0} \frac{(1 + h)^2 - (1)^2}{h} \\ &= \lim_{h \to 0} \frac{1 + 2h + h^2 - 1}{h} = \lim_{h \to 0} \frac{2h + h^2}{h} \\ &= \lim_{h \to 0} \frac{h(2 + h)}{h} = \lim_{h \to 0} (2 + h) = 2.\end{aligned}$$

Thus the tangent line to $y = x^2$ at (1, 1) has a slope of 2 (Fig. 3-7).

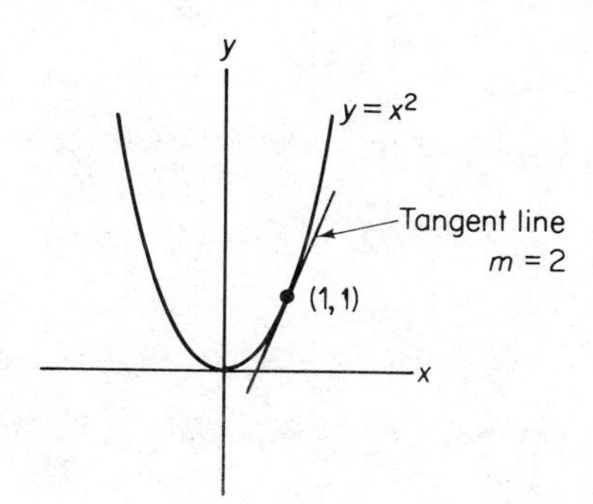

FIG. 3-7

We can generalize the result in (1) to any point $(x, f(x))$ on the curve by replacing x_1 by x. We thus have the following definition, which forms the basis of differential calculus.

DEFINITION. *If $y = f(x)$ defines a function f, the limit*

$$\lim_{h \to 0} \frac{f(x + h) - f(x)}{h},$$

if it exists, is called the ***derivative*** *of f at x and is denoted $f'(x)$, which is read "f prime of x." The process of finding the derivative is called* ***differentiation****.*

Sometimes the derivative of $y = f(x)$ at x is referred to as the derivative *with respect to x*.

EXAMPLE 2

If $f(x) = x^2$, find the derivative of f.

$$f'(x) = \lim_{h \to 0} \frac{f(x+h) - f(x)}{h}$$

$$= \lim_{h \to 0} \frac{(x+h)^2 - x^2}{h} = \lim_{h \to 0} \frac{x^2 + 2xh + h^2 - x^2}{h}$$

$$= \lim_{h \to 0} \frac{2xh + h^2}{h} = \lim_{h \to 0} \frac{h(2x+h)}{h} = \lim_{h \to 0} (2x + h) = 2x.$$

Note that in taking the limit we treated x as a constant because it was h, not x, that was changing. Also note that $f'(x) = 2x$ is a function of x. In all cases, **the derivative of a function f is also a function, f'.**

Ways of denoting the derivative of $y = f(x)$ at x are

$f'(x)$, y' (y prime),

$\frac{d}{dx}[f(x)]$ (pronounced "dee $f(x)$, dee x"), $\frac{dy}{dx}$ (dee y, dee x),

$D_x[f(x)]$ (dee x of $f(x)$), $D_x y$ (dee x of y).

PITFALL. $\frac{dy}{dx}$ *is not a fraction, but is a symbol for a derivative. We have not yet attached any meaning to individual symbols such as dy and dx.*

If the derivative of f can be evaluated at $x = x_1$, the resulting *number* is called the derivative of f at x_1, denoted $f'(x_1)$. In this case we say that f is *differentiable* at x_1. This means that

$f'(x_1)$ is the slope of the tangent to $y = f(x)$ at $(x_1, f(x_1))$.

In addition to the notation $f'(x_1)$ we can also write

$$\left.\frac{dy}{dx}\right|_{x=x_1} \quad \text{and} \quad y'(x_1).$$

EXAMPLE 3

If $f(x) = 2x^2 + 2x + 3$, find $f'(1)$.

We shall find the derivative $f'(x)$ and then evaluate it at $x = 1$.

$$f'(x) = \lim_{h \to 0} \frac{f(x+h) - f(x)}{h}$$

$$= \lim_{h \to 0} \frac{\left[2(x+h)^2 + 2(x+h) + 3\right] - (2x^2 + 2x + 3)}{h}$$

$$= \lim_{h \to 0} \frac{2x^2 + 4xh + 2h^2 + 2x + 2h + 3 - 2x^2 - 2x - 3}{h}$$

$$= \lim_{h \to 0} \frac{4xh + 2h^2 + 2h}{h} = \lim_{h \to 0} (4x + 2h + 2).$$

$$f'(x) = 4x + 2.$$

$$f'(1) = 4(1) + 2 = 6.$$

EXAMPLE 4

a. *Find the slope of the curve $y = x^2$ at the point $(3, 9)$. Then find an equation of the tangent line at $(3, 9)$.*

From Example 2, $y' = 2x$. Thus $y'(3) = 2(3) = 6$. That is, the tangent line to the curve $y = x^2$ at $(3, 9)$ has a slope of 6. A point-slope form of the tangent line is $y - 9 = 6(x - 3)$, from which $y = 6x - 9$ (see Fig. 3-8).

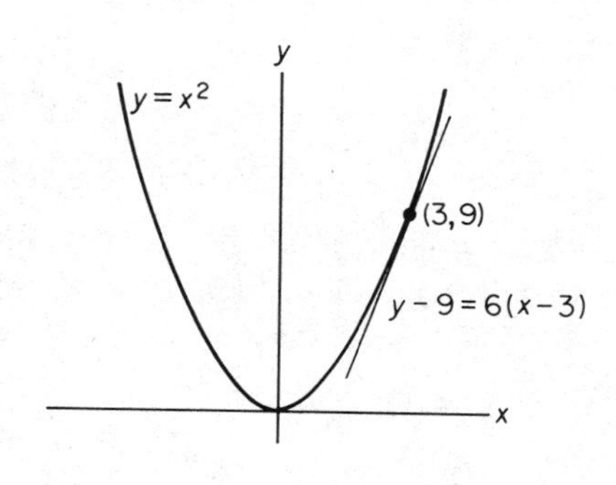

FIG. 3-8

PITFALL. *It is **not** correct to say that since the derivative of $y = x^2$ is $2x$, the tangent at $(3, 9)$ is $y - 9 = 2x(x - 3)$. The derivative must be **evaluated** at the point of tangency to determine the slope of the tangent line.*

b. *Find the slope of the curve $y = 2x + 3$ at the point where $x = 6$.*

Letting $y = f(x) = 2x + 3$, we have

$$y' = \lim_{h\to 0} \frac{f(x+h) - f(x)}{h} = \lim_{h\to 0} \frac{[2(x+h)+3] - (2x+3)}{h}$$

$$= \lim_{h\to 0} \frac{2h}{h} = \lim_{h\to 0} 2 = 2.$$

Since $y' = 2$, the slope when $x = 6$, or at any point, is 2. Note that the curve is a straight line and thus has a constant slope at each point.

EXAMPLE 5

Find $D_x(\sqrt{x})$.

If $f(x) = \sqrt{x}$, then

$$D_x(\sqrt{x}) = \lim_{h\to 0} \frac{f(x+h) - f(x)}{h} = \lim_{h\to 0} \frac{\sqrt{x+h} - \sqrt{x}}{h}.$$

As $h \to 0$, both the numerator and denominator approach zero. This can be avoided by rationalizing the numerator.

$$\frac{\sqrt{x+h} - \sqrt{x}}{h} = \frac{\sqrt{x+h} - \sqrt{x}}{h} \cdot \frac{\sqrt{x+h} + \sqrt{x}}{\sqrt{x+h} + \sqrt{x}} = \frac{(x+h) - x}{h(\sqrt{x+h} + \sqrt{x})}$$

$$= \frac{h}{h(\sqrt{x+h} + \sqrt{x})} = \frac{1}{\sqrt{x+h} + \sqrt{x}}.$$

Thus,

$$D_x(\sqrt{x}) = \lim_{h\to 0} \frac{1}{\sqrt{x+h} + \sqrt{x}} = \frac{1}{\sqrt{x} + \sqrt{x}} = \frac{1}{2\sqrt{x}}.$$

Note that the original function, $\sqrt{x}$, is defined for $x \geqslant 0$. But the derivative, $1/(2\sqrt{x})$, is defined only when $x > 0$. From the graph of $y = \sqrt{x}$ in Fig. 3-9, it is clear that when $x = 0$ the tangent line is vertical and hence does not have a slope.

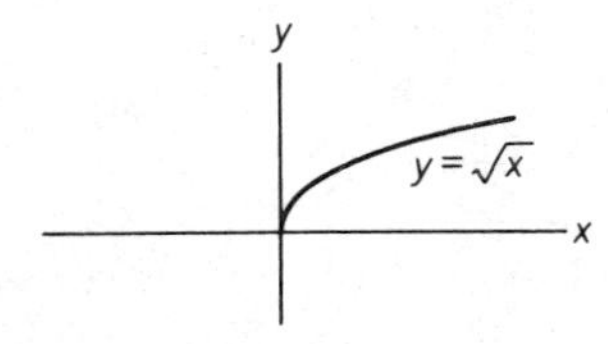

FIG. 3-9

If a variable, say q, is a function of some variable, say p, then we would speak of the derivative of q with respect to p and could write dq/dp.

EXAMPLE 6

If $q = f(p) = \dfrac{1}{2p}$, *find* dq/dp.

$$\frac{dq}{dp} = \lim_{h\to 0} \frac{f(p+h) - f(p)}{h}$$

$$= \lim_{h\to 0} \frac{\dfrac{1}{2(p+h)} - \dfrac{1}{2p}}{h} = \lim_{h\to 0} \frac{p - (p+h)}{h[2p(p+h)]}$$

$$= \lim_{h\to 0} \frac{-h}{h[2p(p+h)]} = \lim_{h\to 0} \frac{-1}{2p(p+h)} = -\frac{1}{2p^2}.$$

Note that when $p = 0$, neither the function nor its derivative exists.

As a final note, we point out that the derivative of $y = f(x)$ at x, namely $f'(x)$, is nothing more than the limit

$$\lim_{h\to 0} \frac{f(x+h) - f(x)}{h}.$$

However, we can interpret f' as a function that gives the slope of the tangent line to the curve $y = f(x)$ at any point $(x, f(x))$. This interpretation is simply a geometric convenience that assists our understanding. The above limit may exist aside from any geometric consideration at all. As you will see later, there are other useful interpretations.

Exercise 3-1

In Problems **1–16**, *use the definition of the derivative to find each of the following.*

1. $\dfrac{d}{dx}[f(x)]$ if $f(x) = x$.

2. $\dfrac{d}{dx}[f(x)]$ if $f(x) = 4x - 1$.

3. $\dfrac{dy}{dx}$ if $y = 2x + 4$.

4. $\dfrac{dy}{dx}$ if $y = -3x$.

5. $\dfrac{d}{dx}(3 - 2x)$.

6. $\dfrac{d}{dx}\left(4 - \dfrac{x}{2}\right)$.

7. $f'(x)$ if $f(x) = 3$.

8. $f'(x)$ if $f(x) = 7.01$.

9. $D_x(x^2 + 4x - 8)$.

10. $D_x y$ if $y = x^2 + 5$.

11. $\frac{dq}{dp}$ if $q = 2p^2 + 5p - 1$.

12. $D_x(x^2 - x - 3)$.

13. $D_x y$ if $y = \frac{1}{x}$.

14. $\frac{dC}{dq}$ if $C = 7 + 2q - 3q^2$.

15. $f'(x)$ if $f(x) = \sqrt{x + 2}$.

16. $g'(x)$ if $g(x) = \frac{2}{x - 3}$.

17. Find the slope of the curve $y = x^2 + 4$ at the point $(-2, 8)$.

18. Find the slope of the curve $y = 2 - 3x^2$ at the point $(1, -1)$.

19. Find the slope of the curve $y = 4x^2 - 5$ when $x = 0$.

In Problems **20–25**, *find an equation of the tangent line to the curve at the given point.*

20. $y = x + 4$; $(3, 7)$.

21. $y = -2x^2$; $(1, -2)$.

22. $y = 2x^2 - 5$; $(-2, 3)$.

23. $y = \frac{3}{x}$; $(3, 1)$.

24. $y = \frac{5}{1 - 3x}$; $(2, -1)$.

25. $y = 3x^2 + 3x - 4$; $(-1, -4)$.

3-2 RULES FOR DIFFERENTIATION

By now you would probably agree that differentiating a function by direct use of the definition of a derivative can be quite tedious. Fortunately there are rules that give us completely mechanical and efficient procedures for differentiation. They also avoid direct use of limits. We shall look at some rules in this section.

To begin with, recall that the graph of the constant function $f(x) = c$ is a horizontal line (Fig. 3-10), which has a slope of zero everywhere. This means that $f'(x) = 0$, which is our first rule. We shall give a formal proof.

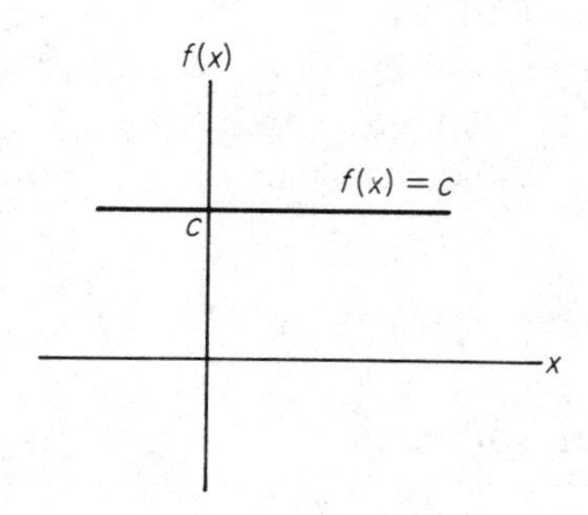

FIG. 3-10

RULE 1. *If $f(x) = c$, where c is a constant, then $f'(x) = 0$. That is, the derivative of a constant function is zero.*

PROOF. If $f(x) = c$, applying the definition of the derivative gives

$$f'(x) = \lim_{h \to 0} \frac{f(x+h) - f(x)}{h} = \lim_{h \to 0} \frac{c - c}{h}$$

$$= \lim_{h \to 0} \frac{0}{h} = \lim_{h \to 0} 0 = 0.$$

Therefore,

$$\boxed{\text{If } f(x) = c, \text{ then } f'(x) = 0.}$$

EXAMPLE 1

a. If $f(x) = 3$, then $f'(x) = 0$.

b. If $g(x) = \sqrt{5}$, then $g'(x) = 0$. For example, the derivative of g when $x = 4$ is $g'(4) = 0$.

c. If $s(t) = (1{,}938{,}623)^{807.4}$, then $ds/dt = 0$ since s is a constant function.

The proof of the next rule requires that we expand a binomial. Recall that

$$(x + h)^2 = x^2 + 2xh + h^2,$$

and

$$(x + h)^3 = x^3 + 3x^2h + 3xh^2 + h^3.$$

In both expansions, notice in reading from left to right that the exponents of x decrease while those of h increase. This is true for the general case $(x + h)^n$, for n a positive integer. That is,

$$(x + h)^n = x^n + nx^{n-1}h + (\quad)x^{n-2}h^2 + \ldots + (\quad)xh^{n-1} + h^n,$$

where the missing numbers inside the parentheses are constants. This formula is used to prove the next rule, which involves the derivative of x raised to a constant power.

RULE 2. *If $f(x) = x^n$, where n is any real number, then $f'(x) = nx^{n-1}$.*

PROOF. We shall give a proof for the case that n is a positive integer. If $f(x) = x^n$, applying the definition of the derivative gives

$$f'(x) = \lim_{h \to 0} \frac{f(x + h) - f(x)}{h} = \lim_{h \to 0} \frac{(x + h)^n - x^n}{h}.$$

By our discussion above,

$$(x + h)^n = x^n + nx^{n-1}h + (\quad)x^{n-2}h^2 + \ldots + h^n.$$

Thus,

$$f'(x) = \lim_{h \to 0} \frac{x^n + nx^{n-1}h + (\quad)x^{n-2}h^2 + \ldots + h^n - x^n}{h}.$$

In the numerator the sum of the first and last terms is 0. Dividing each of the remaining terms by h gives

$$f'(x) = \lim_{h \to 0} \left[nx^{n-1} + (\quad)x^{n-2}h + \ldots + h^{n-1} \right].$$

Each term after the first has h as a factor and must approach 0 as $h \to 0$. Hence,

$$f'(x) = nx^{n-1}.$$

Therefore,

If $f(x) = x^n$, *then* $f'(x) = nx^{n-1}$.

By Rule 2 *the derivative of a constant power of x is the exponent times x raised to a power one less than the given power.*

EXAMPLE 2

a. If $f(x) = x^2$, then by Rule 2,

$$f'(x) = 2x^{2-1} = 2x.$$

b. If $g(w) = w^{9/4}$, then by Rule 2,

$$g'(w) = \tfrac{9}{4}w^{(9/4)-1} = \tfrac{9}{4}w^{5/4}.$$

c. If $F(x) = x = x^1$, then

$$\frac{d}{dx}[F(x)] = \frac{d}{dx}(x) = 1 \cdot x^{1-1} = 1 \cdot x^0 = 1.$$

Thus the derivative of x with respect to x is 1.

d. Suppose $y = x\sqrt{x}$. To find $D_x y$ we first write $x\sqrt{x}$ as $x^{3/2}$. Thus $y = x^{3/2}$ and

$$D_x y = \tfrac{3}{2}x^{(3/2)-1} = \tfrac{3}{2}x^{1/2} = \tfrac{3}{2}\sqrt{x}.$$

e. Suppose $h(x) = \dfrac{1}{x^{3/2}}$. To apply Rule 2 we *must* write $h(x)$ as $h(x) = x^{-3/2}$.

$$D_x\left(\frac{1}{x^{3/2}}\right) = D_x(x^{-3/2}) = -\frac{3}{2}x^{(-3/2)-1} = -\frac{3}{2}x^{-5/2}.$$

PITFALL. $D_x\left(\dfrac{1}{x^{3/2}}\right) \neq \dfrac{1}{\frac{3}{2}x^{1/2}}$. *Do not merely differentiate the denominator.*

RULE 3. *If $g(x) = c\,f(x)$ and $f'(x)$ exists, then $g'(x) = c\,f'(x)$. That is, the derivative of a constant times a function is the constant times the derivative of the function.*

PROOF. If $g(x) = c\,f(x)$, applying the definition of the derivative of g gives

$$g'(x) = \lim_{h\to 0}\frac{g(x+h) - g(x)}{h} = \lim_{h\to 0}\frac{c\,f(x+h) - c\,f(x)}{h}$$

$$= \lim_{h\to 0}\left[c \cdot \frac{f(x+h) - f(x)}{h}\right] = c \cdot \lim_{h\to 0}\frac{f(x+h) - f(x)}{h}.$$

But $\displaystyle\lim_{h\to 0}\frac{f(x+h) - f(x)}{h} = f'(x)$, and thus $g'(x) = c\,f'(x)$. Therefore,

If $g(x) = c\,f(x)$, then $g'(x) = c\,f'(x)$.

EXAMPLE 3

Find the derivative of each of the following functions.

a. $g(x) = 5x^3$.

If we let $f(x) = x^3$, then $g(x) = 5f(x)$ and

$$\begin{aligned} g'(x) &= 5f'(x) && \text{(Rule 3)} \\ &= 5D_x(x^3) = 5(3x^{3-1}) && \text{(Rule 2)} \\ &= 15x^2. \end{aligned}$$

b. $g(p) = \frac{13}{2}p$.

If we let $f(p) = p$, then $g(p) = \frac{13}{2}f(p)$ and by Rule 3,

$$g'(p) = \tfrac{13}{2}f'(p) = \tfrac{13}{2}D_p(p) = \tfrac{13}{2}(1) = \tfrac{13}{2}.$$

c. $y = \dfrac{.702}{x^2\sqrt{x}} = .702x^{-5/2}$.

Note that y can be considered a constant times a function.

$$\begin{aligned} D_x y &= .702D_x(x^{-5/2}) && \text{(Rule 3)} \\ &= .702\left(-\tfrac{5}{2}x^{-7/2}\right) = -1.755x^{-7/2}. && \text{(Rule 2)} \end{aligned}$$

PITFALL. *If $f(x) = (4x)^3$, you may be tempted to write $f'(x) = 3(4x)^2$. This is* ***incorrect****! The reason is that Rule 2 applies to a power of the variable x,* ***not*** *a power of a function such as $4x$. To apply our rules we must get a suitable form for $f(x)$. We can write $(4x)^3$ as 4^3x^3 or $64x^3$. Thus,*

$$f'(x) = 64D_x(x^3) = 64(3x^2) = 192x^2.$$

The next two rules involve derivatives of sums and differences of functions.

RULE 4. *If $F(x) = f(x) + g(x)$ and $f'(x)$ and $g'(x)$ exist, then $F'(x) = f'(x) + g'(x)$.*
That is, the derivative of a sum of two functions is the sum of the derivatives of the functions.

PROOF. If $F(x) = f(x) + g(x)$, applying the definition of the derivative of F gives

$$\begin{aligned}
F'(x) &= \lim_{h\to 0} \frac{F(x+h) - F(x)}{h} \\
&= \lim_{h\to 0} \frac{[f(x+h) + g(x+h)] - [f(x) + g(x)]}{h} \\
&= \lim_{h\to 0} \frac{[f(x+h) - f(x)] + [g(x+h) - g(x)]}{h} \\
&= \lim_{h\to 0} \left[\frac{f(x+h) - f(x)}{h} + \frac{g(x+h) - g(x)}{h}\right].
\end{aligned}$$

Since the limit of a sum is the sum of the limits,

$$F'(x) = \lim_{h\to 0} \frac{f(x+h) - f(x)}{h} + \lim_{h\to 0} \frac{g(x+h) - g(x)}{h}.$$

But these two limits are $f'(x)$ and $g'(x)$. Thus,

$$F'(x) = f'(x) + g'(x).$$

Therefore,

If $F(x) = f(x) + g(x)$, then $F'(x) = f'(x) + g'(x)$.

The proof of the next rule is similar to that of Rule 4.

RULE 5. *If $F(x) = f(x) - g(x)$ and $f'(x)$ and $g'(x)$ exist, then $F'(x) = f'(x) - g'(x)$.*
That is, the derivative of a difference of two functions is the difference of the derivatives of the functions.

If $F(x) = f(x) - g(x)$, then $F'(x) = f'(x) - g'(x)$.

Rules 4 and 5 can be extended to the derivative of any finite number of sums and differences of functions. For example, if $F(x) = f(x) - g(x) + h(x) + k(x)$, then $F'(x) = f'(x) - g'(x) + h'(x) + k'(x)$.

EXAMPLE 4

Differentiate each of the following functions.

a. $F(x) = 3x^5 + \sqrt{x}$.

Let $f(x) = 3x^5$ and $g(x) = \sqrt{x} = x^{1/2}$. Then $F(x) = f(x) + g(x)$, a sum of two functions. Thus,

$$F'(x) = f'(x) + g'(x) = D_x(3x^5) + D_x(x^{1/2}) \qquad \text{(Rule 4)}$$

$$= 3D_x(x^5) + D_x(x^{1/2}) \qquad \text{(Rule 3)}$$

$$= 3(5x^4) + \frac{1}{2}x^{-1/2} = 15x^4 + \frac{1}{2\sqrt{x}}. \qquad \text{(Rule 2)}$$

b. $f(x) = x^5 - \sqrt[3]{x^2}$.

Since f is the difference of two functions,

$$f'(x) = D_x(x^5) - D_x(\sqrt[3]{x^2}) = D_x(x^5) - D_x(x^{2/3}) \qquad \text{(Rule 5)}$$

$$= 5x^4 - \frac{2}{3}x^{-1/3} = 5x^4 - \frac{2}{3\sqrt[3]{x}}. \qquad \text{(Rule 2)}$$

c. $f(z) = \dfrac{z^4}{4} - \dfrac{5}{z^{1/3}}$.

Note that we can write $f(z) = \frac{1}{4}z^4 - 5z^{-1/3}$.

$$\frac{d}{dz}[f(z)] = D_z(\tfrac{1}{4}z^4) - D_z(5z^{-1/3}) \qquad \text{(Rule 5)}$$

$$= \tfrac{1}{4}D_z(z^4) - 5D_z(z^{-1/3}) \qquad \text{(Rule 3)}$$

$$= \tfrac{1}{4}(4z^3) - 5(-\tfrac{1}{3}z^{-4/3}) \qquad \text{(Rule 2)}$$

$$= z^3 + \tfrac{5}{3}z^{-4/3}.$$

d. $y = 6x^3 - 2x^2 + 7x - 8$.

$$D_x y = D_x(6x^3) - D_x(2x^2) + D_x(7x) - D_x(8)$$

$$= 6D_x(x^3) - 2D_x(x^2) + 7D_x(x) - D_x(8)$$

$$= 6(3x^2) - 2(2x) + 7(1) - 0$$

$$= 18x^2 - 4x + 7.$$

EXAMPLE 5

Find the derivative of $f(x) = 2x(x^2 - 5x + 2)$ *when* $x = 2$.

By the distributive property,

$$f(x) = 2x^3 - 10x^2 + 4x.$$

Thus,

$$f'(x) = 2(3x^2) - 10(2x) + 4(1)$$
$$= 6x^2 - 20x + 4,$$

and

$$f'(2) = 6(2)^2 - 20(2) + 4 = -12.$$

EXAMPLE 6

Find an equation of the tangent line to the curve $y = \dfrac{3x^2 - 2}{x}$ *when* $x = 1$.

By writing y as a difference of two functions, we have $y = \dfrac{3x^2}{x} - \dfrac{2}{x} = 3x - 2x^{-1}$. Thus,

$$\frac{dy}{dx} = 3(1) - 2[(-1)x^{-2}] = 3 + \frac{2}{x^2}.$$

The slope of the tangent line to the curve when $x = 1$ is

$$\left.\frac{dy}{dx}\right|_{x=1} = 3 + \frac{2}{1^2} = 5.$$

To find the y-coordinate of the point on the curve where $x = 1$, we substitute this value of x into the *original equation* of the curve. When $x = 1$, then $y = [3(1)^2 - 2]/1 = 1$. Hence the point $(1, 1)$ lies on both the curve and the tangent line. Therefore, an equation of the tangent line is

$$y - 1 = 5(x - 1),$$
$$y = 5x - 4.$$

Exercise 3-2

In Problems **1–58**, *differentiate the functions.*

1. $f(x) = 7$.

2. $f(x) = (\frac{933}{465})^{2/3}$.

3. $f(x) = x^8$.

4. $f(x) = .7x$.

5. $f(x) = 9x^5$.

6. $f(x) = \sqrt{2}\, x^{83/4}$.

7. $g(w) = w^{-7}$.

8. $f(t) = 3t^{-2}$.

9. $f(x) = 4x^{-14/5}$.

10. $v(x) = \dfrac{x^9}{18}$.

11. $f(x) = 3x - 2.$

12. $f(w) = \frac{3w^2}{4}.$

13. $f(p) = \frac{13p}{5} + \frac{7}{3}.$

14. $q(x) = \frac{5x + 2}{8}.$

15. $g(x) = 3x^2 - 5x - 2.$

16. $f(q) = 7q^2 - 5q + 3.$

17. $f(t) = -13t^2 + 14t + 2.$

18. $p(x) = 97x^2 - 383x + 205.$

19. $f(x) = 14x^3 - 6x^2 + 7x - 8.$

20. $f(r) = -8r^3 + 6.$

21. $f(q) = -3q^3 + \frac{9}{2}q^2 + 9q + 9.$

22. $f(x) = 100x^{-3} - 50x^{-1/2} + 10x - 1.$

23. $f(x) = x^8 - 7x^6 + 3x^2 + 9^{2/3}.$

24. $g(u) = 3u^{12} - 8u^8 + 3u^{-3} + 2u^2 - 9.$

25. $f(x) = 2x^{501} - 125x^{100} + .2x^{3.4}.$

26. $f(x) = 17 + 8x^{1/7} - 10x^{12} - 3x^{-15}.$

27. $f(x) = 2(13 - x^4).$

28. $f(s) = 5(s^4 - 3).$

29. $g(x) = \frac{13 - x^4}{3}.$

30. $f(x) = \frac{5(x^4 - 3)}{2}.$

31. $f(x) = x^{-4} - 9x^{1/3} + 5x^{-2/5}.$

32. $f(z) = 3z^{1/4} - 12^2 - 8z^{-3/4}.$

33. $h(x) = -2(27x - 14x^5).$

34. $f(x) = \frac{-(1 + x - x^2 + x^3 + x^4 - x^5)}{2}.$

35. $f(x) = -2x^2 + \frac{3}{2}x + \frac{x^4}{4} + 2.$

36. $p(x) = \frac{x^7}{7} + \frac{x}{2}.$

37. $f(x) = \frac{1}{x}.$

38. $f(x) = \frac{7}{x^3}.$

39. $f(s) = \frac{1}{4s^5}.$

40. $g(w) = \frac{2}{3w^3}.$

41. $f(t) = 4\sqrt{t}.$

42. $f(x) = \frac{4}{\sqrt{x}}.$

43. $q(x) = \frac{1}{\sqrt[5]{x}}.$

44. $f(x) = \frac{3}{\sqrt[4]{x^3}}.$

45. $f(x) = x(3x^2 - 7x + 7).$

46. $f(x) = x^3(3x^6 - 5x^2 + 4).$

47. $g(t) = \frac{t^2}{2} - \frac{2}{t^2}.$

48. $f(x) = x\sqrt{x}.$

49. $f(x) = x^3(3x)^2$.

50. $f(x) = \sqrt{x}\,(5 - 6x + 3\sqrt[4]{x}\,)$.

51. $v(x) = x^{-2/3}(x + 5)$.

52. $f(x) = x^{3/5}(x^2 + 7x + 1)$.

53. $f(q) = \dfrac{4q^3 + 7q - 4}{q}$.

54. $f(w) = \dfrac{w - 5}{w^5}$.

55. $f(x) = (x + 1)(x + 3)$.

56. $f(x) = x^2(x - 2)(x + 4)$.

57. $w(x) = \dfrac{x^2 + x^3}{x^2}$.

58. $f(x) = \dfrac{7x^3 + x}{2\sqrt{x}}$.

For each curve in Problems **59–62**, *find the slope at the indicated points.*

59. $y = 3x^2 + 4x - 8$; $(0, -8), (2, 12), (-3, 7)$.

60. $y = 5 - 6x - 2x^3$; $(0, 5), (\frac{3}{2}, -\frac{43}{4}), (-3, 77)$.

61. $y = 4$; when $x = -4, x = 7, x = 22$.

62. $y = 2x - 3\sqrt{x}$; when $x = 1, x = 16, x = 25$.

In Problems **63** *and* **64**, *find an equation of the tangent line to the curve at the indicated point.*

63. $y = 4x^2 + 5x + 2$; $(1, 11)$.

64. $y = (1 - x^2)/5$; $(4, -3)$.

65. Find an equation of the tangent line to the curve $y = 3 + x - 5x^2 + x^4$ when $x = 0$.

66. Repeat Problem 65 for the curve $y = \dfrac{\sqrt{x}\,(2 - x^2)}{x}$ when $x = 4$.

67. Find all points on the curve $y = \frac{1}{3}x^3 - x^2$ where the tangent line is horizontal.

68. Find all points on the curve $y = x^2 - 5x + 3$ where the slope is 1.

3-3 THE DERIVATIVE AS A RATE OF CHANGE

To denote the change in a variable such as x, the symbol Δx (read "delta x") is sometimes used. For example, if x changes from $x = 1$ to $x = 3$, then the change is $\Delta x = 3 - 1 = 2$. The new value of x $(= 3)$ can be written $1 + \Delta x$. Similarly, if q increases by Δq, the new value is $q + \Delta q$. We shall use Δ-notation in the following discussion.

A manufacturer's **total cost function,** $c = f(q)$, gives the total cost c of producing and marketing q units of a product. Figure 3-11 shows the graph of f, which is called a *total cost curve*. Suppose a manufacturer

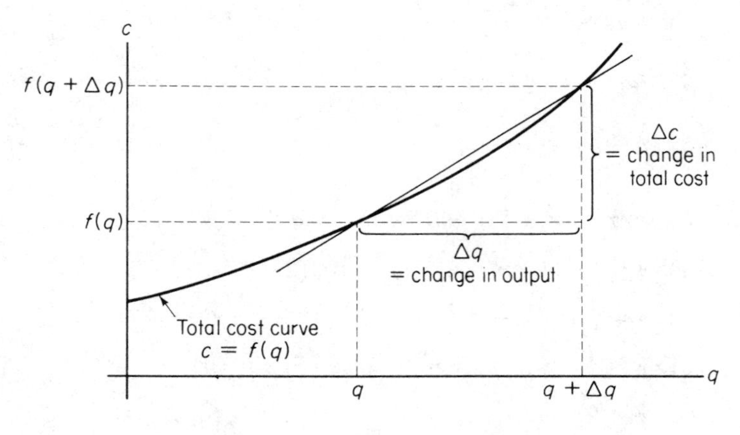

FIG. 3-11

produces q units at a total cost of $f(q)$. If the production level is increased by Δq units to $q + \Delta q$, then the total cost is $f(q + \Delta q)$. The average cost per unit for these Δq additional units is

$$\frac{\text{change in total cost } c}{\text{change in output } q} = \frac{f(q + \Delta q) - f(q)}{\Delta q}.$$

Denoting the change in total cost by Δc, we have

$$\frac{\Delta c}{\Delta q} = \frac{f(q + \Delta q) - f(q)}{\Delta q}, \tag{1}$$

which is called the **average rate of change of cost c with respect to output q** over the interval $[q, q + \Delta q]$. Notice in Fig. 3-11 that $\Delta c/\Delta q$ is the slope of a secant line.

To be more specific, suppose $c = f(q) = .1q^2 + 3$, where c is in dollars and q is in pounds. The cost of 4 lb is $f(4) = 4.6$. If output increases by 2 lb ($\Delta q = 2$), the new level of production is $4 + \Delta q = 6$, and the total cost is $f(4 + \Delta q) = f(6) = 6.6$. Thus the average cost per pound of the additional output on the interval [4, 6] is

$$\frac{\Delta c}{\Delta q} = \frac{f(6) - f(4)}{2} = \frac{6.6 - 4.6}{2} = \$1 \text{ per lb.}$$

Similarly, for changes in output of 1, .1, and .01, we obtain the results given in Table 3-1. Notice that as $\Delta q \to 0$, it appears that $\Delta c/\Delta q \to .80$.

TABLE 3-1

Change in Output Δq	Interval $[4, 4 + \Delta q]$	Average cost per lb of additional output $\Delta c/\Delta q$
2	[4, 6]	1.00
1	[4, 5]	.90
.1	[4, 4.1]	.81
.01	[4, 4.01]	.801

This would indicate that for a small increase in output above 4 lb, the cost per lb of that additional output is approximately \$0.80. We say that $\lim_{\Delta q \to 0} \Delta c/\Delta q$ is the *instantaneous* rate of change of cost c with respect to output q when $q = 4$.

For any cost function $c = f(q)$, we say that $\lim_{\Delta q \to 0} \Delta c/\Delta q$ is the **instantaneous rate of change of c with respect to q**. More simply, this limit is called **marginal cost**. Recall that $\Delta c/\Delta q$ is the slope of a secant line. Thus if $\Delta q \to 0$, then $\Delta c/\Delta q$ approaches the slope of a tangent line; that is, $\lim_{\Delta q \to 0} \Delta c/\Delta q = dc/dq$.† Hence,

$$\frac{dc}{dq} = \left\{ \begin{array}{l} \text{instantaneous rate} \\ \text{of change of } c \\ \text{with respect to } q \end{array} \right\} = \text{marginal cost.}$$

For the case of the previous cost function $c = f(q) = .1q^2 + 3$,

$$\frac{dc}{dq} = .2q.$$

To find marginal cost when 4 lb are produced, we evaluate dc/dq when $q = 4$:

$$\left.\frac{dc}{dq}\right|_{q=4} = .2(4) = \$0.80.$$

†This can also be seen by replacing Δq in Eq. (1) by h:

$$\lim_{\Delta q \to 0} \frac{\Delta c}{\Delta q} = \lim_{\Delta q \to 0} \frac{f(q + \Delta q) - f(q)}{\Delta q} = \lim_{h \to 0} \frac{f(q + h) - f(q)}{h} = \frac{dc}{dq}.$$

From Table 3-1 the actual cost of producing one more pound beyond 4 lb is .90. However, the marginal cost is .80. These numbers are close to each other because the slope of the tangent line to the cost curve at $q = 4$ is a good approximation to the slope of the secant line through $(4, f(4))$ and $(5, f(5))$. See Fig. 3-12. For this reason *we interpret marginal cost as the approximate change in cost resulting from one additional unit of output.*

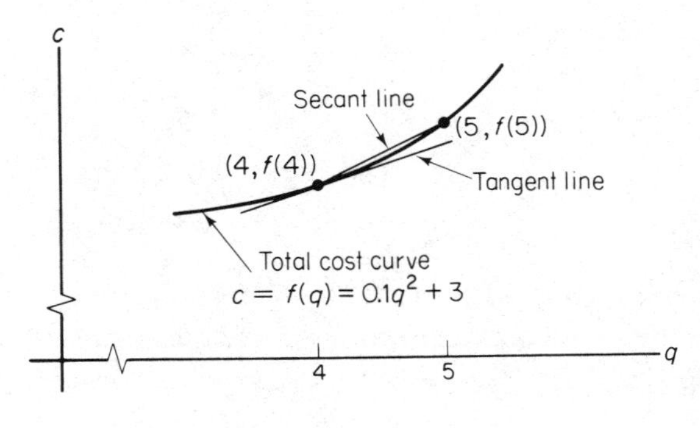

FIG. 3-12

Our discussion concerning rate of change applies not only to cost functions, but also to *any* function $y = f(x)$. In general we can interpret dy/dx as

$$\frac{dy}{dx} = \begin{cases} \text{instantaneous rate of} \\ \text{change of } y \text{ with respect to } x. \end{cases}$$

Thus the instantaneous rate of change of $y = f(x)$ at a point is the slope of the tangent line to the graph of $y = f(x)$ at that point. For convenience we usually refer to instantaneous rate of change simply as *rate of change*.

EXAMPLE 1

Find the (instantaneous) rate of change of $y = x^4$ with respect to x. Evaluate when $x = 2$ and when $x = -1$.

The rate of change is given by dy/dx:

$$\frac{dy}{dx} = 4x^3.$$

Thus the rate at which x^4 changes with respect to x is $4x^3$. When $x = 2$, then $dy/dx = 4(2)^3 = 32$. This means that y is increasing 32 times as fast as x

does. When $x = -1$, then $dy/dx = 4(-1)^3 = -4$. The significance of the minus sign on -4 is that y is *decreasing* four times as fast as x increases.

If c is the total cost of producing q units of a product, then the *average cost* per unit, $\bar{c}$, for producing q units is

$$\bar{c} = \frac{c}{q}. \tag{2}$$

For example, if the total cost of 20 units is \$100, then the average cost per unit is $\bar{c} = 100/20 = \$5$. By multiplying both sides of Eq. (2) by q, we have

$$c = q\bar{c}.$$

That is, total cost is the product of the number of units produced and the average cost per unit.

EXAMPLE 2

If

$$\bar{c} = .0001q^2 - .02q + 5 + \frac{5000}{q}$$

is a manufacturer's average cost equation, find the marginal cost function. What is the marginal cost when 50 units are produced?

We first find total cost c. Since $c = q\bar{c}$, then

$$\begin{aligned} c &= q\bar{c} \\ &= q\left[.0001q^2 - .02q + 5 + \frac{5000}{q}\right]. \\ c &= .0001q^3 - .02q^2 + 5q + 5000. \end{aligned}$$

Differentiating c, we have the marginal cost function:

$$\begin{aligned} \frac{dc}{dq} &= .0001(3q^2) - .02(2q) + 5(1) + 0 \\ &= .0003q^2 - .04q + 5. \end{aligned}$$

The marginal cost when 50 units are produced is

$$\left.\frac{dc}{dq}\right|_{q=50} = .0003(50)^2 - .04(50) + 5 = 3.75.$$

Let us interpret this result. If c is in dollars and production is increased by one unit from $q = 50$ to $q = 51$, then the cost of the additional unit is approximately \$3.75. If production is increased by 1/3 unit, then the cost of the additional output is approximately (1/3)(3.75) = \$1.25.

EXAMPLE 3

A sociologist is studying various suggested programs that can aid in the education of preschool-age children in a certain city. The sociologist believes that x years after the beginning of a particular program, f(x) thousand preschoolers will be enrolled, where

$$f(x) = \tfrac{10}{9}(12x - x^2), \qquad 0 \leqslant x \leqslant 12.$$

At what rate would enrollment change (a) *after three years from the start of this program?* (b) *After nine years?*

The rate of change of $f(x)$ is $f'(x)$:

$$f'(x) = \tfrac{10}{9}(12 - 2x).$$

a. After three years the rate of change is

$$f'(3) = \tfrac{10}{9}[12 - 2(3)] = \tfrac{10}{9} \cdot 6 = \tfrac{20}{3} = 6\tfrac{2}{3}.$$

Thus enrollment would be increasing at the rate of $6\frac{2}{3}$ thousand preschoolers per year.

b. After nine years the rate is

$$f'(9) = \tfrac{10}{9}[12 - 2(9)] = \tfrac{10}{9}[-6] = -\tfrac{20}{3} = -6\tfrac{2}{3}.$$

Thus enrollment would be *decreasing* at the rate of $6\frac{2}{3}$ thousand preschoolers per year.

Suppose $r = f(q) = 2q$ gives the total revenue r (in dollars) that a manufacturer receives for selling q units of his product. The rate of change of revenue with respect to number of units sold is

$$\frac{dr}{dq} = 2.$$

This means that revenue is changing at the rate of \$2 per unit, regardless of the number of units sold. Although this is valuable information, it may be more significant when compared to r itself. For example, if $q = 50$ then $r = 2(50) = \$100$. Thus the rate of change of revenue is $2/100 = .02$ **of** $\boldsymbol{r}$. On the other hand, if $q = 5000$ then $r = 2(5000) = \$10{,}000$, and so the rate of change of r is $2/10{,}000 = .0002$ **of** $\boldsymbol{r}$. Although r changes at the same rate at each level, when compared to r itself this rate is relatively smaller when $r = 10{,}000$ than when $r = 100$. By considering the ratio

$$\frac{dr/dq}{r},$$

we have a means of comparing the rate of change of r with r itself. This ratio is called the *relative rate of change* of the revenue function $r = f(q)$. We have shown above that the relative rate of change when $q = 50$ is

$$\frac{dr/dq}{r} = \frac{2}{100} = .02,$$

and when $q = 5000$, it is

$$\frac{dr/dq}{r} = \frac{2}{10{,}000} = .0002.$$

By multiplying these relative rates by 100, we obtain so-called *percentage rates of change*. Thus the percentage rate of change when $q = 50$ is $(.02)(100) = 2$ percent; when $q = 5000$ it is $(.0002)(100) = .02$ percent.

In general, for any function f we have the following definition.

DEFINITION. *The **relative rate of change of f** is*

$$\frac{f'(x)}{f(x)}.$$

*The **percentage rate of change of f** is*

$$\frac{f'(x)}{f(x)} \cdot 100.$$

EXAMPLE 4

Find the relative and percentage rates of change of

$$y = f(x) = 3x^2 - 5x + 25$$

when $x = 5$.

Here

$$f'(x) = 3(2x) - 5(1) + 0 = 6x - 5.$$

Since $f'(5) = 6(5) - 5 = 25$ and $f(5) = 3(5)^2 - 5(5) + 25 = 75$, the relative rate of change of y when $x = 5$ is

$$\frac{f'(5)}{f(5)} = \frac{25}{75} = .333.$$

Multiplying .333 by 100 gives the percentage rate of change: (.333)(100) = 33.3 percent.

Exercise 3-3

In Problems **1–6**, *find (a) the rate of change of y with respect to x, and (b) the relative rate of change of y. At the given value of x, find (c) the rate of change of y, (d) the relative rate of change of y, and (e) the percentage rate of change of y.*

1. $y = f(x) = x + 4; \quad x = 5.$

2. $y = f(x) = 4 - 2x; \quad x = 3.$

3. $y = 3x^2 + 6; \quad x = 2.$

4. $y = 2 - x^2; \quad x = 0.$

5. $y = 8 - x^3; \quad x = 1.$

6. $y = x^2 + 3x - 4; \quad x = -1.$

In Problems **7–12**, *cost functions are given where c is the cost of producing q units of a product. In each case find the marginal cost function. What is the marginal cost at the given value(s) of q?*

7. $c = 500 + 10q; \quad q = 100.$

8. $c = 5000 + 6q; \quad q = 36.$

9. $c = .3q^2 + 2q + 850; \quad q = 3.$

10. $c = .1q^2 + 3q + 2; \quad q = 3.$

11. $c = q^2 + 50q + 1000; \quad q = 15, q = 16, q = 17.$

12. $c = .03q^3 - .6q^2 + 4.5q + 7700; \quad q = 10, q = 20, q = 100.$

13. The total cost function for a hosiery mill is estimated in [7] by

$$c = -10{,}484.69 + 6.750q - .000328q^2,$$

where q is output in dozens of pairs and c is total cost in dollars. Find the marginal cost function and evaluate it when $q = 5000$.

14. The total cost function for an electric light and power plant is estimated in [34] by

$$c = 32.07 - .79q + .02142q^2 - .0001q^3, \qquad 20 \leqslant q \leqslant 90,$$

where q is eight-hour total output (as percentage of capacity) and c is total fuel cost in dollars. Find the marginal cost function and evaluate it when $q = 70$.

15. For the cost function $c = .2q^2 + 1.2q + 4$, how fast does c change with respect to q when $q = 5$?

16. For the cost function $c = .4q^2 + 4q + 5$, find the rate of change of c with respect to q when $q = 2$. Also, what is $\Delta c/\Delta q$ over the interval [2, 3]?

In Problems **17–20**, $\bar{c}$ *represents average cost, which is a function of the number* q *of units produced. Find the marginal cost function and the marginal cost for the indicated values of* q.

17. $\bar{c} = .01q + 5 + \dfrac{500}{q}$; $q = 50$, $q = 100$.

18. $\bar{c} = 2 + \dfrac{1000}{q}$; $q = 25$, $q = 235$.

19. $\bar{c} = .00002q^2 - .01q + 6 + \dfrac{20{,}000}{q}$; $q = 100$, $q = 500$.

20. $\bar{c} = .001q^2 - .3q + 40 + \dfrac{7000}{q}$; $q = 10$, $q = 20$.

21. The approximate temperature T of the skin in terms of the temperature T_e of the environment is given in **[48]** by

$$T = 32.8 + .27(T_e - 20),$$

where T and T_e are in degrees Celsius. Find the rate of change of T with respect to T_e.

22. The volume V of a spherical cell is given by $V = \frac{4}{3}\pi r^3$, where r is the radius. Find the rate of change of volume with respect to the radius when $r = 6.5 \times 10^{-4}$ cm.

23. For a certain manufacturer, the revenue r obtained from the sale of q units of his product is given by $r = 30q - .3q^2$. (a) How fast does r change with respect to q? When $q = 10$, (b) find the relative rate of change of r, and (c) to the nearest percent, find the percentage rate of change of r.

24. Repeat Problem 23 for the revenue function $r = 20q - .1q^2$ and $q = 100$.

25. In a discussion of contemporary waters of shallow seas, Odum **[36]** claims that in such waters the total organic matter y (in milligrams per liter) is a function of species diversity x (in number of species per thousand individuals). If $y = 100/x$, at what rate is the total organic matter changing with respect to species diversity when $x = 10$? What is the percentage rate of change when $x = 10$?

26. Suppose the 100 largest cities in the U.S. in 1920 are ranked according to magnitude (areas of cities). From [31], the following relation approximately holds:

$$PR^{.93} = 5{,}000{,}000,$$

where P is the population of the city having respective rank R. This relation is called the *law of urban concentration* for 1920. Solve for P in terms of R and then find how fast population is changing with respect to rank.

27. A study of the winter moth was made in Nova Scotia (adapted from [10]). The prepupae of the moth fall onto the ground from host trees. It was found that at a distance of x ft from the base of a host tree, the prepupal density (number of prepupae per square foot of soil) was y, where

$$y = 59.3 - 1.5x - .5x^2, \qquad 1 \leqslant x \leqslant 9.$$

(a) At what rate is the prepupal density changing with respect to distance from the base of the tree when $x = 6$?

(b) For what value of x is the prepupal density decreasing at the rate of 6 prepupae per square foot per foot?

28. Sociologists studied the relation between income and number of years of education for members of a particular urban group. They found that a person with x years of education before seeking regular employment can expect to receive an average yearly income of y dollars per year, where

$$y = 4x^{5/2} + 4900, \qquad 4 \leqslant x \leqslant 16.$$

At what rate does income change with respect to number of years of education when $x = 9$? When $x = 12$?

29. The weight W of a limb of a tree is given by $W = 2t^{.432}$, where t is time. Find the relative rate of change of W with respect to t.

30. A psychological experiment [1] was conducted to analyze human response to electrical shocks (stimuli). The subjects received shocks of various intensities. The response R to a shock of intensity I (in microamperes) was to be a number that indicated the perceived magnitude relative to that of a "standard" shock. The standard shock was assigned a magnitude of 10. Two groups of subjects were tested under slightly different conditions. The responses R_1 and R_2 of the first and second groups to a shock of intensity I were given by

$$R_1 = \frac{I^{1.3}}{1855.24}, \qquad 800 \leqslant I \leqslant 3500,$$

and

$$R_2 = \frac{I^{1.3}}{1101.29}, \qquad 800 \leqslant I \leqslant 3500.$$

(a) For each group determine the relative rate of change of response with respect to intensity.

(b) How do these changes compare?

(c) In general, if $f(x) = C_1x^n$ and $g(x) = C_2x^n$, where C_1 and C_2 are constants, how do the relative rates of change of f and g compare?

3-4 DIFFERENTIABILITY AND CONTINUITY

There is an important relationship between differentiability and continuity, namely

> If f is differentiable at $x = a$, then f is continuous at $x = a$.

Intuitively we can see why this is true in Fig. 3-13. If the derivative exists at $x = a$, then as $h \to 0$ the limiting position of the secant line is

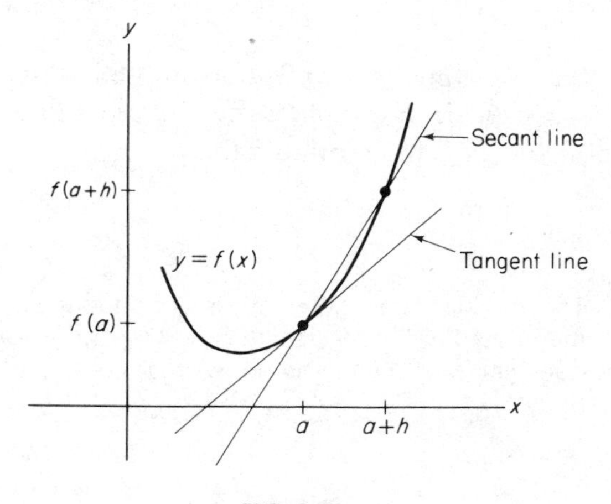

FIG. 3-13

the tangent line. For this position to be approached, the curve must be smooth and connected at a. This is another way of saying that f is continuous at $x = a$.

If a function is not continuous at a point, then it cannot have a derivative there. For example, the function in Fig. 3-14 is discontinuous at $x = a$. The curve has no tangent at that point, and so the function is not differentiable there.

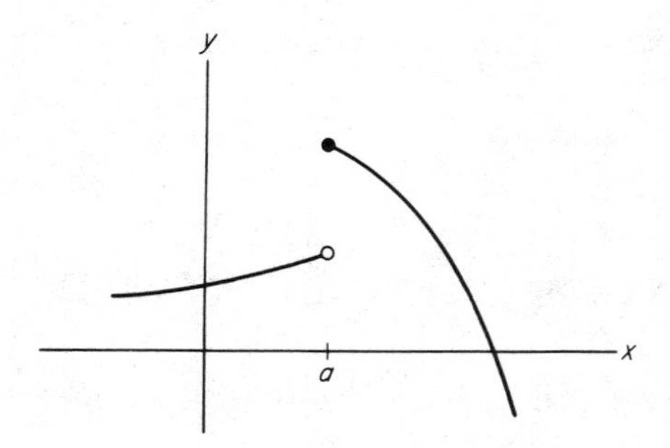

FIG. 3-14

EXAMPLE 1

a. Let $f(x) = x^2$. Since $f'(x) = 2x$ is defined for all values of x, then $f(x) = x^2$ must be continuous for all values of x.

b. The function $f(p) = \dfrac{1}{2p}$ is not continuous at $p = 0$ because f is not defined there. Thus the derivative does not exist at $p = 0$.

The converse of the statement that differentiability implies continuity is *false*. In Example 2 you will see a function that is continuous at a point but not differentiable there.

EXAMPLE 2

The function $y = f(x) = |x|$ is continuous at $x = 0$. See Fig. 3-15. As we mentioned in Sec. 3-1, there is no tangent line at $x = 0$. Thus the derivative does not exist there. This shows that continuity does *not* imply differentiability.

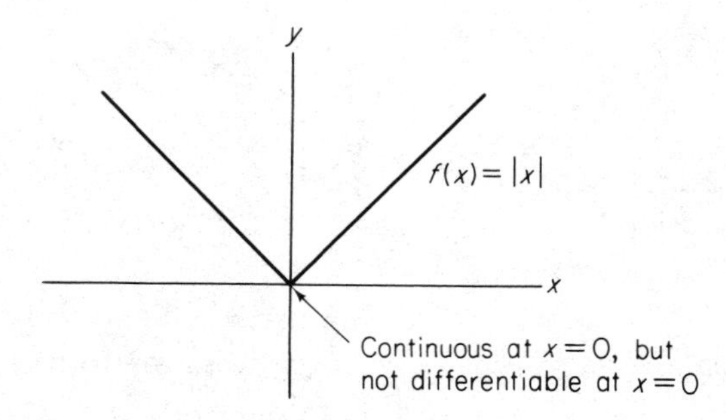

FIG. 3-15

3-5 PRODUCT AND QUOTIENT RULES

The equation $F(x) = (x^2 + 3x)(4x + 5)$ expresses $F(x)$ as a product of two functions: $x^2 + 3x$ and $4x + 5$. To find $F'(x)$ by using only Rules 1–5, we first multiply the functions, which gives $F(x) = 4x^3 + 17x^2 + 15x$. Then we differentiate term by term:

$$F'(x) = 4(3x^2) + 17(2x) + 15(1) = 12x^2 + 34x + 15.$$

In many problems which involve differentiating a product of functions, the multiplication is not so simple as it was here. Often it is not even practical to attempt it. Fortunately, there is a rule for differentiating a product, and the rule avoids such multiplications. Since the derivative of a sum of functions is the sum of their derivatives, you might think that the derivative of a product of two functions is the product of their derivatives. This is **not** the case, as Rule 6 shows (we omit the proof):

RULE 6. Product Rule.

> If $F(x) = f(x) \cdot g(x)$ and $f'(x)$ and $g'(x)$ exist, then
>
> $$F'(x) = f(x) \cdot g'(x) + g(x) \cdot f'(x).$$

The product rule indicates that *the derivative of the product of two functions is the first function times the derivative of the second, plus the second function times the derivative of the first.* To apply the product rule to our previous function, $F(x) = (x^2 + 3x)(4x + 5)$, think of the right side as follows:

$$F(x) = \underbrace{(x^2 + 3x)}_{f(x)}\,\underbrace{(4x + 5)}_{g(x)}.$$

By the product rule,

$$\begin{aligned} F'(x) &= f(x) \cdot g'(x) + g(x) \cdot f'(x) \\ &= (x^2 + 3x)D_x(4x + 5) + (4x + 5)D_x(x^2 + 3x) \\ &= (x^2 + 3x)(4) + (4x + 5)(2x + 3) \\ &= 12x^2 + 34x + 15 \quad \text{(as we saw before).} \end{aligned}$$

PITFALL. *We repeat: The derivative of a product of two functions* **is not** *the product of the derivatives. For example,* $D_x(x^2 + 3x) = 2x + 3$ *and* $D_x(4x + 5) = 4$, *but*

$$D_x[(x^2 + 3x)(4x + 5)] = 12x^2 + 34x + 15 \neq (2x + 3)4.$$

EXAMPLE 1

a. *Find the slope of the graph of* $F(x) = (7x^3 - 5x + 2)(2x^4 + x + 7)$ *when* $x = 1$.

Let $f(x) = 7x^3 - 5x + 2$ and $g(x) = 2x^4 + x + 7$. Then $F(x) = f(x) \cdot g(x)$, a product of two functions. By the product rule,

$$\begin{aligned} F'(x) &= f(x)g'(x) + g(x)f'(x) \\ &= (7x^3 - 5x + 2)D_x(2x^4 + x + 7) + (2x^4 + x + 7)D_x(7x^3 - 5x + 2) \\ &= (7x^3 - 5x + 2)(8x^3 + 1) + (2x^4 + x + 7)(21x^2 - 5). \end{aligned}$$

Evaluating $F'(x)$ at $x = 1$ gives the slope of the graph at that point:

$$F'(1) = 4(9) + 10(16) = 196.$$

Note: **we do not have to simplify $F'(x)$ before evaluating it.**

b. *If* $y = (x^{2/3} + 3)(x^{-1/3} + 5x)$, *find* D_xy.

$$\begin{aligned} D_xy &= (x^{2/3} + 3)D_x(x^{-1/3} + 5x) + (x^{-1/3} + 5x)D_x(x^{2/3} + 3) \\ &= (x^{2/3} + 3)\left(-\tfrac{1}{3}x^{-4/3} + 5\right) + (x^{-1/3} + 5x)\left(\tfrac{2}{3}x^{-1/3}\right) \\ &= \tfrac{25}{3}x^{2/3} + \tfrac{1}{3}x^{-2/3} - x^{-4/3} + 15. \end{aligned}$$

c. *If* $y = (x + 2)(x + 3)(x + 4)$, *find* y'.

By grouping we can consider y to be a product of two functions:

$$y = [(x + 2)(x + 3)](x + 4).$$

The product rule gives

$$\begin{aligned} y' &= [(x + 2)(x + 3)]D_x(x + 4) + (x + 4)D_x[(x + 2)(x + 3)] \\ &= [(x + 2)(x + 3)](1) + (x + 4)D_x[(x + 2)(x + 3)]. \end{aligned}$$

Applying the product rule again, we have

$$\begin{aligned} y' &= [(x + 2)(x + 3)](1) + (x + 4)[(x + 2)D_x(x + 3) + (x + 3)D_x(x + 2)] \\ &= [(x + 2)(x + 3)](1) + (x + 4)[(x + 2)(1) + (x + 3)(1)]. \end{aligned}$$

After simplifying, we have

$$y' = 3x^2 + 18x + 26.$$

Usually we do not use the product rule when simpler ways are obvious. For example, if $f(x) = 2x(x + 3)$, then it is quicker to write $f(x) = 2x^2 + 6x$ from which $f'(x) = 4x + 6$. Similarly, we do not usually use the product rule to differentiate $y = 4(x^2 - 3)$. Since the 4 is a constant multiplier, by Rule 3 we have $y' = 4(2x) = 8x$.

The next rule is used for differentiating a *quotient* of two functions.

RULE 7. Quotient Rule. *Let* $F(x) = \dfrac{f(x)}{g(x)}$ *such that* $g(x) \neq 0$. *If* $f'(x)$ *and* $g'(x)$ *exist, then*

$$F'(x) = \frac{g(x) \cdot f'(x) - f(x) \cdot g'(x)}{[g(x)]^2}.$$

PROOF. Since $F(x) = \dfrac{f(x)}{g(x)}$,

$$F(x) \cdot g(x) = f(x).$$

By the product rule,

$$F(x)g'(x) + g(x)F'(x) = f'(x).$$

Solving for $F'(x)$, we have

$$F'(x) = \frac{f'(x) - F(x)g'(x)}{g(x)}.$$

But $F(x) = f(x)/g(x)$. Thus,

$$F'(x) = \frac{f'(x) - \dfrac{f(x)g'(x)}{g(x)}}{g(x)}$$

$$= \frac{g(x)f'(x) - f(x)g'(x)}{[g(x)]^2}.$$

Therefore,† we have the **quotient rule**:

If $F(x) = f(x)/g(x)$, then

$$F'(x) = \frac{g(x) \cdot f'(x) - f(x) \cdot g'(x)}{[g(x)]^2}.$$

PITFALL. *The derivative of a quotient of two functions* ***is not*** *the quotient of the derivatives of the functions.*

EXAMPLE 2

a. *If* $F(x) = \dfrac{4x^2 - 2x + 3}{2x - 1}$, *find* $F'(x)$.

Let $f(x) = 4x^2 - 2x + 3$ and $g(x) = 2x - 1$. Then $F(x) = f(x)/g(x)$ and by Rule 7, the quotient rule,

$$\begin{aligned} F'(x) &= \frac{g(x) \cdot f'(x) - f(x) \cdot g'(x)}{[g(x)]^2} \\ &= \frac{(2x - 1)D_x(4x^2 - 2x + 3) - (4x^2 - 2x + 3)D_x(2x - 1)}{(2x - 1)^2} \\ &= \frac{(2x - 1)(8x - 2) - (4x^2 - 2x + 3)(2)}{(2x - 1)^2} \\ &= \frac{8x^2 - 8x - 4}{(2x - 1)^2} = \frac{4(2x^2 - 2x - 1)}{(2x - 1)^2}. \end{aligned}$$

b. *If* $y = \dfrac{1}{x^2}$, *find* y'.

Here y can be considered a quotient, and by the quotient rule,

$$\begin{aligned} y' &= \frac{(x^2)D_x(1) - (1)D_x(x^2)}{(x^2)^2} \\ &= \frac{x^2(0) - 1(2x)}{x^4} \\ &= \frac{-2x}{x^4} = -\frac{2}{x^3}. \end{aligned}$$

†You may have observed that this proof assumes the existence of $F'(x)$. However, the quotient rule can be proven without this assumption.

A simpler and more direct method of differentiating y is by writing $y = x^{-2}$ and using Rule 2: $y' = -2x^{-3} = -2/x^3$.

PITFALL. *Note that* $D_x \frac{1}{x^2} \neq \frac{1}{2x}$.

c. *Find an equation of the tangent line to the curve*

$$y = \frac{(x+1)(x^2+2x+5)}{1-x}$$

at (0, 5).

By the quotient rule,

$$y' = \frac{(1-x)D_x[(x+1)(x^2+2x+5)] - [(x+1)(x^2+2x+5)]D_x(1-x)}{(1-x)^2}.$$

Using the product rule to find $D_x[(x+1)(x^2+2x+5)]$, we have

$$y' = \frac{(1-x)[(x+1)(2x+2) + (x^2+2x+5)(1)] - [(x+1)(x^2+2x+5)](-1)}{(1-x)^2}.$$

The slope of the curve at (0, 5) is $y'(0) = 12$. An equation of the tangent line is

$$y - 5 = 12(x - 0),$$

$$y = 12x + 5.$$

Suppose $r = f(q)$ is the **total revenue function** of a manufacturer. The equation $r = f(q)$ states that the total dollar value received for selling q units of his product is r. The **marginal revenue** is defined as the rate of change of the total dollar value received with respect to the total number of units sold. Hence, marginal revenue is just the derivative of r with respect to q:

$$\textbf{marginal revenue} = \frac{dr}{dq}.$$

Marginal revenue indicates the rate at which revenue changes with respect to units sold. We interpret it as the approximate change in revenue that results from selling one additional unit of output.

EXAMPLE 3

Suppose that the demand equation for a manufacturer's product is $p =$

$1000/(q + 5)$, *where* p *is the price per unit (in dollars) when* q *units are demanded. Find the marginal revenue function and evaluate it when* $q = 45$.

The revenue r received for selling q units is

$$\text{revenue} = (\text{price})(\text{quantity}),$$
$$r = pq.$$

Thus the revenue function is

$$r = \left(\frac{1000}{q+5}\right)q,$$

or

$$r = \frac{1000q}{q+5}.$$

To find the marginal revenue function, all we must determine is dr/dq.

$$\frac{dr}{dq} = \frac{(q+5)D_q(1000q) - (1000q)D_q(q+5)}{(q+5)^2}$$

$$= \frac{(q+5)1000 - (1000q)(1)}{(q+5)^2} = \frac{5000}{(q+5)^2}.$$

When $q = 45$,

$$\frac{dr}{dq} = \frac{5000}{(45+5)^2} = \frac{5000}{2500} = 2.$$

This means that selling one additional unit beyond 45 results in approximately \$2 more in revenue.

A function that plays an important role in economic analysis is the **consumption function**. The consumption function $C = f(I)$ expresses a relationship between the total national income I and the total national consumption C. Usually, both I and C are expressed in billions of dollars and I is restricted to some interval. The *marginal propensity to consume* is defined as the rate of change of consumption with respect to income. It is merely the derivative of C with respect to I.

$$\textbf{marginal propensity to consume} = \frac{dC}{dI}.$$

If we assume that the difference between income I and consumption C is savings S, then

$$S = I - C.$$

Differentiating both sides with respect to I gives

$$\frac{dS}{dI} = \frac{d}{dI}(I) - \frac{d}{dI}(C) = 1 - \frac{dC}{dI}.$$

We define dS/dI as the **marginal propensity to save.** Thus, the marginal propensity to save indicates how fast savings change with respect to income.

EXAMPLE 4

If the consumption function is given by

$$C = \frac{5(2\sqrt{I^3} + 3)}{I + 10},$$

determine the marginal propensity to consume and the marginal propensity to save when $I = 100$.

$$\frac{dC}{dI} = \frac{(I + 10)D_I[5(2I^{3/2} + 3)] - 5(2\sqrt{I^3} + 3)D_I[I + 10]}{(I + 10)^2}$$

$$= \frac{(I + 10)[5(3I^{1/2})] - 5(2\sqrt{I^3} + 3)[1]}{(I + 10)^2}.$$

When $I = 100$ the marginal propensity to consume is

$$\left.\frac{dC}{dI}\right|_{I=100} = \frac{6485}{12{,}100} = .536.$$

The marginal propensity to save when $I = 100$ is $1 - .536 = .464$. This means that if a current income of \$100 billion increases by \$1 billion, then the nation will consume approximately 53.6% (536/1000) and save 46.4% (464/1000) of that increase.

Exercise 3-5

In Problems **1–48**, *differentiate the functions.*

1. $f(x) = (3x - 1)(7x + 2)$.

2. $f(x) = (4x + 1)(6x + 3)$.

3. $Q(x) = (5 - 2x)(x^2 + 1)$.

4. $s(t) = (8 - 7t)(t^2 - 2)$.

5. $f(r) = (3r^2 - 4)(r^2 - 5r + 1)$.

6. $C(I) = (2I^2 - 3)(3I^2 - 4I + 1)$.

7. $y = (x^2 + 3x - 2)(2x^2 - x - 3)$.

8. $y = (2 - 3x + 4x^2)(1 + 2x - 3x^2)$.

9. $f(w) = (8w^2 + 2w - 3)(5w^3 + 2)$.

10. $f(x) = (3x - x^2)(3 - x - x^2)$.

11. $g(x) = 3(x^3 - 2x^2 + 5x - 4)(x^4 - 2x^3 + 7x + 1)$.

12. $y = -\frac{3}{2}(2x^4 - 3x + 1)(3x^3 - 6x^2 + 2x - 4)$.

13. $y = (x^2 - 1)(3x^3 - 6x + 5) - (x + 4)(4x^2 + 2x + 1)$.

14. $h(x) = 4(x^5 - 3)(2x^3 + 4) + 3(8x^2 - 5)(3x + 2)$.

15. $f(p) = \frac{3}{2}(\sqrt{p} - 4)(4p - 5)$.

16. $g(x) = (\sqrt{x} - 3x + 1)(\sqrt[4]{x} - 2\sqrt{x})$.

17. $y = (2x^{.45} - 3)(x^{1.3} - 7x)$.

18. $y = (x - 1)(x - 2)(x - 3)$.

19. $y = (2x - 1)(3x + 4)(x + 7)$.

20. $y = \dfrac{x}{x - 3}$.

21. $y = 7 \cdot \frac{2}{3}$.

22. $y = \dfrac{3x - 5}{2x + 1}$.

23. $f(x) = \dfrac{x}{x - 1}$.

24. $f(x) = \dfrac{-2x}{1 - x}$.

25. $y = \dfrac{x - 1}{x + 2}$.

26. $h(w) = \dfrac{3w^2 + 5w - 1}{w - 3}$.

27. $h(z) = \dfrac{5 - 2z}{z^2 - 4}$.

28. $y = \dfrac{x^2 - 4x + 2}{x^2 + x + 1}$.

29. $y = \dfrac{8x^2 - 2x + 1}{x^2 - 5x}$.

30. $f(x) = \dfrac{x^3 - x^2 + 1}{x^2 + 1}$.

31. $y = \dfrac{x^2 - 4x + 3}{2x^2 - 3x + 2}$.

32. $F(z) = \dfrac{z^4 + 4}{3z}$.

33. $g(x) = \dfrac{1}{x^{100} + 1}$.

34. $y = \dfrac{3}{7x^3}$.

35. $u(v) = \dfrac{v^5 - 8}{v}$.

36. $y = \dfrac{x - 5}{2\sqrt{x}}$.

37. $y = \dfrac{3x^2 - x - 1}{\sqrt[3]{x}}$.

38. $y = \dfrac{x^{.3} - 2}{2x^{2.1} + 1}$.

39. $y = 7 - \dfrac{4}{x - 8} + \dfrac{2x}{3x + 1}$.

40. $q(x) = 13x^2 + \dfrac{x - 1}{2x + 3} - \dfrac{4}{x}$.

41. $H(s) = \dfrac{(s + 2)(s - 4)}{s - 5}$.

42. $y = \dfrac{(2x - 1)(3x + 2)}{4 - 5x}$.

43. $y = \dfrac{x - 5}{(x + 2)(x - 4)}$.

44. $y = \dfrac{4 - 5x}{(2x - 1)(3x + 2)}$.

45. $s(t) = \dfrac{t^2 + 3t}{(t^2 - 1)(t^3 + 7)}$.

46. $y = \dfrac{(2x - 3)(x^2 - 4x + 1)}{3x^3 + 1}$.

47. $y = \dfrac{(x - 1)(x - 2)}{(x - 3)(x - 4)}$.

48. $f(s) = \dfrac{17}{s(5s^2 - 10s + 4)}$.

49. Find the slope of the curve $y = (4x^2 + 2x - 5)(x^3 + 7x + 4)$ at $(-1, 12)$.

50. Find the slope of the curve $y = \dfrac{x^3}{x^4 + 1}$ at $(1, \frac{1}{2})$.

In Problems **51–54**, *find an equation of the tangent line to the curve at the given point.*

51. $y = 6/(x - 1)$; $(3, 3)$.

52. $y = \dfrac{4x + 5}{x^2}$; $(-1, 1)$.

53. $y = (2x + 3)[2(x^4 - 5x^2 + 4)]$; $(0, 24)$.

54. $y = \dfrac{x + 1}{x^2(x - 4)}$; $(2, -\frac{3}{8})$.

In Problems **55–58**, *r represents total revenue and is a function of the number q of units sold. Find the marginal revenue function and the marginal revenue for the indicated values of q.*

55. $r = .7q$; $q = 8$, $q = 100$, $q = 200$.

56. $r = q(15 - \frac{1}{30}q)$; $q = 5$, $q = 15$, $q = 150$.

57. $r = 250q + 45q^2 - q^3$; $q = 5$, $q = 10$, $q = 25$.

58. $r = 2q(30 - .1q)$; $q = 10$, $q = 20$.

In Problems **59–62**, *each equation represents a demand function for a certain product where p denotes price per unit for q units. Find the marginal revenue function in each case.*

59. $p = 25 - .02q$.

60. $p = 500/q$.

61. $p = \dfrac{108}{q + 2} - 3$.

62. $p = \dfrac{q + 750}{q + 50}$.

63. For the United States (1922–42), the consumption function is estimated in [16] by

$$C = .672I + 113.1.$$

Find the marginal propensity to consume.

64. Repeat Problem 63 if $C = .712I + 95.05$ for the United States for 1929–41 **[16]**.

In Problems **65–67**, *each equation represents a consumption function. Find the marginal propensity to consume and the marginal propensity to save for the given value of I.*

65. $C = 2 + 2\sqrt{I}\,; \quad I = 9.$

66. $C = 6 + \dfrac{3I}{4} - \dfrac{\sqrt{I}}{3}; \quad I = 25.$

67. $C = \dfrac{20\sqrt{I} + .5\sqrt{I^3} - .4I}{\sqrt{I} + 5}; \quad I = 100.$

68. The persistence of sound in a room after the sound source is turned off is called *reverberation*. The *reverberation time RT* of the room is the time it takes for the intensity level of the sound to fall 60 decibels. In the acoustical design of an auditorium **[8]** the following formula may be used to compute the RT of the room:

$$RT = \frac{.05\,V}{A + xV},$$

where V is the room volume, A is the total room absorption, and x is the air absorption coefficient. Assuming that A and x are positive constants, show that the rate of change of RT with respect to V is always positive. If the total room volume increases by one unit, does reverberation time increase or decrease?

69. In a predator-prey experiment **[19]**, it was statistically determined that the number of prey consumed, y, by an individual predator was a function of prey density x (the number of prey per unit of area), where

$$y = f(x) = \frac{.7355x}{1 + .02744x}.$$

Determine the rate of change of prey consumed with respect to prey density.

70. For a particular host-parasite relationship, it is determined that when the host density (number of hosts per unit of area) is x, then the number of hosts that are parasitized is y, where

$$y = \frac{900x}{10 + 45x}.$$

At what rate is the number of hosts parasitized changing with respect to host density when $x = 2$?

71. If the total cost function for a manufacturer is given by

$$c = \frac{5q^2}{q + 3} + 5000,$$

find the marginal cost function.

3-6 THE CHAIN RULE AND POWER RULE

Our next rule, the chain rule, is one of the most important rules for finding derivatives.

To begin, suppose

$$y = f(u) = 2u^2 - 3u - 2, \tag{1}$$

and $$u = g(x) = x^2 + x. \tag{2}$$

Here y is a function of u and u is a function of x. If we substitute $x^2 + x$ for u in Eq. (1), we can consider y to be a function of x:

$$y = 2(x^2 + x)^2 - 3(x^2 + x) - 2.$$

After expanding we can find dy/dx in the usual way.

$$\begin{aligned} y &= 2x^4 + 4x^3 + 2x^2 - 3x^2 - 3x - 2 \\ &= 2x^4 + 4x^3 - x^2 - 3x - 2. \\ \frac{dy}{dx} &= 8x^3 + 12x^2 - 2x - 3. \end{aligned}$$

From this example you can see that finding dy/dx by first performing a substitution could be quite involved, especially if $2u^2$ in Eq. (1) were $2u^{200}$. The chain rule will allow us to handle such situations with ease.

RULE 8. Chain Rule. *If $y = f(u)$ is a differentiable function of u and u is a differentiable function of x, then y is a differentiable function of x and*

$$\boxed{\frac{dy}{dx} = \frac{dy}{du} \cdot \frac{du}{dx}.}$$

Let us see why the chain rule is reasonable. Suppose $y = 8u + 5$ and $u = 2x - 3$. Let x change by one unit. How does u change? Answer: $du/dx = 2$. But for *each* one-unit change in u there is a change in y of $dy/du = 8$. Therefore, what is the change in y if x changes by one unit, that is, what is dy/dx? Answer: $8 \cdot 2$, which is $\frac{dy}{du} \cdot \frac{du}{dx}$. Thus $\frac{dy}{dx} = \frac{dy}{du} \cdot \frac{du}{dx}$.

EXAMPLE 1

a. *If $y = 2u^2 - 3u - 2$ and $u = x^2 + x$, find dy/dx.*

By Rule 8, the chain rule,

$$\frac{dy}{dx} = \frac{dy}{du} \cdot \frac{du}{dx} = \frac{d}{du}(2u^2 - 3u - 2) \cdot \frac{d}{dx}(x^2 + x)$$

$$= (4u - 3)(2x + 1).$$

We can write our answer exclusively in terms of x by replacing u by $x^2 + x$.

$$\frac{dy}{dx} = [4(x^2 + x) - 3](2x + 1)$$

$$= [4x^2 + 4x - 3](2x + 1)$$

$$= 8x^3 + 12x^2 - 2x - 3 \text{ (as we saw before).}$$

b. *If $y = \sqrt{u}$ and $u = 7 - x^3$, find dy/dx.*

By the chain rule,

$$\frac{dy}{dx} = \frac{dy}{du} \cdot \frac{du}{dx}$$

$$= \frac{d}{du}(\sqrt{u}) \cdot \frac{d}{dx}(7 - x^3)$$

$$= \frac{1}{2\sqrt{u}} \cdot (-3x^2)$$

$$= -\frac{3x^2}{2\sqrt{u}} = -\frac{3x^2}{2\sqrt{7 - x^3}}.$$

c. *If $y = u^{10}$ and $u = 8 - t^2 + t^5$, find dy/dt.*

Here y is a function of u, and u is a function of t. Hence we can consider y as a function of t. By the chain rule,

$$\frac{dy}{dt} = \frac{dy}{du} \cdot \frac{du}{dt} = \frac{d}{du}(u^{10}) \cdot \frac{d}{dt}(8 - t^2 + t^5)$$

$$= (10u^9)(-2t + 5t^4)$$

$$= 10(8 - t^2 + t^5)^9(-2t + 5t^4).$$

d. *If $y = 4u^3 + 10u^2 - 3u - 7$ and $u = 4/(3x - 5)$, find dy/dx when $x = 1$.*

By the chain rule,

$$\frac{dy}{dx} = \frac{dy}{du} \cdot \frac{du}{dx} = \frac{d}{du}(4u^3 + 10u^2 - 3u - 7) \cdot \frac{d}{dx}\left(\frac{4}{3x-5}\right)$$

$$= (12u^2 + 20u - 3) \cdot \frac{(3x-5)D_x(4) - 4D_x(3x-5)}{(3x-5)^2}$$

$$= (12u^2 + 20u - 3) \cdot \frac{-12}{(3x-5)^2}.$$

Even though dy/dx is in terms of x's and u's, we can evaluate it when $x = 1$ if we determine the corresponding value of u. When $x = 1$, then $u = \frac{4}{3(1) - 5} = -2$. Thus,

$$\left.\frac{dy}{dx}\right|_{x=1} = [12(-2)^2 + 20(-2) - 3] \cdot \frac{-12}{[3(1) - 5]^2}$$

$$= 5 \cdot (-3) = -15.$$

Suppose we wanted to differentiate $y = (x^3 - x^2 + 6)^{100}$. We can think of the right side as u^{100} where $x^3 - x^2 + 6$ plays the role of u. This suggests a substitution. Let $x^3 - x^2 + 6$ be u. Then

$$y = u^{100} \text{ where } u = x^3 - x^2 + 6.$$

By the chain rule,

$$\frac{dy}{dx} = \frac{dy}{du} \cdot \frac{du}{dx}$$

$$= (100u^{99})(3x^2 - 2x)$$

$$= 100(x^3 - x^2 + 6)^{99}(3x^2 - 2x).$$

EXAMPLE 2

a. *If $y = \sqrt[5]{8x^2 - 7x}$, find y'.*

Let $u = 8x^2 - 7x$. Then $y = \sqrt[5]{u} = u^{1/5}$.

$$y' = \frac{dy}{du} \cdot \frac{du}{dx}$$

$$= \left(\tfrac{1}{5}u^{-4/5}\right)(16x - 7)$$

$$= \frac{(16x - 7)(8x^2 - 7x)^{-4/5}}{5}.$$

b. *If* $y = \dfrac{1}{(x^2 - 2)^4}$, *find* y'.

Let $u = x^2 - 2$. Then $y = 1/u^4 = u^{-4}$. Thus,

$$\frac{dy}{dx} = \frac{dy}{du} \cdot \frac{du}{dx} = (-4u^{-5})(2x)$$

$$= -\frac{8x}{u^5} = -\frac{8x}{(x^2 - 2)^5}.$$

In Example 2 we used the chain rule to differentiate a power of a *function* of x. In (a) we differentiated $y = (8x^2 - 7x)^{1/5}$, and in (b) we differentiated $y = (x^2 - 2)^{-4}$. The following rule generalizes our results and is known as the *power rule.*

RULE 9. Power Rule. *If* $y = u^n$, *where* n *is any real number and* u *is a differentiable function of* x, *then*

$$\frac{dy}{dx} = nu^{n-1}\frac{du}{dx}.$$

PROOF. By the chain rule,

$$\frac{dy}{dx} = \frac{dy}{du} \cdot \frac{du}{dx}.$$

But by Rule 2, $$\frac{dy}{du} = \frac{d}{du}(u^n) = nu^{n-1},$$

and so $$\frac{dy}{dx} = nu^{n-1}\frac{du}{dx}.$$

Therefore we have the ***power rule:***

> If $y = u^n$ where u is a function of x, then
> $$\frac{dy}{dx} = nu^{n-1}\frac{du}{dx}.$$

Another way of writing the power rule is

> $$\frac{d}{dx}([u(x)]^n) = n[u(x)]^{n-1}u'(x).$$

EXAMPLE 3

a. *If* $y = (x^3 - 1)^7$, *find* y'.

Since y is a power of a *function* of x, the power rule applies. Letting $u(x) = x^3 - 1$ and $n = 7$, we have

$$\begin{aligned} y' &= n[u(x)]^{n-1}u'(x) \\ &= 7(x^3 - 1)^{7-1}\frac{d}{dx}(x^3 - 1) \\ &= 7(x^3 - 1)^6(3x^2) = 21x^2(x^3 - 1)^6. \end{aligned}$$

b. *If* $y = \sqrt{4x^2 + 3x - 1}$, *find* dy/dx *when* $x = -2$.

Since $y = (4x^2 + 3x - 1)^{1/2}$, we use the power rule and choose $u(x) = 4x^2 + 3x - 1$ and $n = \frac{1}{2}$.

$$\begin{aligned} \frac{dy}{dx} &= \frac{1}{2}(4x^2 + 3x - 1)^{(1/2)-1}\frac{d}{dx}(4x^2 + 3x - 1) \\ &= \frac{8x + 3}{2\sqrt{4x^2 + 3x - 1}}. \end{aligned}$$

$$\left.\frac{dy}{dx}\right|_{x=-2} = \frac{-13}{2\sqrt{9}} = -\frac{13}{6}.$$

c. *If* $z = \left(\dfrac{2s + 5}{s^2 + 1}\right)^4$, *find* $\dfrac{dz}{ds}$.

Since z is a power of a function, we first use the power rule.

$$\frac{dz}{ds} = 4\left(\frac{2s + 5}{s^2 + 1}\right)^{4-1}\frac{d}{ds}\left(\frac{2s + 5}{s^2 + 1}\right).$$

By the quotient rule,

$$\frac{dz}{ds} = 4\left(\frac{2s + 5}{s^2 + 1}\right)^3\frac{(s^2 + 1)(2) - (2s + 5)(2s)}{(s^2 + 1)^2}.$$

Simplifying, we have

$$\begin{aligned} \frac{dz}{ds} &= 4 \cdot \frac{(2s + 5)^3}{(s^2 + 1)^3} \cdot \frac{(-2s^2 - 10s + 2)}{(s^2 + 1)^2} \\ &= \frac{-8(s^2 + 5s - 1)(2s + 5)^3}{(s^2 + 1)^5}. \end{aligned}$$

d. *If* $y = (x^2 - 4)^5(3x + 5)^4$, *find* y'.

Since y is a product, we first apply the product rule.

$$y' = (x^2 - 4)^5 D_x[(3x + 5)^4] + (3x + 5)^4 D_x[(x^2 - 4)^5].$$

Now we can use the power rule.

$$y' = (x^2 - 4)^5[4(3x + 5)^3(3)] + (3x + 5)^4[5(x^2 - 4)^4(2x)].$$

Simplifying, we have

$$\begin{aligned} y' &= 12(x^2 - 4)^5(3x + 5)^3 + 10x(3x + 5)^4(x^2 - 4)^4 \\ &= 2(x^2 - 4)^4(3x + 5)^3[6(x^2 - 4) + 5x(3x + 5)] \qquad \text{(factoring)} \\ &= 2(x^2 - 4)^4(3x + 5)^3(21x^2 + 25x - 24). \end{aligned}$$

Usually, the power rule should be used to differentiate $y = [u(x)]^n$. Although a function such as $y = (x^2 + 2)^2$ may be written as $y = x^4 + 4x^2 + 4$ and differentiated easily, this method is impractical for a function such as $y = (x^2 + 2)^{1000}$. Since $y = (x^2 + 2)^{1000}$ is of the form $y = [u(x)]^n$, we have

$$y' = 1000(x^2 + 2)^{999}(2x).$$

As an aid to you, Table 3-2 lists the rules of differentiation that were discussed in this chapter. Not only should you be totally familiar

TABLE 3-2 Differentiation Rules

$$\frac{d}{dx}(c) = 0, \text{ where } c \text{ is any constant.}$$

$$\frac{d}{dx}(x^n) = nx^{n-1}, \text{ where } n \text{ is any real number.}$$

$$\frac{d}{dx}[c\,f(x)] = c\,f'(x).$$

$$\frac{d}{dx}[f(x) + g(x)] = f'(x) + g'(x).$$

$$\frac{d}{dx}[(f(x) - g(x)] = f'(x) - g'(x).$$

$$\frac{d}{dx}[f(x) \cdot g(x)] = f(x)g'(x) + g(x)f'(x).$$

$$\frac{d}{dx}\left[\frac{f(x)}{g(x)}\right] = \frac{g(x)f'(x) - f(x)g'(x)}{[g(x)]^2}.$$

$$\frac{dy}{dx} = \frac{dy}{du} \cdot \frac{du}{dx}, \text{ where } y \text{ is a function of } u \text{ and } u \text{ is a function of } x.$$

$$\frac{d}{dx}(u^n) = nu^{n-1}\frac{du}{dx}.$$

with these formulas and the mechanics involved in applying them, but you should also know both the definition and interpretations of a derivative.

Exercise 3-6

In Problems **1–8**, *use the chain rule.*

1. If $y = u^2 - 2u$ and $u = x^2 - x$, find dy/dx.

2. If $y = 2u^3 - 8u$ and $u = 7x - x^3$, find dy/dx.

3. If $y = \dfrac{1}{w^2}$ and $w = 2 - x$, find dy/dx.

4. If $y = \sqrt[3]{z}$ and $z = x^6 - x^2 + 1$, find dy/dx.

5. If $w = u^2$ and $u = \dfrac{t+1}{t-1}$, find dw/dt when $t = 3$.

6. If $z = u^2 + \sqrt{u} + 9$ and $u = 2s^2 - 1$, find dz/ds when $s = -1$.

7. If $y = 3w^2 - 8w + 4$ and $w = 3x^2 + 1$, find dy/dx when $x = 0$.

8. If $y = 3u^3 - u^2 + 7u - 2$ and $u = 3x - 2$, find dy/dx when $x = 1$.

In Problems **9–46**, *find* y'.

9. $y = (7x + 4)^8$.

10. $y = (4 - 3x)^{25}$.

11. $y = (3 - 2p^2)^{14}$.

12. $y = (x^2 - 8x)^{40}$.

13. $y = \dfrac{(4x^3 - 8x + 2)^{10}}{3}$.

14. $y = \dfrac{(7 - q^2 + q)^{12}}{9}$.

15. $y = (4r^2 - 10r + 3)^{-15}$.

16. $y = (t^2 - 5)^{-4}$.

17. $y = \dfrac{7}{(x^3 - x^2 + 2)^7}$.

18. $y = \dfrac{6}{(2 - x^2 + x)^4}$.

19. $y = 15(4z^3 - z^2 + 2)^{1/5}$.

20. $y = 2(8x - 2)^{2/3}$.

21. $y = \sqrt{2x^2 - x + 3}$.

22. $y = \sqrt[3]{8s^2 - 1}$.

23. $y = \sqrt[5]{(x^2 + 1)^3}$.

24. $y = \dfrac{1}{(3x^2 - x)^{2/3}}$.

25. $y = (x^5 + 7x^3 - 8x + 4)^{-6.5}$.

26. $y = (4x^3 - 6x^2 + 7)^{3.8}$.

27. $y = \left(\frac{x-7}{x+4}\right)^{10}$.

28. $y = \left(\frac{2w}{w+2}\right)^4$.

29. $y = 2\left(\frac{q^3 - 2q + 4}{5q^2 + 1}\right)^5$.

30. $y = 3\left(\frac{x^2 + 2x - 2}{x^3 + x}\right)^8$.

31. $y = \sqrt{\frac{x-2}{x+3}}$.

32. $y = \sqrt[3]{\frac{8x^2 - 3}{x^2 + 2}}$.

33. $y = (x^2 + 2x - 1)^3(5x + 7)$.

34. $y = (8x^3 - 1)^3(2x^2 + 1)^2$.

35. $y = [(4x + 3)(6x^2 + x + 8)]^8$.

36. $y = \frac{2t - 5}{(t^2 + 4)^3}$.

37. $y = \frac{(2w + 3)^3}{w^2 + 4}$.

38. $y = \sqrt{(x - 1)(x + 2)^3}$.

39. $y = 6(5x^2 + 2)\sqrt{x^4 + 5}$.

40. $y = \sqrt[3]{\frac{8x - 7}{5x^2 + 6}}$.

41. $y = (4 - 3x^2)^2(2 - 3x)^3$.

42. $y = 6 + 3x - 4x(7x + 1)^2$.

43. $y = 8t + \frac{t - 1}{t + 4} - \left(\frac{8t - 7}{4}\right)^2$.

44. $y = 4[(3p - 8)(3p^2 - 2p + 1)^3]^4$.

45. $y = \frac{(8x - 1)^5}{(3x - 1)^3}$.

46. $y = \frac{(4x^2 - 2)(8x - 1)}{(3x - 1)^2}$.

47. If $y = (5u + 6)^3$ and $u = (x^2 + 1)^4$, find dy/dx when $x = 0$.

48. If $z = 2y^2 - 4y + 5$, $y = 6x - 5$, and $x = 2t$, find dz/dt when $t = 1$.

49. Find the slope of the curve $y = (x^2 - 7x - 8)^3$ at the point (8, 0).

50. Find the slope of the curve $y = \sqrt{x + 1}$ at the point (8, 3).

In Problems **51–54**, *find an equation of the tangent line to the curve at the given point.*

51. $y = \sqrt[3]{(x^2 - 8)^2}$; (3, 1).

52. $y = (2x + 3)^2$; (−2, 1).

53. $y = \frac{\sqrt{7x + 2}}{x + 1}$; $(1, \frac{3}{2})$.

54. $y = \frac{-3}{(3x^2 + 1)^3}$; (0, −3).

55. Suppose the cost c of producing q units of a product is given by $c = 4000 + 10q + .1q^2$. If the price per unit p is given by the equation $q = 800 - 2.5p$, use the chain rule to find the rate of change of cost with respect to price per unit when $p = 80$.

56. The volume V of a spherical cell is given by $V = \frac{4}{3}\pi r^3$, where r is the radius. At time t seconds, the radius r (in centimeters) is given by $r = 10^{-8}t^2 + 10^{-7}t$. Use the chain rule to find dV/dt when $t = 10$.

57. For a certain population, if E is the number of years of a person's education and S represents a numerical value of the person's status based on that educational level, then $S = 4\left(\frac{E}{4} + 1\right)^2$. (a) How fast is status changing with respect to education when $E = 16$? (b) At what level of education does the rate of change of status equal 8?

58. Under certain conditions, the pressure p developed in body tissue by ultrasonic beams is given in **[48]** by a function of the intensity I:

$$p = (2\rho VI)^{1/2},$$

where ρ (a Greek letter read "rho") is density, V is velocity of propagation, and I is intensity. Here, ρ and V are constants. (a) Find the rate of change of p with respect to I. (b) Find the relative rate of change.

59. Suppose that for a certain group of 20,000 births, the number l_x of people surviving to age x years is

$$l_x = 2000\sqrt{100 - x}, \qquad 0 \leq x \leq 100.$$

(a) Find the rate of change of l_x with respect to x and evaluate your answer for $x = 36$. (b) Find the relative rate of change of l_x when $x = 36$.

60. Muscles have the ability to shorten when a load, such as a weight, is imposed on it. The equation

$$(P + a)(v + b) = k$$

is called the "fundamental equation of muscle contraction" **[48]**. Here P is the load imposed on the muscle, v is the velocity of the shortening of the muscle fibres, and a, b, and k are positive constants. Express v as a function of P. Use your result to find dv/dP.

61. A governmental health agency examined the records of a group of individuals who were hospitalized with a particular illness. It was found that the total proportion who had been discharged at the end of t days of hospitalization is given by $f(t)$, where

$$f(t) = 1 - \left(\frac{300}{300 + t}\right)^3.$$

Find $f'(300)$ and interpret your answer.

62. If the total cost function of a manufacturer is given by

$$c = \frac{5q^2}{\sqrt{q^2 + 3}} + 5000,$$

find the marginal cost function.

63. If $p = 100 - \sqrt{q^2 + 20}$ is a demand equation, find the marginal revenue function.

In Problems **64** *and* **65**, *each equation represents a consumption function. Find the marginal propensity to consume and the marginal propensity to save for the given value of I.*

64. $C = \dfrac{20\sqrt{I} + .5\sqrt{I^3} - .4I}{\sqrt{I + 100}}$; $\quad I = 100.$

65. $C = \dfrac{5(2I + \sqrt{I + 9}\,)}{\sqrt{I + 9}}$; $\quad I = 135.$

3-7 IMPLICIT DIFFERENTIATION

To introduce implicit differentiation, we shall find the slope of a tangent line to a circle. Let us take the circle of radius 2 whose center is at the origin (Fig. 3-16). Its equation is

$$x^2 + y^2 = 4,$$

$$x^2 + y^2 - 4 = 0. \tag{1}$$

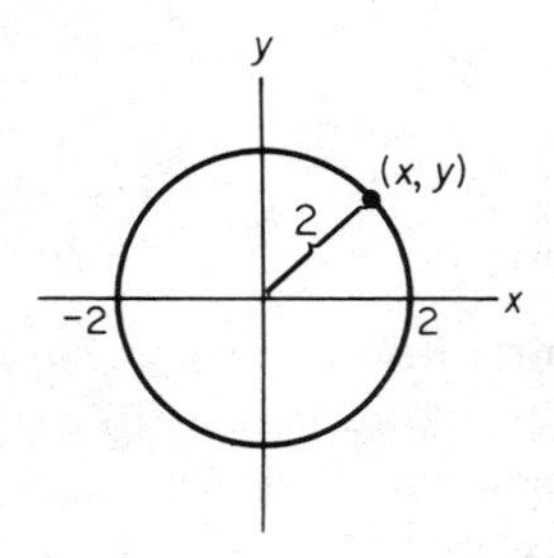

FIG. 3-16

The point $(\sqrt{2}, \sqrt{2}\,)$ lies on the circle. To find the slope at this point, we need to find dy/dx there. Until now we have always had y

given explicitly (directly) in terms of x before determining y'; that is, in the form $y = f(x)$. In Eq. (1) this is not so. We say that Eq. (1) has the form $F(x, y) = 0$, where $F(x, y)$ denotes a function of two variables. The obvious thing to do is solve Eq. (1) for y in terms of x:

$$x^2 + y^2 - 4 = 0,$$

$$y^2 = 4 - x^2,$$

$$y = \pm\sqrt{4 - x^2}. \qquad (2)$$

A problem now occurs—Eq. (2) may give two values of y for a value of x. It does not define y explicitly as a function of x. We can, however, "consider" Eq. (1) as defining y as two different functions of x:

$$y = +\sqrt{4 - x^2} \quad \text{and} \quad y = -\sqrt{4 - x^2},$$

whose graphs are given in Fig. 3-17. Since $(\sqrt{2}, \sqrt{2})$ lies on the graph

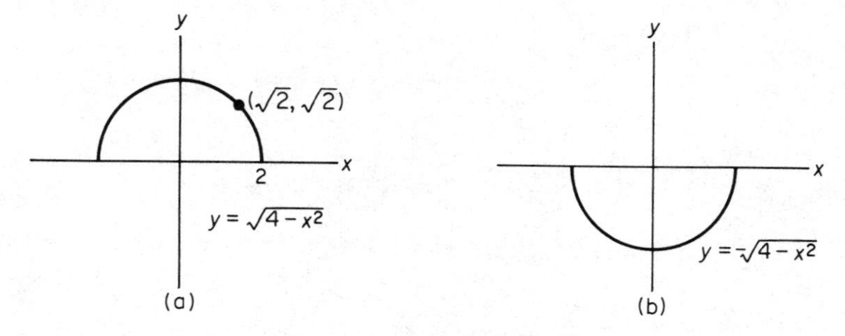

FIG. 3-17

of $y = \sqrt{4 - x^2}$, we should differentiate that function:

$$y = \sqrt{4 - x^2}.$$

$$\frac{dy}{dx} = \frac{1}{2}(4 - x^2)^{-1/2}(-2x)$$

$$= -\frac{x}{\sqrt{4 - x^2}}.$$

$$\left.\frac{dy}{dx}\right|_{x=\sqrt{2}} = -\frac{\sqrt{2}}{\sqrt{4 - 2}} = -1.$$

Thus the slope of the circle $x^2 + y^2 - 4 = 0$ at the point $(\sqrt{2}, \sqrt{2})$ is -1.

Let us summarize the difficulties we had. First, y was not originally given explicitly in terms of x. Second, after we tried to find such a relation, we ended up with more than one function of x. In fact, depending on the equation given, it may be very complicated or even impossible to find an explicit expression for y. For example, it would be difficult to solve $y + y^3 = x$ for y. We shall now consider a method that avoids such difficulties.

An equation of the form $F(x, y) = 0$, such as we had originally, is said to express y *implicitly* as a function of x. The word "implicitly" is used since y is not given *explicitly* as a function of x. However, it is assumed or *implied* that the equation defines y as at least one differentiable function of x. Thus we assume that the former equation $x^2 + y^2 - 4 = 0$ defines at least one function of x, say $y = f(x)$. Hence, to find dy/dx we treat y as a function of x and differentiate both sides of the equation with respect to x.

$$\frac{d}{dx}(x^2 + y^2 - 4) = \frac{d}{dx}(0),$$

$$\frac{d}{dx}(x^2) + \frac{d}{dx}(y^2) - \frac{d}{dx}(4) = \frac{d}{dx}(0).$$

We know that $\frac{d}{dx}(x^2) = 2x$ and that both $\frac{d}{dx}(4)$ and $\frac{d}{dx}(0)$ are 0. But $\frac{d}{dx}(y^2)$ is **not** $2y$, because we are differentiating with respect to x, not y. That is, y is not the independent variable. Since y is assumed to be a function of x, the y^2-term has the form u^n, where y plays the role of u. Just as the power rule states that $\frac{d}{dx}(u^2) = 2u\frac{du}{dx}$, we have $\frac{d}{dx}(y^2) = 2y\frac{dy}{dx} = 2yy'$. Hence the above equation becomes

$$2x + 2yy' = 0.$$

Solving for y', we obtain

$$2yy' = -2x,$$

$$y' = -\frac{x}{y}. \tag{3}$$

Notice that the expression for y' involves the variable y as well as x. This means that to find y' at a point, both coordinates of the point must be substituted into y'. Thus,

$$y'\big|_{(\sqrt{2},\,\sqrt{2})} = -\frac{\sqrt{2}}{\sqrt{2}} = -1 \text{ as before.}$$

This method of finding dy/dx is called **implicit differentiation.** We note that Eq. (3) is not defined when $y = 0$. Geometrically this is clear, since the tangent line to the circle at either $(2, 0)$ or $(-2, 0)$ is vertical and the slope is not defined.

EXAMPLE 1

For each of the following, find y′ by implicit differentiation.

a. $y + y^3 = x$.

We treat y as a function of x and differentiate both sides with respect to x.

$$D_x(y) + D_x(y^3) = D_x(x).$$

Here $D_x(y)$ is merely y', and $D_x(x) = 1$. By the power rule, $D_x(y^3) = 3y^2\frac{dy}{dx} = 3y^2y'$. Thus,

$$y' + 3y^2y' = 1.$$

Solving for y', we get

$$y'(1 + 3y^2) = 1,$$

$$y' = \frac{1}{1 + 3y^2}.$$

b. $x^2 + 5xy - 3y^3 - 5 = 0$.

We assume that y is a function of x and differentiate both sides with respect to x.

$$D_x(x^2) + 5D_x(xy) - 3D_x(y^3) - D_x(5) = D_x(0).$$

To find $D_x(xy)$ we use the product rule.

$$[2x] + 5[xD_x(y) + yD_x(x)] - 3[3y^2D_x(y)] - 0 = 0,$$

$$[2x] + 5[xy' + y] - 3[3y^2y'] = 0,$$

$$y'(5x - 9y^2) = -2x - 5y,$$

$$y' = \frac{2x + 5y}{9y^2 - 5x}.$$

c. $\sqrt{x} + \sqrt{y} = 4$.

Since $x^{1/2} + y^{1/2} = 4$, then $D_x(x^{1/2}) + D_x(y^{1/2}) = D_x(4)$:

$$\frac{1}{2x^{1/2}} + \frac{1}{2y^{1/2}}(y') = 0,$$

$$y' = -\frac{y^{1/2}}{x^{1/2}} = -\frac{\sqrt{y}}{\sqrt{x}}.$$

Using the original equation, we can express y' in terms of x:

$$y' = -\frac{4 - \sqrt{x}}{\sqrt{x}}.$$

d. $x^3 = (y - x^2)^2$.

$$D_x(x^3) = D_x[(y - x^2)^2],$$

$$3x^2 = 2(y - x^2)(y' - 2x),$$

$$3x^2 = 2(yy' - 2xy - x^2y' + 2x^3),$$

$$3x^2 + 4xy - 4x^3 = 2y'(y - x^2),$$

$$y' = \frac{3x^2 + 4xy - 4x^3}{2(y - x^2)}.$$

If we want to find the slope of the curve $x^3 = (y - x^2)^2$ at the point $(1, 2)$, we have

$$y'|_{(1, 2)} = \frac{3(1)^2 + 4(1)(2) - 4(1)^3}{2[2 - (1)^2]} = \frac{7}{2}.$$

EXAMPLE 2

If $p = 500 - 3q^2$, find dq/dp, the rate of change of q with respect to p.

Differentiating both sides with respect to p and treating q as a function of p, we have

$$D_p(p) = D_p(500) - D_p(3q^2),$$

$$1 = 0 - 6q\frac{dq}{dp}.$$

Solving for dq/dp gives

$$\frac{dq}{dp} = -\frac{1}{6q}.$$

Exercise 3-7

In Problems **1–18** *find dy/dx by implicit differentiation*

1. $x^2 + 4y^2 = 4$.

2. $3x^2 + 6y^2 = 1$.

3. $xy = 4$.

4. $x + xy - 2 = 0$.

5. $xy - y - 4x = 5$.

6. $x^2 + y^2 = 2xy + 3$.

7. $x^3 + y^3 - 12xy = 0$.

8. $2x^2 - 3y^2 = 4$.

9. $x^{3/4} + y^{3/4} = 7$.

10. $y^3 = 4x$.

11. $3y^4 - 5x = 0$.

12. $x^{1/5} + y^{1/5} = 4$.

13. $\sqrt{x} + \sqrt{y} = 3$.

14. $2x^3 + 3xy + y^3 = 0$.

15. $x = \sqrt{y} + \sqrt[3]{y}$.

16. $x^3y^3 + x = 9$.

17. $3x^2y^3 - x + y = 25$.

18. $ax^2 - by^2 = c$.

19. If $x + xy + y^2 = 7$, find y' at $(1, 2)$.

20. Find the slope of the curve $4x^2 + 9y^2 = 1$ at the point $(0, \frac{1}{3})$; at the point (x_0, y_0).

21. Find an equation of the tangent line to the curve $x^3 + y^2 = 3$ at the point $(-1, 2)$.

22. Repeat Problem 21 for the curve $y^2 + xy - x^2 = 5$ at the point $(4, 3)$.

For the demand equations in Problems **23–25**, *find the rate of change of q with respect to p.*

23. $p = 100 - q^2$.

24. $p = 20/(q^2 + 5)$.

25. $p = 20/(q + 5)^2$.

26. The equation $(P + a)(v + b) = k$ is called the "fundamental equation of muscle contraction" [48]. Here P is the load imposed on the muscle, v is the velocity of the shortening of the muscle fibres, and a, b, and k are positive constants. Use implicit differentiation to show that dv/dP, in terms of P, is given by

$$\frac{dv}{dP} = -\frac{k}{(P + a)^2}.$$

3-8 HIGHER-ORDER DERIVATIVES

Since the derivative of a function is itself a function, it too may be differentiated. When this is done, the result (being a function itself) may also be differentiated. Continuing in this way, we obtain *higher-order derivatives*.

If $y = f(x)$, then $f'(x)$ is called the **first derivative** of f with respect to x. The derivative of $f'(x)$, denoted $f''(x)$, is called the **second derivative** of f with respect to x, etc. Some of the ways in which higher-order derivatives may be written are given in Table 3-3.

TABLE 3-3

first derivative:	y',	$f'(x)$,	$\frac{dy}{dx}$,	$\frac{d}{dx}[f(x)]$,	$D_x y$
second derivative:	y'',	$f''(x)$,	$\frac{d^2y}{dx^2}$,	$\frac{d^2}{dx^2}[f(x)]$,	$D_x^2 y$
third derivative:	y''',	$f'''(x)$,	$\frac{d^3y}{dx^3}$,	$\frac{d^3}{dx^3}[f(x)]$,	$D_x^3 y$
fourth derivative:	$y^{(4)}$,	$f^{(4)}(x)$,	$\frac{d^4y}{dx^4}$,	$\frac{d^4}{dx^4}[f(x)]$,	$D_x^4 y$

PITFALL. *The symbol $D_x^2 y$ represents the second derivative of y. It is not the same as $[D_x y]^2$, the square of the first derivative of y.*

$$D_x^2 y \neq [D_x y]^2.$$

EXAMPLE 1

a. *If $y = 2x^4 + 6x^3 - 12x^2 + 6x - 2$, find y'''.*

Differentiating y, we obtain

$$y' = 8x^3 + 18x^2 - 24x + 6.$$

Differentiating y', we obtain

$$y'' = 24x^2 + 36x - 24.$$

Differentiating y'', we obtain

$$y''' = 48x + 36.$$

b. *If* $f(x) = 7$, *find* $f''(x)$.

$$f'(x) = 0.$$
$$f''(x) = 0.$$

c. *If* $y = (2x^2 + 3)^4$, *find* $\frac{d^2y}{dx^2}$.

$$\frac{dy}{dx} = 4(2x^2 + 3)^3(4x) = 16x(2x^2 + 3)^3.$$

$$\frac{d^2y}{dx^2} = 16\left[x(3)(2x^2 + 3)^2(4x) + (2x^2 + 3)^3(1)\right]$$
$$= 16(2x^2 + 3)^2[12x^2 + (2x^2 + 3)]$$
$$= 16(2x^2 + 3)^2(14x^2 + 3).$$

d. *If* $y = f(x) = \frac{x^2}{x + 4}$, *find* $\frac{d^2y}{dx^2}$ *and evaluate it when* $x = 4$.

$$\frac{dy}{dx} = \frac{(x + 4)(2x) - (x^2)(1)}{(x + 4)^2} = \frac{x^2 + 8x}{(x + 4)^2}.$$

$$\frac{d^2y}{dx^2} = \frac{(x + 4)^2(2x + 8) - (x^2 + 8x)(2)(x + 4)}{(x + 4)^4}$$
$$= \frac{(x + 4)[(x + 4)(2x + 8) - (x^2 + 8x)(2)]}{(x + 4)^4}$$
$$= \frac{32}{(x + 4)^3}.$$

$$\left.\frac{d^2y}{dx^2}\right|_{x=4} = \frac{1}{16}.$$

The second derivative evaluated at $x = 4$ can also be denoted $f''(4)$ or $y''(4)$.

e. *If* $f(x) = \frac{1}{\sqrt{3x - 1}}$, *find the rate of change of* $f''(x)$.

To find the rate of change of any function, we must find its derivative. Thus we want $D_x[f''(x)]$, which is $f'''(x)$. Since $f(x) = (3x - 1)^{-1/2}$,

$$f'(x) = -\tfrac{1}{2}(3x - 1)^{-3/2}(3) = -\tfrac{3}{2}(3x - 1)^{-3/2}.$$

$$f''(x) = -\tfrac{3}{2}\left[\left(-\tfrac{3}{2}\right)(3x - 1)^{-5/2}(3)\right] = \tfrac{27}{4}(3x - 1)^{-5/2}.$$

$$f'''(x) = \tfrac{27}{4}\left[\left(-\tfrac{5}{2}\right)(3x - 1)^{-7/2}(3)\right] = -\tfrac{405}{8}(3x - 1)^{-7/2}.$$

We shall now find a higher-order derivative by means of implicit differentiation. Keep in mind that we shall assume y to be a function of x.

EXAMPLE 2

Find y'' if $x^2 + 4y^2 = 4$.

$$x^2 + 4y^2 = 4.$$

Differentiating both sides with respect to x, we obtain

$$2x + 8yy' = 0.$$

$$y' = \frac{-x}{4y}. \tag{1}$$

$$y'' = \frac{4yD_x(-x) - (-x)D_x(4y)}{(4y)^2}$$

$$= \frac{4y(-1) - (-x)(4y')}{16y^2}.$$

$$y'' = \frac{-4y + 4xy'}{16y^2}.$$

$$y'' = \frac{-y + xy'}{4y^2}. \tag{2}$$

Since $y' = \frac{-x}{4y}$ from Eq. (1), by substituting into Eq. (2) we have

$$y'' = \frac{-y + x\left(\frac{-x}{4y}\right)}{4y^2} = \frac{-4y^2 - x^2}{16y^3}$$

$$= -\frac{4y^2 + x^2}{16y^3}.$$

Since $x^2 + 4y^2 = 4$,

$$y'' = -\frac{4}{16y^3} = -\frac{1}{4y^3}.$$

Exercise 3-8

In Problems **1–18**, *find the indicated derivatives.*

1. $y = 4x^3 - 12x^2 + 6x + 2, y'''$.

2. $y = 2x^4 - 6x^2 + 7x - 2, y'''$.

3. $y = 7 - x, \frac{d^2y}{dx^2}$.

4. $y = -x - x^2, \frac{d^2y}{dx^2}$.

5. $y = \dfrac{1}{\sqrt{x}}$, $y^{(4)}$.

6. $f(q) = 8q^{1/4} - 8q^{-1/4}$, $f'''(q)$.

7. $f(p) = \dfrac{1}{6p^3}$, $f'''(p)$.

8. $y = \dfrac{1}{x}$, y'''.

9. $f(r) = \sqrt{1 - r}$, $f''(r)$.

10. $f(x) = \sqrt{x}$, $D_x^2[f(x)]$.

11. $y = \dfrac{1}{5x - 6}$, $\dfrac{d^2y}{dx^2}$.

12. $y = \sqrt{8 + 3^3}$, y''.

13. $y = (x^2 - 4)^{10}$, y''.

14. $y = (2x + 1)^4$, y''.

15. $y = \dfrac{x + 1}{x - 1}$, y''.

16. $y = 2x^{1/2} + (2x)^{1/2}$, y''.

17. $y = x(1 - 2x)^3$, y''.

18. $y = \dfrac{3}{(4x - 3)^4}$, y''.

In Problems **19–26**, *find* y''.

19. $x^2 + 4y^2 - 16 = 0$.

20. $x^2 - y^2 = 16$.

21. $y^2 = 4x$.

22. $4x^2 + 3y^2 = 4$.

23. $\sqrt{x} + 4\sqrt{y} = 4$.

24. $y^2 - 6xy = 4$.

25. $xy + y - x = 4$.

26. $xy + y^2 = 1$.

27. Find the rate of change of $f'(x)$ if $f(x) = (5x - 3)^4$.

28. Find the rate of change of $f''(x)$ if $f(x) = 6\sqrt{x} + \dfrac{1}{6\sqrt{x}}$.

29. If $c = .3q^2 + 2q + 850$ is a cost function, how fast is marginal cost changing when $q = 100$?

30. If $p = 1000 - 45q - q^2$ is a demand equation, how fast is marginal revenue changing when $q = 10$?

3-9 REVIEW

IMPORTANT TERMS AND SYMBOLS IN CHAPTER 3

tangent line *(p. 83)*

slope of a curve *(p. 84)*

derivative *(p. 85)*

y', $f'(x)$, D_xy *(p. 86)*

$\lim\limits_{h \to 0} \dfrac{f(x + h) - f(x)}{h}$ *(p. 85)*

$\dfrac{dy}{dx}$, $\dfrac{d^3y}{dx^3}$, $\dfrac{d^4}{dx^4}[f(x)]$, D_x^2y *(p. 136)*

Δx *(p. 99)*

rate of change *(p. 102)*

relative rate of change *(p. 105)*

product rule *(p. 111)*

marginal revenue *(p. 115)*

marginal propensity to consume *(p. 116)*

chain rule *(p. 121)*

implicit differentiation *(p. 133)*

marginal cost *(p. 101)*

average cost *(p. 103)*

percentage rate of change *(p. 105)*

quotient rule *(p.113)*

consumption function *(p. 116)*

marginal propensity to save *(p. 117)*

power rule *(p. 124)*

higher-order derivatives *(p. 136)*

REVIEW SECTION

1. In terms of a limit, the definition of the derivative of $f(x)$ with respect to x is ________.

   Ans. $\lim\limits_{h \to 0} \dfrac{f(x+h) - f(x)}{h}$.

2. Geometrically, the derivative $f'(x)$ evaluated at $x = a$ is the ________ of the tangent line to the graph of $y = f(x)$ at $x = a$.

   Ans. slope.

3. If $f'(x) = 2x^3 + 7$, then $f''(x) =$ ________.

   Ans. $6x^2$.

4. The slope of the tangent line to $y = x^2$ at $(1, 1)$ is ________.

   Ans. 2.

5. If $f(x) = \frac{1}{3}$, then $f'''(4) =$ ________.

   Ans. 0.

6. If $y = f(x)$, then dy/dx denotes an (average) (instantaneous) rate of change of y with respect to x.

   Ans. instantaneous.

7. Does the derivative of $y = |x|$ exist at $x = 0$? ________

   Ans. No.

8. True or false: If a function is continuous at a point, then it is differentiable there. __(a)__. If a function is differentiable at a point, then it is continuous there. __(b)__.

   Ans. (a) false; (b) true.

9. The expression $f'(x)/f(x)$ is called the ________ rate of change of $f(x)$.

   Ans. relative.

10. The slope of the curve $y = (x + 2)^2$ at $x = 1$ is ________.

    Ans. 6.

11. If $y = \frac{x}{7}$, then $D_x^2 y =$ ________.

    Ans. 0.

12. If u is a function of x, then the power rule asserts that $D_x(u^n) =$ ________.

    Ans. $nu^{n-1}\frac{du}{dx}$.

13. If $f'(x) = 7$, then $D_x(8f(x)) =$ ________.

    Ans. 56.

14. If 27 percent is the percentage rate of change of $f(x)$, then the relative rate of change is ________.

    Ans. .27.

15. The derivative of a function (is) (is not) a function.

    Ans. is.

16. True or false: If $y = f(x) \cdot g(x)$, then $y' = f'(x) \cdot g'(x)$. __(a)__. If $y = f(x)/g(x)$, then $y' = f'(x)/g'(x)$. __(b)__.

    Ans. (a) false; (b) false.

17. If $f(x) = 7x + 1$, then $f'(1) =$ ________.

    Ans. 7.

18. If the marginal propensity to consume is .6, then the marginal propensity to save is ________.

    Ans. .4.

19. True or false: $D_x(x^2 + 7)^3 = 3(x^2 + 7)^2$. ________.

    Ans. false.

20. The rate of change of total revenue with respect to the number of units sold is called __(a)__, and the rate of change of total cost with respect to the number of units produced is called __(b)__.

    Ans. (a) marginal revenue; (b) marginal cost.

REVIEW PROBLEMS

In Problems **1–38**, *differentiate.*

1. $y = 6^3$.

2. $y = x$.

3. $y = 7x^4 - 6x^3 + 5x^2 + 1$.

4. $y = 4(x^2 + 5) - 7x$.

5. $f(s) = s^2(s^2 + 2)$.

6. $y = \sqrt{x + 3}$.

7. $y = \dfrac{x^2 + 3}{5}$.

8. $y = \dfrac{1}{x^3}$.

9. $y = (x^2 + 6x)(x^3 - 6x^2 + 4)$.

10. $y = (x^2 + 1)^{100}(x - 6)$.

11. $f(x) = (2x^2 + 4x)^{100}$.

12. $f(w) = w\sqrt{w} + w^2$.

13. $y = \dfrac{1}{2x + 1}$.

14. $y = \dfrac{x^4 + 4x^2}{2x}$.

15. $y = (8 + 2x)(x^2 + 1)^4$.

16. $g(z) = (2z)^{3/5} + 5$.

17. $f(z) = \dfrac{z^2 - 1}{z^2 + 1}$.

18. $y = \dfrac{x - 5}{(x + 2)^2}$.

19. $y = \sqrt[3]{4x - 1}$.

20. $h(t) = (1 + 2^2)^0$.

21. $2xy + y^2 = 6$ (find y').

22. $4x^2 - 9y^2 = 4$ (find y').

23. $y = \dfrac{1}{\sqrt{1 - x}}$.

24. $y = \dfrac{x(x + 1)}{2x^2 + 3}$.

25. $h(x) = (x - 6)^4(x + 5)^3$.

26. $y + xy + y^2 = 1$ (find y').

27. $y = \dfrac{5x - 4}{x + 1}$.

28. $y = \dfrac{(x + 3)^5}{x}$.

29. $y = 2x^{-3/8} + (2x)^{-3/8}$.

30. $f(x) = 5x\sqrt{1 - 2x}$.

31. $y = \dfrac{x^2 + 6}{\sqrt{x^2 + 5}}$.

32. $y = \sqrt{\dfrac{x}{2}} + \sqrt{\dfrac{2}{x}}$.

33. $y = (x^3 + 6x^2 + 9)^{3/5}$.

34. $y = \sqrt[3]{(1 - 3x^2)^2}$.

35. $x^2y^2 = 1$ (find y').

36. $y = .5x(x + 1)^{-2} + .3$.

37. $g(z) = \dfrac{-7z}{(z - 1)^{-1}}$.

38. $g(z) = \dfrac{-3}{4(z^5 + 2z - 5)^4}$.

In Problems **39–46**, *find the indicated derivative at the given point. It is not necessary to simplify the derivative before substituting the coordinates.*

39. $y = x^4 - 2x^3 + 6x$; y''', $(1, 5)$.

40. $y = x^2\sqrt{x}$; y''', $(1, 1)$.

41. $y = \dfrac{x}{\sqrt{x - 1}}$; y'', $(5, \frac{5}{2})$.

42. $y = \dfrac{2}{1 - x}$; y'', $(-2, \frac{2}{3})$.

43. $x + xy + y = 5$; y'', $(2, 1)$.

44. $xy + y^2 = 2$; y'', $(1, 1)$.

45. $y = \dfrac{4x}{x^2 + 4}$; y'', $(2, 1)$.

46. $y = (x + 1)^3(x - 1)$; y'', $(-1, 0)$.

In Problems **47–52**, *find an equation of the tangent line to the curve at the point corresponding to the given value of x.*

47. $y = x^2 - 6x + 4$, $x = 1$.

48. $y = -2x^3 + 6x + 1$, $x = 2$.

49. $y = \sqrt[3]{x}$, $x = 8$.

50. $y = \dfrac{x}{1 - x}$, $x = 3$.

51. $x^2 - y^2 = 9$, $x = 7$, $y > 0$.

52. $xy = 6$, $x = 1$.

53. If $f(x) = 4x^2 + 2x + 8$, find the relative and percentage rates of change of $f(x)$ when $x = 1$.

54. If $f(x) = x/(x + 4)$, find the relative and percentage rates of change of $f(x)$ when $x = 1$.

55. If $r = q(20 - .1q)$ is a total revenue function, find the marginal revenue function.

56. If $c = .0001q^3 - .02q^2 + 3q + 6000$ is a total cost function, find the marginal cost when $q = 100$.

57. The total cost function for an electric light and power plant is estimated in [34] by

$$c = 16.68 + .125q + .00439q^2, \qquad 20 \leqslant q \leqslant 90,$$

where q is eight-hour total output (as percentage of capacity) and c is total fuel cost in dollars. Find the marginal cost function and evaluate it when $q = 70$.

58. Bacteria are growing in a culture. The time t (in hours) for the number of bacteria to double in number (generation time) is a function of the temperature T (in degrees Celsius) of the culture and is given by

$$t = f(T) = \begin{cases} \frac{1}{24}T + \frac{11}{4}, & \text{if } 30 \leqslant T \leqslant 36, \\ \frac{4}{3}T - \frac{175}{4}, & \text{if } 36 < T \leqslant 39. \end{cases}$$

Find dt/dT when (a) $T = 38$, (b) $T = 35$.

59. In a study (adapted from [10]) of the winter moth in Nova Scotia, it was determined that the average number of eggs, y, in a female moth was a function of the female's abdominal width x (in millimeters), where

$$y = f(x) = 14x^3 - 17x^2 - 16x + 34, \qquad 1.5 \leqslant x \leqslant 3.5.$$

At what rate does the number of eggs change with respect to abdominal width when $x = 2$?

60. For a particular host-parasite relationship, it is found that when the host density (number of hosts per unit of area) is x, then the number of hosts that are parasitized is y, where

$$y = 10\left(1 - \frac{1}{1 + 2x}\right), \qquad x \geqslant 0.$$

For what value of x does dy/dx equal $1/5$?

61. If $C = 7 + .6I - .25\sqrt{I}$ is a consumption function, find the marginal propensity to consume and the marginal propensity to save when $I = 16$.

62. If $p = (q + 14)/(q + 4)$ is a demand equation, find the rate of change of price p with respect to quantity q.

63. If $p = -.5q + 450$ is a demand equation, find the marginal revenue function.

64. If $\bar{c} = .03q + 1.2 + (3/q)$ is an average cost function, find marginal cost when $q = 100$.

CHAPTER 4

Applications of Differentiation

4-1 INTERCEPTS AND SYMMETRY

Examining the graphical behavior of equations is a basic part of mathematics and has applications to other areas of study. In this section we shall examine equations to determine whether their graphs have certain features. Specifically, we shall consider *intercepts* and *symmetry*.

A point where a graph intersects the x-axis is called an x-*intercept* of the graph and has the form $(x, 0)$. A y-*intercept* is a point $(0, y)$ where the graph intersects the y-axis.

EXAMPLE 1

Find the x- and y-intercepts of the graphs of the following equations.

a. $x^2 + y^2 = 25$.

If $(x, 0)$ is an x-intercept, its coordinates must satisfy $x^2 + y^2 = 25$. Setting

$y = 0$ and solving for x will give the first-coordinates of the x-intercepts.

$$x^2 + 0^2 = 25.$$
$$x = \pm 5.$$

The x-intercepts are thus $(5, 0)$ and $(-5, 0)$. Similarly, to determine the second-coordinates of the y-intercepts we set $x = 0$ in $x^2 + y^2 = 25$ and solve for y.

$$0^2 + y^2 = 25.$$
$$y = \pm 5.$$

Thus the y-intercepts are $(0, 5)$ and $(0, -5)$. See Fig. 4-1.

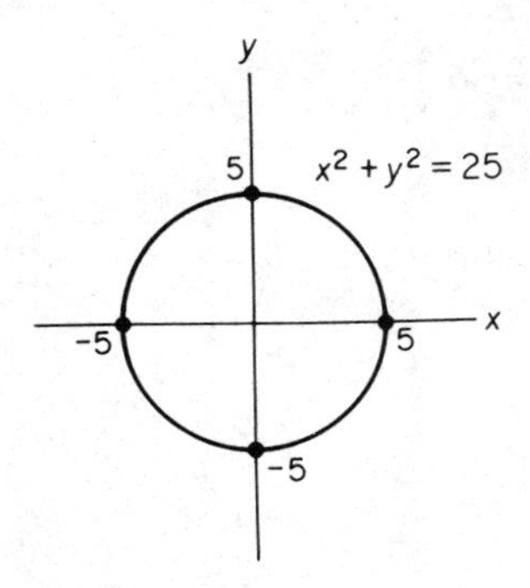

FIG. 4-1

b. $y = \dfrac{1}{x}$.

Since x cannot be 0, the graph has no y-intercept. If $y = 0$, then $0 = 1/x$ and this equation has no solution. Thus no x-intercepts exist either (see Fig. 4-2).

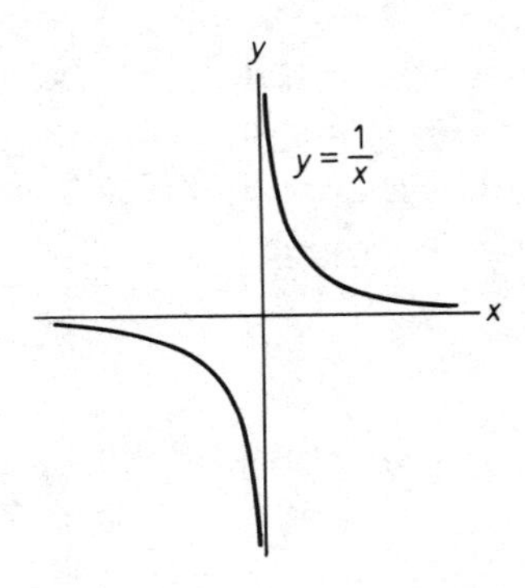

FIG. 4-2

At times it may be quite difficult or even impossible to find intercepts. For example, it would be difficult to find the x-intercepts of the graph of $y - \sqrt{2}x^5 - 4x^2 - 7 = 0$, although the y-intercept is easily found to be $(0, 7)$. In cases such as this, we settle for those intercepts that we can find conveniently.

Some graphs may have *symmetry*. For example, consider the graph of $y = x^2$ in Fig. 4-3. The portion for which $x \leqslant 0$ is the reflection (or mirror image) through the y-axis of that portion for which $x \geqslant 0$, and vice versa. More precisely, if (x_0, y_0) is any point on this graph, then the point $(-x_0, y_0)$ must also lie on the graph. We say that this graph is *symmetric with respect to the y-axis*.

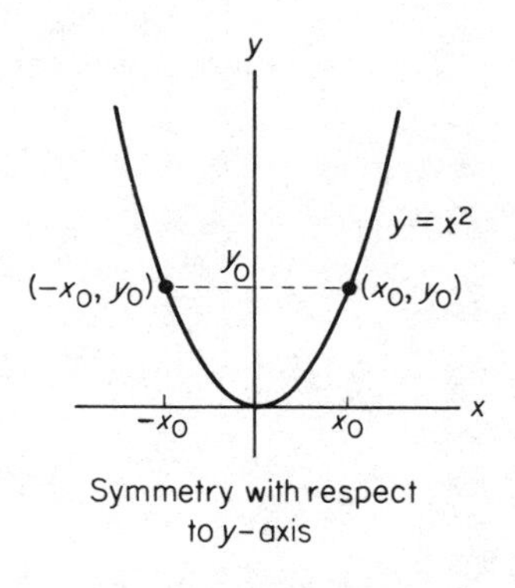

FIG. 4-3

DEFINITION. *A graph is **symmetric with respect to the y-axis** if and only if $(-x_0, y_0)$ lies on the graph when (x_0, y_0) does.*

EXAMPLE 2

Use the definition above to show that the graph of $y = x^2$ is symmetric with respect to the y-axis.

Suppose (x_0, y_0) is *any* point on the graph of $y = x^2$. Then

$$y_0 = x_0^2.$$

We must show that the coordinates of $(-x_0, y)$ satisfy $y = x^2$:

$$y_0 = (-x_0)^2\ ?$$

$$y_0 = x_0^2\ ?$$

But from above we know that $y_0 = x_0^2$. Thus the graph *is* symmetric with respect to the y-axis.

When testing for symmetry in Example 2, (x_0, y_0) can be any point on the graph. In the future, for convenience we shall drop the subscripts. This means that a graph is symmetric with respect to the y-axis if replacing x by $-x$ in its equation results in an equivalent equation.

The graph of $x = y^2$ appears in Fig. 4-4. Here the portion for which $y \leqslant 0$ is the reflection through the x-axis of that portion for which $y \geqslant 0$, and vice versa. If (x, y) lies on the graph, then $(x, -y)$ also lies on it. We say that this graph is *symmetric with respect to the x-axis*.

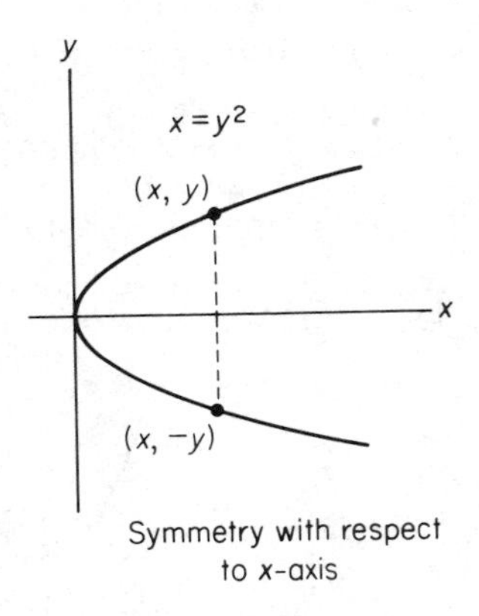

Symmetry with respect to x-axis

FIG. 4-4

DEFINITION. *A graph is* ***symmetric with respect to the x-axis*** *if and only if $(x, -y)$ lies on the graph when (x, y) does.*

A third type of symmetry, *symmetry with respect to the origin*, is illustrated by the graph of $y = x^3$ (Fig. 4-5). Whenever (x, y) lies on the

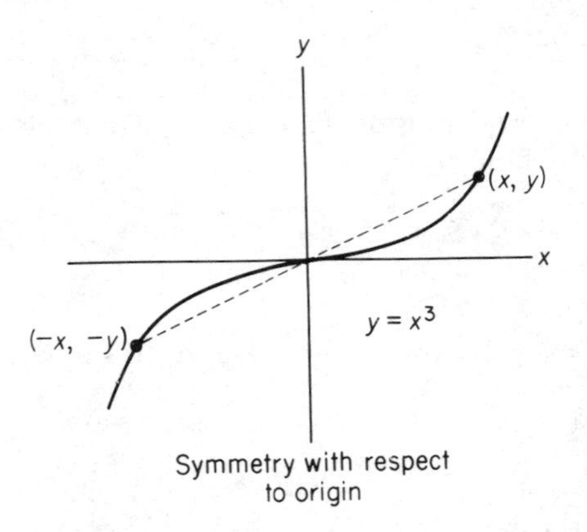

Symmetry with respect to origin

FIG. 4-5

graph, then $(-x, -y)$ also lies on it. Note that the line segment joining (x, y) and $(-x, -y)$ is bisected by the origin.

DEFINITION. *A graph is **symmetric with respect to the origin** if and only if $(-x, -y)$ lies on the graph when (x, y) does.*

A summary of tests for symmetry is given in Table 4-1.

TABLE 4-1 Tests for Symmetry

x-axis	Replace y by $-y$ in given equation. Symmetric if equivalent equation is obtained.
y-axis	Replace x by $-x$ in given equation. Symmetric if equivalent equation is obtained.
origin	Replace x by $-x$ and y by $-y$ in given equation. Symmetric if equivalent equation is obtained.

EXAMPLE 3

Determine whether or not the graph of $y = f(x) = 1 - x^4$ is symmetric with respect to the x-axis, the y-axis, or the origin. Then find the intercepts and sketch the graph.

Symmetry. *x-axis*: Replacing y by $-y$ in $y = 1 - x^4$ gives

$$-y = 1 - x^4,$$
$$y = -1 + x^4,$$

which is not equivalent to the given equation. Thus the graph is *not* symmetric with respect to the x-axis.

y-axis: Replacing x by $-x$ in $y = 1 - x^4$ gives

$$y = 1 - (-x)^4,$$
$$y = 1 - x^4,$$

which is equivalent to the given equation. Thus the graph *is* symmetric with respect to the y-axis.

origin: Replacing x by $-x$ and y by $-y$ in $y = 1 - x^4$ gives

$$-y = 1 - (-x)^4,$$
$$-y = 1 - x^4,$$
$$y = -1 + x^4,$$

which is not equivalent to the given equation. Thus the graph is *not* symmetric with respect to the origin.

Intercepts. Testing for x-intercepts, we set $y = 0$ in $y = 1 - x^4$. Then

$$\begin{aligned} 1 - x^4 &= 0, \\ (1 - x^2)(1 + x^2) &= 0, \\ (1 - x)(1 + x)(1 + x^2) &= 0, \\ x = 1 \text{ or } x &= -1. \end{aligned}$$

The x-intercepts are thus $(1, 0)$ and $(-1, 0)$. Testing for y-intercepts, we set $x = 0$. Then $y = 1$ and so $(0, 1)$ is the only y-intercept.

Discussion. If the intercepts and some points (x, y) where $x > 0$ are plotted, we can sketch the entire graph by using the property of symmetry with respect to the y-axis (Fig. 4-6).

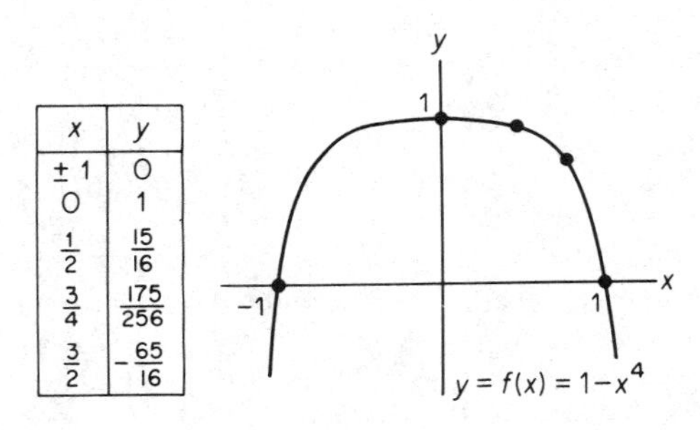

x	y
± 1	0
0	1
$\frac{1}{2}$	$\frac{15}{16}$
$\frac{3}{4}$	$\frac{175}{256}$
$\frac{3}{2}$	$-\frac{65}{16}$

FIG. 4-6

In Example 3 you saw that the graph of $y = f(x) = 1 - x^4$ does not have x-axis symmetry. With the exception of the constant function $f(x) = 0$, *the graph of any* ***function*** *$y = f(x)$ cannot be symmetric with respect to the x-axis.* To see why, let (x_0, y_0) be any point on the graph. Then $y_0 = f(x_0)$. If the graph has x-axis symmetry, then $(x_0, -y_0)$ also lies on it. Thus $-y_0 = f(x_0)$. Hence, with the input number x_0 the function f associates two output numbers, y_0 and $-y_0$. But a function can associate only one output with a given input. Thus $y_0 = -y_0$, which implies $y_0 = 0$. Since (x_0, y_0) is *any* point on the graph, then $y = f(x) = 0$.

EXAMPLE 4

Test the graph of $4x^2 + 9y^2 = 36$ for intercepts and symmetry. Sketch the graph.

Intercepts. *If* $y = 0$, then $4x^2 = 36$, and so $x = \pm 3$. Thus the x-intercepts are $(3, 0)$ and $(-3, 0)$. If $x = 0$, then $9y^2 = 36$, and so $y = \pm 2$. Thus the y-intercepts are $(0, 2)$ and $(0, -2)$.

Symmetry. Testing for x-axis symmetry, we replace y by $-y$:

$$4x^2 + 9(-y)^2 = 36,$$
$$4x^2 + 9y^2 = 36.$$

Since we get the original equation, the graph is symmetric with respect to the x-axis.

Testing for y-axis symmetry, we replace x by $-x$:

$$4(-x)^2 + 9y^2 = 36,$$
$$4x^2 + 9y^2 = 36.$$

Again we have the original equation, and so the graph is also symmetric with respect to the y-axis.

Testing for symmetry with respect to the origin, we replace x by $-x$ and y by $-y$:

$$4(-x)^2 + 9(-y)^2 = 36,$$
$$4x^2 + 9y^2 = 36.$$

Since this is the original equation, the graph is also symmetric with respect to the origin.

Discussion. In Fig. 4-7 the intercepts and some points in the first quadrant are plotted. The points in that quadrant are then connected by a smooth curve. By symmetry with respect to the x-axis, the points in the fourth

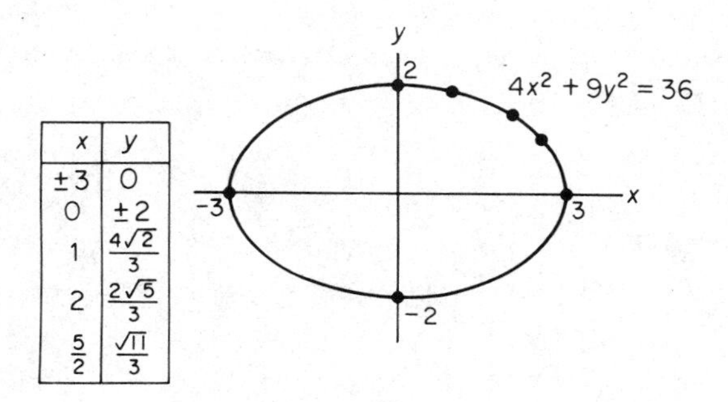

x	y
± 3	0
0	± 2
1	$\frac{4\sqrt{2}}{3}$
2	$\frac{2\sqrt{5}}{3}$
$\frac{5}{2}$	$\frac{\sqrt{11}}{3}$

FIG. 4-7

quadrant are obtained. Then by symmetry with respect to the y-axis the complete graph is found. There are other ways of graphing the equation by using symmetry. For example, after plotting the intercepts and some points in the first quadrant, then by symmetry with respect to the origin we can obtain the points in the third quadrant. By symmetry with respect to the x-axis (or y-axis) we can then obtain the entire graph.

In Example 4 the graph of $4x^2 + 9y^2 = 36$ was symmetric with respect to the x-axis, the y-axis, and the origin. **For any graph, if any two of the three types of symmetry that we have discussed exist, then the third type must also exist.**

Exercise 4-1

In Problems **1–14**, *find the x- and y-intercepts of the graphs of the equations. Also determine whether or not the graphs are symmetric with respect to the x-axis, the y-axis, or the origin. Do not sketch the graphs.*

1. $y = 5x$.

2. $y = f(x) = x^2 - 4$.

3. $2x^2 + y^2x^4 = 8 - y$.

4. $x = y^3$.

5. $4x^2 - 9y^2 = 36$.

6. $y = 7$.

7. $x = -2$.

8. $y = |2x| - 2$.

9. $x = -y^{-4}$.

10. $y = \sqrt{x^2 - 4}$.

11. $x - 4y - y^2 + 21 = 0$.

12. $x^3 - xy + y^2 = 0$.

13. $y = f(x) = x^3/(x^2 + 5)$.

14. $x^2 + xy + y^2 = 0$.

In Problems **15–22**, *find the x- and y-intercepts of the graphs of the equations. Also determine whether or not the graphs are symmetric with respect to the x-axis, the y-axis, or the origin. Then sketch the graphs.*

15. $|x| - |y| = 0$.

16. $x = y^4$.

17. $2x + y^2 = 4$.

18. $y = x - x^3$.

19. $y = f(x) = x^3 - 4x$.

20. $x^2 + y^2 = 16$.

21. $4x^2 + y^2 = 16$.

22. $x^2 - y^2 = 1$.

4-2 ASYMPTOTES

In Example 1(b) of the last section, we showed that the graph of the function $y = 1/x$ has no intercepts. Although this graph is symmetric with respect to the origin (as you may verify), it has other distinguishing features (see Fig. 4-8). As x approaches zero from the right, $1/x$ becomes

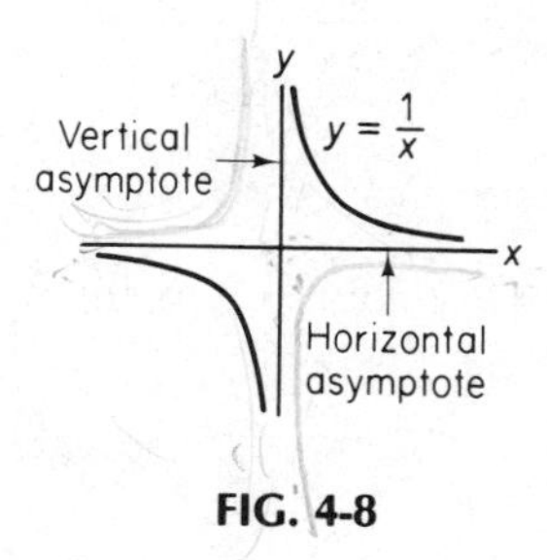

FIG. 4-8

positively infinite; as x approaches zero from the left, $1/x$ becomes negatively infinite. In terms of limits,

$$\lim_{x\to 0^+} \frac{1}{x} = \infty \quad \text{and} \quad \lim_{x\to 0^-} \frac{1}{x} = -\infty.$$

We say that the line $x = 0$ (the y-axis) is a *vertical asymptote* of the graph of $y = 1/x$. This means it is a vertical line near which the graph "blows up" (that is, the graph rises or falls, so to speak, without bound).

On the other hand, as x approaches ∞, as well as $-\infty$, $1/x$ approaches 0. That is,

$$\lim_{x\to \infty} \frac{1}{x} = 0 \quad \text{and} \quad \lim_{x\to -\infty} \frac{1}{x} = 0.$$

We say that the line $y = 0$ (the x-axis) is a *horizontal asymptote* of the graph of $y = 1/x$. This means that it is a horizontal line near which the graph "settles down" as $x \to \infty$ or $x \to -\infty$.

DEFINITION. *The line $x = a$ is a* ***vertical asymptote*** *of the graph of the function f if and only if*

$$\lim_{x\to a^+} f(x) = \infty \text{ (or } -\infty)$$

or $$\lim_{x\to a^-} f(x) = \infty \text{ (or } -\infty).$$

The line $y = b$ is a ***horizontal asymptote*** *of the graph of f if and only if*

$$\lim_{x\to\infty} f(x) = b \quad \textit{or} \quad \lim_{x\to -\infty} f(x) = b.$$

Note that if $x = a$ is a vertical asymptote, the function cannot be continuous at a—in fact, it will have an infinite discontinuity at a.

EXAMPLE 1

Determine the horizontal and vertical asymptotes for the graphs of the following functions.

a. $y = \dfrac{1}{x-2} + 3.$

To test for horizontal asymptotes, we find the limits of y as $x \to \infty$ and as $x \to -\infty$.

$$\lim_{x\to\infty}\left[\frac{1}{x-2} + 3\right] = \lim_{x\to\infty}\frac{1}{x-2} + \lim_{x\to\infty} 3.$$

As x increases without bound, so does $x - 2$. Hence $1/(x-2)$ approaches zero. Thus,

$$\lim_{x\to\infty}\left[\frac{1}{x-2} + 3\right] = 3,$$

and so the line $y = 3$ is a horizontal asymptote. Also,

$$\lim_{x\to-\infty}\left[\frac{1}{x-2} + 3\right] = \lim_{x\to-\infty}\frac{1}{x-2} + \lim_{x\to-\infty} 3$$
$$= 0 + 3 = 3.$$

Hence the graph will settle down near the line $y = 3$ both as $x \to \infty$ and $x \to -\infty$.

To determine vertical asymptotes, we must find where $\dfrac{1}{x-2} + 3$ blows up. Note that the denominator of $1/(x-2)$ is 0 when $x = 2$. If x is slightly larger than 2, then $x - 2$ is both close to 0 and positive. Thus $1/(x-2)$ is very large and so

$$\lim_{x\to 2^+}\left[\frac{1}{x-2} + 3\right] = \infty.$$

Hence, the line $x = 2$ is a vertical asymptote. If x is slightly less than 2, then $x - 2$ is very close to 0 but negative. Thus $1/(x-2)$ is "very negative" and so

$$\lim_{x\to 2^-}\left[\frac{1}{x-2} + 3\right] = -\infty.$$

We conclude that the function increases without bound as $x \to 2^+$ and decreases without bound as $x \to 2^-$. The graph appears in Fig. 4-9. The dotted lines indicate the asymptotes and are not part of the graph. However, they can be useful guides in sketching it.

Usually, good places to look for vertical asymptotes of a function $y = f(x)$ are at those values of x for which a denominator of $f(x)$ is 0.

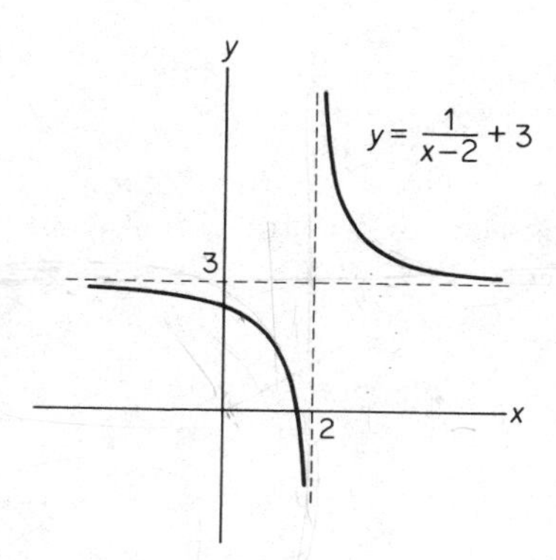

FIG. 4-9

b. $y = f(x) = x^3 + 2x$.

Testing for horizontal asymptotes, we have

$$\lim_{x \to \infty} (x^3 + 2x) = \infty \quad \text{and} \quad \lim_{x \to -\infty} (x^3 + 2x) = -\infty.$$

Thus the graph does not settle down as $x \to \infty$ or $x \to -\infty$. Hence, there are no horizontal asymptotes.

We previously said that if the line $x = a$ is a vertical asymptote of the graph of a function f, then f must have an infinite discontinuity at a. But $y = f(x) = x^3 + 2x$ is a polynomial function and is continuous everywhere. Thus its graph has no vertical asymptotes. See Fig. 4-10.

In general, **any polynomial function that is not a constant has neither horizontal nor vertical asymptotes.**

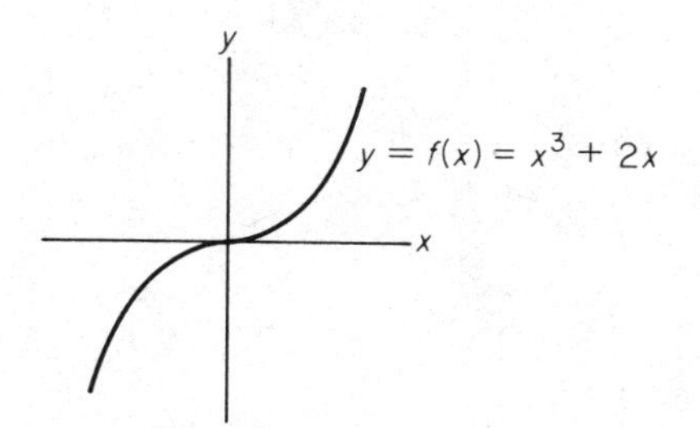

FIG. 4-10

EXAMPLE 2

Sketch the graph of $y = \dfrac{x^2}{x^2 - 1}$ *with the aid of intercepts, symmetry, and asymptotes.*

Intercepts. If $x = 0$, then $y = 0$; if $y = 0$, then $x = 0$. Therefore, the x-intercept, as well as the y-intercept, is $(0, 0)$.

Symmetry. Since y is a function of x and it is not the zero function, then the graph is *not* symmetric with respect to the x-axis.
Testing for y-axis symmetry, we replace x by $-x$:

$$y = \frac{(-x)^2}{(-x)^2 - 1},$$

$$y = \frac{x^2}{x^2 - 1}.$$

The graph *is* symmetric with respect to the y-axis. *Symmetry with respect to exactly one axis implies that the graph cannot be symmetric with respect to the origin.*

Asymptotes.

Horizontal:

$$\lim_{x\to\infty} \frac{x^2}{x^2 - 1} = \lim_{x\to\infty} \frac{1}{1 - \frac{1}{x^2}} \qquad \text{(dividing both numerator and denominator by } x^2\text{)}$$

$$= \frac{1}{1 - 0} = 1.$$

Therefore, as $x \to \infty$ the graph approaches the line $y = 1$, a horizontal asymptote. By symmetry, as $x \to -\infty$ the graph again approaches the line $y = 1$.

Vertical:
Since

$$\frac{x^2}{x^2 - 1} = \frac{x^2}{(x - 1)(x + 1)},$$

it is easy to see from the denominator that the graph will blow up when x is close to 1 or -1. In fact,

$$\lim_{x\to 1^-} \frac{x^2}{(x - 1)(x + 1)} = -\infty \text{ and } \lim_{x\to 1^+} \frac{x^2}{(x - 1)(x + 1)} = \infty.$$

By symmetry,

$$\lim_{x\to -1^+} \frac{x^2}{(x - 1)(x + 1)} = -\infty \quad \text{and} \quad \lim_{x\to -1^-} \frac{x^2}{(x - 1)(x + 1)} = \infty.$$

Thus the lines $x = 1$ and $x = -1$ are vertical asymptotes.

Discussion. By plotting the intercept and other points on the graph when $x > 0$, and by using properties of symmetry and asymptotes, it is relatively easy to sketch the graph (see Fig. 4-11).

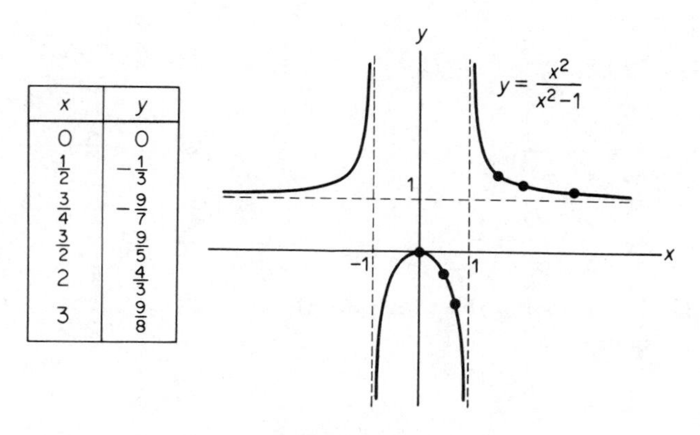

x	y
0	0
$\frac{1}{2}$	$-\frac{1}{3}$
$\frac{3}{4}$	$-\frac{9}{7}$
$\frac{3}{2}$	$\frac{9}{5}$
2	$\frac{4}{3}$
3	$\frac{9}{8}$

FIG. 4-11

Exercise 4-2

In Problems **1–12**, *determine the horizontal and vertical asymptotes of the graphs of the functions. Do not sketch the curves.*

1. $y = \dfrac{4}{x}$.

2. $y = -\dfrac{4}{x^2}$.

3. $y = \dfrac{4}{x-6} + 4$.

4. $y = \dfrac{2x+1}{2x-1}$.

5. $y = x^3 - 5x + 8$.

6. $y = \dfrac{x^3}{x^2-9}$.

7. $f(x) = \dfrac{x-1}{2x+3}$.

8. $f(x) = \dfrac{x^2}{5}$.

9. $f(x) = \dfrac{2x^2}{x^2+x-6}$.

10. $f(x) = \dfrac{5}{2x^2-9x+4}$.

11. $f(x) = \sqrt[3]{x^2}$.

12. $y = \dfrac{x^2(x^2-9)}{x^2}$.

In Problems **13–22**, *find the x- and y-intercepts of the graphs of the functions; determine whether the graphs are symmetric with respect to the x-axis, y-axis, or origin; determine horizontal and vertical asymptotes; sketch the graphs.*

13. $y = \dfrac{3}{x-1}$.

14. $y = \dfrac{x}{4-x}$.

15. $f(x) = \dfrac{8}{x^3}$.

16. $f(x) = \dfrac{1}{x^4}$.

17. $f(x) = \dfrac{1}{x^2 - 1}$.

18. $f(x) = \dfrac{x^2}{x^2 - 4}$.

19. $y = \dfrac{x^2 - 1}{x^2 - 4}$.

20. $y = \dfrac{x^3 - x}{x}$.

21. $y = \dfrac{x^2(x^2 - 9)}{x^2}$.

22. $f(x) = \begin{cases} 1/x, & \text{if } x > 0, \\ (x + 1)/x, & \text{if } x < 0. \end{cases}$

23. In discussing the time pattern of purchasing, Mantell and Sing [32] use the curve

$$y = \frac{x}{a + bx}$$

as a mathematical model. They claim that $y = 1/b$ is an asymptote. Verify this.

4-3 RELATIVE MAXIMA AND MINIMA

In curve sketching, plotting points at random usually is not good enough to properly determine a curve's shape. For example, the points $(-1, 0)$, $(0, -1)$, and $(1, 0)$ satisfy the equation $y = (x + 1)^3(x - 1)$. You might hastily conclude that its graph should appear as in Fig. 4-12, but in fact the actual shape is given in Fig. 4-13. In this section, as well as in the

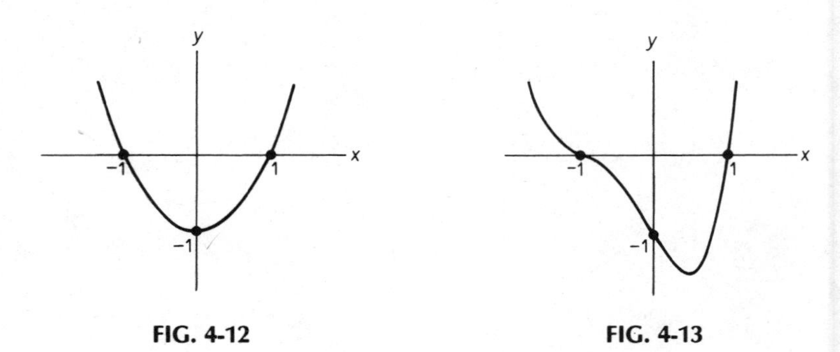

FIG. 4-12 FIG. 4-13

following one, we shall explore the powerful role that differentiation plays in analyzing a function so that we may determine the true shape and behavior of its graph.

Suppose Fig. 4-14 gives the graph of a function $y = f(x)$. As x increases (goes from left to right) on the interval I_1 determined by a and b, the corresponding values of y increase and the curve is rising. Symbolically, if x_1 and x_2 are any two points in I_1 such that $x_1 < x_2$, then

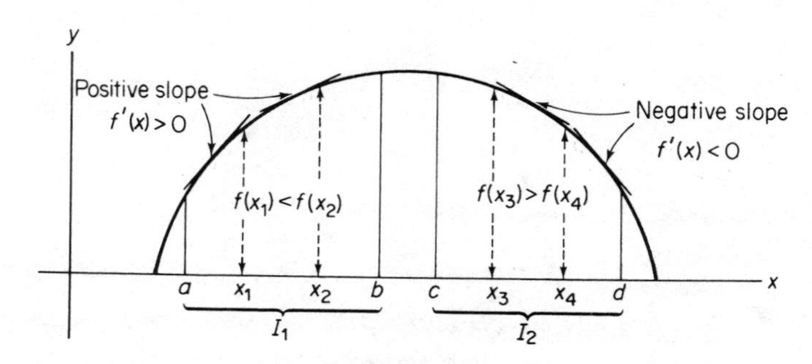

FIG. 4-14

$f(x_1) < f(x_2)$ and f is said to be an *increasing function* on I_1. Moreover, for this portion of the curve, any tangent line will have a positive slope, and thus $f'(x)$ must be positive for all x in I_1. On the other hand, as x increases on the interval I_2 determined by c and d, the curve is falling. Here $x_3 < x_4$ implies $f(x_3) > f(x_4)$ and f is said to be a *decreasing function* on I_2. In this case any tangent line has a negative slope, and thus $f'(x) < 0$ for all x in I_2.

DEFINITION. *A function f is an* ***increasing*** *[****decreasing****]* ***function*** *on the interval I if and only if for any two points x_1, x_2, in I such that $x_1 < x_2$, then $f(x_1) < f(x_2)$ $[f(x_1) > f(x_2)]$.*

RULE 1. *If $f'(x) > 0$ on an interval I, then f is an increasing function on I. If $f'(x) < 0$ on I, then f is a decreasing function on I.*

A function f is said to be increasing at a *point* x_0 if there is an open interval containing x_0 on which f is increasing. Thus in Fig. 4-14, f is increasing at x_1 and is decreasing at x_3.

To illustrate these notions we shall use Rule 1 to find when $y = 18x - \frac{2x^3}{3}$ is increasing or decreasing. If $y = f(x)$, we must determine when $f'(x)$ is positive and when $f'(x)$ is negative.

$$f'(x) = 18 - 2x^2 = 2(9 - x^2) = 2(3 + x)(3 - x).$$

Using the technique of Sec. 2-4, we can find the signs of $f'(x)$ by considering the intervals determined by the roots of $2(3 + x)(3 - x) = 0$, namely 3 and -3 (see Fig. 4-15). In each interval the sign of $f'(x)$ is

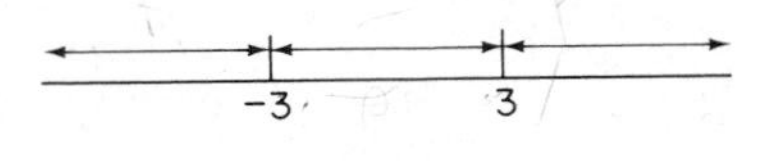

FIG. 4-15

determined by the signs of its factors:

if $x < -3$, then $f'(x) = 2(-)(+) = (-)$ and f is decreasing;
if $-3 < x < 3$, then $f'(x) = 2(+)(+) = (+)$ and f is increasing;
if $x > 3$, then $f'(x) = 2(+)(-) = (-)$ and f is decreasing [see Fig. 4-16(a)].

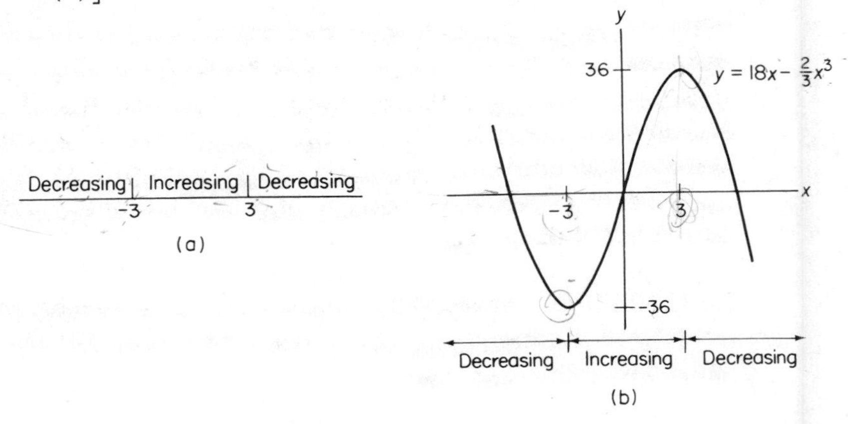

FIG. 4-16

Thus f is decreasing on $(-\infty, -3)$ and $(3, \infty)$, and is increasing on $(-3, 3)$, as seen in Fig. 4-16(b). These results could be sharpened. Actually, by definition, f is decreasing on $(-\infty, -3]$ and $[3, \infty)$, and increasing on $[-3, 3]$. However, for our purposes open intervals are sufficient. *It will be our practice to determine* ***open*** *intervals on which a function is increasing or decreasing.*

Look now at the graph of $y = f(x)$ in Fig. 4-17. Three observations can be made.

First, there is something special about the points P_1, P_2, and P_3. Notice that P_1 is *higher* than any other "nearby" points on the curve—likewise for P_3. The point P_2 is *lower* than any other "nearby" points on the curve. Since P_1, P_2, and P_3 may not necessarily be the highest or lowest points on the *entire* curve, we simply say that f has a *relative maximum* when $x = x_1$ and when $x = x_3$, and a *relative minimum* when $x = x_2$. Actually, there is an *absolute maximum* (highest point) when

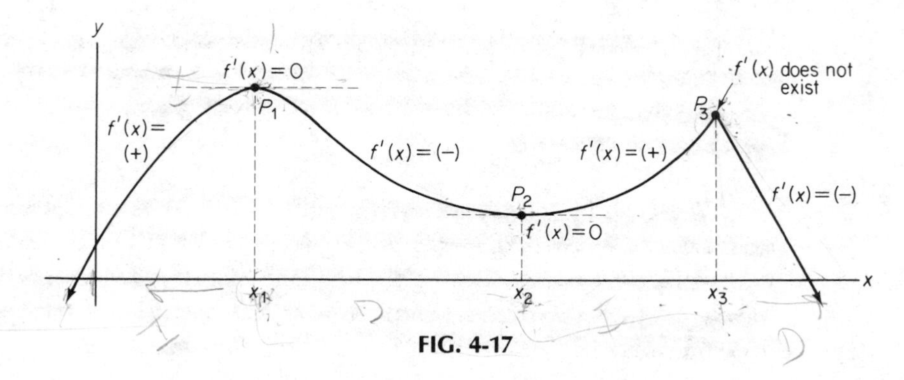

FIG. 4-17

$x = x_1$, but there is no *absolute minimum* (lowest point), since the curve is assumed to extend downward indefinitely. We define these new terms as follows.

DEFINITION. *A function f has a* ***relative maximum*** *[**relative minimum**] when $x = x_0$ if there is an open interval containing x_0 on which $f(x_0) \geqslant f(x)$ $[f(x_0) \leqslant f(x)]$ for all x in the interval.*

DEFINITION. *A function f has an* ***absolute maximum*** *[**absolute minimum**] when $x = x_0$ if $f(x_0) \geqslant f(x)$ $[f(x_0) \leqslant f(x)]$ for all x in the domain of f.*

We refer to either a relative maximum or a relative minimum as a **relative extremum** (plural: *relative extrema*). Similarly, we speak of **absolute extrema**.

Our second observation is that at a relative extremum the derivative may not be defined (as when $x = x_3$). But whenever it is defined, it is 0 (as when $x = x_1$ and $x = x_2$), and hence the tangent line is horizontal as shown in Fig. 4-17. We may state:

RULE 2. *If f has a relative extremum when $x = x_0$, then $f'(x_0) = 0$ or $f'(x_0)$ is not defined.*

Third, each relative extremum occurs at a point at which the sign of $f'(x)$ is changing, regardless of whether or not the derivative is defined at the point. For the relative maximum when $x = x_1$, $f'(x)$ goes from $(+)$ for $x < x_1$ to $(-)$ for $x > x_1$, *as long as x is near x_1*. At the relative minimum when $x = x_2$, $f'(x)$ goes from $(-)$ to $(+)$, and at the relative maximum when $x = x_3$, it again goes from $(+)$ to $(-)$. Thus, *around relative maxima, f is increasing and then decreasing, and the reverse holds for relative minima.*

RULE 3. *If x_0 is in the domain of f and $f'(x)$ changes from positive to negative as x increases through x_0, then f has a relative maximum when $x = x_0$. If $f'(x)$ changes from negative to positive as x increases through x_0, then f has a relative minimum when $x = x_0$.*

From Rules 1, 2, and 3, it should be clear that relative extrema may occur at values of x for which $f'(x) = 0$ or is not defined, for it is there that $f'(x)$ may change sign. These values of x are called *critical values*. To find relative extrema, you should examine the signs of $f'(x)$ over the intervals determined by the critical values.

DEFINITION. *If $f'(x_0) = 0$ or $f'(x_0)$ is not defined, x_0 is called a* ***critical value*** *of f. If x_0 is a critical value and is in the domain of f, then $(x_0, f(x_0))$ is called a* ***critical point.***

PITFALL. *Not every critical value corresponds to a relative extremum.* For example, if $y = f(x) = x^3$, then $f'(x) = 3x^2$. Since $f'(0) = 0$, then 0 is a critical value. But if $x < 0$, then $3x^2 > 0$. If $x > 0$, then $3x^2 > 0$. Since $f'(x)$ does not change sign, no relative maximum or minimum exists. Indeed, since $f'(x) \geqslant 0$ for all x, the graph of f never falls and f is said to be *nondecreasing* (see Fig. 4-18). On the other hand, if $y = f(x) = 1/x^2$, then $y' = -2/x^3$. Since y' is not defined when $x = 0$, then 0 is a critical value. If $x < 0$, then $y' > 0$. If $x > 0$, then $y' < 0$. Although a change in sign of y' occurs around $x = 0$, no relative maximum exists there since 0 is not in the domain of f. Nevertheless, this critical value is important in determining the intervals over which f is increasing or decreasing. Here f is increasing on $(-\infty, 0)$ and decreasing on $(0, \infty)$ [see Fig. 4-19].

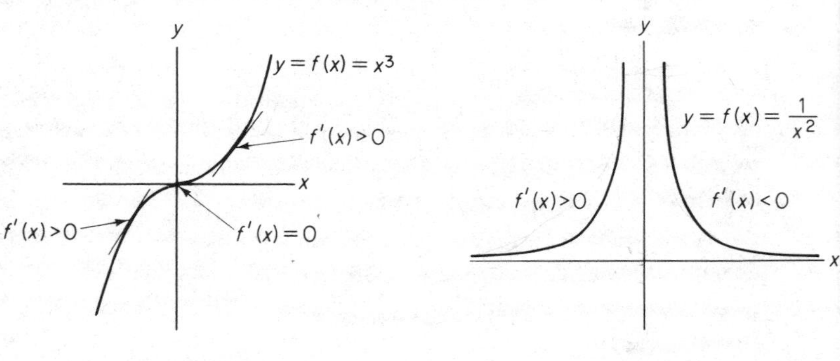

FIG. 4-18

FIG. 4-19

From our discussions and the "Pitfall" above, we can conclude that a critical value is only a "candidate" for a relative extremum. It may be a relative maximum, a relative minimum, or neither.

Summarizing the results of this section, we have the *first-derivative test* for the relative extrema of $y = f(x)$:

FIRST-DERIVATIVE TEST.

1. **Find $f'(x)$.**
2. **Determine critical values.**
3. **On the intervals suggested by the critical values, determine whether f is increasing ($f'(x) > 0$) or decreasing ($f'(x) < 0$).**
4. **For each critical value x_0 in the domain of f, determine whether $f'(x)$ changes sign as x increases through x_0. There is a relative maximum when $x = x_0$ if $f'(x)$ changes from $(+)$ to $(-)$, and a relative minimum if $f'(x)$ changes from $(-)$ to $(+)$. If $f'(x)$ does not change sign, there is no relative maximum or minimum when $x = x_0$.**

EXAMPLE 1

If $y = f(x) = x + \frac{4}{x+1}$, use the first-derivative test to determine intervals on which f is increasing or decreasing and locate all relative extrema.

1. $f'(x) = 1 - \frac{4}{(x+1)^2} = \frac{(x+1)^2 - 4}{(x+1)^2} = \frac{(x+3)(x-1)}{(x+1)^2}$.

2. Setting $f'(x) = 0$ gives the critical values $x = -3, 1$. Since $f'(-1)$ does not exist, $x = -1$ is also a critical value.

3. There are four intervals to consider (Fig. 4-20):

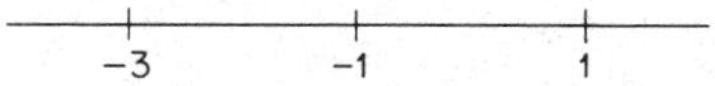

$-3 \qquad -1 \qquad 1$

FIG. 4-20

$$\text{if } x < -3, \text{ then } f'(x) = \frac{(-)(-)}{(+)} = (+) \text{ and } f \text{ is increasing;}$$

$$\text{if } -3 < x < -1, \text{ then } f'(x) = \frac{(+)(-)}{(+)} = (-) \text{ and } f \text{ is decreasing;}$$

$$\text{if } -1 < x < 1, \text{ then } f'(x) = \frac{(+)(-)}{(+)} = (-) \text{ and } f \text{ is decreasing;}$$

$$\text{if } x > 1, \text{ then } f'(x) = \frac{(+)(+)}{(+)} = (+) \text{ and } f \text{ is increasing (Fig. 4-21).}$$

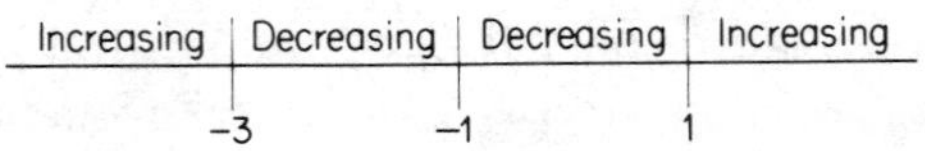

Increasing	Decreasing	Decreasing	Increasing
	−3	−1	1

FIG. 4-21

Thus f is increasing on $(-\infty, -3)$ and $(1, \infty)$ and is decreasing on $(-3, -1)$ and $(-1, 1)$.

4. When $x = -3$, there is a relative maximum since $f'(x)$ changes from $(+)$ to $(-)$. When $x = 1$, there is a relative minimum since $f'(x)$ changes from $(-)$ to $(+)$. We ignore $x = -1$ since -1 is not in the domain of f (Fig. 4-22).

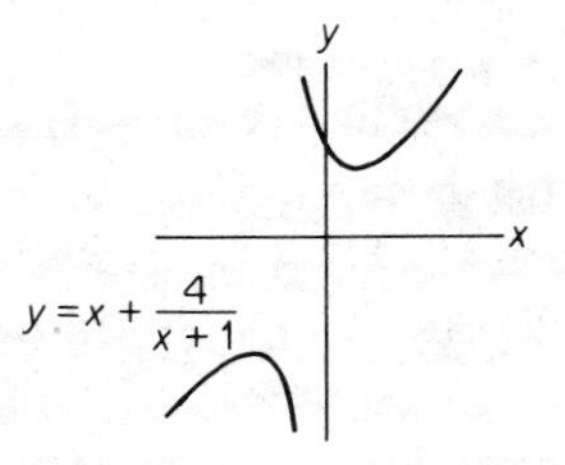

FIG. 4-22

EXAMPLE 2

Test $y = f(x) = x^{2/3}$ for relative extrema.

We have $f'(x) = \frac{2}{3}x^{-1/3} = 2/(3\sqrt[3]{x})$. When $x = 0$, then $f'(x)$ is not defined and thus $x = 0$ is a critical value. If $x < 0$, then $f'(x) < 0$. If $x > 0$, then $f'(x) > 0$. Since 0 is also in the domain of f, there is a relative (as well as an absolute) minimum when $x = 0$ (Fig. 4-23).

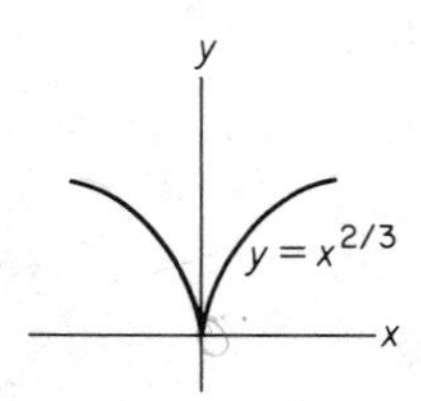

FIG. 4-23

EXAMPLE 3

Sketch the graph of $y = f(x) = 2x^2 - x^4$.

Intercepts. If $x = 0$, then $y = 0$. If $y = 0$, then

$$0 = 2x^2 - x^4 = x^2(2 - x^2) = x^2(\sqrt{2} + x)(\sqrt{2} - x),$$

and thus $x = 0, \pm\sqrt{2}$. The intercepts are $(0, 0)$, $(\sqrt{2}, 0)$, and $(-\sqrt{2}, 0)$.

Symmetry. Testing for y-axis symmetry, we have

$$y = 2(-x)^2 - (-x)^4,$$
$$y = 2x^2 - x^4.$$

Since this is the original equation, there is y-axis symmetry. It can be shown that there is no x-axis symmetry and hence no symmetry with respect to the origin.

Asymptotes. No horizontal or vertical asymptotes exist since f is a nonconstant polynomial function.

First-derivative Test.

1. $y' = 4x - 4x^3 = 4x(1 - x^2) = 4x(1 + x)(1 - x)$.
2. Setting $y' = 0$ gives the critical values $x = 0, \pm 1$. The critical points are $(-1, 1)$, $(0, 0)$, and $(1, 1)$. The y-coordinates of these points were found by substituting $x = 0, \pm 1$ into the *original* equation, $y = 2x^2 - x^4$.
3. There are four intervals to consider in Fig. 4-24:

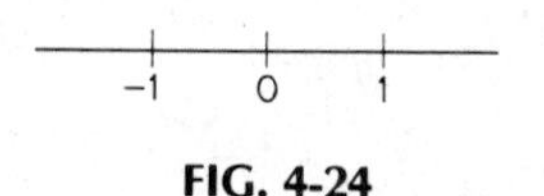

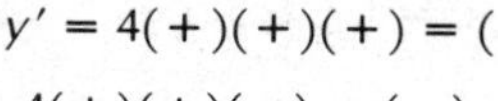

FIG. 4-24

if $x < -1$, then $y' = 4(-)(-)(+) = (+)$ and f is increasing;

if $-1 < x < 0$, then $y' = 4(-)(+)(+) = (-)$ and f is decreasing;

if $0 < x < 1$, then $y' = 4(+)(+)(+) = (+)$ and f is increasing;

if $x > 1$, then $y' = 4(+)(+)(-) = (-)$ and f is decreasing (Fig. 4-25).

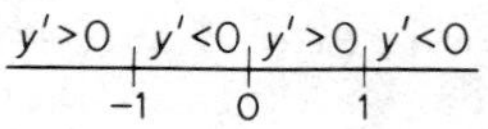

$y'>0$ | $y'<0$ | $y'>0$ | $y'<0$
-1 0 1

FIG. 4-25

4. Relative maxima occur at $(-1, 1)$ and $(1, 1)$; a relative minimum occurs at $(0, 0)$.

Discussion. In Fig. 4-26(a) we have plotted the intercepts and the horizontal tangents at the relative maximum and minimum points. We know the curve rises from the left, has a relative maximum, then falls, has a relative minimum, then rises to a relative maximum, and falls thereafter. A sketch is shown in Fig. 4-26(b).

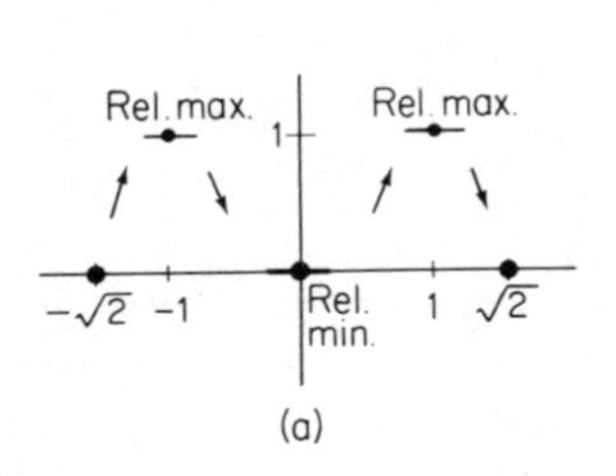

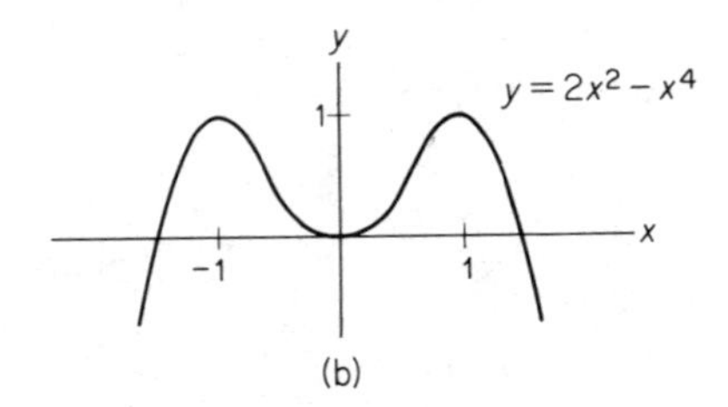

FIG. 4-26

In Example 3 relative maxima, as well as absolute maxima, occur at $x = \pm 1$ [see Fig. 4-26(b)]. Although there is a relative minimum, there is no absolute minimum.

If the domain of a function is a closed interval, to determine absolute extrema we must not only examine the function for relative extrema, but we must also take into consideration the values of $f(x)$ at the endpoints. Although endpoints are not considered to be relative maxima or minima, they may yield *absolute* maxima or minima. Example 4 will illustrate.

EXAMPLE 4

Find all extrema (relative and absolute) for $y = f(x) = x^2 - 4x + 5$ *on the closed interval* $[1, 4]$.

1. $f'(x) = 2x - 4 = 2(x - 2)$.

2. Setting $f'(x) = 0$ gives the critical value $x = 2$, which is in the domain of f.

3. The intervals to consider are when $x < 2$ and when $x > 2$.

4. If $x < 2$, then $f'(x) < 0$ and f is decreasing; if $x > 2$, then $f'(x) > 0$ and f is increasing. Thus there is a relative minimum when $x = 2$. It occurs on the graph at the point $(2, 1)$ [see Fig. 4-27].

5. Since f is decreasing for $x < 2$, then an absolute maximum may *possibly* occur at the left-hand endpoint of the domain of f, that is, when $x = 1$. Similarly, since f is increasing for $x > 2$, an absolute maximum may

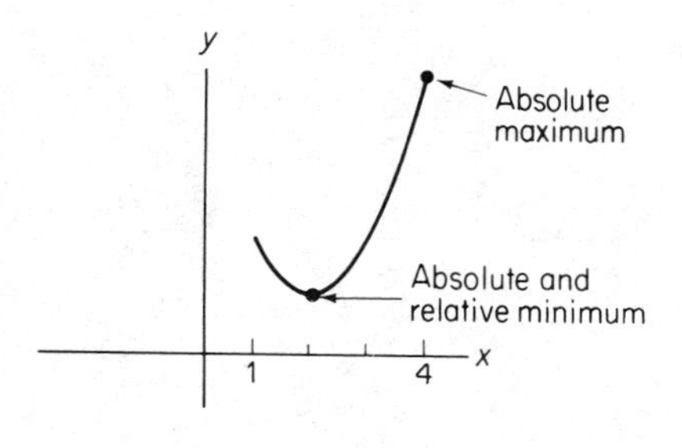

FIG. 4-27

possibly occur at the right-hand endpoint, that is, when $x = 4$. Testing the endpoints, we have $f(1) = 2$ and $f(4) = 5$. Noting that $f(4) > f(1)$, we conclude that an absolute maximum occurs when $x = 4$. When $x = 2$ there is an absolute, as well as a relative, minimum.

If f is a function such that $f'(x) > 0$ on an interval I, then f is increasing there. In fact, f' gives the rate of increase of f. However, this rate itself might be increasing or decreasing. That is, $[f'(x)]'$, or more simply $f''(x)$, may be positive or negative. If $f''(x) > 0$ on I, then we say that f is increasing *at an increasing rate* there. If $f''(x) < 0$, then f is increasing *at a decreasing rate*. On the other hand, if $f'(x) < 0$ on I, then f is decreasing there. We say that f is decreasing at an increasing or decreasing rate depending, respectively, on whether we have $f''(x) > 0$ or $f''(x) < 0$ on I. In the next section you will see how these concepts relate to the graph of a function.

EXAMPLE 5

Show that $y = f(x) = \sqrt{x}$ is increasing for $x > 0$ and determine whether it is increasing at an increasing or decreasing rate there.

Since $f'(x) = \frac{1}{2}x^{-1/2} = 1/(2\sqrt{x})$, then $f'(x) > 0$ for $x > 0$. Thus f is increasing on $(0, \infty)$. To determine if f is increasing at an increasing or decreasing rate, we must consider the sign of $f''(x)$ on $(0, \infty)$.

$$f''(x) = -\frac{1}{4}x^{-3/2} = -\frac{1}{4\sqrt{x^3}}.$$

Since $f''(x) < 0$ for $x > 0$, then f is increasing at a *decreasing* rate on $(0, \infty)$.

Exercise 4-3

In Problems **1–20** *determine when the function is increasing or decreasing and locate all relative maxima and minima. Do not sketch the graph.*

1. $y = x^2 + 2$.

2. $y = x^2 + 4x + 3$.

3. $y = x - x^2 + 2$.

4. $y = 4x - x^2$.

5. $y = -\dfrac{x^3}{3} - 2x^2 + 5x - 2$.

6. $y = 4x^3 - 3x^4$.

7. $y = x^4 - 2x^2$.

8. $y = -2 + 12x - x^3$.

9. $y = x^3 - 6x^2 + 9x$.

10. $y = x^3 - 6x^2 + 12x - 6$.

11. $y = 3x^5 - 5x^3$.

12. $y = 5x - x^5$.

13. $y = -x^5 - 5x^4 + 200$.

14. $y = 3x^4 - 4x^3 + 1$.

15. $y = \dfrac{1}{x - 1}$.

16. $y = \dfrac{3}{x}$.

17. $y = \dfrac{10}{\sqrt{x}}$.

18. $y = \dfrac{x}{x + 1}$.

19. $y = \dfrac{x^2}{1 - x}$.

20. $y = x + \dfrac{4}{x}$.

In Problems **21–32**, *determine: intervals on which the functions are increasing or decreasing; relative maxima and minima; symmetry; horizontal and vertical asymptotes; those intercepts that can be obtained conveniently. Then sketch the graphs.*

21. $y = x^2 - 6x - 7$.

22. $y = 2x^2 - 5x - 12$.

23. $y = 3x - x^3$.

24. $y = x^4 - 16$.

25. $y = 2x^3 - 9x^2 + 12x$.

26. $y = x^3 - 9x^2 + 24x - 19$.

27. $y = x^4 + 4x^3 + 4x^2$.

28. $y = x^5 - \frac{5}{4}x^4$.

29. $y = \dfrac{x + 1}{x - 1}$.

30. $y = \dfrac{x^2}{x^2 + 1}$.

31. $y = \dfrac{x^2}{x + 3}$.

32. $y = x + \dfrac{1}{x}$.

In Problems **33–38**, *find when absolute maxima and minima occur for the given function on the given interval.*

33. $f(x) = x^2 - 2x + 3$; $[-1, 2]$.

34. $f(x) = -2x^2 - 6x + 5$; $[-2, 3]$.

35. $f(x) = \frac{1}{3}x^3 - x^2 - 3x + 1$; $[0, 2]$.

36. $f(x) = \frac{1}{4}x^4 - \frac{3}{2}x^2$; $[0, 1]$.

37. $f(x) = 4x^3 + 3x^2 - 18x + 3$; $[\frac{1}{2}, 3]$.

38. $f(x) = x^{4/3}$; $[-8, 8]$.

39. A manufacturer's *fixed costs* are those costs that under normal conditions do not depend on the level of production. Some examples are rent, officers' salaries, and normal maintenance. If fixed costs are $c_f = 25{,}000$, show that the average fixed cost function $\bar{c}_f = c_f/q$ is a decreasing function for $q > 0$. Thus, as output q increases, each unit's portion of fixed cost declines.

40. If $c = 4q - q^2 + 2q^3$ is a cost function of a manufacturer, when is marginal cost increasing?

41. Given the demand function $p = 400 - 2q$ for a product, find when marginal revenue is increasing.

42. For a manufacturer's cost function $c = \sqrt{q}$, show that marginal and average costs are always decreasing for $q > 0$.

43. A psychological experiment (adapted from [1]) was conducted to analyze human response to electrical shocks (stimuli). The subjects received shocks of various intensities. The response R to a shock of intensity I (in microamperes) was to be a number that indicated the perceived magnitude relative to that of a "standard" shock. The standard shock was assigned a magnitude of 10. If

$$R = \frac{I^{4/3}}{2500}, \qquad 500 \leqslant I \leqslant 3500,$$

show that R increases with respect to I and that it is increasing at an increasing rate.

44. In a discussion of thermal pollution in [43], the efficiency E of a power plant is given by

$$E = .71\left(1 - \frac{T_c}{T_h}\right),$$

where T_c and T_h are the respective absolute temperatures of the hotter and colder reservoirs. Assume that T_c is a positive constant and that T_h is positive. Using calculus, show that as T_h increases, the efficiency increases. Also show that E increases with respect to T_h at a decreasing rate.

45. To study a predator-prey relationship, Holling [19] conducted an experiment in which a blindfolded subject, the "predator," stood in front of a 3-foot square table on which uniform sandpaper discs, the "prey," were placed. For one minute the "predator" searched for the discs by tapping

with a finger. As a disc was found, it was removed and searching resumed. The experiment was repeated at various disc densities (number of discs per 9 sq ft). It was estimated that y discs were picked up in one minute when x discs were on the table, where

$$y = \frac{.70x}{1 + .03x}, \qquad x \geqslant 0.$$

Holling states that y increases at a decreasing rate as the density of the discs (x) increases. (a) Verify this claim. (b) Sketch the graph of the equation.

46. For the predator-prey relationship of Problem 45, y/x gives the proportion of prey "destroyed" at a given prey density ($x > 0$). Show that y/x is a decreasing function of x, and sketch the graph of this function.

47. For a certain population, if E is the number of years of a person's education and S represents a numerical value of status based on that educational level, then $S = .25E^2$. (a) Determine the rate of change and the relative rate of change of status with respect to education when $E = 10$. (b) Show that as E increases, then the rate of change of status with respect to education increases, but the relative rate of change decreases.

4-4 CONCAVITY

We have seen how the first derivative is used to determine when a function is increasing or decreasing and to locate relative maxima and minima. However, for us to know with assurance the actual shape of a curve, we must have additional information. For example, consider the curve $y = f(x) = x^2$. Since $f'(x) = 2x$ and $f'(0) = 0$, then $x = 0$ is a critical value. If $x < 0$, then $f'(x) < 0$ and f is decreasing; if $x > 0$, then $f'(x) > 0$ and f is increasing. Thus there is a relative minimum when $x = 0$. Figures 4-28(a) and (b) both satisfy the preceding conditions. But

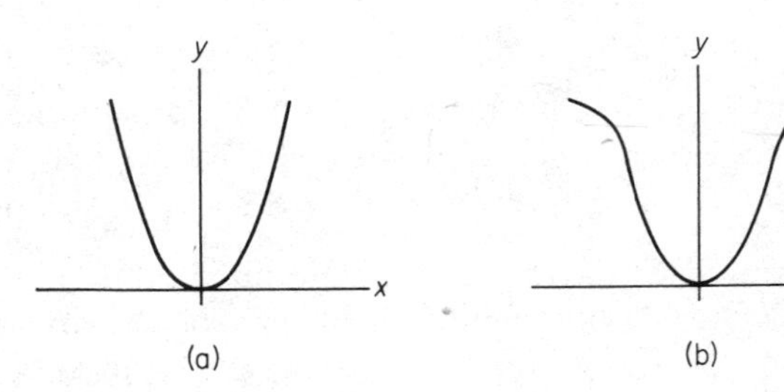

FIG. 4-28

which one truly describes the curve? This question will easily be settled by using the second derivative and the notion of *concavity*.

In Fig. 4-29, note that in each case the curve $y = f(x)$ "bends" (or opens) upward. Moreover, for each curve, if tangent lines are drawn their slopes increase in value as x increases. In (a) the slopes go from small positive values to larger values; in (b) they are negative and

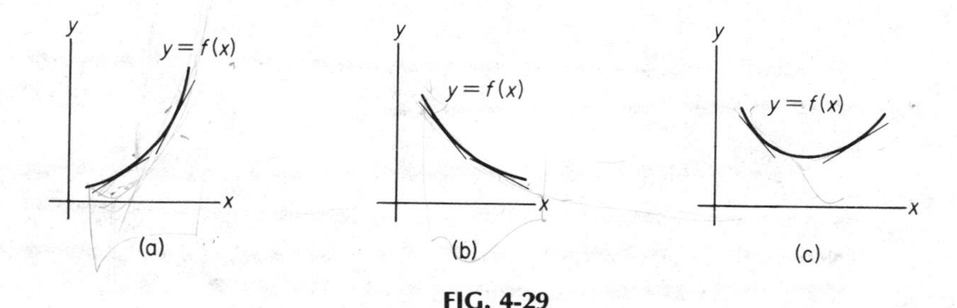

FIG. 4-29

approaching zero (thus increasing); in (c) they pass from negative values to positive values. Since $f'(x)$ gives the slope at a point, f' is an increasing function here. In each case we say that the curve (or function f) is *concave up*. By a similar analysis, in Fig. 4-30 it can be seen in each case that as x increases, the slopes of the tangent lines are decreasing

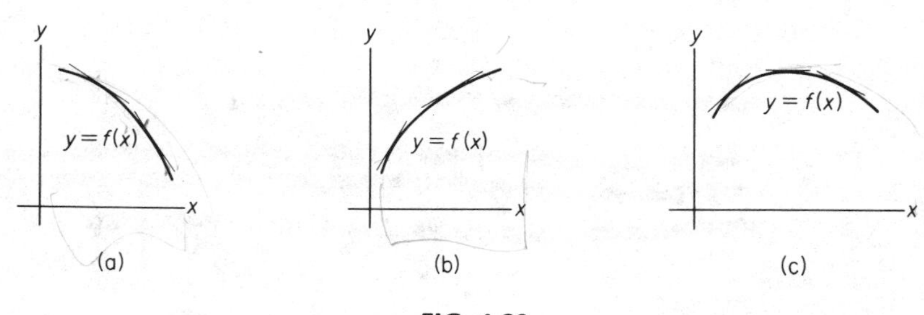

FIG. 4-30

and the curves are bending downward. Thus f' is a decreasing function here, and we say f is *concave down*.

DEFINITION. *A function f is said to be* ***concave up*** *[**concave down**] on an interval I if f' is an increasing [decreasing] function on I.*

Remember: If f is concave up on an interval I, then its graph is bending upward there. If f is concave down, then its graph is bending downward.

PITFALL. *Concavity relates to whether f', not f, is increasing or decreasing. Thus in Fig. 4-29(b), note that f is concave up and decreasing, but in Fig. 4-30(a) note that f is concave down and decreasing.*

Since f' is increasing when its derivative $D_x[f'(x)] = f''(x)$ is positive, and f' is decreasing when $f''(x)$ is negative, we can state the following rule:

RULE 4. *A function f is concave up on an interval I if $f''(x) > 0$ on I. It is concave down on I if $f''(x) < 0$ on I.*

A function f is also said to be concave up at a *point* x_0 if there exists an interval around x_0 on which f is concave up. In fact, for the functions that we shall consider, if $f''(x_0) > 0$, then f is concave up at x_0.† Similarly, f is concave down at x_0 if $f''(x_0) < 0$.

From the previous section we know that if $f'(x) > 0$ and $f''(x) > 0$ on an interval, then f is increasing at an increasing rate there. Geometrically this means that the graph of f is rising from left to right and bending upward (concave up), such as in Fig. 4-29(a). In Fig. 4-29(b), f is decreasing at an increasing rate. What can you say about the functions in Figures 4-30(a) and (b)?

EXAMPLE 1

a. *Test $y = f(x) = (x - 1)^3 + 1$ for concavity.*

We must find y''. Since $y' = 3(x - 1)^2$, then $y'' = 6(x - 1)$. Thus f is concave up when $6(x - 1) > 0$; that is, when $x > 1$. Also, f is concave down when $6(x - 1) < 0$; that is, when $x < 1$. See Fig. 4-31.

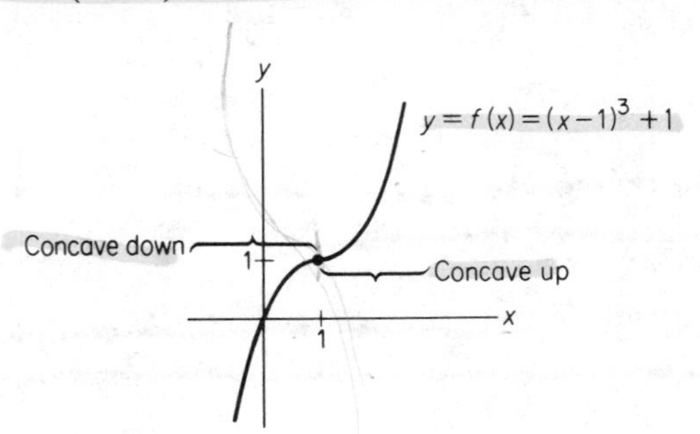

FIG. 4-31

†This is guaranteed for functions f such that f'' is continuous.

b. *Test $y = x^2$ for concavity.*

Since $y' = 2x$, then $y'' = 2 > 0$. Thus the graph of $y = x^2$ must always be concave up, as in Fig. 4-28(a). The graph of $y = x^2$ cannot appear as in Fig. 4-28(b), for in that situation there are intervals on which the curve is concave down.

A point on a graph, such as (1, 1) in Fig. 4-31, where concavity changes from downward to upward, or vice versa, is called a **point of inflection**. For this to occur, the sign of $f''(x)$ must go from (−) to (+) or from (+) to (−). *Possible* inflection points are points where $f''(x) = 0$ or is not defined. In fact, such points determine intervals that should be examined when testing for concavity. We use the same method as that for determining when a function is increasing or decreasing.

EXAMPLE 2

Test $y = 6x^4 - 8x^3 + 1$ for concavity and points of inflection.

$$y' = 24x^3 - 24x^2.$$
$$y'' = 72x^2 - 48x = 24x(3x - 2) = 72x\left(x - \tfrac{2}{3}\right).$$

Setting $y'' = 0$ gives $x = 0, \frac{2}{3}$ as *possible* points of inflection. There are three intervals to consider (Fig. 4-32):

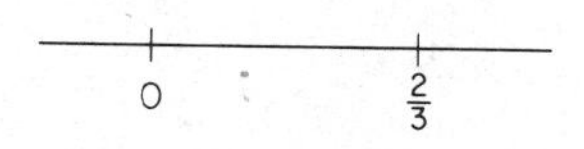

FIG. 4-32

if $x < 0$, then $y'' = 72(-)(-) = (+)$ and the curve is concave up;

if $0 < x < \frac{2}{3}$, then $y'' = 72(+)(-) = (-)$ and the curve is concave down;

if $x > \frac{2}{3}$, then $y'' = 72(+)(+) = (+)$ and the curve is concave up (see Fig. 4-33).

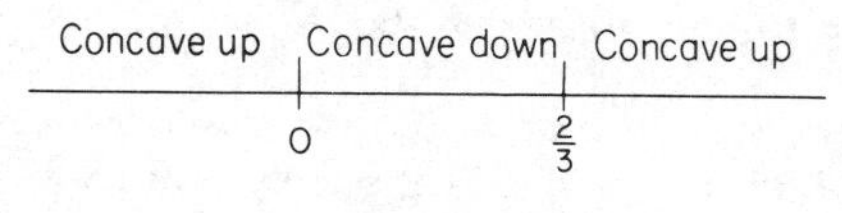

FIG. 4-33

Since concavity changes when $x = 0$ and $x = \frac{2}{3}$, inflection points occur for these values of x. See Fig. 4-34. Summarizing, the curve is concave up on

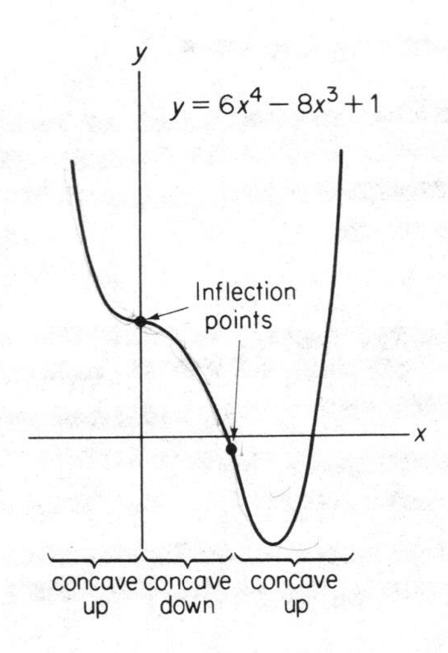

FIG. 4-34

$(-\infty, 0)$ and $(\frac{2}{3}, \infty)$ and is concave down on $(0, \frac{2}{3})$. Inflection points occur when $x = 0$ or $x = \frac{2}{3}$.

PITFALL. *If $f''(x_0) = 0$, this does not prove that the graph of f has an inflection point when $x = x_0$. For if $f(x) = x^4$, then $f''(x) = 12x^2$ and $f''(0) = 0$. But $x < 0$ implies $f''(x) > 0$, and $x > 0$ implies $f''(x) > 0$. Thus concavity does not change and there are no inflection points. See Fig.* 4-35.

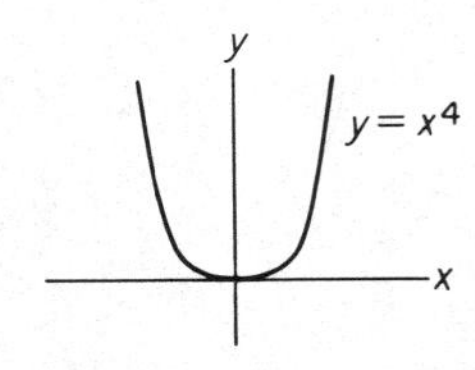

FIG. 4-35

EXAMPLE 3

Sketch the graph of $y = 2x^3 - 9x^2 + 12x$.

Intercepts. When $x = 0$, then $y = 0$. Setting $y = 0$ gives $0 = x(2x^2 - 9x + 12)$, which has $x = 0$ as the only real root. Thus the only intercept is $(0, 0)$.

Symmetry. None.

Asymptotes. None.

Letting $y = f(x)$, we have

$$f'(x) = 6x^2 - 18x + 12 = 6(x^2 - 3x + 2) = 6(x-1)(x-2).$$
$$f''(x) = 12x - 18 = 12\left(x - \tfrac{3}{2}\right).$$

Maxima and Minima. From $f'(x)$ the critical values are $x = 1, 2$. See Fig. 4-36.

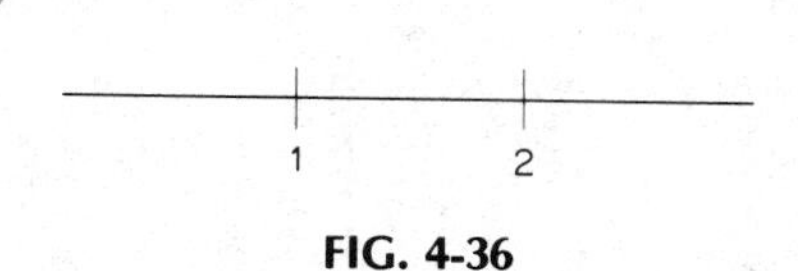

FIG. 4-36

If $x < 1$, then $f'(x) = 6(-)(-) = (+)$ and f is increasing;
if $1 < x < 2$, then $f'(x) = 6(+)(-) = (-)$ and f is decreasing;
if $x > 2$, then $f'(x) = 6(+)(+) = (+)$ and f is increasing (see Fig. 4-37).

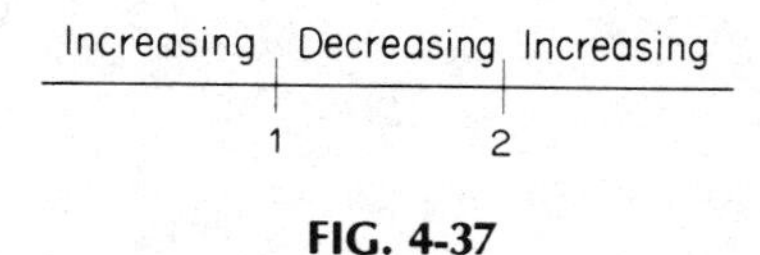

FIG. 4-37

There is a relative maximum when $x = 1$ and a relative minimum when $x = 2$.

Concavity. Setting $f''(x) = 0$ gives a possible inflection point at $x = \frac{3}{2}$. When $x < \frac{3}{2}$, then $f''(x) < 0$ and f is concave down. When $x > \frac{3}{2}$, then $f''(x) > 0$ and f is concave up. See Fig. 4-38.

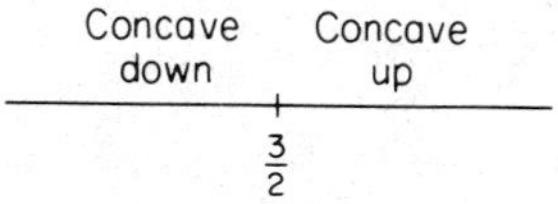

Concave down | Concave up

$\frac{3}{2}$

FIG. 4-38

Since concavity changes, there is a point of inflection when $x = \frac{3}{2}$.

Discussion. We now find the coordinates of the important points on the graph (and any other points if there is doubt as to the behavior of the

x	0	1	$\frac{3}{2}$	2
y	0	5	$\frac{9}{2}$	4

curve). As x increases, the function is first concave down and increases to a relative maximum at $(1, 5)$; it then decreases to $(\frac{3}{2}, \frac{9}{2})$; it then becomes concave up but continues to decrease until it reaches a relative minimum at $(2, 4)$; thereafter it increases and is still concave up. See Fig. 4-39.

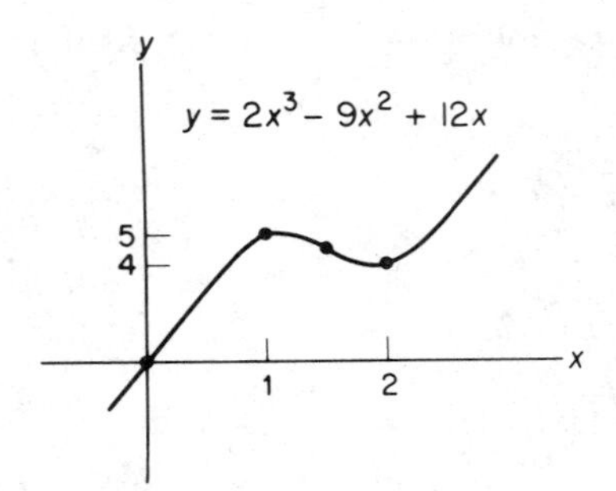

FIG. 4-39

EXAMPLE 4

Sketch the graph of $y = \dfrac{4x}{x^2 + 1}$.

Intercepts. When $x = 0$, then $y = 0$; when $y = 0$, then $x = 0$. Thus, $(0, 0)$ is the only intercept.

Symmetry. There is symmetry only with respect to the origin: replacing x by $-x$ and y by $-y$ gives

$$-y = \frac{4(-x)}{(-x)^2 + 1},$$

which is equivalent to

$$y = \frac{4x}{x^2 + 1}.$$

Asymptotes. Testing for horizontal asymptotes, we have

$$\lim_{x \to \infty} \frac{4x}{x^2 + 1} = \lim_{x \to \infty} \frac{\frac{4}{x}}{1 + \frac{1}{x^2}} = \frac{0}{1} = 0,$$

and

$$\lim_{x \to -\infty} \frac{4x}{x^2 + 1} = 0.$$

The x-axis $(y = 0)$ is a horizontal asymptote. There are no vertical asymptotes because the given function is never discontinuous.

Letting $y = f(x)$, we have

$$f'(x) = \frac{(x^2+1)(4) - 4x(2x)}{(x^2+1)^2} = \frac{4-4x^2}{(x^2+1)^2} = \frac{4(1+x)(1-x)}{(x^2+1)^2}.$$

$$f''(x) = \frac{(x^2+1)^2(-8x) - (4-4x^2)(2)(x^2+1)(2x)}{(x^2+1)^4}$$

$$= \frac{8x(x^2+1)(x^2-3)}{(x^2+1)^4} = \frac{8x(x+\sqrt{3})(x-\sqrt{3})}{(x^2+1)^3}.$$

Maxima and Minima. From $f'(x)$, the critical values are $x = \pm 1$.

If $x < -1$, then $f'(x) = \dfrac{4(-)(+)}{(+)} = (-)$ and f is decreasing;

if $-1 < x < 1$, then $f'(x) = \dfrac{4(+)(+)}{(+)} = (+)$ and f is increasing;

if $x > 1$, then $f'(x) = \dfrac{4(+)(-)}{(+)} = (-)$ and f is decreasing (see Fig. 4-40).

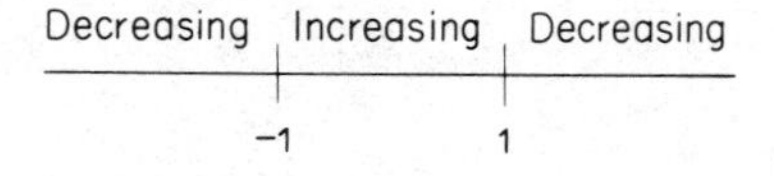

FIG. 4-40

There is a relative minimum when $x = -1$ and a relative maximum when $x = 1$.

Concavity. Setting $f''(x) = 0$, we see that the possible points of inflection are when $x = \pm\sqrt{3}, 0$.

If $x < -\sqrt{3}$, then $f''(x) = \dfrac{8(-)(-)(-)}{(+)} = (-)$ and f is concave down;

if $-\sqrt{3} < x < 0$, then $f''(x) = \dfrac{8(-)(+)(-)}{(+)} = (+)$ and f is concave up;

if $0 < x < \sqrt{3}$, then $f''(x) = \dfrac{8(+)(+)(-)}{(+)} = (-)$ and f is concave down;

if $x > \sqrt{3}$, then $f''(x) = \dfrac{8(+)(+)(+)}{(+)} = (+)$ and f is concave up (see Fig. 4-41).

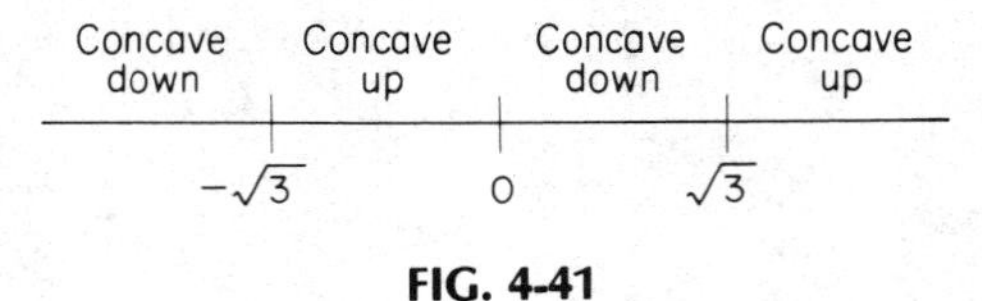

FIG. 4-41

Inflection points occur when $x = 0, \pm\sqrt{3}$.

Discussion. After considering all of the above information, the graph of $y = 4x/(x^2 + 1)$ is given in Fig. 4-42 together with a table of coordinates of important points.

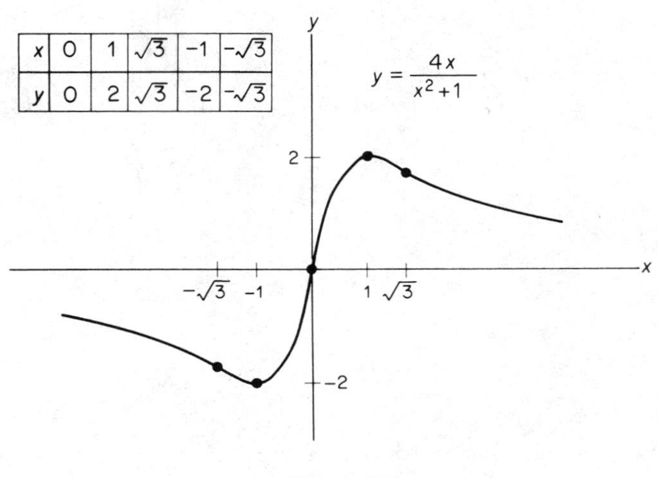

x	0	1	$\sqrt{3}$	−1	$-\sqrt{3}$
y	0	2	$\sqrt{3}$	−2	$-\sqrt{3}$

FIG. 4-42

EXAMPLE 5

Sketch the graph of $y = \dfrac{1}{4 - x^2}$.

Intercepts. $(0, \frac{1}{4})$.

Symmetry. Symmetric to the y-axis.

Asymptotes. As $x \to \infty$, then $y \to 0$; as $x \to -\infty$, then $y \to 0$. Thus $y = 0$ (the x-axis) is a horizontal asymptote. Since $1/(4 - x^2)$ blows up near $x = \pm 2$, the vertical asymptotes are the lines $x = 2$ and $x = -2$.

Maxima and Minima. Since $y = (4 - x^2)^{-1}$,

$$y' = -1(4 - x^2)^{-2}(-2x) = \frac{2x}{(4 - x^2)^2}.$$

The critical values are $x = 0, \pm 2$. If $x < -2$, then $y' < 0$; if $-2 < x < 0$, then $y' < 0$; if $0 < x < 2$, then $y' > 0$; if $x > 2$, then $y' > 0$. The function is decreasing on $(-\infty, -2)$ and $(-2, 0)$ and increasing on $(0, 2)$ and $(2, \infty)$. See Fig. 4-43. There is a relative minimum when $x = 0$.

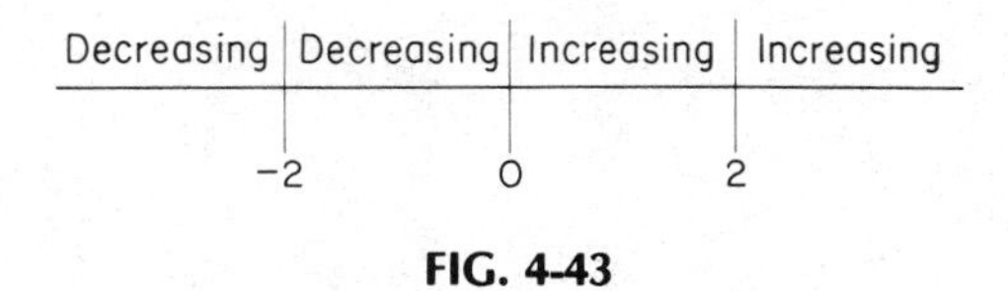

FIG. 4-43

Concavity.

$$y'' = \frac{(4 - x^2)^2(2) - (2x)2(4 - x^2)(-2x)}{(4 - x^2)^4}$$

$$= \frac{2(4 - x^2)(4 + 3x^2)}{(4 - x^2)^4} = \frac{8 + 6x^2}{(4 - x^2)^3}.$$

Setting $y'' = 0$, we get no real roots. However, y'' is undefined when $x = \pm 2$. Thus concavity may change around these values. If $x < -2$, then $y'' < 0$; if $-2 < x < 2$, then $y'' > 0$; if $x > 2$, then $y'' < 0$. The graph is concave up on $(-2, 2)$ and concave down on $(-\infty, -2)$ and $(2, \infty)$. See Fig. 4-44. Although concavity changes around $x = \pm 2$, these values of x do not give points of inflection since y *itself* is not defined at $x = \pm 2$.

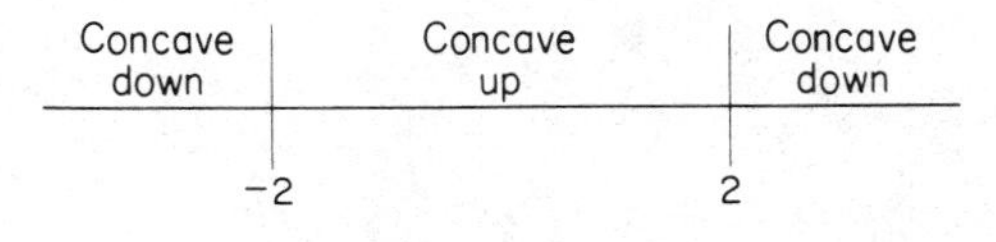

FIG. 4-44

Discussion. Plotting the points in the table in Fig. 4-45, some arbitrarily chosen, and using the above information, we get the indicated graph. Due to symmetry our table has only $x > 0$.

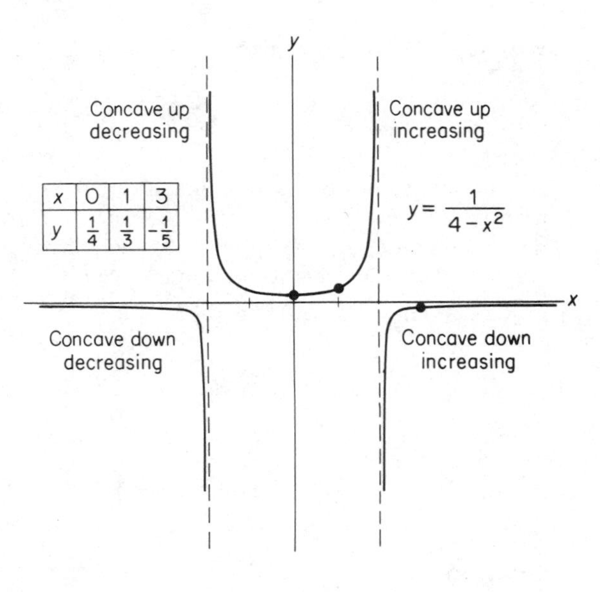

FIG. 4-45

Exercise 4-4

In Problems **1–10**, *determine concavity and where points of inflection occur. Do not sketch the graphs.*

1. $y = -2x^2 + 4x$.

2. $y = 3x^2 - 6x + 5$.

3. $y = 4x^3 + 12x^2 - 12x$.

4. $y = x^3 - 6x^2 + 9x + 1$.

5. $y = x^4 - 6x^2 + 5x - 6$.

6. $y = -\dfrac{x^4}{4} + \dfrac{9x^2}{2} + 2x$.

7. $y = \dfrac{x + 1}{x - 1}$.

8. $y = x + \dfrac{1}{x}$.

9. $y = \dfrac{x^2}{x^2 + 1}$.

10. $y = \dfrac{x^2}{x + 3}$.

In Problems **11–40** *sketch each curve. Determine: intervals on which the function is increasing, decreasing, concave up, concave down; relative maxima and minima; inflection points; symmetry; horizontal and vertical asymptotes; those intercepts which can be obtained conveniently.*

11. $y = x^2 + 4x + 3$.

12. $y = x^2 + 2$.

13. $y = 4x - x^2$.

14. $y = x - x^2 + 2$.

15. $y = 2x^2 - 5x - 12.$

16. $y = x^2 - 6x - 7.$

17. $y = x^3 - 9x^2 + 24x - 19.$

18. $y = 3x - x^3.$

19. $y = \frac{x^3}{3} - 3x.$

20. $y = x^3 - 6x^2 + 9x.$

21. $y = x^3 - 3x^2 + 3x - 3.$

22. $y = 2x^3 - 9x^2 + 12x.$

23. $y = 4x^2 - x^4.$

24. $y = -\frac{x^3}{3} - 2x^2 + 5x - 2.$

25. $y = 4x^3 - 3x^4.$

26. $y = x^4 - 2x^2.$

27. $y = -2 + 12x - x^3.$

28. $y = (3 + 2x)^3.$

29. $y = x^3 - 6x^2 + 12x - 6.$

30. $y = 3x^5 - 5x^3.$

31. $y = 5x - x^5.$

32. $y = \frac{x^5}{100} - \frac{x^4}{20}.$

33. $y = 3x^4 - 4x^3 + 1.$

34. $y = x(1 - x)^3.$

35. $y = \frac{3}{x}.$

36. $y = \frac{1}{x - 1}.$

37. $y = \frac{x}{x + 1}.$

38. $y = \frac{10}{\sqrt{x}}.$

39. $y = x^2 + \frac{1}{x^2}.$

40. $y = \frac{x^2}{1 - x}.$

41. Show that the graph of the demand equation $p = 100/(q + 2)$ for a product is decreasing and concave up for $q > 0$.

42. If the cost function for a manufacturer's product is $c = 3q^2 + 5q + 6$, show that the graph of the average cost function $\bar{c}$ is always concave up for $q > 0$.

43. When researchers count the number of species of plants on a plot, this number may depend on the size of the plot. For example, in Fig. 4-46 we see that on 1-square-meter plots there are three species (A, B, and C on the

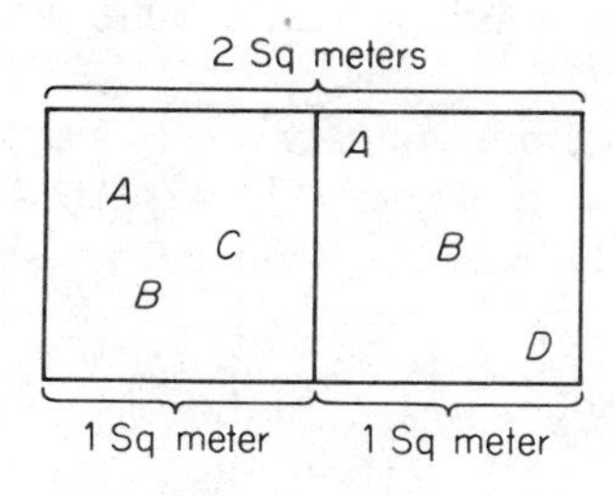

FIG. 4-46

left plot; A, B, and D on the right plot), and on a 2-square-meter plot there are four species (A, B, C, and D).

In a study (adapted from [39]) of rooted plants in a certain geographic region, it was determined that the average number of species, S, occurring on plots of size A (in square meters) is given by

$$S = f(A) = 12\sqrt[4]{A}\,, \qquad 0 \leqslant A \leqslant 625.$$

Sketch the graph of f. (*Note*: Your graph should be rising and concave down. Thus, the number of species is increasing with respect to area, but at a decreasing rate.)

44. In a model of the effect of contraception on birth rate [28], the equation

$$R = f(x) = \frac{x}{4.4 - 3.4x}, \qquad 0 \leqslant x \leqslant 1$$

gives the proportional reduction R in the birth rate as a function of the efficiency x of a contraception method. An efficiency of .2 (or 20 percent) means that the probability of becoming pregnant is 80 percent of the probability of becoming pregnant without the contraceptive. Find the reduction (as a percentage) when efficiency is (a) 0, (b) .5, and (c) 1. Find dR/dx and d^2R/dx^2 and sketch the graph of the equation.

45. If you were to recite members of a category, such as four-legged animals, the words that you utter would probably occur in "chunks" with distinct pauses between such chunks. For example, you might say the following for the category of four-legged animals:

dog, cat, mouse, rat,
(pause)

horse, donkey, mule,
(pause)

cow, pig, goat, lamb
etc.

The pauses may occur because one may have to mentally search for subcategories (animals around the house, beasts of burden, farm animals, etc.)

The elapsed time between onsets of successive words is called *interresponse time*. A function [15] has been used to analyze the length of time for pauses and chunk size (number of words in a chunk). This function f is such that

$$f(t) = \begin{cases} \text{the average number of words} \\ \text{that occur in succession with} \\ \text{interresponse times less than } t. \end{cases}$$

The graph of f has a shape similar to that in Fig. 4-47 and is best fit by a

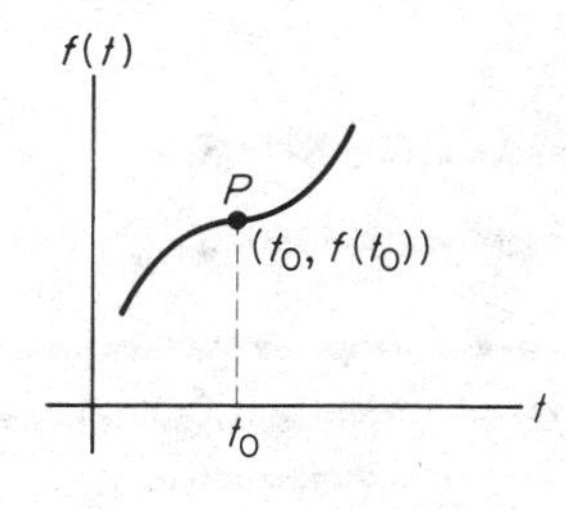

FIG. 4-47

third-degree polynomial:

$$f(t) = At^3 + Bt^2 + Ct + D.$$

The point P has special meaning. It is such that the value t_0 separates interresponse times *within* chunks from those *between* chunks. Mathematically, P is a critical point that is also a point of inflection. Assume these two conditions and show that (a) $t_0 = -B/(3A)$ and (b) $B^2 = 3AC$.

4-5 THE SECOND-DERIVATIVE TEST

The second derivative may be used to test certain critical values for relative extrema. Observe in Fig. 4-48 that when $x = x_0$ there is a horizontal tangent; that is, $f'(x_0) = 0$. This suggests a relative maximum

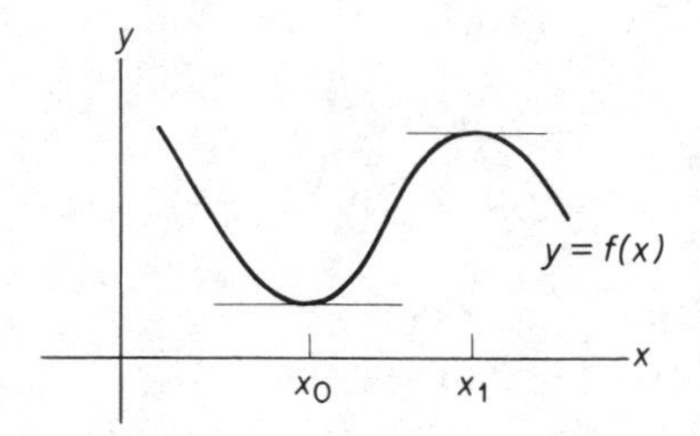

FIG. 4-48

or minimum. However, we see also that the curve is bending upward there (that is, $f''(x_0) > 0$). This leads us to conclude that there is a relative minimum at x_0. On the other hand, $f'(x_1) = 0$ but the curve is bending downward at x_1 (that is, $f''(x_1) < 0$). From this we conclude that a relative maximum exists there. This technique of examining the second derivative at points where $f'(x) = 0$ is called the *second-derivative test* for relative maxima and minima.

SECOND-DERIVATIVE TEST.

Suppose $f'(x_0) = 0$.

If $f''(x_0) < 0$, then f has a relative maximum at x_0;
if $f''(x_0) > 0$, then f has a relative minimum at x_0;
if $f''(x_0) = 0$, the test fails (use the first-derivative test).

PITFALL. *In the case where $f'(x_0) = f''(x_0) = 0$, some students conclude that there is no relative maximum or minimum at x_0. This conclusion may be* ***false****, as Example 1(c) will show.*

EXAMPLE 1

Use the second-derivative test to examine the following for relative maxima and minima.

a. $y = 18x - \frac{2}{3}x^3$.

$$y' = 18 - 2x^2 = 2(9 - x^2) = 2(3 + x)(3 - x).$$
$$y'' = -4x.$$

Solving $y' = 0$ gives the critical values $x = \pm 3$. If $x = 3$, then we have $y'' = -4(3) = -12 < 0$ and so there is a relative maximum when $x = 3$. If $x = -3$, then $y'' > 0$ and so there is a relative minimum when $x = -3$. See Fig. 4-16.

b. $y = 6x^4 - 8x^3 + 1$.

$$y' = 24x^3 - 24x^2 = 24x^2(x - 1).$$
$$y'' = 72x^2 - 48x.$$

Solving $y' = 0$ gives the critical values $x = 0, 1$. Since $y'' > 0$ if $x = 1$, there is a relative minimum when $x = 1$. For $x = 0$, then $y'' = 0$ and the second-derivative test fails. We now turn to the first-derivative test. If $x < 0$, then $y' < 0$; if $0 < x < 1$, then $y' < 0$. Thus no relative maximum or minimum exists when $x = 0$. See Fig. 4-34.

c. $y = x^4$.

$$y' = 4x^3.$$
$$y'' = 12x^2.$$

Solving $y' = 0$ gives the critical value $x = 0$. But if $x = 0$, then $y'' = 0$ and the second-derivative test fails. Since $y' < 0$ for $x < 0$, and $y' > 0$ for $x > 0$, by the first-derivative test there is a relative minimum when $x = 0$. See Fig. 4-35.

If a continuous function has *exactly one* relative extremum on an interval, it can be shown that the relative extremum must also be an *absolute* extremum on the interval. To illustrate, in Example 1(c), $y = x^4$ has a relative minimum when $x = 0$ and there are no other relative extrema. Since $y = x^4$ is continuous, this relative minimum is also an absolute minimum for the function.

Exercise 4-5

In Problems **1–10**, *test for relative maxima and minima by using the second-derivative test. In Problems* **1–4**, *state whether the relative extrema are also absolute extrema.*

1. $y = x^2 - 5x + 6.$

2. $y = -2x^2 + 6x + 12.$

3. $y = -4x^2 + 2x - 8.$

4. $y = 3x^2 - 5x + 6.$

5. $y = x^3 - 27x + 1.$

6. $y = x^3 - 12x + 1.$

7. $y = -x^3 + 3x^2 + 1.$

8. $y = x^4 - 2x^2 + 4.$

9. $y = 2x^4 + 2.$

10. $y = -x^7.$

4-6 APPLIED MAXIMA AND MINIMA

By using techniques of the previous sections, we can examine situations which require determining the value of a variable that will maximize or minimize a function. For example, we might want to maximize profit or minimize cost. The crucial part is setting up the function to be investigated. Then we find its derivative and test the resulting critical values. For this the first-derivative test or second-derivative test may be used, although it is often obvious from the nature of the problem whether or not a critical value represents an appropriate answer. For the problems that we shall consider, our interest is in *absolute* maxima and minima. In some cases they occur at endpoints.

Read each of the examples carefully so that you may gain insight into setting up the function to be analyzed.

EXAMPLE 1

A manufacturer determines that his total cost function is $c = \frac{q^2}{4} + 3q + 400$, *where q is the number of units produced. At what level of output will average cost per unit be a minimum? What is this minimum?*

The quantity to be minimized is average cost $\bar{c}$. The average cost function is given by

$$\bar{c} = \frac{c}{q} = \frac{\frac{q^2}{4} + 3q + 400}{q} = \frac{q}{4} + 3 + \frac{400}{q}. \quad (1)$$

$$D_q\bar{c} = \frac{1}{4} - \frac{400}{q^2} = \frac{q^2 - 1600}{4q^2}.$$

Setting $D_q\bar{c} = 0$, we get

$$q^2 - 1600 = 0,$$

$$(q - 40)(q + 40) = 0,$$

$$q = 40 \quad [\text{since we assume } q > 0].$$

To determine if this level of output gives a relative minimum, we shall use the second-derivative test.

$$D_q^2\bar{c} = \frac{800}{q^3},$$

which is positive for $q = 40$. Thus $\bar{c}$ has a relative minimum when $q = 40$. We note that $\bar{c}$ is continuous for $q > 0$. Since $q = 40$ is the only relative extrema for $q > 0$, we conclude that this relative minimum is indeed an absolute minimum. Substituting $q = 40$ in (1) gives the minimum average cost $\bar{c} = 23$.

EXAMPLE 2

An enzyme is a protein which acts as a catalyst for increasing the rate of a chemical reaction that occurs in cells. In a certain reaction, an enzyme is converted to another enzyme called the product. The product acts as a catalyst for its own formation. The rate R at which the product is formed (with respect to time) is given by

$$R = kp(I - p),$$

where I is the total initial amount of both enzymes, p is the amount of the product enzyme, and k is a positive constant. For what value of p will R be a maximum?

We can write $R = k(pl - p^2)$. Setting $dR/dp = 0$ and solving for p gives

$$\frac{dR}{dp} = k(l - 2p) = 0.$$

$$p = \frac{l}{2}.$$

Now, $d^2R/dp^2 = -2k$. Since $k > 0$, the second derivative is always negative. Hence, $p = l/2$ gives a relative maximum. Moreover, since R is a continuous function of p, we conclude that we indeed have an absolute maximum when $p = l/2$.

Calculus can be applied to inventory decisions as the following example shows.

EXAMPLE 3

A manufacturer annually produces and sells 10,000 units of a product. Sales are uniformly distributed throughout the year. He wishes to determine the number of units to be manufactured in each production run in order to minimize annual set-up costs and carrying costs. The size of such production runs is referred to as the ***economic lot size*** *or* ***economic order quantity.*** *The production cost of each unit is \$20 and carrying costs (insurance, interest, storage, etc.) are estimated to be 10 percent of the value of the average inventory. Set-up costs per production run are \$40. Find the economic lot size.*

Let q be the number of units in a production run. Since sales are distributed at a uniform rate, we shall assume that inventory varies uniformly from q to 0 between production runs. Thus we take the average inventory to be $q/2$ units. The production costs are \$20 per unit, and so the value of the average inventory is $20(q/2)$. Carrying costs are 10 percent of this value:

$$.10(20)\left(\frac{q}{2}\right).$$

The number of production runs per year is $10{,}000/q$. Thus the total set-up costs are

$$40\left(\frac{10{,}000}{q}\right).$$

Hence the total annual carrying costs and set-up costs C are

$$C = .10(20)\left(\frac{q}{2}\right) + 40\left(\frac{10{,}000}{q}\right)$$

$$= q + \frac{400{,}000}{q}.$$

$$\frac{dC}{dq} = 1 - \frac{400{,}000}{q^2} = \frac{q^2 - 400{,}000}{q^2}.$$

Setting $dC/dq = 0$, we get

$$q^2 = 400{,}000.$$

Since $q > 0$, we choose

$$q = \sqrt{400{,}000} = 200\sqrt{10} \approx 632.5.$$

The symbol "$\approx$" means "approximately equals." To determine if this value of q gives minimum cost, we shall examine the first derivative. If $0 < q < \sqrt{400{,}000}$, then $dC/dq < 0$. If $q > \sqrt{400{,}000}$, then $dC/dq > 0$. Since C always decreases to the left of $q = \sqrt{400{,}000}$ and always increases to the right of $q = \sqrt{400{,}000}$, we conclude that there is an *absolute* minimum at $q = 632.5$. The number of production runs is $10{,}000/632.5 \approx 15.8$. For practical purposes, there would be 16 lots, each having an economic lot size of 625 units.

EXAMPLE 4

The Vista TV Cable Co. currently has 2000 subscribers who are paying a monthly rate of \$5. A survey reveals that there will be 50 more subscribers for each \$.10 decrease in the rate. At what rate will maximum revenue be obtained and how many subscribers will there be at this rate?

Let x be the rate. Then the total decrease in the rate is $5 - x$ and the number of \$.10 decreases is $\frac{5 - x}{.10}$. For *each* of these decreases there will be 50 more subscribers. Thus the total of *new* subscribers is $50\left(\frac{5 - x}{.10}\right)$, and the total of all subscribers is

$$2000 + 50\left(\frac{5 - x}{.10}\right). \qquad (2)$$

The revenue r is given by $r =$ (rate) (number of subscribers):

$$r = x\left[2000 + 50\left(\frac{5 - x}{.10}\right)\right]$$

$$= 4500x - 500x^2.$$

Setting $r' = 0$, we have

$$r' = 4500 - 1000x = 0.$$

$$x = 4.50.$$

Since $r'' = -1000 < 0$, we have a relative maximum when $x = 4.50$. Since r is a continuous function of x, we conclude that when $x = 4.50$ there is an absolute maximum. Substituting $x = 4.50$ in (2) gives 2250 subscribers.

EXAMPLE 5

A sociologist determined that if a particular health-care program for the elderly were initiated, then t years after its start, n thousand elderly people would receive direct benefits, where

$$n = \frac{t^3}{3} - 6t^2 + 32t, \qquad 0 \leqslant t \leqslant 12.$$

For what value of t does the maximum number receive benefits?

Setting $dn/dt = 0$, we have

$$\frac{dn}{dt} = t^2 - 12t + 32 = 0.$$
$$(t-4)(t-8) = 0.$$
$$t = 4 \quad \text{or} \quad t = 8.$$

Now, $d^2n/dt^2 = 2t - 12$, which is negative for $t = 4$ and positive for $t = 8$. Thus there is a relative maximum when $t = 4$. This gives $n = 53\frac{1}{3}$. To determine whether this is an absolute maximum, we must find n at the endpoints of the domain. If $t = 0$, then $n = 0$. If $t = 12$, then $n = 96$. Thus an absolute maximum occurs when $t = 12$. A graph of the function is given in Fig. 4-49.

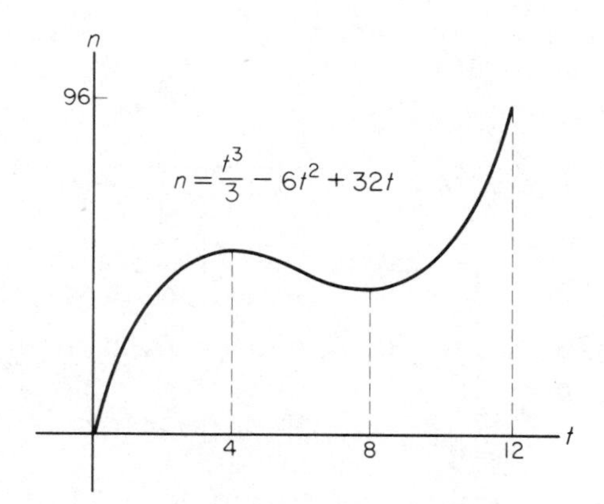

FIG. 4-49

Suppose $p = f(q)$ is the demand function for a firm where p is price per unit and q is the number of units produced and sold. Then the total revenue $r = qp = qf(q)$ is a function of q. Let the total cost c of producing q units be given by the cost function $c = g(q)$. Thus, the total profit P, which is total revenue − total cost, is also a function of q:

$$P = r - c = qf(q) - g(q).$$

Let us consider the most profitable output for the firm. Ignoring special cases, profit is maximized when $dP/dq = 0$ and $d^2P/dq^2 < 0$.

$$\frac{dP}{dq} = \frac{d}{dq}(r - c) = \frac{dr}{dq} - \frac{dc}{dq} = 0.$$

Thus,

$$\frac{dr}{dq} = \frac{dc}{dq}.$$

That is, at the level of maximum profit the slope of the tangent to the total revenue curve must equal the slope of the tangent to the total cost curve (Fig. 4-50). But dr/dq is marginal revenue MR, and dc/dq is

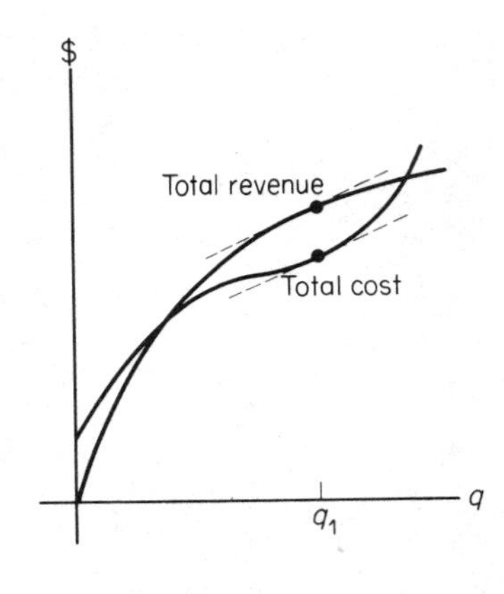

FIG. 4-50

marginal cost MC. Thus, under typical conditions, *to maximize profit it is necessary that*

$$MR = MC.$$

For this to indeed correspond to a maximum, it is necessary that $d^2P/dq^2 < 0$.

$$\frac{d^2P}{dq^2} = \frac{d^2r}{dq^2} - \frac{d^2c}{dq^2} < 0, \quad \text{or} \quad \frac{d^2r}{dq^2} < \frac{d^2c}{dq^2}.$$

That is, when $MR = MC$, in order to insure maximum profit the slope of the marginal revenue curve must be less than the slope of the marginal cost curve.

In the next example we use the word *monopolist*. Under a situation of monopoly, there is only one seller of a product for which there are no similar substitutes, and he, the monopolist, controls the market. By considering the demand equation for his product, he may set the price (or volume of output) so that maximum profit will be obtained.

EXAMPLE 6

Suppose that the demand equation for a monopolist's product is $p = 400 - 2q$ and the average cost function is $\bar{c} = .2q + 4 + (400/q)$, where q is number of units, and both p and $\bar{c}$ are expressed in dollars.

a. *Determine the level of output at which profit is maximized.*

b. *Determine the price at which maximum profit occurs.*

c. *Determine the maximum profit.*

d. *If, as a regulatory device, the government imposes a tax of \$22 per unit on the monopolist, what is the new price for profit maximization?*

Since revenue $r = pq = 400q - 2q^2$ and total cost $c = q\bar{c} = .2q^2 + 4q + 400$, profit P is

$$P = r - c = 400q - 2q^2 - (.2q^2 + 4q + 400).$$

$$P = 396q - 2.2q^2 - 400. \tag{3}$$

a. Setting $dP/dq = 0$, we have

$$\frac{dP}{dq} = 396 - 4.4q = 0.$$

$$q = 90.$$

Since $d^2P/dq^2 = -4.4 < 0$, we conclude that $q = 90$ gives a maximum.

b. From the demand equation, $p = 400 - 2(90) = 220$.

c. Substituting $q = 90$ in (3) gives $P = 17{,}420$.

d. The tax of \$22 per unit means that for q units the total cost increases by $22q$. The new cost function is $c_1 = .2q^2 + 4q + 400 + 22q$, and the profit P_1 is given by

$$P_1 = 400q - 2q^2 - (.2q^2 + 4q + 400 + 22q).$$

$$P_1 = 374q - 2.2q^2 - 400.$$

Setting $dP_1/dq = 0$ gives

$$\frac{dP_1}{dq} = 374 - 4.4q = 0.$$

$$q = 85.$$

Thus to maximize profit, the monopolist restricts output to 85 units at a higher price of $p_1 = 400 - 2(85) = 230$. Since this is only \$10 more than before, only part of the tax has been shifted to the consumer, and the monopolist must bear the cost of the balance. His new profit is \$15,495, which is lower than the former profit.

EXAMPLE 7

For insurance purposes a manufacturer plans to fence in a 10,800 sq ft rectangular storage area adjacent to a building by using the building as one side of the enclosed area (see Fig. 4-51). The fencing parallel to the building

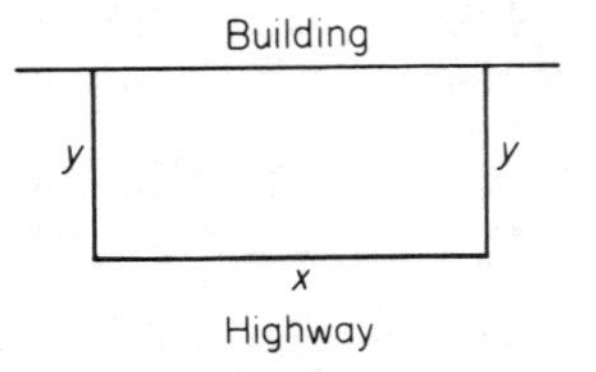

FIG. 4-51

faces a highway and will cost \$3 per ft installed, while the fencing for the other two sides costs \$2 per ft installed. Find the amount of each type of fence so that the total cost of the fence will be a minimum. What is the minimum cost?

In Fig. 4-51 we have labeled the length of the side parallel to the building as x and the lengths of the other two sides as y, where x and y are in feet. The cost (in dollars) of the fencing along the highway is $3x$, and along each of the other sides it is $2y$. Thus the total cost C of the fencing is

$$C = 3x + 2y + 2y = 3x + 4y.$$

We wish to minimize C. In order to differentiate, we first express C in terms of one variable only. To do this we find a relationship between x and y. Since the storage area xy must be 10,800,

$$xy = 10{,}800,$$

$$\text{or} \qquad y = \frac{10{,}800}{x}.$$

By substitution we have

$$C = 3x + 4\left(\frac{10{,}800}{x}\right) = 3x + \frac{43{,}200}{x}.$$

To minimize C we set $dC/dx = 0$ and solve for x:

$$\frac{dC}{dx} = 3 - \frac{43{,}200}{x^2} = 0,$$

from which

$$x^2 = \frac{43{,}200}{3} = 14{,}400.$$

$$x = 120 \qquad (\text{since } x > 0).$$

Now, $d^2C/dx^2 = 86{,}400/x^3 > 0$ for $x = 120$, and we conclude that $x = 120$ indeed gives the minimum value of C. When $x = 120$, then $y = 10{,}800/120 = 90$. Thus 120 ft of the \$3 fencing and 180 ft of the \$2 fencing are needed. This gives a cost of \$720.

Exercise 4-6

In each of the following, p is price per unit and q is output per unit of time. Fixed costs refer to costs that remain constant at all levels of production in a given time period. (An example is rent.)

1. A manufacturer finds that the total cost c of producing his product is given by the cost function $c = .05q^2 + 5q + 500$. At what level of output will average cost per unit be at a minimum?

2. For a monopolist's product, the revenue function is given by $r = 240q + 57q^2 - q^3$. Determine the output for maximum revenue.

3. The demand equation for a monopolist's product is $p = -5q + 30$. At what price will revenue be maximized?

4. In his model for storage and shipping costs of materials for a manufacturing process, Lancaster [27] derives the following cost function:

$$C(k) = 100\left[100 + 9k + \frac{144}{k}\right], \qquad 1 \leqslant k \leqslant 100,$$

where $C(k)$ is the total cost (in dollars) of storage and transportation for 100 days of operation if a load of k tons of material is moved every k days. (a) Find $C(1)$. (b) For what value of k does $C(k)$ have a minimum? (c) What is the minimum value?

5. A monopolist estimates the demand equation for his product to be $p = 72 - .04q$. If his cost function is $c = 500 + 30q$, at what level of output will his profit be maximized? At what price does this occur and what is the profit?

6. For a monopolist, the demand function is $p = 50/\sqrt{q}$ and his average cost function is $\bar{c} = .50 + (1000/q)$. Find the profit-maximizing price and output. At this level, show that marginal revenue is equal to marginal cost.

7. For XYZ Manufacturing Co., total fixed costs are \$1200, material and labor costs combined are \$2 per unit, and the demand equation is $p = 100/\sqrt{q}$. What level of output will maximize profit? Show that this occurs when marginal revenue is equal to marginal cost. What is the price at profit maximization?

8. The severity R of the reaction of the human body to an initial dose D of a drug is given in [52] by

$$R = f(D) = D^2\left(\frac{C}{2} - \frac{D}{3}\right),$$

where the constant C denotes the maximum amount of the drug that may be given. Show that R has a maximum *rate of change* when $D = C/2$.

9. In a laboratory an experimental antibacterial agent is applied to a population of 100 bacteria. Data indicate that the number N of bacteria, t hours after the agent is introduced, is given by

$$N = \frac{14{,}400 + 120t + 100t^2}{144 + t^2}.$$

For what value of t does the maximum number of bacteria in the population occur? What is this maximum number?

10. The cost per hour C (in dollars) of operating an automobile is given by

$$C = .12s - .0012s^2 + .08, \qquad 0 \leqslant s \leqslant 60,$$

where s is the speed in miles per hour. At what speed is the cost per hour a minimum?

11. A group of biologists studied the nutritional effects on rats that were fed a diet containing 10 percent protein (adapted from [4]). The protein consisted of yeast and cottonseed flour. By varying the percent p of yeast in the protein mix, the group found that the (average) weight gain (in grams) of a rat over a period of time was

$$f(p) = 160 - p - \frac{900}{p + 10}, \qquad 0 \leqslant p \leqslant 100.$$

Find (a) the maximum weight gain and (b) the minimum weight gain.

12. A real estate firm owns 70 garden-type apartments. At \$125 per month

each apartment can be rented. However, for each \$5 per month increase, there will be two vacancies with no possibility of filling them. What rent per apartment will maximize monthly revenue?

13. A TV cable company has 1000 subscribers who are paying \$5 per month. It can get 100 more subscribers for each \$.10 decrease in the monthly fee. What rate will yield maximum revenue, and what will this revenue be?

14. A manufacturer finds that for the first 500 units of his product that are produced and sold, the profit is \$50 per unit. The profit on each of the units beyond 500 is decreased by \$.10 times the number of additional units produced. For example, the total profit when 502 units are produced and sold is 500(50) + 2(49.80). What level of output will maximize his profit?

15. Find two numbers whose sum is 40 and whose product is a maximum.

16. Find two nonnegative numbers whose sum is 20 such that the product of twice one number and the square of the other number will be a maximum.

17. A company has set aside \$3000 to fence in a rectangular portion of land adjacent to a stream by using the stream for one side of the enclosed area. The cost of the fencing parallel to the stream is \$5 per foot installed, and the fencing for the remaining two sides is \$3 per foot installed. Find the dimensions of the maximum enclosed area.

18. The owner of the Laurel Nursery Garden Center wants to fence in 1000 square feet of land in a rectangular plot to be used for different types of shrubs. The plot is to be divided into four equal plots with three fences parallel to the same pair of sides as shown in Fig. 4-52. What is the least number of feet of fence needed?

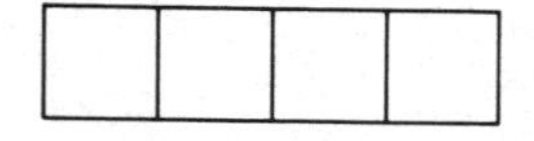

FIG. 4-52

19. A container manufacturer is designing a rectangular box, open at the top and with a square base, which is to have a volume of 32 cubic feet. If the box is to require the least amount of material, what must be the dimensions of the box? See Fig. 4-53.

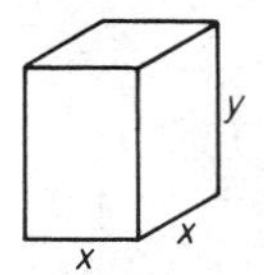

FIG. 4-53

20. An open top box with a square base is to be constructed from 192 square feet of material. What should be the dimensions of the box if the volume is to be a maximum? What is the maximum volume? See Fig. 4-53.

21. A rectangular cardboard poster is to have 150 square inches for printed matter. It is to have a 3-inch margin at the top and bottom and a 2-inch margin on each side. Find the dimensions of the poster so that the amount of cardboard used is minimized. See Fig. 4-54.

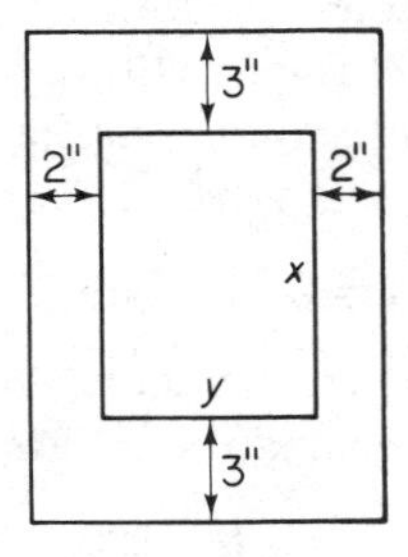

FIG. 4-54

22. An open box is to be made by cutting equal squares from each corner of a 12-inch square piece of cardboard and then folding up the sides. Find the length of the side of the square that must be cut out if the volume of the box is to be maximized. What is the maximum volume? See Fig. 4-55.

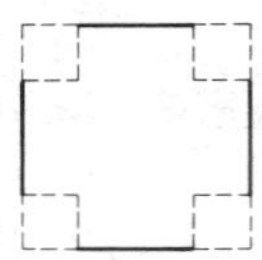

FIG. 4-55

23. A cylindrical can, open at the top, is to have a fixed volume of K. Show that if the least amount of material is to be used, then both the radius and the height are equal to $\sqrt[3]{K/\pi}$. See Fig. 4-56.

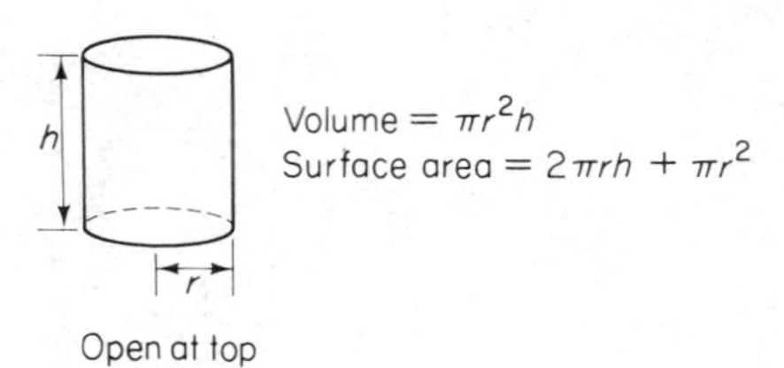

FIG. 4-56

24. A cylindrical can, open at the top, is to be made from a fixed amount of material, K. If the volume is to be a maximum, show that both the radius and the height are equal to $\sqrt{K/(3\pi)}$. See Fig. 4-56.

25. The demand equation for a monopolist is $p = 600 - 2q$ and his total cost function is $c = .2q^2 + 28q + 200$. Find the profit-maximizing output and price, and determine the corresponding profit. If the government were to impose a tax of \$22 per unit on the manufacturer, what would be the new profit-maximizing output and price? What is his profit now?

26. Use the *original* data in Problem 25 and assume that the government imposes a license fee of \$100 on the manufacturer. This is a lump-sum amount without regard to output. Show that marginal revenue and marginal cost do not change and, hence, the profit maximizing price and output remain the same. Show, however, that he will have less profit.

27. A manufacturer has to produce annually 1000 units of a product that is sold at a uniform rate during the year. The production cost of each unit is \$10 and carrying costs (insurance, interest, storage, etc.) are estimated to be 12.8 percent of the value of average inventory. Set-up costs per production run are \$40. Find the economic lot size.

28. In a model by Smith [46] for power output P of an animal at a given speed as a function of its movement or *gait* j, the following relation is derived:

$$P(j) = Aj\frac{L^4}{V} + B\frac{V^3L^2}{1+j}.$$

Here A and B are constants, j is a measure of the "jumpiness" of the gait, L is a constant representing linear dimension, and V is a constant forward speed. Assume that P is a minimum when $dP/dj = 0$. Show that when this occurs, then

$$(1+j)^2 = \frac{BV^4}{AL^2}.$$

As a passing comment, Smith indicates, " . . . at top speed, j is zero for an elephant, .3 for a horse, and 1 for a greyhound, approximately."

29. Imperial Educational Services (I.E.S.) is considering offering a workshop in resource allocation to key personnel at Acme Corp. To make the offering economically feasible, I.E.S. feels that at least thirty persons must attend at a cost of \$50 each. Moreover, I.E.S. will agree to reduce the charge for *everybody* by \$1.25 for each person over the thirty who attends. How many people should be in the group for I.E.S. to maximize revenue? Assume that the maximum allowable number in the group is forty.

30. For a manufacturer, the cost of making a part is \$3 per unit for labor and \$1 per unit for materials; overhead is fixed at \$2000 per week. If more than 5000 units are made each week, he must pay for labor \$4.50 per unit for those units in excess of 5000. At what level of production will average cost per unit be at a minimum?

31. The cost of operating a truck on a throughway (excluding the salary of the driver) is $.11 + (s/600)$ dollars per mile, where s is the (steady) speed of the truck in miles per hour. The truck driver's salary is \$6 per hour. At what speed should the truck driver operate the truck to make a 700-mile trip most economical?

32. A company produces daily x tons of chemical $A (x \leqslant 4)$ and y tons of chemical B, where $y = (24 - 6x)/(5 - x)$. The profit on chemical A is \$2000 per ton and on B it is \$1000 per ton. How much of chemical A should be produced per day to maximize profit? Answer the same question if the profit on A is P per ton and that on B is $P/2$ per ton.

33. To erect an office building, fixed costs are \$250,000 and include land, architect's fee, basement, foundation, etc. If x floors are to be constructed, the cost (excluding fixed costs) is $c = (x/2)[100{,}000 + 5000(x - 1)]$. The revenue per month is \$5000 per floor. Find the number of floors that will yield a maximum rate of return on investment (rate of return = total revenue/total cost).

34. In a model of traffic flow on a lane of a freeway, the number N of cars the lane can carry per unit time is given in [43] by

$$N = \frac{-2a}{-2at_r + v - \dfrac{2al}{v}},$$

where a is acceleration of a car when stopping $(a < 0)$, t_r is reaction time to begin braking, v is average speed of the cars, and l is length of a car. Assume that a, t_r, and l are constant. To find at most how many cars a lane can carry, we want to find the speed v that maximizes N. To maximize N it suffices to minimize the denominator $-2at_r + v - (2al/v)$. (a) Find the value of v that minimizes the denominator. (b) Evaluate your answer in (a) when $a = -19.6(\text{ft}/\text{sec}^2)$, $l = 20(\text{ft})$ and $t_r = .5(\text{sec})$. Your answer will be in feet per second. (c) Find the corresponding value of N to one decimal place. Your answer will be in cars per second. Convert your answer to cars per hour.

4-7 DIFFERENTIALS

We shall now give a reason for using the symbol dy/dx to denote the derivative of y with respect to x. To do this we introduce the notion of the *differential* of a function.

DEFINITION. *Suppose that $y = f(x)$ is a differentiable function of x. Then the* ***differential of y****, denoted dy or $d[f(x)]$, is*

$$dy = f'(x) \cdot h,$$

where h is any real number. Note that dy is a function of two variables, x and h.

EXAMPLE 1

Find the differential of $y = x^3 - 2x^2 + 3x - 4$ and evaluate it when $x = 1$ and $h = .04$.

$$dy = D_x(x^3 - 2x^2 + 3x - 4) \cdot h,$$

$$dy = (3x^2 - 4x + 3)h.$$

If $x = 1$ and $h = .04$, then

$$dy = [3(1)^2 - 4(1) + 3](.04) = .08.$$

If $f(x) = x$, then $d[f(x)] = d[x] = D_x[x]h = 1h = h$. Hence the differential of x is h. From now on we shall use the symbol dx for $d[x] = h$. For example,

$$d(x^2 + 5) = D_x(x^2 + 5) \cdot dx = 2x\, dx.$$

Summarizing, if $y = f(x)$ defines a differentiable function of x, we have

$$\boxed{dy = f'(x)\, dx.}$$

EXAMPLE 2

a. If $f(x) = \sqrt{x}$, then

$$d(\sqrt{x}) = D_x(\sqrt{x}) \cdot dx$$

$$= \frac{1}{2} x^{-1/2}\, dx = \frac{1}{2\sqrt{x}}\, dx.$$

b. If $u = (x^2 + 3)^5$, then

$$du = 5(x^2 + 3)^4 (2x)\, dx = 10x(x^2 + 3)^4\, dx.$$

c. If $u = x$, then $du = 1\, dx = dx$.

If $y = f(x)$, then $dy = f'(x)\,dx$. Provided that $dx \neq 0$, we can divide both sides by dx:

$$\frac{dy}{dx} = f'(x).$$

That is, dy/dx can be viewed either as the quotient of two differentials, dy divided by dx, or as the derivative of f at x. It is for this reason that we introduced the symbol dy/dx to denote the derivative.

The differential can be interpreted geometrically. In Fig. 4-57 the point $P(x, f(x))$ is on the curve $y = f(x)$. Suppose x changes by dx, a real number. Then its new value is $x + dx$ and the corresponding point on the curve is $Q(x + dx, f(x + dx))$. Through P and Q are horizontal and

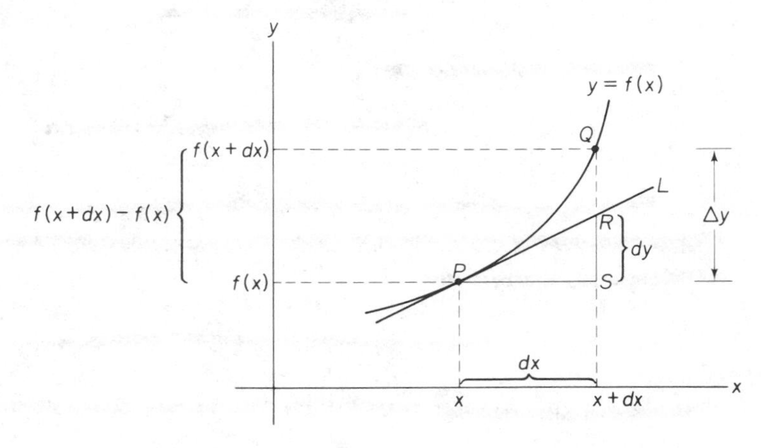

FIG. 4-57

vertical lines that intersect at S. A line L tangent to the curve at P intersects QS at R, forming the right triangle PRS. The slope of L is $f'(x)$ or, equivalently, it is $\overline{SR}/\overline{PS}$:

$$f'(x) = \frac{\overline{SR}}{\overline{PS}}.$$

Since $dy = f'(x)\,dx$, and $dx = \overline{PS}$,

$$\begin{aligned} dy &= f'(x)\,dx \\ &= \frac{\overline{SR}}{\overline{PS}} \cdot \overline{PS} \\ &= \overline{SR}. \end{aligned}$$

Thus, if dx is a change in x at P, dy is the corresponding vertical change in the **tangent line** at P. Note that for the same dx, the vertical change in the **curve** is $\Delta y = \overline{SQ} = f(x + dx) - f(x)$. Do not confuse Δy with dy. However, it is apparent from Fig. 4-57 that

when dx is *small*, dy is an approximation to Δy.

Symbolically we write $\Delta y \approx dy$ to indicate that Δy is approximately equal to dy. Observe that the graph of f near P is approximated by the tangent line at P.

EXAMPLE 3

A governmental health agency examined the records of a group of individuals who were hospitalized with a particular illness. It was found that the total proportion P who were discharged at the end of t days of hospitalization is given by

$$P = P(t) = 1 - \left(\frac{300}{300 + t}\right)^3.$$

Use differentials to find the approximate change in the proportion discharged if t changes from 300 to 305. Also, find the actual change.

We want to approximate the change in P, that is, approximate ΔP, when t goes from 300 to 305. We approximate ΔP by dP when $t = 300$ and $dt = 5$.

$$\Delta P \approx dP = P'\,dt = -3\left(\frac{300}{300 + t}\right)^2\left[-\frac{300}{(300 + t)^2}\right] dt$$

$$= -3\left(\frac{300}{600}\right)^2\left[-\frac{300}{(600)^2}\right]5$$

$$= -3\left(\frac{1}{2}\right)^2\left[-\frac{1}{2(600)}\right]5 = \frac{1}{320} \approx .0031.$$

The actual change ΔP is $P(305) - P(300) \approx .87807 - .87500 = .00307$.

We said that if $y = f(x)$, then $\Delta y \approx dy$ if dx is small. Thus

$$\Delta y = f(x + dx) - f(x) \approx dy,$$

or

$$\boxed{f(x + dx) \approx f(x) + dy.} \qquad (1)$$

This gives us a way of approximating a functional value. For example, suppose we approximate $(1.06)^3$. If we let $y = f(x) = x^3$, then we want to approximate $f(1.06)$. By (1) and the fact that $d(x^3) = 3x^2\,dx$, we have

$$f(x + dx) \approx f(x) + dy,$$
$$(x + dx)^3 \approx x^3 + 3x^2\,dx.$$

Note that $1.06 = 1 + .06$. Since .06 is small and we know the exact value of $1^3 = 1$, we shall let $x = 1$ and $dx = .06$:

$$(1.06)^3 = (1 + .06)^3 \approx 1^3 + 3(1)^2(.06) = 1.18.$$

The actual value of $(1.06)^3$ to two decimal places is 1.19.

EXAMPLE 4

The demand function for a product is given by $p = f(q) = 20 - \sqrt{q}$, where p is the price per unit in dollars for q units. By using differentials, approximate the price when 99 units are demanded.

We want to approximate $f(99)$. By (1),

$$f(q + dq) \approx f(q) + dp,$$

where
$$dp = -\frac{1}{2\sqrt{q}}\,dq.$$

We choose $q = 100$ because 100 is near 99 and it is easy to compute $f(100) = 20 - \sqrt{100} = 10$. Choosing $dq = -1$ (note the negative sign), we have

$$f(99) = f[100 + (-1)] \approx f(100) - \frac{1}{2\sqrt{100}}(-1).$$
$$f(99) \approx 10 + .05 = 10.05.$$

The price per unit when 99 units are demanded is approximately \$10.05.

The equation $y = x^3 + 4x + 5$ defines y as a function of x. However, it also defines x implicitly as a function of y. Thus we can look at the derivative of x with respect to y, dx/dy. Since dx/dy can be considered a quotient of differentials, we are motivated to write (and it is indeed true)

$$\boxed{\frac{dx}{dy} = \frac{1}{\frac{dy}{dx}}, \quad dy/dx \neq 0.}$$

But dy/dx is the derivative of y with respect to x and equals $3x^2 + 4$. Thus,

$$\frac{dx}{dy} = \frac{1}{3x^2 + 4}.$$

This is the *reciprocal* of dy/dx.

EXAMPLE 5

Find dp/dq *if* $q = \sqrt{2500 - p^2}$.

Since $q = (2500 - p^2)^{1/2}$, then

$$\frac{dq}{dp} = \frac{1}{2}(2500 - p^2)^{-1/2}(-2p) = -\frac{p}{\sqrt{2500 - p^2}}.$$

Hence,

$$\frac{dp}{dq} = \frac{1}{\frac{dq}{dp}} = -\frac{\sqrt{2500 - p^2}}{p}.$$

Exercise 4-7

In Problems **1–10**, *find the differentials of the functions in terms of x and dx.*

1. $y = 3x - 4$.

2. $y = 2$.

3. $f(x) = \sqrt{x^4 + 2}$.

4. $f(x) = (4x^2 - 5x + 2)^3$.

5. $u = \dfrac{1}{x^2}$.

6. $u = \dfrac{1}{\sqrt{x}}$.

7. $p = x^2(4x + 3)$.

8. $p = x\sqrt{x}$.

9. $g(x) = \dfrac{2x + 1}{x^2 + 3}$.

10. $g(x) = (x + 4)(x^2 - 5)^4$.

In Problems **11–14**, *evaluate* $d[f(x)]$ *for the indicated values of x and dx.*

11. $f(x) = 4 - 7x$; $x = 3$, $dx = .02$.

12. $f(x) = 4x^2 - 3x + 10$; $x = -1$, $dx = .25$.

13. $f(x) = \sqrt{25 - x^2}$; $x = 4$, $dx = -.1$.

14. $f(x) = (3x + 2)^2$; $x = -1$, $dx = -.03$.

In Problems **15–18**, *approximate each expression by using differentials.*

15. $\sqrt{101}$.

16. $\sqrt{120}$.

17. $\sqrt[3]{63}$.

18. $\sqrt[4]{17}$.

In Problems **19–24**, *find dx/dy or dp/dq.*

19. $y = 2x - 1$.

20. $y = 5x^2 + 3x + 2$.

21. $q = (p^2 + 5)^3$.

22. $q = \sqrt{p + 5}$.

23. $q = \dfrac{1}{p}$.

24. $q = \dfrac{p + 3}{p - 3}$.

In Problems **25** *and* **26**, *find the rate of change of q with respect to p for the indicated value of q.*

25. $p = \dfrac{500}{q + 2}$; $q = 18$.

26. $p = 50 - \sqrt{q}$; $q = 100$.

27. Suppose the profit P (in dollars) of producing q units of a product is

$$P = 396q - 2.2q^2 - 400.$$

Using differentials, find the approximate change in profit if the level of production changes from $q = 80$ to $q = 81$. Find the actual change.

28. Given the revenue function

$$r = 250q + 45q^2 - q^3,$$

use differentials to find the approximate change in revenue if the number of units increases from $q = 40$ to $q = 41$. Find the actual change.

29. The demand equation for a monopolist is $p = 10/\sqrt{q}$. Using differentials, approximate the price when 24 units are demanded.

30. Answer the same question in Problem 29 if 101 units are demanded.

31. The volume V of a spherical cell is given by $V = \frac{4}{3}\pi r^3$, where r is the radius. Estimate the change in volume when the radius changes from 6.5×10^{-4} cm to 6.6×10^{-4} cm.

32. The equation $(P + a)(v + b) = k$ is called the "fundamental equation of muscle contraction" **[48]**. Here P is the load imposed on the muscle, v is the velocity of the shortening of the muscle fibers, and a, b, and k are positive constants. Find v in terms of P and then use the differential to approximate the change in v due to a small change in P.

33. In a study of rooted plants in a certain region (adapted from [39]), it was determined that the average number of species S occurring on plots of size A (in square meters) was given by

$$S = 12\sqrt[4]{A}, \qquad 0 \leqslant A \leqslant 900.$$

Use differentials to approximate the average number of species in an 80-square-meter plot.

34. Suppose S is a numerical value of status attributable to a person's annual income I (in thousands of dollars). For a certain population, suppose $S = 20\sqrt{I}$. Use differentials to approximate S for a person with an annual income of $15,000, that is, $I = 15$.

35. If $y = f(x)$, then the *proportional change* in y is defined to be $\Delta y/y$, which can be approximated with differentials by dy/y. Use this last form to approximate the proportional change in the cost function $c = f(q) = (q^4/2) + 3q + 400$ when $q = 10$ and $dq = 2$. Give answer correct to one decimal place.

4-8 REVIEW

IMPORTANT TERMS AND SYMBOLS IN CHAPTER 4

x-intercept *(p. 145)*
y-intercept *(p. 145)*
x-axis symmetry *(p. 148)*
y-axis symmetry *(p. 147)*
symmetry about origin *(p. 149)*
horizontal asymptote *(p. 153)*
vertical asymptote *(p. 153)*
relative maximum *(p. 161)*
relative minimum *(p. 161)*
relative extrema *(p. 161)*
increasing function *(p. 159)*
decreasing function *(p. 159)*
concave up *(p. 171)*
concave down *(p. 171)*
critical value *(p. 162)*
first-derivative test *(p. 163)*
second-derivative test *(p. 184)*
differential *(p. 199)*
dy, dx *(p. 199)*
absolute extrema *(p. 161)*
inflection point *(p. 173)*

REVIEW SECTION

1. The graph of $y = x^2 + 1$ has no (a) (x) (y)-intercept but has one (b) (x) (y)-intercept.

   Ans. (a) x; (b) y.

2. The y-intercept of the graph of $y = 8x^3 - 7x + 4$ is ________.

   Ans. (0, 4).

3. The graph of $y = x^4 - x^2 + 3$ is symmetric with respect to the (a) axis but not the (b) -axis.

   Ans. (a) y; (b) x.

4. If a graph is symmetric with respect to both the x-axis and y-axis, then it (is) (is not) symmetric with respect to the origin.

   Ans. is.

5. In the graph of $y = \dfrac{6}{x+2} + 3$ the line $x =$ (a) is a vertical asymptote and the line $y =$ (b) is a horizontal asymptote.

   Ans. (a) -2; (b) 3.

6. The graph of $y = \dfrac{1}{x^2+1}$ has a (a) (horizontal) (vertical) but no (b) (horizontal) (vertical) asymptote.

   Ans. (a) horizontal; (b) vertical.

7. Suppose that f is a function defined on an interval I. If $f'(x) < 0$ on I, then f is (a) (increasing) (decreasing) on I and the graph of f is (b) (rising) (falling). If $f'(x) > 0$ on I, then f is (c) (increasing) (decreasing) on I. If $f'(x_1) = 0$ or is not defined, then $x = x_1$ is called a (d) value. At such a point, f may have a relative (e) or (f) or neither.

   Ans. (a) decreasing; (b) falling; (c) increasing; (d) critical; (e) maximum; (f) minimum.

8. If $f''(x) > 0$ on I, then f is concave (up) (down) on I.

   Ans. up.

9. If $f''(x_1) = 0$ or is not defined, then there may be a point of (a) when $x = x_1$. If $f'(x_1) = 0$ and $f''(x_1) > 0$, then f has a relative (b) when $x = x_1$.

   Ans. (a) inflection; (b) minimum.

10. If $f(x_1) \geqslant f(x)$ for all x in the domain of f, then when $x = x_1$ there occurs a(n) ________ maximum.

Ans. absolute.

11. Under typical conditions, profit is maximized at the level of output for which marginal revenue = ________.

Ans. marginal cost.

12. If f has a relative maximum when $x = x_1$, then as x increases through x_1, $f'(x)$ changes from (a) (+) (−) to (b) (+) (−). At a relative minimum, $f'(x)$ changes from (c) (+) (−) to (d) (+) (−).

Ans. (a) +; (b) −; (c) −; (d) +.

13. True or false: If $f'(x_1) = 0$, then f has a relative maximum or minimum when $x = x_1$, __(a)__. If $f''(x_2) = 0$, then f has a point of inflection when $x = x_2$, __(b)__.

Ans. (a) false; (b) false.

14. In the graph of $y = f(x)$ in Fig. 4-58, an absolute maximum occurs at point(s) __(a)__, an absolute minimum at __(b)__, a relative maximum at __(c)__, a relative minimum at __(d)__, and inflection point(s) at __(e)__.

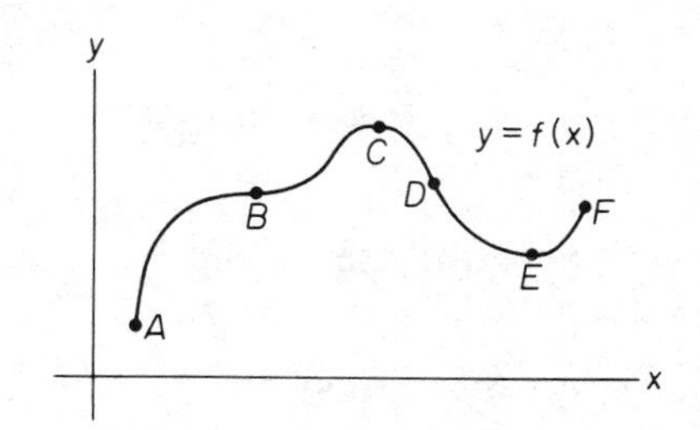

FIG. 4-58

Ans. (a) C; (b) A; (c) C; (d) E; (e) B, D.

15. If the graph of $y = f(x)$ is falling from left to right over the interval I, and f is concave up there, then f is (a) (increasing) (decreasing) at a(n) (b) (increasing) (decreasing) rate.

Ans. (a) decreasing; (b) increasing.

16. The differential of $y = x^2 + 3$ is ________.

Ans. $2x\,dx$.

17. If $\dfrac{dy}{dx} = \dfrac{2}{3}$, then $\dfrac{dx}{dy} =$ ________.

Ans. $\frac{3}{2}$.

18. The differential of $y = x$ when $x = 6$ and $dx = .001$ is ________.

Ans. .001.

REVIEW PROBLEMS

In Problems **1–8** *sketch the graphs of the functions. Indicate intervals on which the function is increasing, decreasing, concave up, concave down; indicate relative maximum points, relative minimum points, points of inflection, horizontal asymptotes, vertical asymptotes, symmetry, and those intercepts that can be obtained conveniently.*

1. $y = x^2 - 2x - 24.$

2. $y = x^3 - 27x.$

3. $y = x^3 - 12x + 20.$

4. $y = x^4 - 4x^3 - 20x^2 + 150.$

5. $y = x^3 + x.$

6. $y = \dfrac{x+2}{x-3}.$

7. $f(x) = \dfrac{100(x+5)}{x^2}.$

8. $f(x) = \dfrac{1}{x^2 - 5}.$

9. Over what interval is $y = f(x) = x^3 - x^2 - x + 10$ decreasing at an increasing rate?

10. Over what interval is $y = f(x) = (x^2 - 1)/(x^2 + 1)$ increasing at an increasing rate?

11. A manufacturer determines that m employees on a certain production line will produce q units per month where $q = 80m^2 - .1m^4$. To obtain maximum monthly production, how many employees should be assigned to the production line?

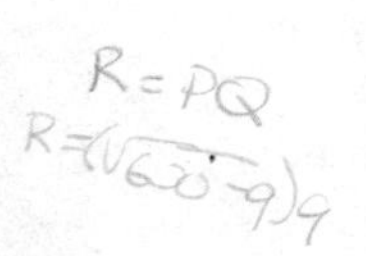

12. The demand function for a monopolists's product is $p = \sqrt{600 - q}$. If he wants to produce at least 100 units but not more than 300 units, how many units should he produce to maximize total revenue?

13. The demand function for a monopolist's product is $p = 400 - 2q$; p is given in dollars. If the average cost per unit for producing q units is $\bar{c} = q + 160 + (2000/q)$, find the maximum profit that the monopolist can achieve.

14. The Kiddie Toy Company plans to lease an electric motor to be used 90,000 horsepower-hours per year in manufacturing. One horsepower-hour is the work done in one hour by a one-horsepower motor. The annual cost to lease a suitable motor is \$150 plus \$.60 per horsepower. The cost per horsepower-hour of operating the motor is $\$.006/N$, where N is the horsepower. What size motor, in horsepower, should be leased in order to minimize cost?

15. If $c = .01q^2 + 5q + 100$ is a cost equation, find the average cost equation. At what level of production q is there a minimum average cost? Show algebraically that the graph of the marginal cost function intersects the graph of the average cost function at this point.

16. A rectangular box is to be made by cutting out equal squares from each corner of a piece of cardboard 10 in. by 16 in. and then folding up the sides. What must be the length of the side of the square cut out if the volume of the box is to be a maximum?

17. A rectangular poster having an area of 500 square inches is to have a 4-inch margin at each side and at the bottom and a 6-inch margin at the top. The remainder of the poster is for printed matter. Find the dimensions of the poster so that the area for the printed matter is maximized.

In Problems **18** *and* **19**, *determine the differentials of the functions in terms of x and dx.*

18. $f(x) = (1 - 3x)^4$.

19. $u = (x^2 + 5)/(x - 7)$.

Approximate the expressions in Problems **20** *and* **21** *by use of differentials.*

20. $\sqrt{25.5}$.

21. $\sqrt{143}$.

22. Fahrenheit temperature F and Celsius temperature C are related by $F = \frac{9}{5}C + 32$. Using differentials, find how much F would change due to a change in C of $\frac{1}{2}°$.

CHAPTER 5

Integration

Chapters 3 and 4 dealt with differential calculus. We differentiated a function and got another function called its derivative. *Integral calculus* is concerned with the opposite process. We are given the derivative of a function and must find the original function. The need for doing this arises in a natural way. For example, we may have a marginal revenue function and want to find the revenue function from it. However, the real power of integral calculus is that it allows us to take the limit of a special kind of sum as the number of terms in the sum becomes infinite. With this notion we may find such things as areas of regions which cannot be found by any other convenient method. Furthermore, you will see that the limit concept involved with integral calculus is related to the limit concept used in differential calculus.

5-1 THE INDEFINITE INTEGRAL

If F is a function such that

$$F'(x) = f(x), \tag{1}$$

then F is called an *antiderivative* of f. Thus an antiderivative of f is merely a function F which when differentiated gives f. Multiplying both sides of Eq. (1) by the differential dx gives $F'(x)\,dx = f(x)\,dx$. However, $F'(x)\,dx$ is the differential dF:

$$dF = f(x)\,dx.$$

Thus we can look at an antiderivative of f as a function whose differential is $f(x)\,dx$.

DEFINITION. *An* ***antiderivative*** *of a function f is a function F such that*

$$F'(x) = f(x)$$

or equivalently, in differential notation,

$$dF = f(x)\,dx.$$

For example, since

$$D_x(x^2) = 2x,$$

x^2 is an antiderivative of $2x$. Is this the only antiderivative of $2x$? The answer is "no"! Since

$$D_x(x^2 + 1) = 2x$$

and

$$D_x(x^2 - 5) = 2x,$$

then $x^2 + 1$ and $x^2 - 5$ are also antiderivatives of $2x$. In fact, any antiderivative of $2x$ must have the form $x^2 + C$, where C is a constant. Thus *any two antiderivatives of $2x$ differ only by a constant.*

The symbol we shall use for an arbitrary antiderivative of $2x$ is $\int 2x\,dx$, which is read "the *indefinite integral* of $2x$ with respect to x." Since all antiderivatives of $2x$ have the form $x^2 + C$, we write

$$\int 2x\,dx = x^2 + C.$$

The symbol $\int$ is called the **integral sign**, $2x$ is the **integrand**, and C is the **constant of integration**. The "dx" is part of the integral notation and indicates the variable involved. We say here that x is the **variable of integration**.

More generally, the **indefinite integral** of f with respect to x is written $\int f(x)\,dx$ and denotes an arbitrary antiderivative of f. It can be shown that all antiderivatives of f differ at most by a constant. Thus if F is an antiderivative of f, then

$$\int f(x)\,dx = F(x) + C, \text{ where } C \text{ is a constant.}$$

To *integrate* f, that is, to find $\int f(x)\,dx$, we just determine an antiderivative of f and add on the constant of integration. Thus

$$\boxed{\int f(x)\,dx = F(x) + C \text{ if and only if } F'(x) = f(x).}$$

EXAMPLE 1

Find $\int 5\,dx$.

First we must find (perhaps better words are "guess at") a function whose derivative is 5. Since $D_x(5x) = 5$, then $5x$ is an antiderivative of 5. Thus

$$\int 5\,dx = 5x + C.$$

PITFALL. *It is incorrect to write* $\int 5\,dx = 5x$. ***Do not forget the constant of integration.***

Using differentiation formulas from Chapter 3, we have compiled a list of basic integration formulas in Table 5-1. These formulas are easily verified. For example, formula (2) is true since the derivative of $x^{n+1}/(n+1)$ is x^n for $n \neq -1$. To verify formula (3) we must show that k times an antiderivative of f is an antiderivative of kf. To demonstrate this, let F be an antiderivative of f. Then $F'(x) = f(x)$, and so $D_x[kF(x)] = k\,f(x)$. Thus kF is an antiderivative of kf and the validity of formula (3) is established. You should verify the other formulas. Formula (4) can be extended to any finite number of sums and/or differences.

We point out that formula (2) states that the indefinite integral of most powers of x is obtained by increasing the exponent of x by one,

TABLE 5-1

BASIC INTEGRATION FORMULAS
1. $\int k\,dx = kx + C$, k a constant.
2. $\int x^n\,dx = \frac{x^{n+1}}{n+1} + C$, $n \neq -1$.
3. $\int k\,f(x)\,dx = k\int f(x)\,dx$, k a constant.
4. $\int [f(x) \pm g(x)]\,dx = \int f(x)\,dx \pm \int g(x)\,dx$.

dividing by the new exponent, and adding on the constant of integration. The case when $n = -1$ would result in division by zero and will be discussed in Chapter 6.

EXAMPLE 2

Find the following indefinite integrals.

a. $\int 1\,dx$.

By formula (1) in Table 5–1 with $k = 1$,

$$\int 1\,dx = 1x + C = x + C.$$

Usually we write $\int 1\,dx$ as $\int dx$. Thus, $\int dx = x + C$.

b. $\int x^5\,dx$.

By formula (2) with $n = 5$,

$$\int x^5\,dx = \frac{x^{5+1}}{5+1} + C = \frac{x^6}{6} + C.$$

c. $\int 7x\,dx$.

By formula (3) with $k = 7$ and $f(x) = x$,

$$\int 7x\,dx = 7\int x\,dx.$$

Since $\int x\,dx = \int x^1\,dx$, by formula (2) with $n = 1$ we have

$$\int x^1\,dx = \frac{x^{1+1}}{1+1} + C_1 = \frac{x^2}{2} + C_1,$$

where C_1 is the constant of integration. Therefore,

$$\int 7x\,dx = 7\int x\,dx = 7\left[\frac{x^2}{2} + C_1\right] = \frac{7}{2}x^2 + 7C_1.$$

Replacing the constant $7C_1$ by C, we have

$$\int 7x\,dx = \frac{7}{2}x^2 + C.$$

It is not necessary to write all intermediate steps when integrating. More simply we write

$$\int 7x\,dx = 7\int x\,dx = (7)\frac{x^2}{2} + C = \frac{7}{2}x^2 + C.$$

PITFALL. *Only a constant factor can "jump" in front of an integral sign. It is correct to write* $\int 7x\,dx = 7\int x\,dx = (7/2)x^2 + C$, *but it is* ***incorrect*** *to write* $\int 7x\,dx = 7x\int dx = (7x)(x + C) = 7x^2 + 7Cx$.

EXAMPLE 3

Find the following indefinite integrals.

a. $\int \frac{1}{\sqrt{t}}\,dt.$

Here t is the *variable of integration.* We rewrite the integrand so that one of the basic forms can be used.

$$\int \frac{1}{\sqrt{t}}\,dt = \int t^{-1/2}\,dt.$$

By formula (2) in Table 5–1 with $n = -\frac{1}{2}$,

$$\int \frac{1}{\sqrt{t}}\,dt = \int t^{-1/2}\,dt = \frac{t^{(-1/2)+1}}{-\frac{1}{2}+1} + C = \frac{t^{1/2}}{\frac{1}{2}} + C = 2\sqrt{t} + C.$$

b. $\int \frac{1}{6x^3}\,dx.$

$$\int \frac{1}{6x^3}\,dx = \frac{1}{6}\int x^{-3}\,dx = \left(\frac{1}{6}\right)\frac{x^{-3+1}}{-3+1} + C$$

$$= -\frac{x^{-2}}{12} + C = -\frac{1}{12x^2} + C.$$

EXAMPLE 4

Find the following indefinite integrals.

a. $\int (x^2 + 2x - 4)\, dx.$

By formula (4) in Table 5–1,

$$\int (x^2 + 2x - 4)\, dx = \int x^2\, dx + \int 2x\, dx - \int 4\, dx.$$

Now,

$$\int x^2\, dx = \frac{x^{2+1}}{2+1} + C_1 = \frac{x^3}{3} + C_1,$$

$$\int 2x\, dx = 2\int x\, dx = (2)\frac{x^{1+1}}{1+1} + C_2 = x^2 + C_2,$$

and

$$\int 4\, dx = 4x + C_3.$$

Thus,

$$\int (x^2 + 2x - 4)\, dx = \frac{x^3}{3} + x^2 - 4x + C_1 + C_2 - C_3.$$

Letting $C = C_1 + C_2 - C_3$, we have

$$\int (x^2 + 2x - 4)\, dx = \frac{x^3}{3} + x^2 - 4x + C.$$

Omitting intermediate steps, we simply write

$$\begin{aligned}\int (x^2 + 2x - 4)\, dx &= \int x^2\, dx + 2\int x\, dx - \int 4\, dx \\ &= \frac{x^3}{3} + (2)\frac{x^2}{2} - 4x + C \\ &= \frac{x^3}{3} + x^2 - 4x + C.\end{aligned}$$

b. $\int (2\sqrt[5]{x^4} - 7x^3 - 1)\, dx.$

$$\begin{aligned}\int \left(2\sqrt[5]{x^4} - 7x^3 - 1\right) dx &= 2\int x^{4/5}\, dx - 7\int x^3\, dx - \int 1\, dx \\ &= (2)\frac{x^{9/5}}{\frac{9}{5}} - (7)\frac{x^4}{4} - x + C \\ &= \frac{10}{9}x^{9/5} - \frac{7}{4}x^4 - x + C.\end{aligned}$$

EXAMPLE 5

For a particular urban group, sociologists studied the average yearly income y (in dollars) that a person can expect to receive with x years of education before seeking regular employment. They estimated that the rate at which income changes with respect to education is given by

$$\frac{dy}{dx} = 10x^{3/2}, \qquad 4 \leqslant x \leqslant 16,$$

where $y = 5872$ *when* $x = 9$. *Find y.*

Since $dy/dx = 10x^{3/2}$, y is an antiderivative of $10x^{3/2}$. Thus

$$\begin{aligned} y &= \int 10x^{3/2}\,dx = 10\int x^{3/2}\,dx \\ &= (10)\frac{x^{5/2}}{\frac{5}{2}} + C = 4x^{5/2} + C. \end{aligned}$$

Thus, *any* function of the form

$$y = 4x^{5/2} + C \tag{2}$$

satisfies $dy/dx = 10x^{3/2}$. To determine the *specific* one for which $y = 5872$ when $x = 9$, we must find the right C. Substituting $y = 5872$ and $x = 9$ into Eq. (2), we obtain

$$\begin{aligned} 5872 &= 4(9)^{5/2} + C \\ &= 4(243) + C. \\ 5872 &= 972 + C. \end{aligned}$$

Therefore $C = 4900$ and

$$y = 4x^{5/2} + 4900.$$

EXAMPLE 6

Given that $y'' = x^2 - 6$, $y'(0) = 2$, *and* $y(1) = -1$, *find y.*

Since $y'' = \frac{d}{dx}(y') = x^2 - 6$, then y' is an antiderivative of $x^2 - 6$. Thus

$$y' = \int (x^2 - 6)\,dx = \frac{x^3}{3} - 6x + C_1. \tag{3}$$

Since $y'(0) = 2$ means that $y' = 2$ when $x = 0$, from Eq. (3) we have

$$2 = \frac{0^3}{2} - 6(0) + C_1.$$

Hence $C_1 = 2$ and

$$y' = \frac{x^3}{3} - 6x + 2.$$

By integration we can find y:

$$y = \int \left(\frac{x^3}{3} - 6x + 2\right) dx$$

$$= \left(\frac{1}{3}\right)\frac{x^4}{4} - (6)\frac{x^2}{2} + 2x + C_2.$$

$$y = \frac{x^4}{12} - 3x^2 + 2x + C_2. \quad (4)$$

Since $y = -1$ when $x = 1$, from Eq. (4) we have

$$-1 = \frac{1^4}{12} - 3(1)^2 + 2(1) + C_2.$$

Therefore $C_2 = -\frac{1}{12}$ and

$$y = \frac{x^4}{12} - 3x^2 + 2x - \frac{1}{12}.$$

EXAMPLE 7

The president of a company determined that the marginal revenue function for his U.S. subsidiary is

$$dr/dq = 2000 - 20q - 3q^2.$$

Determine the demand function.

Since dr/dq is the derivative of total revenue r,

$$r = \int (2000 - 20q - 3q^2)\, dq$$

$$= 2000q - (20)\frac{q^2}{2} - (3)\frac{q^3}{3} + C.$$

$$r = 2000q - 10q^2 - q^3 + C. \quad (5)$$

When no units are sold, total revenue is 0; that is, $r = 0$ when $q = 0$. From Eq. (5),

$$0 = 2000(0) - 10(0)^2 - 0^3 + C.$$

Hence $C = 0$ and

$$r = 2000q - 10q^2 - q^3.$$

But $r = pq$ where p is the price per unit. Solving for p gives the demand function:

$$p = \frac{r}{q} = \frac{2000q - 10q^2 - q^3}{q}.$$

$$p = 2000 - 10q - q^2.$$

EXAMPLE 8

The chief executive officer of a corporation asked his vice president of production to determine the total cost of producing 10,000 pounds of product in one week. In his files the vice president found that the marginal cost function dc/dq is

$$\frac{dc}{dq} = \frac{1}{1{,}000{,}000}\left[\frac{1}{500}q^2 - 25q\right] + .2,$$

where q is pounds of product per week, and fixed costs per week are \$400. Fixed costs refer to costs, such as rent, that remain constant at all levels of production in a given time period. What reply should the vice president give?

Since dc/dq is the derivative of total cost c,

$$c = \int\left\{\frac{1}{1{,}000{,}000}\left[\frac{1}{500}q^2 - 25q\right] + .2\right\} dq$$

$$= \frac{1}{1{,}000{,}000}\int\left[\frac{1}{500}q^2 - 25q\right] dq + \int .2\, dq.$$

$$c = \frac{1}{1{,}000{,}000}\left[\frac{q^3}{1500} - \frac{25q^2}{2}\right] + .2q + C.$$

Fixed costs are constant regardless of output. Therefore, when $q = 0$, then $c = 400$ and so $C = 400$. Thus

$$c = \frac{1}{1{,}000{,}000}\left[\frac{q^3}{1500} - \frac{25q^2}{2}\right] + .2q + 400. \qquad (6)$$

From Eq. (6), when $q = 10{,}000$ then $c = 1816\frac{2}{3}$. The vice president should reply that the total cost for producing 10,000 pounds of product in one week is \$1,816.67.

Exercise 5-1

In Problems **1–52**, *find the indefinite integrals.*

1. $\int 5\, dx.$

2. $\int \frac{1}{2}\, dx.$

3. $\int x^8\,dx.$

4. $\int 2x^{25}\,dx.$

5. $\int t^{13/2}\,dt.$

6. $\int \frac{1}{2}x^{5/3}\,dx.$

7. $\int x^{-7}\,dx.$

8. $\int \frac{z^{-3}}{3}\,dz.$

9. $\int \frac{1}{x^{10}}\,dx.$

10. $\int \frac{7}{x^4}\,dx.$

11. $\int \frac{1}{y^{11/5}}\,dy.$

12. $\int \frac{7}{2x^{9/4}}\,dx.$

13. $\int \sqrt[5]{x^6}\,dx.$

14. $\int -\frac{3}{2}\sqrt{x}\,dx.$

15. $\int \frac{1}{\sqrt[8]{x^7}}\,dx.$

16. $\int \frac{3}{4\sqrt[8]{x}}\,dx.$

17. $\int (7+4)\,dx.$

18. $\int (5-6)\,dx.$

19. $\int x^{\sqrt{2}}\,dx.$

20. $\int x^{1.2}\,dx.$

21. $\int 3x^7\,dx.$

22. $\int 5x^4\,dx.$

23. $\int (8+u)\,du.$

24. $\int (r^3+2r)\,dr.$

25. $\int (y^5+5y)\,dy.$

26. $\int (7-3w-2w^2)\,dw.$

27. $\int (3t^2-4t-5)\,dt.$

28. $\int (1+u+u^2+u^3)\,du.$

29. $\int \left(\frac{x}{7}-\frac{3}{4}x^4\right)dx.$

30. $\int \left(\frac{2x^2}{7}-\frac{8}{3}x^4\right)dx.$

31. $\int \frac{2\sqrt{x}}{3}\,dx.$

32. $\int \frac{3}{x^{-4}}\,dx.$

33. $\int \left(\frac{3}{2y^4}-\frac{3y^4}{2}\right)dy.$

34. $\int \frac{1}{12}\left(\frac{1}{3}x^5\right)dx.$

35. $\int (p^{2.1}-p^{-2.1})\,dp.$

36. $\int dw.$

37. $\int (x^{-2}-5x^{-3}+2x^{-4})\,dx.$

38. $\int (-3x^{-2}-2x^{-3})\,dx.$

39. $\int (x^{8.3} - 9x^6 + 3x^{-4} + x^{-3})\, dx.$

40. $\int (.3y^4 - 8y^{-3} + 2)\, dy.$

41. $\int \left(\frac{x^3}{3} - \frac{3}{x^3}\right) dx.$

42. $\int \left(\frac{1}{2x^3} - \frac{1}{x^4}\right) dx.$

43. $\int \left(\frac{3w^2}{2} - \frac{2}{3w^2}\right) dw.$

44. $\int \frac{2}{7^{-1}}\, ds.$

45. $\int (\sqrt[3]{x} - \sqrt[4]{x} + \sqrt[5]{x})\, dx.$

46. $\int \left(\frac{\sqrt[5]{x}}{2} - \frac{2}{3}\sqrt[3]{x}\right) dx.$

47. $\int (2\sqrt{x} - 3\sqrt[4]{x})\, dx.$

48. $\int 0\, dx.$

49. $\int 2x^{-6/5}\, dx.$

50. $\int (x^{-4/5} + 2)\, dx.$

51. $\int \left(-\frac{\sqrt[3]{x^2}}{5} - \frac{7}{2\sqrt{x}} + 6x\right) dx.$

52. $\int \left(\sqrt[3]{x} - \frac{1}{\sqrt[3]{x}}\right) dx.$

In Problems 53–56, *find* y *subject to the given conditions.*

53. $dy/dx = 3x - 4$; $y(-1) = \frac{13}{2}$.

54. $dy/dx = x^2 - x$; $y(3) = 4$.

55. $y'' = -x^2 - 2x$; $y'(1) = 0, y(1) = 1$.

56. $y'' = x + 1$; $y'(0) = 0, y(0) = 5$.

In Problems 57–60, dc/dq *is a marginal cost function and fixed costs are indicated in braces. For Problems* 57 *and* 58 *find the total cost function. For Problems* 59 *and* 60, *find the total cost for the indicated value of* q.

57. $dc/dq = 1.35$; $\{200\}$.

58. $dc/dq = 2q + 50$; $\{1000\}$.

59. $dc/dq = .09q^2 - 1.2q + 4.5$; $\{7700\}$; $q = 10$.

60. $dc/dq = .000102q^2 - .034q + 5$; $\{10{,}000\}$; $q = 100$.

In Problems 61–63, dr/dq *is a marginal revenue function. Find the demand function.*

61. $dr/dq = .7$.

62. $dr/dq = 15 - \frac{1}{15}q$.

63. $dr/dq = 275 - q - .3q^2$.

64. A manufacturer of manikins has determined that their marginal cost function is $dc/dq = .003q^2 - .4q + 40$, where q is the number of manikins

produced. If their marginal cost when 50 manikins are produced is \$27.50 and fixed costs are \$5000, what is the *average* cost of producing 100 manikins?

65. A study of the winter moth was made in Nova Scotia (adapted from **[10]**). The prepupae of the moth fall onto the ground from host trees. It was found that the (approximate) rate at which prepupal density y (number of prepupae per square foot of soil) changes with respect to distance x (in feet) from the base of a host tree is

$$\frac{dy}{dx} = -1.5 - x, \qquad 1 \leqslant x \leqslant 9.$$

If $y = 57.3$ when $x = 1$, find y.

66. A group of biologists studied the nutritional effects on rats that were fed a diet containing 10 percent protein (adapted from **[4]**). The protein consisted of yeast and corn flour. Over a period of time, the group found that the (approximate) rate of change of the average weight gain G (in grams) of a rat with respect to the percentage P of yeast in the protein mix is

$$\frac{dG}{dP} = -\frac{P}{25} + 2, \qquad 0 \leqslant P \leqslant 100.$$

If $G = 38$ when $P = 10$, find G.

67. In the study of flow of fluid in a tube of constant radius R, such as blood flow in portions of the body, one can think of the tube as consisting of concentric tubes of radius r, where $0 \leqslant r \leqslant R$. The velocity v of the fluid is a function of r and is given in **[48]** by

$$v = \int -\frac{(P_1 - P_2)r}{2l\eta}\, dr,$$

where P_1 and P_2 are pressures at the ends of the tube, η (a Greek letter read "eta") is fluid viscosity, and l is the length of the tube. If $v = 0$ when $r = R$, show that

$$v = \frac{(P_1 - P_2)(R^2 - r^2)}{4l\eta}.$$

5-2 TECHNIQUES OF INTEGRATION

Sometimes, in order to apply the basic integration formulas it is necessary to first perform algebraic manipulations on an integrand. Example 1 shows some techniques.

EXAMPLE 1

Find the following indefinite integrals.

a. $\int y^2\left(y + \frac{2}{3}\right) dy.$

By multiplying the integrand we get

$$\int y^2\left(y + \frac{2}{3}\right) dy = \int \left(y^3 + \frac{2}{3}y^2\right) dy$$
$$= \frac{y^4}{4} + \left(\frac{2}{3}\right)\frac{y^3}{3} + C = \frac{y^4}{4} + \frac{2y^3}{9} + C.$$

b. $\int \frac{(3x + \sqrt{x})(\sqrt[3]{x} - 2)}{6} dx.$

By factoring out $\frac{1}{6}$ and multiplying the binomials, we get

$$\int \frac{(3x + x^{1/2})(x^{1/3} - 2)}{6} dx$$
$$= \frac{1}{6}\int (3x^{4/3} + x^{5/6} - 6x - 2x^{1/2}) dx$$
$$= \frac{1}{6}\left[(3)\frac{x^{7/3}}{\frac{7}{3}} + \frac{x^{11/6}}{\frac{11}{6}} - (6)\frac{x^2}{2} - (2)\frac{x^{3/2}}{\frac{3}{2}}\right] + C$$
$$= \frac{3x^{7/3}}{14} + \frac{x^{11/6}}{11} - \frac{x^2}{2} - \frac{2x^{3/2}}{9} + C.$$

c. $\int \frac{x^4 - 1}{x^2} dx.$

We can rewrite the integrand by dividing each term in the numerator by the denominator.

$$\int \frac{x^4 - 1}{x^2} dx = \int \left[x^2 - \frac{1}{x^2}\right] dx$$
$$= \frac{x^3}{3} - \int x^{-2} dx$$
$$= \frac{x^3}{3} + \frac{1}{x} + C.$$

The formula

$$\int x^n \, dx = \frac{x^{n+1}}{n + 1} + C, \text{ if } n \neq -1,$$

can be generalized to handle not only a power of x but also a power of a *function* of x. Let u be a differentiable function of x. By the power rule, if $n \neq -1$ then

$$\frac{d}{dx}\left(\frac{[u(x)]^{n+1}}{n+1}\right) = \frac{(n+1)[u(x)]^{n}u'(x)}{n+1}$$
$$= [u(x)]^{n}u'(x).$$

Thus,

$$\int [u(x)]^{n} \cdot u'(x)\,dx = \frac{[u(x)]^{n+1}}{n+1} + C, \qquad n \neq -1. \quad (1)$$

We call this the **power rule for integration.** Since $u'(x)\,dx$ is the differential of u, namely du, using mathematical shorthand we can write this rule as

$$\boxed{\int u^{n}\,du = \frac{u^{n+1}}{n+1} + C.}$$

EXAMPLE 2

Use the power rule for integration to find the following indefinite integrals.

a. $\int (x+1)^{20}\,dx.$

Since we have a power of $x+1$, we shall set $u = x + 1$. Then $du = dx$ and $\int (x+1)^{20}\,dx$ has the form $\int u^{20}\,du$. Hence,

$$\int (x+1)^{20}\,dx = \int u^{20}\,du = \frac{u^{21}}{21} + C = \frac{(x+1)^{21}}{21} + C.$$

Note that we give our answer not in terms of u but explicitly in terms of x.

b. $\int 3x^{2}(x^{3}+7)^{3}\,dx.$

Let $u = x^{3} + 7$. Then $du = 3x^{2}\,dx$ and

$$\int 3x^{2}(x^{3}+7)^{3}\,dx = \int (x^{3}+7)^{3}[3x^{2}\,dx] = \int u^{3}\,du$$
$$= \frac{u^{4}}{4} + C = \frac{(x^{3}+7)^{4}}{4} + C.$$

EXAMPLE 3

Find the following indefinite integrals.

a. $\int x\sqrt{x^2+5}\ dx$.

We can write this integral as

$$\int x(x^2+5)^{1/2}\,dx.$$

If $u = x^2 + 5$, then $du = 2x\,dx$. Since the *constant* factor 2 in du does *not* appear in the integrand, this integral does not have the form $\int u^n\,du$. However, we can put it in this form by first multiplying and dividing the integrand by 2. This does not change its value. Thus,

$$\int x(x^2+5)^{1/2}\,dx = \int \frac{2}{2}x(x^2+5)^{1/2}\,dx = \int \frac{1}{2}(x^2+5)^{1/2}[2x\,dx].$$

Since the integrand now has 1/2 as a *constant* factor, we have

$$\int x(x^2+5)^{1/2}\,dx = \frac{1}{2}\int (x^2+5)^{1/2}[2x\,dx]$$

$$= \frac{1}{2}\int u^{1/2}\,du = \frac{1}{2}\left[\frac{u^{3/2}}{\frac{3}{2}}\right] + C.$$

Going back to x, we have

$$\int x\sqrt{x^2+5}\ dx = \frac{(x^2+5)^{3/2}}{3} + C.$$

In summary, to find $\int x\sqrt{x^2+5}\ dx$ we multiplied the integrand by the constant 2 so that we would have the form $\int u^n\,du$. This step was adjusted for by inserting the factor $\frac{1}{2}$ in front of the integral sign.

b. $\int \sqrt[3]{6y}\ dy$.

If we set $u = 6y$, then $du = 6\,dy$. Thus we insert a factor of 6 and adjust for it with a factor of $\frac{1}{6}$.

$$\int \sqrt[3]{6y}\ dy = \int (6y)^{1/3}\,dy = \frac{1}{6}\int (6y)^{1/3}[6\,dy] = \frac{1}{6}\int u^{1/3}\,du$$

$$= \left(\frac{1}{6}\right)\frac{u^{4/3}}{\frac{4}{3}} + C = \frac{(6y)^{4/3}}{8} + C.$$

c. $\int \frac{2x^3 + 3x}{(x^4 + 3x^2 + 7)^4}\, dx.$

We can write this as $\int (x^4 + 3x^2 + 7)^{-4}(2x^3 + 3x)\, dx$. If $u = x^4 + 3x^2 + 7$, then $du = (4x^3 + 6x)\, dx$ which is two times the quantity $(2x^3 + 3x)\, dx$ in the integral. Thus we insert a factor of 2 and adjust for it with a factor of $\frac{1}{2}$.

$$\int (x^4 + 3x^2 + 7)^{-4}(2x^3 + 3x)\, dx$$

$$= \frac{1}{2}\int (x^4 + 3x^2 + 7)^{-4}[2(2x^3 + 3x)\, dx]$$

$$= \frac{1}{2}\int (x^4 + 3x^2 + 7)^{-4}[(4x^3 + 6x)\, dx]$$

$$= \frac{1}{2}\int u^{-4}\, du = \frac{1}{2} \cdot \frac{u^{-3}}{-3} + C = -\frac{1}{6u^3} + C$$

$$= -\frac{1}{6(x^4 + 3x^2 + 7)^3} + C.$$

d. $\int \frac{1}{\sqrt{x}\,(\sqrt{x} - 2)^3}\, dx.$

We can write this integral as

$$\int \frac{(\sqrt{x} - 2)^{-3}}{\sqrt{x}}\, dx.$$

Let $u = \sqrt{x} - 2$. Then $du = \frac{1}{2\sqrt{x}}\, dx$ and

$$\int \frac{(\sqrt{x} - 2)^{-3}}{\sqrt{x}}\, dx = 2\int (\sqrt{x} - 2)^{-3}\left[\frac{1}{2\sqrt{x}}\, dx\right]$$

$$= 2\int u^{-3}\, du$$

$$= 2\left(\frac{u^{-2}}{-2}\right) + C$$

$$= -u^{-2} + C = -(\sqrt{x} - 2)^{-2} + C.$$

e. $\int \frac{1}{(1 - w)^2}\, dw.$

$$\int \frac{1}{(1 - w)^2}\, dw = \int (1 - w)^{-2}\, dw$$

$$= -1\int (1 - w)^{-2}[-dw].$$

The integral has the form $\int u^{-2}\,du$, where $u = 1 - w$. Thus,

$$\int \frac{1}{(1-w)^2}\,dw = -\frac{(1-w)^{-1}}{-1} + C$$
$$= \frac{1}{1-w} + C.$$

When using the power rule for integration, care must be taken when making your choice for u. For example, in Example 3(c) you would *not* be able to proceed very far if, for instance, you let $u = 2x^3 + 3x$. At times you may find it necessary to try many different choices. **Skill at integration comes only after many hours of practice and conscientious study.**

EXAMPLE 4

Find $\int 4x^2(x^4 + 1)^2\,dx$.

If we set $u = x^4 + 1$, then $du = 4x^3\,dx$. Here we are not able to get the form for du in the integral because we need an additional factor of the *variable* x. Remember: you can insert only a **constant** factor in an integral and can adjust for it in front of the integral sign. In our situation we can find the integral by first expanding the integrand.

$$\int 4x^2(x^4+1)^2\,dx = 4\int x^2(x^8 + 2x^4 + 1)\,dx$$
$$= 4\int (x^{10} + 2x^6 + x^2)\,dx$$
$$= 4\left(\frac{x^{11}}{11} + \frac{2x^7}{7} + \frac{x^3}{3}\right) + C.$$

EXAMPLE 5

The Minister of Economic Affairs of a country determined that the marginal propensity to consume for his country is given by

$$\frac{dC}{dI} = \frac{3}{4} - \frac{1}{2\sqrt{3I}},$$

where consumption C is a function of national income I, written $C = C(I)$. Here I is expressed in billions of slugs (50 slugs = \$.01). Determine the consumption function for the country if it is known that consumption is 10 billion slugs ($C = 10$) when $I = 12$.

Since the marginal propensity to consume is the derivative of the consumption function C, we have

$$C(I) = \int \left(\frac{3}{4} - \frac{1}{2\sqrt{3I}}\right) dI = \int \frac{3}{4}\, dI - \frac{1}{2}\int (3I)^{-1/2}\, dI$$

$$= \frac{3}{4} I - \frac{1}{2}\int (3I)^{-1/2}\, dI.$$

If we let $u = 3I$, then $du = 3\, dI$ and

$$C(I) = \frac{3}{4} I - \left(\frac{1}{2}\right)\frac{1}{3}\int (3I)^{-1/2}[3\, dI]$$

$$= \frac{3}{4} I - \frac{1}{6}\frac{(3I)^{1/2}}{\frac{1}{2}} + C_1.$$

$$C(I) = \frac{3}{4} I - \frac{\sqrt{3I}}{3} + C_1.$$

When $I = 12$, then $C(I) = 10$ and thus

$$10 = \frac{3}{4}(12) - \frac{\sqrt{3(12)}}{3} + C_1.$$

$$10 = 9 - 2 + C_1.$$

$$C_1 = 3.$$

The consumption function is

$$C(I) = \frac{3}{4} I - \frac{\sqrt{3I}}{3} + 3.$$

Exercise 5-2

In Problems **1–50**, *find the indefinite integrals.*

1. $\int (x^2 + 5)(x - 3)\, dx.$

2. $\int x^4(x^3 + 3x^2 + 7)\, dx.$

3. $\int \sqrt{x}\,(x + 3)\, dx.$

4. $\int (z + 2)^2\, dz.$

5. $\int (2u + 1)^2\, du.$

6. $\int \left(\frac{1}{\sqrt[3]{x}} + 1\right)^2 dx.$

7. $\int v^{-2}(2v^4 + 3v^2 - 2v^{-3})\, dv.$

8. $\int \frac{\sqrt{x}\,(x^5 - \sqrt[3]{x} + 2)}{3}\, dx.$

9. $\int \frac{3x^3 + x^2 - 1}{x^2}\, dx.$

10. $\int \frac{3x^2 - 7x}{4x}\, dx.$

11. $\int (x + 4)^8\, dx.$

12. $\int 2(x + 3)^3\, dx.$

13. $\int 2x(x^2 + 16)^3\, dx.$

14. $\int (3x^2 + 14x)(x^3 + 7x^2 + 1)\, dx.$

15. $\int (3y^2 + 6y)(y^3 + 3y^2 + 1)^{2/3}\, dy.$

16. $\int (-12z^2 - 12z + 1)(-4z^3 - 6z^2 + z)^{18}\, dz.$

17. $\int \frac{3}{(3x - 1)^3}\, dx.$

18. $\int \frac{4x}{(2x^2 - 7)^{10}}\, dx.$

19. $\int \sqrt{x + 10}\, dx.$

20. $\int \frac{1}{\sqrt{x - 2}}\, dx.$

21. $\int (7x - 6)^4\, dx.$

22. $\int x^2(3x^3 + 7)^3\, dx.$

23. $\int x(x^2 + 3)^{12}\, dx.$

24. $\int x\sqrt{1 + 2x^2}\, dx.$

25. $\int x^4(27 + x^5)^{1/3}\, dx.$

26. $\int (3 - 2x)^{10}\, dx.$

27. $\int \frac{6z}{(z^2 - 6)^5}\, dz.$

28. $\int \frac{1}{(8y - 3)^3}\, dy.$

29. $\int \sqrt{5x}\, dx.$

30. $\int \frac{1}{(4x)^7}\, dx.$

31. $\int \frac{x}{\sqrt{x^2 - 4}}\, dx.$

32. $\int (t^2 + 4t)(t^3 + 6t^2)^6\, dt.$

33. $\int (x + 1)(3 - 3x^2 - 6x)^3\, dx.$

34. $\int \sqrt{4x - 3}\, dx.$

35. $\int 4\sqrt[3]{y + 1}\, dy.$

36. $\int \frac{x^2}{\sqrt[3]{2x^3 + 9}}\, dx.$

37. $\int x\sqrt{(7 - 5x^2)^3}\, dx.$

38. $\int -(x^2 - 2x^5)(x^3 - x^6)^{-10}\, dx.$

39. $\int \frac{dx}{\sqrt{2x}}.$

40. $\int (2x^3 + x)(x^4 + x^2)\, dx.$

41. $\int (x^2 + 1)^2 \, dx.$

42. $\int \frac{18 + 12x}{(4 - 9x - 3x^2)^5} \, dx.$

43. $\int (r^3 + 5)^2 \, dr.$

44. $\int (w^3 - 8w^7 + 1)(w^4 - 4w^8 + 4w)^{-6} \, dw.$

45. $\int \frac{8x^3 - 6x - x^4}{3x^3} \, dx.$

46. $\int \frac{2x^4 - 8x^3 - 6x + 4}{x^3} \, dx.$

47. $\int \frac{(\sqrt{x} + 2)^2}{3\sqrt{x}} \, dx.$

48. $\int \frac{4}{\sqrt{2 - 3x}} \, dx.$

49. $\int \sqrt{t}\,(5 - t\sqrt{t}\,)^4 \, dt.$

50. $\int \sqrt{x}\,\sqrt{(8x)^{3/2} + 3} \, dx.$

In Problems **51–54**, *find y subject to the given conditions.*

51. $D_x y = (3 - 2x)^2; \quad y(0) = 1.$

52. $D_x y = x/(x^2 + 4)^2; \quad y(1) = 0.$

53. $y'' = 1/x^3; \quad y'(-1) = 1, y(1) = 0.$

54. $y'' = \sqrt{x + 2}\,; \quad y'(2) = \frac{1}{3}, y(2) = -\frac{7}{15}.$

In Problems **55** *and* **56**, *dr/dq is a marginal revenue function. Find the demand function.*

55. $\frac{dr}{dq} = \frac{200}{(q + 2)^2}.$

56. $\frac{dr}{dq} = \frac{900}{(2q + 3)^3}.$

In Problems **57–59**, *dC/dI represents marginal propensity to consume. Find the consumption function subject to the given condition.*

57. $\frac{dC}{dI} = \frac{1}{\sqrt{I}}; \quad C(9) = 8.$

58. $\frac{dC}{dI} = \frac{3}{4} - \frac{1}{2\sqrt{3I}}; C(3) = \frac{11}{4}.$

59. $\frac{dC}{dI} = \frac{3}{4} - \frac{1}{6\sqrt{I}}; \quad C(25) = 23.$

5-3 SUMMATION

To prepare you for further applications of integration, we need to discuss certain sums.

Consider finding the sum S of the first n positive integers:

$$S = 1 + 2 + \ldots + (n - 1) + n. \tag{1}$$

Writing the right side of Eq. (1) in reverse order, we have

$$S = n + (n - 1) + \ldots + 2 + 1. \tag{2}$$

Adding the corresponding sides of Eqs. (1) and (2) gives

$$\begin{array}{rccccccccc} S = & 1 & + & 2 & + \ldots + & (n-1) & + & n \\ S = & n & + & (n-1) & + \ldots + & 2 & + & 1 \\ \hline 2S = & (n+1) & + & (n+1) & + \ldots + & (n+1) & + & (n+1). \end{array}$$

On the right side of the last equation the term $(n + 1)$ occurs n times. Thus $2S = n(n + 1)$, and so

$$S = \frac{n(n+1)}{2} \qquad \left[\text{the sum of the first } n \text{ positive integers}\right]. \tag{3}$$

For example, the sum of the first 100 positive integers corresponds to $n = 100$ and is $100(100 + 1)/2$ or 5050.

For convenience, to indicate a sum we shall introduce *sigma notation*, so named because the Greek letter Σ (sigma) is used. For example,

$$\sum_{k=1}^{3} (2k + 5)$$

denotes the sum of those numbers obtained from the expression $(2k + 5)$ by first replacing k by 1, then by 2, and finally by 3. Thus

$$\begin{aligned} \sum_{k=1}^{3} (2k + 5) &= [2(1) + 5] + [2(2) + 5] + [2(3) + 5] \\ &= 7 + 9 + 11 = 27. \end{aligned}$$

The letter k is called the *index of summation*; the numbers 1 and 3 are the *limits of summation* (1 is the *lower limit* and 3 is the *upper limit*). The symbol used for the index is a "dummy" symbol in the sense that it does not affect the sum of the terms. Any other letter can be used. For example,

$$\sum_{j=1}^{3} (2j + 5) = 7 + 9 + 11 = \sum_{k=1}^{3} (2k + 5).$$

EXAMPLE 1

Evaluate each of the following.

a. $\displaystyle\sum_{k=4}^{7} \frac{k^2 + 3}{2}$.

Here the sum begins with $k = 4$.

$$\sum_{k=4}^{7} \frac{k^2 + 3}{2} = \frac{4^2 + 3}{2} + \frac{5^2 + 3}{2} + \frac{6^2 + 3}{2} + \frac{7^2 + 3}{2}$$

$$= \frac{19}{2} + \frac{28}{2} + \frac{39}{2} + \frac{52}{2} = 69.$$

b. $\displaystyle\sum_{j=0}^{2} (-1)^{j+1}(j - 1)^2$.

$$\sum_{j=0}^{2} (-1)^{j+1}(j - 1)^2$$

$$= (-1)^{0+1}(0 - 1)^2 + (-1)^{1+1}(1 - 1)^2 + (-1)^{2+1}(2 - 1)^2$$

$$= (-1) + 0 + (-1) = -2.$$

To express the sum of the first n positive integers in sigma notation, we can write

$$\sum_{k=1}^{n} k = 1 + 2 + \ldots + n.$$

By Eq. (3),

$$\boxed{\sum_{k=1}^{n} k = \frac{n(n + 1)}{2}.} \tag{4}$$

Note in (4) that $\displaystyle\sum_{k=1}^{n} k$ is a function of n alone, not of k.

EXAMPLE 2

Evaluate each of the following.

a. $\sum_{k=1}^{60} k$.

Here we must find the sum of the first sixty positive integers. By Eq. (4) with $n = 60$,

$$\sum_{k=1}^{60} k = \frac{60(60+1)}{2} = 1830.$$

b. $\sum_{k=1}^{n-1} k$.

Here we must add the first $n-1$ positive integers. Replacing n by $n-1$ in Eq. (4), we obtain

$$\sum_{k=1}^{n-1} k = \frac{(n-1)[(n-1)+1]}{2} = \frac{(n-1)n}{2}.$$

Another useful formula is that for the sum of the *squares* of the first n positive integers. We shall use it in the next section.

$$\sum_{k=1}^{n} k^2 = \frac{n(n+1)(2n+1)}{6}. \tag{5}$$

EXAMPLE 3

Evaluate $1 + 4 + 9 + 16 + 25 + 36$.

This sum can be written as $\sum_{k=1}^{6} k^2$. By Eq. (5) with $n = 6$,

$$\sum_{k=1}^{6} k^2 = \frac{6(6+1)[2(6)+1]}{6} = 91.$$

We conclude with a property of sigma. If $x_1, x_2, \ldots, x_n$ are real numbers and c is a constant, then

$$\sum_{i=1}^{n} cx_i = cx_1 + cx_2 + \ldots + cx_n$$

$$= c(x_1 + x_2 + \ldots + x_n) = c\sum_{i=1}^{n} x_i.$$

Thus,

$$\sum_{i=1}^{n} cx_i = c \sum_{i=1}^{n} x_i.$$

This means that a constant factor can "jump" before sigma. For example,

$$\sum_{i=1}^{5} 3i^2 = 3 \sum_{i=1}^{5} i^2.$$

By Eq. (5) we have

$$\sum_{i=1}^{5} 3i^2 = 3 \sum_{i=1}^{5} i^2 = 3\left[\frac{5(6)(11)}{6}\right] = 165.$$

PITFALL. *Although constant factors can "jump" before sigma, nothing else can.*

Exercise 5-3

In Problems **1–10**, *evaluate the given sum.*

1. $\sum_{k=1}^{5} (k + 4)$.

2. $\sum_{k=12}^{15} (5 - 2k)$.

3. $\sum_{j=1}^{10} (-1)^j$.

4. $\sum_{j=0}^{5} 2^j$.

5. $\sum_{n=2}^{3} (3n^2 - 7)$.

6. $\sum_{n=2}^{4} \frac{n+1}{n-1}$.

7. $\sum_{k=3}^{4} \frac{(-1)^k(k+1)}{2^k}$.

8. $\sum_{n=1}^{5} 1$.

9. $\sum_{k=1}^{3} \frac{(-1)^{k-1}(1-k^2)}{k}$.

10. $\sum_{n=1}^{4} (n^2 + n)$.

In Problems **11–16**, *express the given sums in sigma notation.*

11. $1 + 2 + 3 + \ldots + 15$.

12. $7 + 8 + 9 + 10$.

13. $1 + 3 + 5 + 7$.

14. $2 + 4 + 6 + 8$.

15. $1^2 + 2^2 + 3^2 + \ldots + 12^2$.

16. $3 + 6 + 9 + 12$.

In Problems **17–22**, *by using Eqs.* (4) *and* (5) *evaluate the sums.*

17. $\sum_{k=1}^{450} k$.

18. $\sum_{k=1}^{10} k^2$.

19. $\sum_{j=1}^{6} 4j$.

20. $\sum_{i=1}^{40} \frac{i}{2}$.

21. $\sum_{i=1}^{6} 3i^2$.

22. $\sum_{j=1}^{8} \left(\frac{j}{2}\right)^2$.

23. A company has an asset whose original value is \$3200 and which has no salvage value. The maintenance cost each year is \$100 and increases by \$100 each year. Show that the average annual total cost C over a period of n years is

$$C = \frac{3200}{n} + 50(n + 1).$$

Find the value of n that minimizes C. What is the average annual cost at this value of n?

5-4 THE DEFINITE INTEGRAL

Figure 5-1 shows the region bounded by the lines $y = f(x) = 2x$, $y = 0$ (the x-axis), and $x = 1$. It is simply a right triangle. If b and h are the length of the base and the height, respectively, then from geometry the area A of the triangle is $A = \frac{1}{2}bh = \frac{1}{2}(1)(2) = 1$ square unit. We shall now find this area by another method which, as you will see later, applies to more complex regions. This method involves summation of areas of rectangles.

Let us divide the interval $[0, 1]$ on the x-axis into four subintervals of equal length by means of the equally spaced points $x_0 = 0$, $x_1 = \frac{1}{4}$, $x_2 = \frac{2}{4}$, $x_3 = \frac{3}{4}$, and $x_4 = \frac{4}{4} = 1$ (see Fig. 5-2). Each subinterval has length $\Delta x = \frac{1}{4}$. These subintervals determine four subregions: R_1, R_2, R_3, and R_4, as indicated.

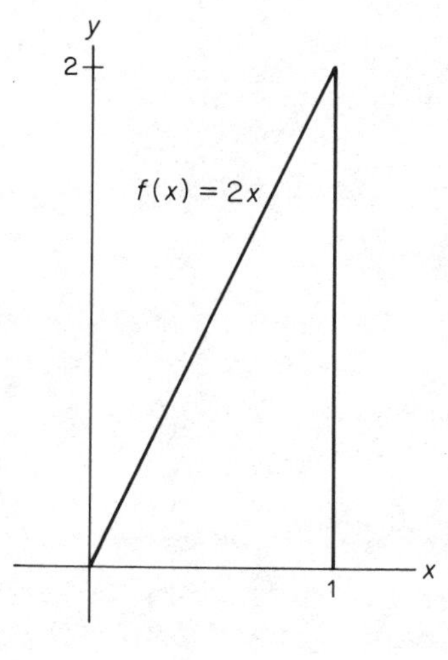

FIG. 5-1

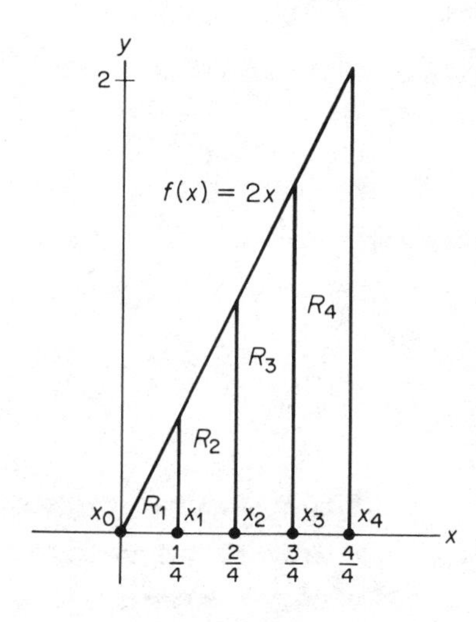

FIG. 5-2

With each subregion we can associate a *circumscribed* rectangle (Fig. 5-3), that is, a rectangle whose base is the corresponding subinterval and whose height is the *maximum* value of $f(x)$ on that subinterval. Since f is

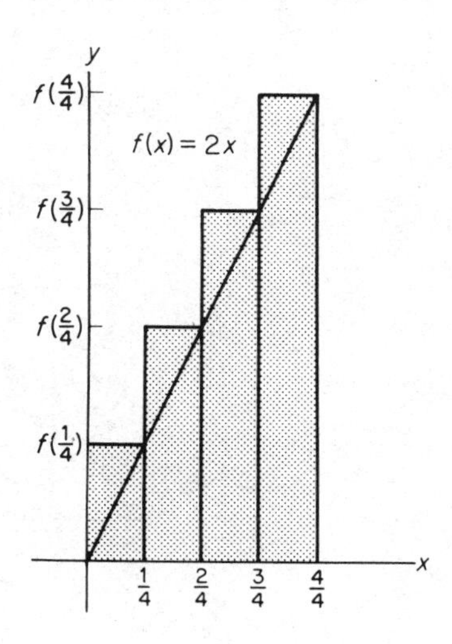

FIG. 5-3

an increasing function, the maximum value of $f(x)$ on each subinterval occurs when x is the right-hand endpoint. Thus the areas of the circumscribed rectangles associated with regions R_1, R_2, R_3, and R_4 are

$\frac{1}{4}f(\frac{1}{4})$, $\frac{1}{4}f(\frac{2}{4})$, $\frac{1}{4}f(\frac{3}{4})$, and $\frac{1}{4}f(\frac{4}{4})$, respectively. The area of each rectangle is an approximation to the area of its corresponding subregion. Thus the sum of the areas of these rectangles, denoted by $\bar{S}_4$, (read "S sub 4 upper bar" or "the fourth upper sum"), approximates the area A of the triangle.

$$\bar{S}_4 = \tfrac{1}{4}f(\tfrac{1}{4}) + \tfrac{1}{4}f(\tfrac{2}{4}) + \tfrac{1}{4}f(\tfrac{3}{4}) + \tfrac{1}{4}f(\tfrac{4}{4})$$
$$= \tfrac{1}{4}\left[2(\tfrac{1}{4}) + 2(\tfrac{2}{4}) + 2(\tfrac{3}{4}) + 2(\tfrac{4}{4})\right] = \tfrac{5}{4}.$$

You may verify that we can write $\bar{S}_4$ as $\bar{S}_4 = \sum_{i=1}^{4} f(x_i)\,\Delta x$. The fact that $\bar{S}_4$ is greater than the actual area of the triangle might have been expected, since $\bar{S}_4$ includes areas of shaded regions that are not in the triangle (see Fig. 5-3).

On the other hand, with each subregion we can also associate an *inscribed* rectangle (see Fig. 5-4), that is, a rectangle whose base is the corresponding subinterval but whose height is the *minimum* value of $f(x)$

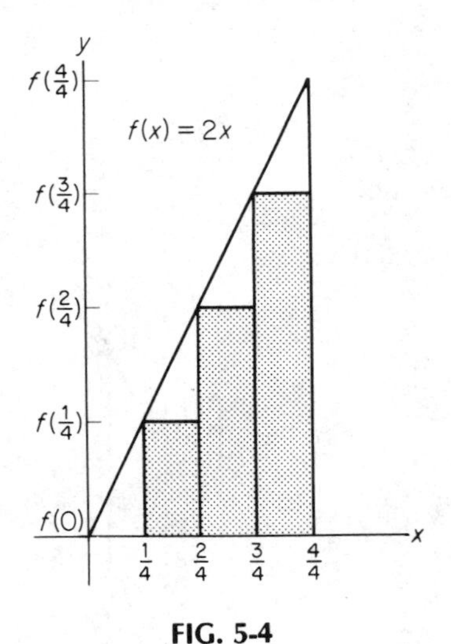

FIG. 5-4

on that subinterval. Since f is an increasing function, the minimum value of $f(x)$ on each subinterval will occur when x is the left-hand endpoint. Thus the areas of the four inscribed rectangles associated with R_1, R_2, R_3, and R_4 are $\frac{1}{4}f(0)$, $\frac{1}{4}f(\frac{1}{4})$, $\frac{1}{4}f(\frac{2}{4})$, and $\frac{1}{4}f(\frac{3}{4})$, respectively. Their sum, denoted $\underline{S}_4$ (read "S sub 4 lower bar" or "the fourth lower sum"), is also an approximation to the area A of the triangle.

$$\underline{S}_4 = \tfrac{1}{4}f(0) + \tfrac{1}{4}f\left(\tfrac{1}{4}\right) + \tfrac{1}{4}f\left(\tfrac{2}{4}\right) + \tfrac{1}{4}f\left(\tfrac{3}{4}\right)$$

$$= \tfrac{1}{4}\left[2(0) + 2\left(\tfrac{1}{4}\right) + 2\left(\tfrac{2}{4}\right) + 2\left(\tfrac{3}{4}\right)\right] = \tfrac{3}{4}.$$

Using sigma notation, we can write $\underline{S}_4$ as $\underline{S}_4 = \sum_{i=1}^{4} f(x_{i-1})\,\Delta x$. Note that $\underline{S}_4$ is less than the area of the triangle because the rectangles do not account for that portion of the triangle which is not shaded in Fig. 5-4.

Since $\frac{3}{4} = \underline{S}_4 \leqslant A \leqslant \bar{S}_4 = \frac{5}{4}$, we say that $\underline{S}_4$ is an approximation to A from *below* and $\bar{S}_4$ is an approximation to A from *above*.

If $[0, 1]$ is divided into more subintervals, we expect that better approximations to A will occur. To test this out, let us use six subintervals of equal length $\Delta x = \frac{1}{6}$. Then $\bar{S}_6$, the total area of six circumscribed

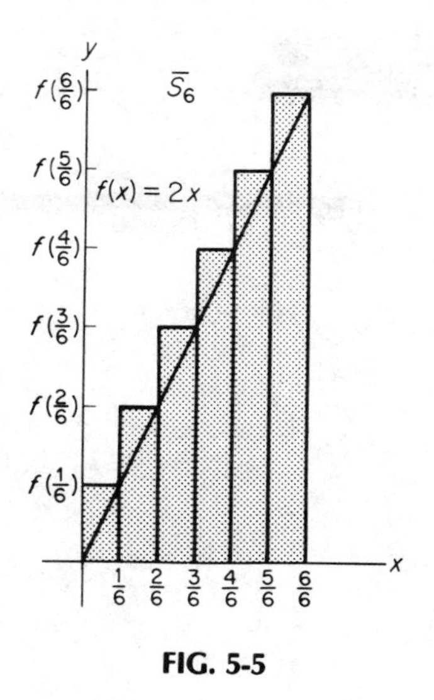

FIG. 5-5

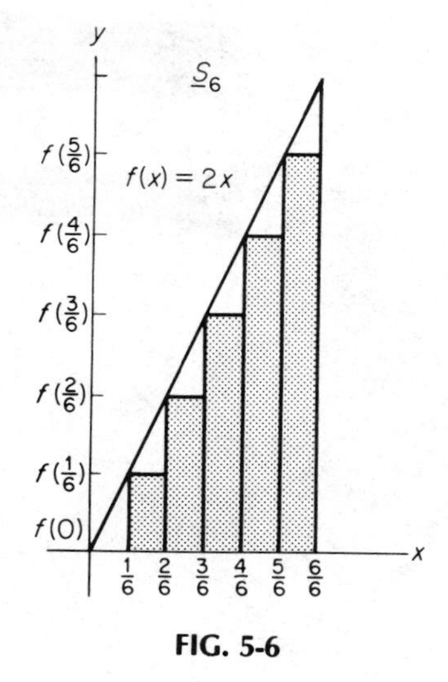

FIG. 5-6

rectangles (see Fig. 5-5), and $\underline{S}_6$, the total area of six inscribed rectangles (see Fig. 5-6), are

$$\bar{S}_6 = \tfrac{1}{6}f\left(\tfrac{1}{6}\right) + \tfrac{1}{6}f\left(\tfrac{2}{6}\right) + \tfrac{1}{6}f\left(\tfrac{3}{6}\right) + \tfrac{1}{6}f\left(\tfrac{4}{6}\right) + \tfrac{1}{6}f\left(\tfrac{5}{6}\right) + \tfrac{1}{6}f\left(\tfrac{6}{6}\right)$$

$$= \tfrac{1}{6}\left[2\left(\tfrac{1}{6}\right) + 2\left(\tfrac{2}{6}\right) + 2\left(\tfrac{3}{6}\right) + 2\left(\tfrac{4}{6}\right) + 2\left(\tfrac{5}{6}\right) + 2\left(\tfrac{6}{6}\right)\right] = \tfrac{7}{6}$$

and

$$\underline{S}_6 = \tfrac{1}{6}f(0) + \tfrac{1}{6}f\left(\tfrac{1}{6}\right) + \tfrac{1}{6}f\left(\tfrac{2}{6}\right) + \tfrac{1}{6}f\left(\tfrac{3}{6}\right) + \tfrac{1}{6}f\left(\tfrac{4}{6}\right) + \tfrac{1}{6}f\left(\tfrac{5}{6}\right)$$

$$= \tfrac{1}{6}\left[2(0) + 2\left(\tfrac{1}{6}\right) + 2\left(\tfrac{2}{6}\right) + 2\left(\tfrac{3}{6}\right) + 2\left(\tfrac{4}{6}\right) + 2\left(\tfrac{5}{6}\right)\right] = \tfrac{5}{6}.$$

Note that $\underline{S}_6 \leqslant A \leqslant \bar{S}_6$ and, with appropriate labelling, both $\bar{S}_6$ and $\underline{S}_6$ will be of the *form* $\Sigma f(x)\,\Delta x$. Using six subintervals gave better approximations to the area than did four subintervals, as expected.

More generally, if we divide $[0, 1]$ into n subintervals of equal length Δx, then $\Delta x = 1/n$ and the endpoints of the subintervals are $x = 0, 1/n, 2/n, \ldots, (n-1)/n$, and $n/n = 1$. See Fig. 5-7. The total area of n *circumscribed* rectangles is

$$\bar{S}_n = \frac{1}{n}f\left(\frac{1}{n}\right) + \frac{1}{n}f\left(\frac{2}{n}\right) + \ldots + \frac{1}{n}f\left(\frac{n}{n}\right) \tag{1}$$

$$= \frac{1}{n}\left[2\left(\frac{1}{n}\right) + 2\left(\frac{2}{n}\right) + \ldots + 2\left(\frac{n}{n}\right)\right]$$

$$= \frac{2}{n^2}[1 + 2 + \ldots + n] \qquad \left(\text{by factoring } \frac{2}{n} \text{ from each term}\right).$$

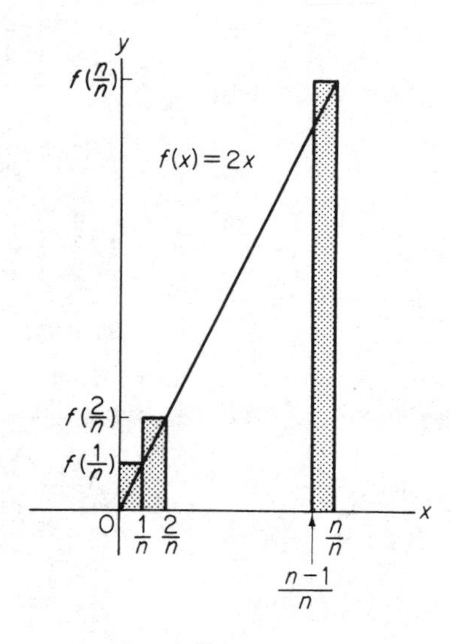

FIG. 5-7

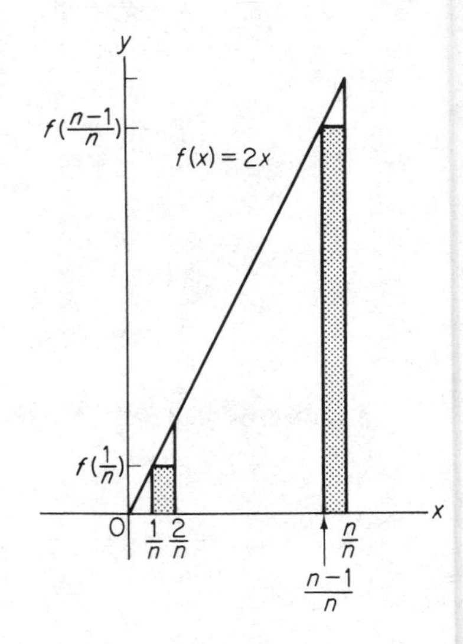

FIG. 5-8

From Sec. 5-3, the sum of the first n positive integers is $\dfrac{n(n+1)}{2}$. Thus

$$\bar{S}_n = \left(\frac{2}{n^2}\right)\frac{n(n+1)}{2} = \frac{n+1}{n}.$$

For n *inscribed* rectangles, the total area determined by the subintervals (see Fig. 5-8) is

$$\begin{aligned} \underline{S}_n &= \frac{1}{n}f(0) + \frac{1}{n}f\left(\frac{1}{n}\right) + \ldots + \frac{1}{n}f\left(\frac{n-1}{n}\right) \qquad (2) \\ &= \frac{1}{n}\left[2(0) + 2\left(\frac{1}{n}\right) + \ldots + 2\left(\frac{n-1}{n}\right)\right] \\ &= \frac{2}{n^2}[1 + \ldots + (n-1)]. \end{aligned}$$

Summing the first $n - 1$ positive integers as we did in Example 2(b) of Sec. 5-3, we obtain

$$\underline{S}_n = \left(\frac{2}{n^2}\right)\frac{(n-1)n}{2} = \frac{n-1}{n}.$$

From Equations (1) and (2) we again see that both $\bar{S}_n$ and $\underline{S}_n$ are sums of the *form* $\Sigma f(x)\,\Delta x$.

From the nature of $\bar{S}_n$ and $\underline{S}_n$, it seems reasonable and it is indeed true that

$$\underline{S}_n \leqslant A \leqslant \bar{S}_n.$$

As n becomes larger, $\underline{S}_n$ and $\bar{S}_n$ become better approximations to A. In fact, let us take the limit of $\underline{S}_n$ and $\bar{S}_n$ as n approaches ∞ through positive integral values.

$$\lim_{n\to\infty} \underline{S}_n = \lim_{n\to\infty} \frac{n-1}{n} = \lim_{n\to\infty}\left(1 - \frac{1}{n}\right) = 1.$$

$$\lim_{n\to\infty} \bar{S}_n = \lim_{n\to\infty} \frac{n+1}{n} = \lim_{n\to\infty}\left(1 + \frac{1}{n}\right) = 1.$$

Since $\bar{S}_n$ and $\underline{S}_n$ have the same common limit, namely

$$\lim_{n\to\infty} \bar{S}_n = \lim_{n\to\infty} \underline{S}_n = 1, \qquad (3)$$

and since

$$\underline{S}_n \leqslant A \leqslant \overline{S}_n,$$

we shall take this limit to be the area of the triangle. Thus $A = 1$ square unit, which agrees with our prior finding.

Mathematically, the sums $\overline{S}_n$ and $\underline{S}_n$, as well as their common limit, have a meaning that is independent of area. With that in mind, we define the common limit of $\overline{S}_n$ and $\underline{S}_n$, namely 1, to be the **definite integral** of $f(x) = 2x$ from $x = 0$ to $x = 1$, and we abbreviate this symbolically by writing

$$\int_0^1 2x\, dx = 1. \tag{4}$$

The reason for using the term "definite integral" and the symbolism in Eq. (4) will become apparent in the next section. The numbers 0 and 1 appearing with the integral sign $\int$ in Eq. (4) are called the ***limits of integration***; 0 is the ***lower limit*** and 1 is the ***upper limit***.

Two points must be made about the definite integral: first, aside from any geometrical interpretation, it is nothing more than a real number; second, the definite integral is a limit of a sum of the form $\Sigma f(x)\,\Delta x$. In fact, one can think of the integral sign as an elongated "S", the first letter of "Summation."

In general, the definite integral of a function f over the interval from $x = a$ to $x = b$, where $a \leqslant b$, is the common limit of $\overline{S}_n$ and $\underline{S}_n$, if it exists, and is written

$$\int_a^b f(x)\, dx.$$

The symbol x is the ***variable of integration*** and $f(x)$ is the ***integrand***. In terms of a limiting process we have

$$\Sigma f(x)\,\Delta x \to \int_a^b f(x)\, dx.$$

As you saw in (3), $\lim_{n\to\infty} \underline{S}_n$ is equal to $\lim_{n\to\infty} \overline{S}_n$. For an arbitrary function this is not always true. However, for the functions that we shall consider, these limits will be equal and the definite integral will always exist. To save time we shall just use the **right-hand endpoint** of each subinterval in computing a sum. For our functions this will correspond to either $\underline{S}_n$ or $\overline{S}_n$ and we shall denote this sum by S_n.

EXAMPLE 1

a. *Find the area of the region in the first quadrant bounded by the curve* $y = f(x) = 4 - x^2$ *and the lines* $x = 0$ *and* $y = 0$.

A sketch of the region appears in Fig. 5-9. The interval over which x varies in this region is seen to be [0, 2] which we divide into n subintervals of equal length Δx. Since the length of [0, 2] is 2, we take $\Delta x = 2/n$. The endpoints of the subintervals are $x = 0, 2/n, 2(2/n), \ldots, (n-1)(2/n)$, and $n(2/n) = 2$ (see Fig. 5-10).

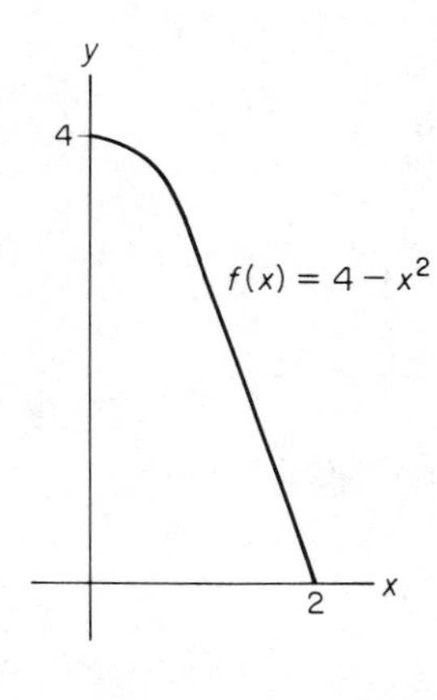

FIG. 5-9

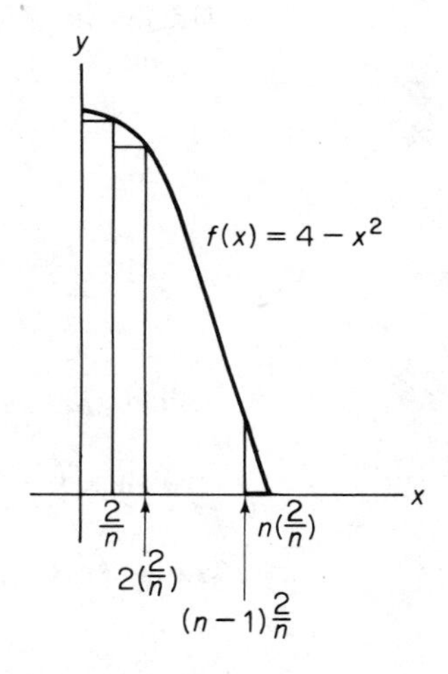

FIG. 5-10

Using the right-hand endpoints, we get

$$\begin{aligned} S_n &= \frac{2}{n} f\left(\frac{2}{n}\right) + \frac{2}{n} f\left[2\left(\frac{2}{n}\right)\right] + \ldots + \frac{2}{n} f\left[n\left(\frac{2}{n}\right)\right] \\ &= \frac{2}{n}\left[f\left(\frac{2}{n}\right) + f\left[2\left(\frac{2}{n}\right)\right] + \ldots + f\left[n\left(\frac{2}{n}\right)\right]\right] \\ &= \frac{2}{n}\left[\left\{4 - \left[\frac{2}{n}\right]^2\right\} + \left\{4 - \left[2\left(\frac{2}{n}\right)\right]^2\right\} + \ldots + \left\{4 - \left[n\left(\frac{2}{n}\right)\right]^2\right\}\right]. \end{aligned}$$

Since the number 4 occurs n times in the sum, we can simplify S_n.

$$\begin{aligned} S_n &= \frac{2}{n}\left[4n - \left(\frac{2}{n}\right)^2 - 2^2\left(\frac{2}{n}\right)^2 - \ldots - n^2\left(\frac{2}{n}\right)^2\right] \\ &= \frac{2}{n}\left[4n - \left(\frac{2}{n}\right)^2 \{1^2 + 2^2 + \ldots + n^2\}\right]. \end{aligned}$$

From Sec. 5-3, $\sum_{k=1}^{n} k^2 = \frac{n(n+1)(2n+1)}{6}$ and thus

$$
\begin{aligned}
S_n &= \frac{2}{n}\left[4n - \left(\frac{2}{n}\right)^2 \frac{n(n+1)(2n+1)}{6}\right] \\
&= 8 - \frac{4(n+1)(2n+1)}{3n^2} \\
&= 8 - \frac{4}{3}\left(\frac{2n^2+3n+1}{n^2}\right).
\end{aligned}
$$

Finally we take the limit of S_n as $n \to \infty$.

$$
\begin{aligned}
\lim_{n\to\infty} S_n &= \lim_{n\to\infty}\left[8 - \frac{4}{3}\left(\frac{2n^2+3n+1}{n^2}\right)\right] \\
&= \lim_{n\to\infty}\left[8 - \frac{4}{3}\left(2 + \frac{3}{n} + \frac{1}{n^2}\right)\right] \\
&= 8 - \frac{8}{3} = \frac{16}{3}.
\end{aligned}
$$

Hence the area of the region is $\frac{16}{3}$ square units.

b. *Evaluate* $\int_0^2 (4 - x^2)\, dx$.

Since $\int_0^2 (4 - x^2)\, dx = \lim_{n\to\infty} S_n$, from part (a) we conclude that

$$\int_0^2 (4 - x^2)\, dx = \frac{16}{3}.$$

EXAMPLE 2

Integrate $f(x) = x - 5$ *from* $x = 0$ *to* $x = 3$; *that is, evaluate* $\int_0^3 (x - 5)\, dx$.

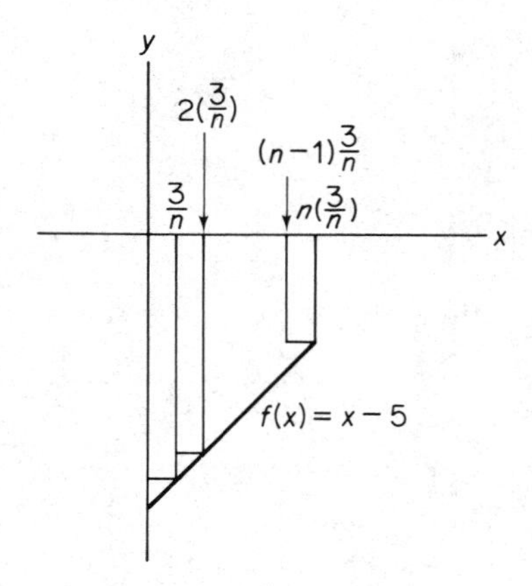

FIG. 5-11

A sketch of $f(x) = x - 5$ over $[0, 3]$ appears in Fig. 5-11. We divide $[0, 3]$ into n subintervals that are of equal length $\Delta x = 3/n$. The endpoints are $x = 0, 3/n, 2(3/n), \ldots, (n-1)(3/n)$, and $n(3/n) = 3$. Note that $f(x)$ is negative at each endpoint. We form the sum

$$S_n = \frac{3}{n} f\left(\frac{3}{n}\right) + \frac{3}{n} f\left[2\left(\frac{3}{n}\right)\right] + \ldots + \frac{3}{n} f\left[n\left(\frac{3}{n}\right)\right].$$

Since all terms are negative, they do *not* represent areas of rectangles; in fact, they are the negatives of areas of rectangles. Simplifying, we have

$$\begin{aligned} S_n &= \frac{3}{n}\left[\left\{\frac{3}{n} - 5\right\} + \left\{2\left(\frac{3}{n}\right) - 5\right\} + \ldots + \left\{n\left(\frac{3}{n}\right) - 5\right\}\right] \\ &= \frac{3}{n}\left[-5n + \frac{3}{n}\{1 + 2 + \ldots + n\}\right] \\ &= \frac{3}{n}\left[-5n + \left(\frac{3}{n}\right)\frac{n(n+1)}{2}\right] \\ &= -15 + \frac{9}{2} \cdot \frac{n+1}{n} \\ &= -15 + \frac{9}{2}\left(1 + \frac{1}{n}\right). \end{aligned}$$

Taking the limit, we obtain

$$\lim_{n\to\infty} S_n = \lim_{n\to\infty}\left[-15 + \frac{9}{2}\left(1 + \frac{1}{n}\right)\right] = -15 + \frac{9}{2} = -\frac{21}{2}.$$

Thus

$$\int_0^3 (x - 5)\, dx = -\frac{21}{2}.$$

This definite integral is **not** the area of the region bounded by $f(x) = x - 5$, $y = 0$, $x = 0$, and $x = 3$. It represents the negative of that area.

In Example 2 it was shown that *the definite integral does not have to represent area*. In fact, there the definite integral was negative. However, if f is continuous and $f(x) \geqslant 0$ on $[a, b]$, then $S_n \geqslant 0$ for all n. Hence $\lim_{n\to\infty} S_n \geqslant 0$ and so $\int_a^b f(x)\, dx \geqslant 0$. Furthermore, this definite integral gives the area of the region bounded by $y = f(x)$, $y = 0$, $x = a$ and $x = b$ (see Fig. 5-12).

Although the approach that we took to discuss the definite integral is sufficient for our purposes, it is by no means rigorous. **The important thing to remember about the definite integral is that it is the limit of a sum.**

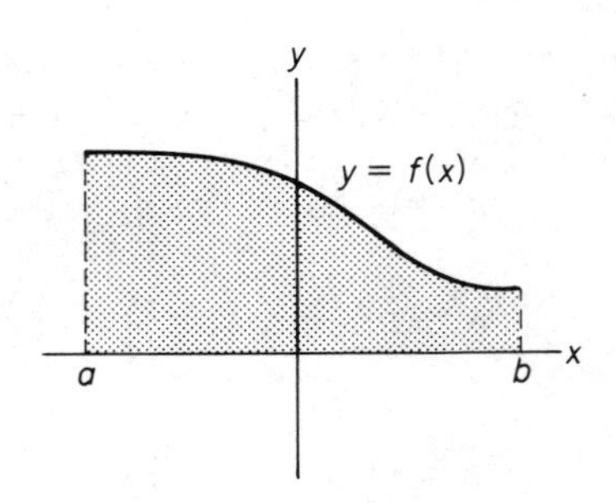

FIG. 5-12

Exercise 5-4

In Problems **1–4**, *sketch the region in the first quadrant which is bounded by the given curves. Approximate the area of the region by the indicated sum. Use the right-hand endpoint of each subinterval.*

1. $f(x) = x, y = 0, x = 1; S_3$.

2. $f(x) = 3x, y = 0, x = 1; S_5$.

3. $f(x) = x^2, y = 0, x = 1; S_3$.

4. $f(x) = x^2 + 1, y = 0, x = 0, x = 1; S_2$.

In Problems **5–10**, *sketch the region in the first quadrant which is bounded by the given curves. Determine the exact area of the region by considering the limit of S_n as $n \to \infty$. Use the right-hand endpoint of each subinterval.*

5. Region as described in Problem 1.

6. Region as described in Problem 2.

7. Region as described in Problem 3.

8. Region as described in Problem 4.

9. $f(x) = 2x^2, y = 0, x = 2$.

10. $f(x) = 9 - x^2, y = 0, x = 0$.

For each of the following problems, evaluate the given definite integral by taking the limit of S_n. Use the right-hand endpoint of each subinterval. Sketch the graph, over the given interval, of the function to be integrated.

11. $\int_0^4 9\,dx$.

12. $\int_0^2 3x\,dx$.

13. $\int_0^3 -4x\,dx$.

14. $\int_0^3 (2x - 9)\,dx$.

15. $\int_0^1 (x^2 + x)\,dx$.

5-5 THE FUNDAMENTAL THEOREM OF INTEGRAL CALCULUS

Until now the limiting processes of both the derivative and the definite integral have been considered separately. We shall now bring these fundamental ideas together and develop the important relationship that exists between them. As a result, we may evaluate definite integrals more efficiently.

The graph of a function f is given in Fig. 5-13. Assume that f is continuous on the interval $[a, b]$ and its graph does not fall below the x-axis. That is, $f(x) \geqslant 0$. From the last section we know that the area of

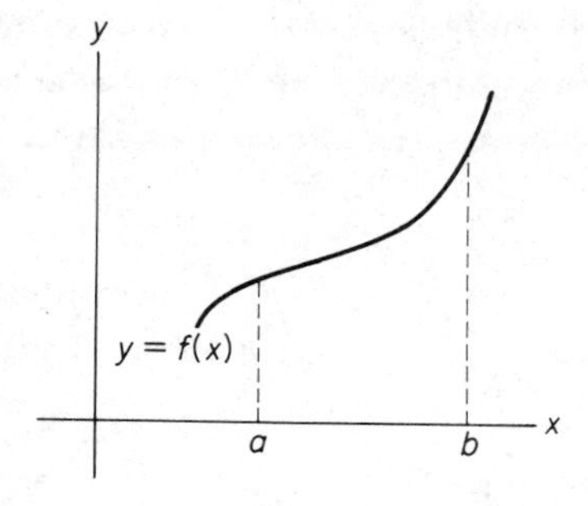

FIG. 5-13

the region below the graph and above the x-axis from $x = a$ to $x = b$ is given by $\int_a^b f(x)\,dx$. We shall now consider another way to determine this area.

Suppose there is a function $A = A(x)$, which we shall refer to as an "area" function, that gives the area of the region below the graph of f and above the x-axis from a to x, where $a \leqslant x \leqslant b$. This region is shaded in Fig. 5-14.

From its definition we can state two properties of A immediately:

(1) $A(a) = 0$ since there is no area from a to a;

(2) $A(b)$ is the area from a to b; that is,

$$A(b) = \int_a^b f(x)\,dx.$$

If x is increased by h units, then $A(x + h)$ is the area of the shaded region in Fig. 5-15. Hence $A(x + h) - A(x)$ will be the difference of the

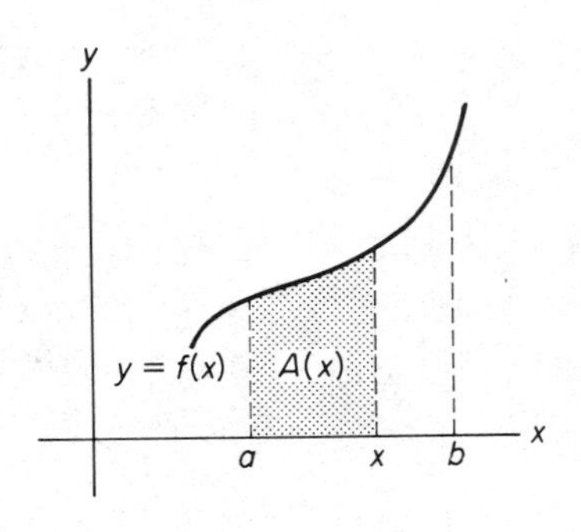

FIG. 5-14

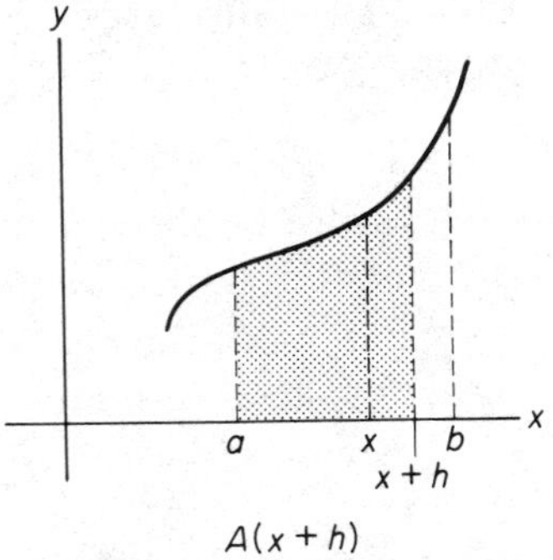

A(x + h)

FIG. 5-15

areas in Figs. 5-15 and 5-14; namely, the area of the shaded region in Fig. 5-16. For h sufficiently close to zero, the area of this region is the same as the area of a rectangle (Fig. 5-17) whose base is h and whose

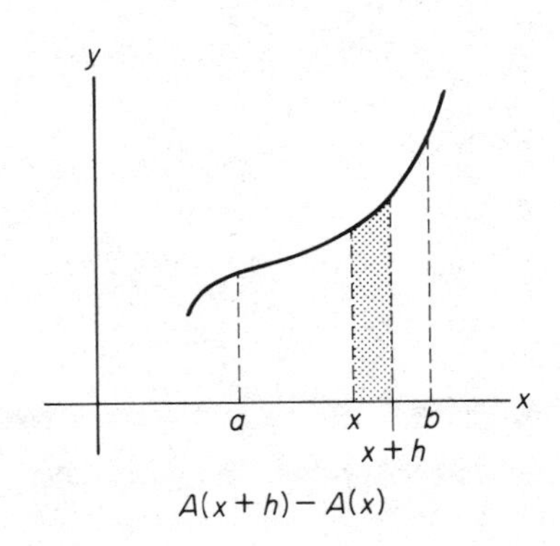

FIG. 5-16

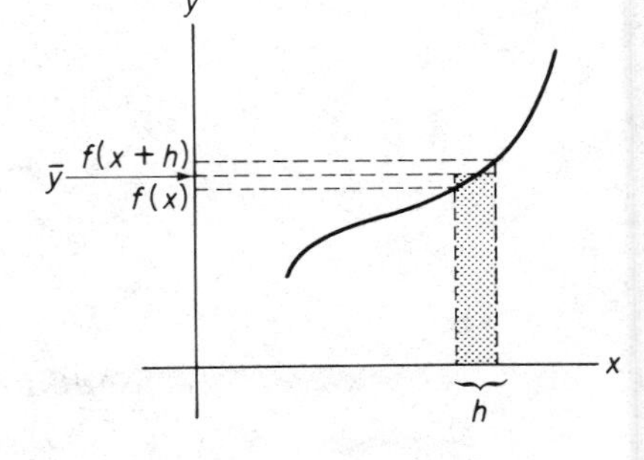

FIG. 5-17

height is some value $\bar{y}$ between $f(x)$ and $f(x + h)$. Here $\bar{y}$ is a function of h. Thus the area of the rectangle is, on the one hand, $A(x + h) - A(x)$, and on the other hand it is $h\bar{y}$:

$$A(x + h) - A(x) = h\bar{y},$$

or

$$\frac{A(x + h) - A(x)}{h} = \bar{y} \qquad \text{[dividing by } h\text{]}.$$

As $h \to 0$, then $\bar{y}$ approaches the number $f(x)$ and so

$$\lim_{h \to 0} \frac{A(x + h) - A(x)}{h} = f(x). \tag{1}$$

But the left side is merely the derivative of A. Thus Eq. (1) becomes

$$A'(x) = f(x).$$

We conclude that the area function A has the additional property that its derivative A' is f. That is, A is an antiderivative of f. Now, suppose that F is *any* antiderivative of f. Since both A and F are antiderivatives of the same function, they must differ by a constant C:

$$A(x) = F(x) + C. \qquad (2)$$

Recall that $A(a) = 0$. Evaluating both sides of Eq. (2) when $x = a$ gives

$$0 = F(a) + C$$

or

$$C = -F(a).$$

Thus Eq. (2) becomes

$$A(x) = F(x) - F(a). \qquad (3)$$

If $x = b$, then from Eq. (3)

$$A(b) = F(b) - F(a). \qquad (4)$$

But recall that

$$A(b) = \int_a^b f(x)\,dx. \qquad (5)$$

From Eqs. (4) and (5) we get

$$\int_a^b f(x)\,dx = F(b) - F(a).$$

Thus a relationship between a definite integral and antidifferentiation has become clear. To find $\int_a^b f(x)\,dx$ it suffices to find an antiderivative of f, say F, and subtract its value at the lower limit a from its value at the upper limit b. We assumed here that f was continuous and $f(x) \geqslant 0$ so that we could appeal to the "area" concept. However, our result is true for any continuous function† and is known as the *fundamental theorem of integral calculus*.

†If f is continuous on $[a, b]$, it can be shown that $\int_a^b f(x)\,dx$ does indeed exist.

FUNDAMENTAL THEOREM OF INTEGRAL CALCULUS

If f is continuous on the interval $[a, b]$ and F is any antiderivative of f there, then

$$\int_a^b f(x)\,dx = F(b) - F(a).$$

It is crucial that you understand the difference between a definite integral and an indefinite integral. The **definite integral** $\int_a^b f(x)\,dx$ **is a number** defined to be the limit of a sum. The Fundamental Theorem says that the **indefinite integral** $\int f(x)\,dx$ (an antiderivative of f), which **is a function** of x and is related to the differentiation process, can be used to determine this limit.

Suppose we apply the Fundamental Theorem to evaluate the definite integral $\int_0^2 (4 - x^2)\,dx$. Here $f(x) = 4 - x^2$, $a = 0$, and $b = 2$. Since an antiderivative of $4 - x^2$ is $F(x) = 4x - (x^3/3)$, then

$$\int_0^2 (4 - x^2)\,dx = F(2) - F(0) = \left(8 - \frac{8}{3}\right) - (0) = \frac{16}{3}.$$

This confirms our result in Example 1(b) of Sec. 5-4. If we had chosen $F(x)$ to be $4x - (x^3/3) + C$, then $F(2) - F(0) = [(8 - \frac{8}{3}) + C] - [0 + C] = \frac{16}{3}$ as before. Since the choice of the value of C is immaterial, for convenience we shall always choose it to be 0, as originally done. Usually, $F(b) - F(a)$ is abbreviated by writing

$$F(x)\Big|_a^b.$$

Hence we have

$$\int_0^2 (4 - x^2)\,dx = \left(4x - \frac{x^3}{3}\right)\Bigg|_0^2 = \left(8 - \frac{8}{3}\right) - 0 = \frac{16}{3}.$$

For a definite integral, we have the following convention:

$$\int_b^a f(x)\,dx = -\int_a^b f(x)\,dx.$$

That is, interchanging the limits of integration changes the integral's sign. For example,

$$\int_2^0 (4 - x^2)\, dx = -\int_0^2 (4 - x^2)\, dx.$$

Two properties of the definite integral deserve mention:

(1) $\int_a^b f(x)\, dx = \int_a^b f(t)\, dt$. The variable of integration x used in $\int_a^b f(x)\, dx$ is a "dummy variable" in the sense that any other variable would produce the same result, that is, the same number. You may verify, for example, that $\int_0^2 x^2\, dx = \int_0^2 t^2\, dt$.

(2) If f is continuous on an interval I and a, b, and c are in I, then

$$\int_a^c f(x)\, dx = \int_a^b f(x)\, dx + \int_b^c f(x)\, dx.$$

This means that you may subdivide the interval over which a definite integral is to be evaluated. Thus,

$$\int_0^2 (4 - x^2)\, dx = \int_0^1 (4 - x^2)\, dx + \int_1^2 (4 - x^2)\, dx.$$

We shall look at some examples of definite integration now and compute some areas in the next section.

EXAMPLE 1

Evaluate each of the following definite integrals.

a. $\int_{-1}^3 (3x^2 - x + 6)\, dx$.

$$\int_{-1}^3 (3x^2 - x + 6)\, dx = \left(x^3 - \frac{x^2}{2} + 6x\right)\Big|_{-1}^3$$

$$= \left[3^3 - \frac{3^2}{2} + 6(3)\right] - \left[(-1)^3 - \frac{(-1)^2}{2} + 6(-1)\right]$$

$$= \left(\frac{81}{2}\right) - \left(-\frac{15}{2}\right) = 48.$$

b. $\int_0^1 \frac{x^3}{\sqrt{1+x^4}}\,dx.$

$$\int_0^1 \frac{x^3}{\sqrt{1+x^4}}\,dx = \int_0^1 x^3(1+x^4)^{-1/2}\,dx$$

$$= \frac{1}{4}\int_0^1 (1+x^4)^{-1/2}[4x^3\,dx] = \left(\frac{1}{4}\right)\frac{(1+x^4)^{1/2}}{\frac{1}{2}}\Bigg|_0^1$$

$$= \frac{1}{2}(1+x^4)^{1/2}\Big|_0^1 = \frac{1}{2}(2)^{1/2} - \frac{1}{2}(1)^{1/2}$$

$$= \frac{1}{2}(\sqrt{2}-1).$$

PITFALL. *In part (b), the result of evaluating the antiderivative* $\frac{1}{2}(1+x^4)^{1/2}$ *at the lower limit* 0 *was* $\frac{1}{2}(1)^{1/2}$. **Do not** *assume that an evaluation at the limit zero will yield* 0.

c. $\int_1^2 [4t^{1/3} + t(t^2+1)^3]\,dt.$

$$\int_1^2 \left[4t^{1/3} + t(t^2+1)^3\right] dt = 4\int_1^2 t^{1/3}\,dt + \frac{1}{2}\int_1^2 (t^2+1)^3[2t\,dt]$$

$$= (4)\frac{t^{4/3}}{\frac{4}{3}}\Bigg|_1^2 + \left(\frac{1}{2}\right)\frac{(t^2+1)^4}{4}\Bigg|_1^2$$

$$= 3(2^{4/3}-1) + \frac{1}{8}(5^4 - 2^4)$$

$$= 3\cdot 2^{4/3} - 3 + \frac{609}{8}$$

$$= 6\sqrt[3]{2} + \frac{585}{8}.$$

EXAMPLE 2

Evaluate $\int_{-2}^1 x^3\,dx.$

$$\int_{-2}^1 x^3\,dx = \frac{x^4}{4}\Bigg|_{-2}^1 = \frac{1^4}{4} - \frac{(-2)^4}{4} = \frac{1}{4} - \frac{16}{4} = -\frac{15}{4}.$$

The reason the result is negative is clear from the graph of $y = x^3$ on the interval $[-2, 1]$ (see Fig. 5-18). For $-2 \leqslant x < 0$, $f(x)$ is negative. Since a definite integral is a limit of a sum of the form $\Sigma f(x)\Delta x$, then $\int_{-2}^0 x^3\,dx$ is not only a negative number, but it is also the negative of the area of the shaded

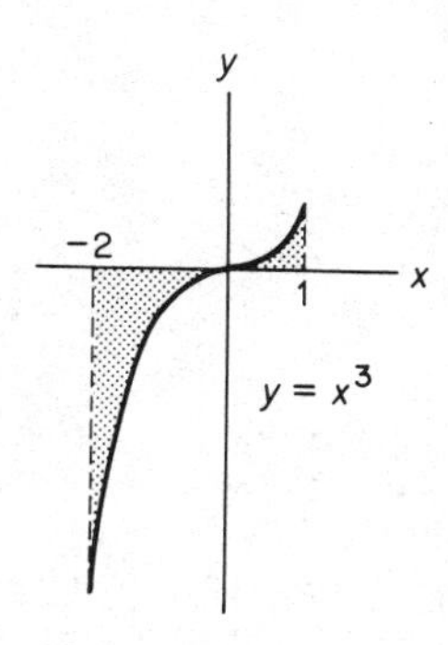

FIG. 5-18

region in the third quadrant. On the other hand, $\int_0^1 x^3\,dx$ is the area of the shaded region in the first quadrant. However, the definite integral over the entire interval $[-2, 1]$ is the *algebraic* sum of these numbers since

$$\int_{-2}^{1} x^3\,dx = \int_{-2}^{0} x^3\,dx + \int_0^1 x^3\,dx.$$

Thus $\int_{-2}^{1} x^3\,dx$ does not represent the area between the curve and the x-axis. However, the area can be given in the form

$$\left|\int_{-2}^{0} x^3\,dx\right| + \int_0^1 x^3\,dx.$$

PITFALL. *Remember that* $\int_a^b f(x)\,dx$ *is a limit of a sum. In some cases this limit represents area. In others it does not. Do not attach units of area to every definite integral.*

Since f is an antiderivative of f', by the Fundamental Theorem we have

$$\int_a^b f'(x)\,dx = f(b) - f(a).$$

But $f'(x)$ is the rate of change of f with respect to x. Thus, if we know the rate of change of f and want to find the difference in functional values $f(b) - f(a)$, it suffices to evaluate $\int_a^b f'(x)\,dx$.

EXAMPLE 3

A manufacturer's marginal cost function is

$$\frac{dc}{dq} = .6q + 2.$$

If production is presently set at $q = 80$ units per week, how much more would it cost to increase production to 100 units per week?

Since $c = c(q)$ is the total cost function, we want to find the difference $c(100) - c(80)$. However, the rate of change of c is dc/dq, and thus

$$\begin{aligned} c(100) - c(80) &= \int_{80}^{100} \frac{dc}{dq}\, dq = \int_{80}^{100} (.6q + 2)\, dq \\ &= \left[\frac{.6q^2}{2} + 2q\right]\Bigg|_{80}^{100} = [.3q^2 + 2q]\Bigg|_{80}^{100} \\ &= [.3(100)^2 + 2(100)] - [.3(80)^2 + 2(80)] \\ &= 3200 - 2080 = 1120. \end{aligned}$$

If c is in dollars, then the cost of increasing production from 80 units to 100 units is \$1120.

Exercise 5-5

In Problems **1–36**, *evaluate the definite integral.*

1. $\int_0^3 4\, dx.$

2. $\int_1^3 (2 + .2)\, dx.$

3. $\int_1^2 3x\, dx.$

4. $\int_0^2 -5x\, dx.$

5. $\int_{-1}^3 \frac{5}{3}x^3\, dx.$

6. $\int_0^{10} .04x^3\, dx.$

7. $\int_{-2}^1 (4x - 6)\, dx.$

8. $\int_{-1}^1 (5y + 2)\, dy.$

9. $\int_0^2 (t^2 + t)\, dt.$

10. $\int_1^3 (2w^2 + 1)\, dw.$

11. $\int_2^3 (y^2 - 2y + 1)\, dy.$

12. $\int_3^2 (2t - t^2)\, dt.$

13. $\int_{-2}^{-1} (3w^2 - w - 1)\, dw.$

14. $\int_8^9 dt.$

15. $\int_1^2 -4t^{-4}\,dt$.

16. $\int_1^2 \frac{x^{-2}}{2}\,dx$.

17. $\int_{-1}^1 \sqrt[3]{x^5}\,dx$.

18. $\int_{1/2}^{3/2} (x^2 + x + 1)\,dx$.

19. $\int_{1/2}^3 \frac{1}{x^2}\,dx$.

20. $\int_4^9 \left(\frac{1}{\sqrt{x}} - 2\right) dx$.

21. $\int_{-1}^1 (z + 1)^5\,dz$.

22. $\int_1^8 (x^{1/3} - x^{-1/3})\,dx$.

23. $\int_0^1 2x^2(x^3 - 1)^3\,dx$.

24. $\int_1^3 (x + 3)^3\,dx$.

25. $\int_1^2 (x + 1)(x^2 + 2x - 5)^3\,dx$.

26. $\int_0^1 (3x^2 + 4x)(x^3 + 2x^2)^4\,dx$.

27. $\int_4^5 \frac{2}{(q - 3)^3}\,dq$.

28. $\int_0^6 \sqrt{2x + 4}\,dx$.

29. $\int_{1/3}^2 \sqrt{10 - 3p}\,dp$.

30. $\int_{-1}^1 q\sqrt{q^2 + 3}\,dq$.

31. $\int_0^1 x^2\sqrt[3]{7x^3 + 1}\,dx$.

32. $\int_0^{\sqrt{7}} \left[2x - \frac{x}{(x^2 + 1)^{5/3}}\right] dx$.

33. $\int_1^2 (x + x^{-2} - x^{-3})\,dx$.

34. $\int_a^b (m + ny)\,dy$.

35. $\int_0^1 \frac{6x^3 + 3x}{(x^2 + x^4 + 1)^2}\,dx$.

36. $\int_1^2 \left(6\sqrt{x} - \frac{1}{\sqrt{2x}}\right) dx$.

37. In a discussion of gene mutation [11] the following integral occurs:

$$\int_0^{10^{-4}} x^{-1/2}\,dx.$$

Evaluate.

38. Imagine a "one-dimensional" country of length $2R$ (see Fig. 5-19) [50]. Suppose the production of goods for this country is continuously distributed

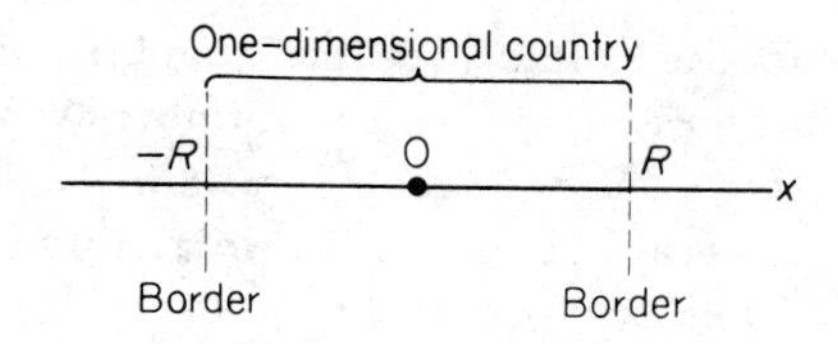

FIG. 5-19

from border to border. If the amount produced each year per unit of distance is $f(x)$, then the country's total yearly production is given by

$$G = \int_{-R}^{R} f(x)\, dx.$$

Evaluate G if $f(x) = i$ where i is constant.

39. In discussing traffic safety, Shonle **[43]** considers how much acceleration a person can tolerate in a crash so that there is no major injury. The *severity index* is defined as follows:

$$\text{severity index} = \int_{0}^{T} \alpha^{5/2}\, dt,$$

where α (a Greek letter read "alpha") is considered a constant involved with a weighted average acceleration, and T is the duration of the crash. Find the severity index.

40. In statistics, the mean μ (a Greek letter read "mu") of the continuous probability density function f defined on the interval $[a, b]$ is

$$\mu = \int_{a}^{b} (x \cdot f(x))\, dx,$$

and the variance σ^2 (σ is a Greek letter read "sigma") is

$$\sigma^2 = \int_{a}^{b} (x - \mu)^2 f(x)\, dx.$$

Compute μ and then σ^2 if $a = 0$, $b = 1$, and $f(x) = 1$.

41. For a certain population, suppose l is a function such that $l(x)$ is the number of persons who reach the age of x in any year of time. This function is called a *life table function*. Under appropriate conditions, the integral

$$\int_{x}^{x+n} l(t)\, dt$$

gives the expected number of people in the population between the exact ages of x and $x + n$, inclusive. If $l(x) = 10{,}000\sqrt{100 - x}$, determine the number of people between the exact ages of 36 and 64 inclusive. Give your answer to the nearest integer, since fractional answers make no sense.

42. The economist Pareto **[54]** has stated an empirical law of distribution of higher incomes that gives the number N of persons receiving x or more dollars. If $dN/dx = -Ax^{-B}$, where A and B are constants, set up a definite integral that gives the total number of persons having incomes between a and b where $a < b$.

43. A manufacturer's marginal cost function is $dc/dq = .2q + 3$. If c is in dollars, determine the cost involved to increase production from 60 to 70 units.

44. Repeat Problem 43 if $dc/dq = .003q^2 - .6q + 40$ and production increases from 100 to 200 units.

45. A manufacturer's marginal revenue function is $dr/dq = 1000/\sqrt{100q}$. If r is in dollars, find the change in the manufacturer's total revenue if production is increased from 400 to 900 units.

46. Repeat Problem 45 if $dr/dq = 250 + 90q - 3q^2$ and production is increased from 10 to 20 units.

47. A sociologist is studying the crime rate in a certain city. He estimates that t months after the beginning of next year, the total number of crimes committed will increase at the rate of $8t + 10$ crimes per month. Determine the total number of crimes that can be expected to be committed next year. How many crimes can be expected to be committed during the last six months of that year?

48. For a group of hospitalized individuals, suppose the discharge rate is given by

$$f(t) = \frac{81 \times 10^6}{(300 + t)^4},$$

where $f(t)$ is the proportion of the group discharged per day at the end of t days. What proportion has been discharged by the end of 700 days?

49. In studying the flow of a fluid in a tube of constant radius R, such as blood flow in portions of the body, one can think of the tube as consisting of concentric tubes of radius r, where $0 \leqslant r \leqslant R$. The velocity v of the fluid is a function of r and is given in **[48]** by

$$v = \frac{(P_1 - P_2)(R^2 - r^2)}{4\eta l},$$

where P_1 and P_2 are pressures at the ends of the tube, η (a Greek letter read "eta") is the fluid viscosity, and l is the length of the tube. The volume rate of flow, Q, through the tube is given by

$$Q = \int_0^R 2\pi r v \, dr.$$

Show that $Q = \dfrac{\pi R^4(P_1 - P_2)}{8\eta l}$. Note that R occurs as a factor to the fourth power. Thus, doubling the radius of the tube has the effect of increasing the flow by a factor of 16. The formula that you derived for the volume rate of flow is called Poiseuille's law, after the French physiologist Jean Poiseuille.

5-6 AREA

SKIP

In Sec. 5-4 we found the area of a region by evaluating the limit of a sum of the form $\Sigma f(x)\,\Delta x$. Since this limit also is a definite integral, we can use the Fundamental Theorem to evaluate the limit.

When using the definite integral to determine area, you should make a rough sketch of the region involved. Let us consider the area of the region bounded by $y = f(x)$ and the x-axis from $x = a$ to $x = b$, as shown in Fig. 5-20. Since we are summing areas of rectangles, a sample rectangle should be included in the sketch. This will help you understand the integration process. Such a rectangle (see Fig. 5-20) is called a **vertical element of area** (or a **vertical strip**). In the diagram the width of the vertical element is Δx. The length is the y-value of the curve. Hence the rectangle has area $y\,\Delta x$ or $f(x)\,\Delta x$. We want to add the areas of all such elements between $x = a$ and $x = b$ by means of definite integration:

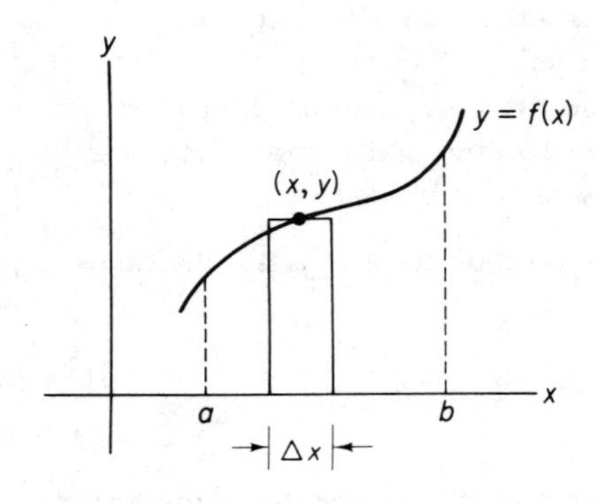

FIG. 5-20

$$\Sigma f(x)\,\Delta x \to \int_a^b f(x)\,dx.$$

For example, let us find the area of the region in the first quadrant bounded by the curve $y = x^2 - 1$, the x-axis, and the line $x = 2$ (see Fig. 5-21). The width of the sample element is Δx and its length is y. Thus the area of the element is $y\,\Delta x$. All such areas of elements between $x = 1$ and $x = 2$ are to be summed. Hence the limits of integration are $x = 1$ and $x = 2$.

$$\Sigma y\,\Delta x \to \int_1^2 y\,dx.$$

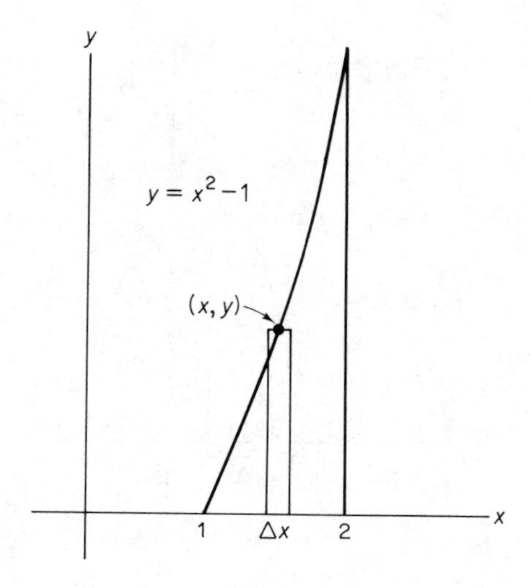

FIG. 5-21

To evaluate this integral we must express the integrand in terms of the variable of integration x. Since $y = x^2 - 1$,

$$\text{area} = \int_1^2 (x^2 - 1)\, dx = \left(\frac{x^3}{3} - x\right)\Bigg|_1^2$$

$$= \left(\frac{8}{3} - 2\right) - \left(\frac{1}{3} - 1\right)$$

$$= \frac{4}{3} \text{ sq units.}$$

EXAMPLE 1

Find the area of the region bounded by the curve $y = 6 - x - x^2$ and the x-axis.

First we must sketch the curve so that we can visualize the region. Since $y = -(x^2 + x - 6) = -(x - 2)(x + 3)$, the x-intercepts are $(2, 0)$ and $(-3, 0)$. Using techniques of graphing that were previously discussed, we obtain the graph shown in Fig. 5-22. Note that for this region, it is crucial that the x-intercepts of the curve be found in order to determine the values of x over which the areas of the vertical elements must be summed. For the vertical element shown, the width is Δx and the length is y. Hence the area of the element is $(6 - x - x^2)\,\Delta x$. Summing these from $x = -3$ to $x = 2$

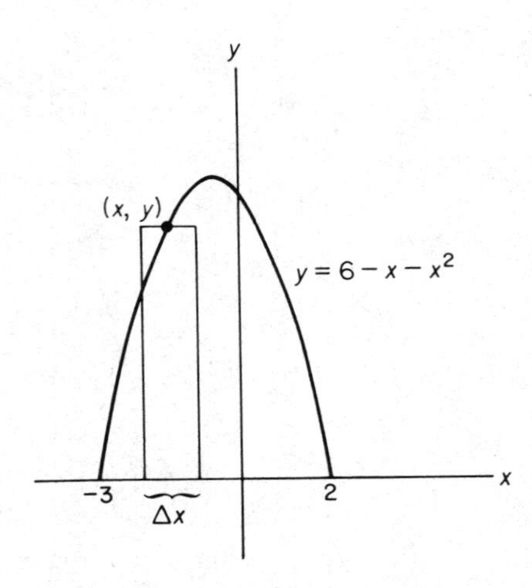

FIG. 5-22

gives

$$\text{area} = \int_{-3}^{2} (6 - x - x^2)\,dx = \left(6x - \frac{x^2}{2} - \frac{x^3}{3}\right)\Big|_{-3}^{2}$$

$$= \left(12 - \frac{4}{2} - \frac{8}{3}\right) - \left(-18 - \frac{9}{2} - \frac{-27}{3}\right)$$

$$= \frac{125}{6} \text{ sq units.}$$

EXAMPLE 2

Find the area of the region bounded by $y = x^2 + 2x + 2$, the x-axis, and the lines $x = -2$ and $x = 1$.

A sketch of the region is given in Fig. 5-23.

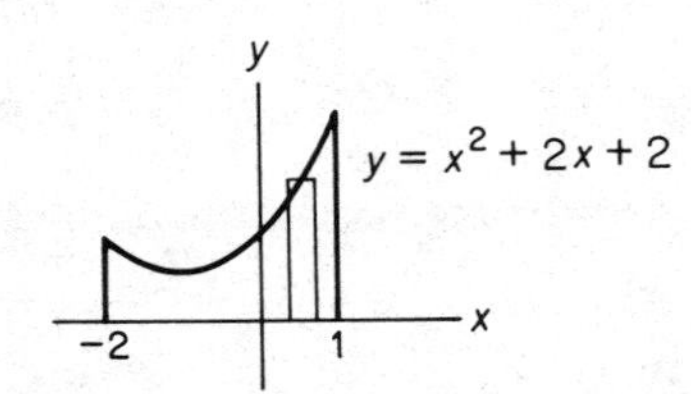

FIG. 5-23

$$\text{area} = \int_{-2}^{1} y\,dx = \int_{-2}^{1} (x^2 + 2x + 2)\,dx$$

$$= \left(\frac{x^3}{3} + x^2 + 2x\right)\Big|_{-2}^{1} = \left(\frac{1}{3} + 1 + 2\right) - \left(-\frac{8}{3} + 4 - 4\right)$$

$$= 6 \text{ sq units.}$$

EXAMPLE 3

Find the area of the region bounded by the curves $y = x^2 - x - 2$ and $y = 0$ (the x-axis) from $x = -2$ to $x = 2$.

A sketch of the region is given in Fig. 5-24. Notice that the x-intercepts are $(-1, 0)$ and $(2, 0)$.

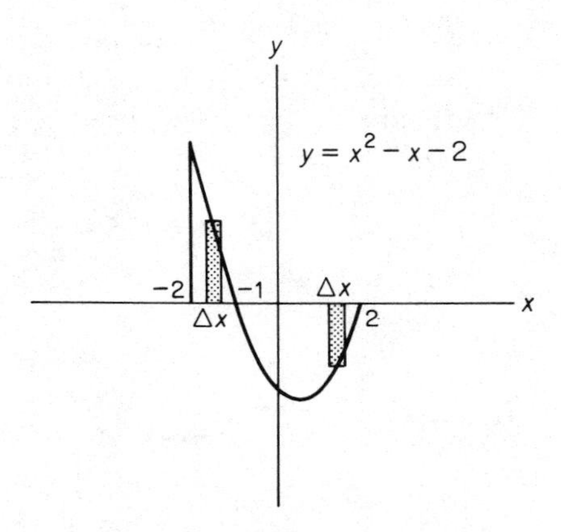

FIG. 5-24

PITFALL. *It is wrong to write hastily that the area is $\int_{-2}^{2} (x^2 - x - 2)\, dx$ for the following reason. For the left rectangle the length is y. However, for the rectangle on the right, y is negative and so the rectangle has length $-y$. This points out the importance of sketching the region.*

On the interval $[-2, -1]$, the area of the element is

$$y\,\Delta x = (x^2 - x - 2)\,\Delta x.$$

On $[-1, 2]$ it is

$$-y\,\Delta x = -(x^2 - x - 2)\,\Delta x.$$

Thus,

$$\begin{aligned}
\text{area} &= \int_{-2}^{-1} (x^2 - x - 2)\, dx + \int_{-1}^{2} -(x^2 - x - 2)\, dx \\
&= \left(\frac{x^3}{3} - \frac{x^2}{2} - 2x\right)\Bigg|_{-2}^{-1} - \left(\frac{x^3}{3} - \frac{x^2}{2} - 2x\right)\Bigg|_{-1}^{2} \\
&= \left[\left(-\frac{1}{3} - \frac{1}{2} + 2\right) - \left(-\frac{8}{3} - \frac{4}{2} + 4\right)\right] - \\
&\qquad \left[\left(\frac{8}{3} - \frac{4}{2} - 4\right) - \left(-\frac{1}{3} - \frac{1}{2} + 2\right)\right] = \frac{19}{3} \text{ sq units.}
\end{aligned}$$

The next example shows the use of area as a probability in statistics.

EXAMPLE 4

In statistics, a (probability) density function f of a continuous variable x defined on the interval [a, b] has the following properties:

1. $f(x) \geqslant 0$.
2. $\int_a^b f(x)\,dx = 1$.
3. *The probability that x assumes a value between c and d, written $P(c \leqslant x \leqslant d)$, is represented by the area of the region bounded by the graph of f and the x-axis between x = c and x = d. Hence, (see Fig. 5-25)*

$$P(c \leqslant x \leqslant d) = \int_c^d f(x)\,dx.$$

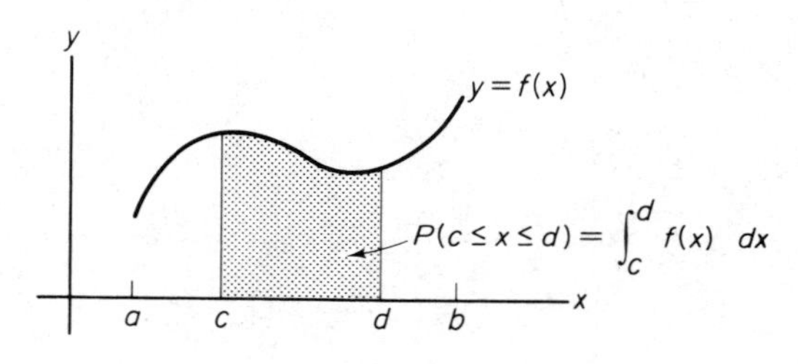

FIG. 5-25

The (cumulative) distribution function F is given by $F(t) = \int_a^t f(x)\,dx$, where $a \leqslant t \leqslant b$. This integral gives the area under the graph of the density function to the left of x = t and represents the probability that x will assume a value less than or equal to t.

Suppose $f(x) = k(x - x^2)$, where k is a constant and $0 \leqslant x \leqslant 1$. If f is a density function, find (a) the value of k, (b) $P(0 \leqslant x \leqslant \frac{1}{4})$, and (c) the distribution function F.

a. We must have $\int_0^1 f(x)\,dx = 1$. Thus,

$$\int_0^1 k(x - x^2)\,dx = 1,$$

$$k\int_0^1 (x - x^2)\,dx = 1,$$

$$k\left(\frac{x^2}{2} - \frac{x^3}{3}\right)\Bigg|_0^1 = 1,$$

$$k\left(\frac{1}{2} - \frac{1}{3}\right) - k(0) = 1,$$

$$k \cdot \frac{1}{6} = 1,$$

$$k = 6.$$

b. Since $f(x) = 6(x - x^2)$,

$$P\left(0 \leq x \leq \frac{1}{4}\right) = \int_0^{1/4} 6(x - x^2)\,dx = 6\int_0^{1/4} (x - x^2)\,dx$$

$$= 6\left(\frac{x^2}{2} - \frac{x^3}{3}\right)\Big|_0^{1/4}$$

$$= 6\left(\frac{\left[\frac{1}{4}\right]^2}{2} - \frac{\left[\frac{1}{4}\right]^3}{3}\right) - 6(0)$$

$$= 6\left(\frac{1}{32} - \frac{1}{192}\right) = 6\left(\frac{5}{192}\right) = \frac{5}{32}.$$

c. $F(t) = \int_0^t 6(x - x^2)\,dx = 6\left(\frac{x^2}{2} - \frac{x^3}{3}\right)\Big|_0^t$

$$= 6\left(\frac{t^2}{2} - \frac{t^3}{3}\right) - 6(0)$$

$$= 3t^2 - 2t^3, \qquad \text{where } 0 \leq t \leq 1.$$

Exercise 5-6

In Problems **1–30**, *find the area of the region bounded by the given curves, the x-axis, and the given lines. In each case first sketch the region.*

1. $y = 4x$, $x = 2$.

2. $y = 3x + 1$, $x = 0$, $x = 4$.

3. $y = 3x + 2$, $x = 2$, $x = 3$.

4. $y = x + 5$, $x = 2$, $x = 4$.

5. $y = x - 1$, $x = 5$.

6. $y = 2x^2$, $x = 1$, $x = 2$.

7. $y = x^2$, $x = 2$, $x = 3$.

8. $y = 2x^2 - x$, $x = -2$, $x = -1$.

9. $y = x^2 + 2$, $x = -1$, $x = 2$.

10. $y = 2x + x^3$, $x = 1$.

11. $y = x^2 - 2x$, $x = -3$, $x = -1$.

12. $y = 3x^2 - 4x$, $x = -2$, $x = -1$.

13. $y = 9 - x^2$.

14. $y = 3$, $x = -1$, $x = 1$.

15. $y = 1 - x - x^3$, $x = -2$, $x = 0$.

16. $y = 8 - 2x^2$.

17. $y = 3 + 2x - x^2$.

18. $y = \frac{1}{x^2}$, $x = 2$, $x = 3$.

19. $y = \sqrt{x + 9}$, $x = -9$, $x = 0$.

20. $y = x^2 - 2x$, $x = 1$, $x = 3$.

21. $y = \sqrt{2x - 1}$, $x = 1$, $x = 5$.

22. $y = x^3 + 3x^2$, $x = -2$, $x = 2$.

23. $y = \sqrt[3]{x}$, $x = 2$.

24. $y = x^2 - 4$, $x = -2$, $x = 2$.

25. $y = x^2 - 2x - 8$.

26. $y = |x|$, $x = -2$, $x = 2$.

27. $y = x^3$, $x = -2$, $x = 4$.

28. $y = \sqrt{x-2}$, $x = 2$, $x = 6$.

29. $y = 2x - x^2$, $x = 1$, $x = 3$.

30. $y = x^2 - x + 1$, $x = 0$, $x = 1$.

31. Under conditions of a continuous uniform distribution, a topic in statistics, the proportion of persons with incomes between a and t, where $a \leqslant t \leqslant b$, is the area of the region between the curve $y = 1/(b - a)$ and the x-axis from $x = a$ to $x = t$. Sketch the graph of the curve and determine the area of the given region.

32. Given

$$f(x) = \begin{cases} 3x^2, \text{ if } 0 \leqslant x \leqslant 2, \\ 16 - 2x, \text{ if } x \geqslant 2, \end{cases}$$

find the area of the region bounded by the graph of $y = f(x)$, the x-axis, and the line $x = 3$. Include a sketch of the region.

33. Suppose $f(x) = kx$, where k is a constant and $0 \leqslant x \leqslant 4$. If f is a density function (see Example 4), find (a) the value of k, (b) $P(2 \leqslant x \leqslant 4)$, and (c) the distribution function F.

34. Suppose $f(x) = k(1 - x)^2$, where k is a constant and $0 \leqslant x \leqslant 1$. If f is a density function (see Example 4), find (a) the value of k, (b) $P(\frac{1}{3} \leqslant x \leqslant \frac{1}{2})$, and (c) the distribution function F.

5-7 AREA BETWEEN CURVES

We shall now consider finding the area of a region enclosed by several curves. As before, our procedure will be to draw a sample element of area and use the definite integral to "add together" the areas of all such elements.

EXAMPLE 1

Find the area of the region bounded by the curves $y = \sqrt{x}$ and $y = x$.

A sketch of the region appears in Fig. 5-26. To determine where the curves intersect, we solve the system formed by the equations $y = \sqrt{x}$ and $y = x$. Eliminating y by substitution, we obtain

$$\sqrt{x} = x.$$

$$x = x^2 \quad \text{[squaring both sides]}.$$

$$0 = x^2 - x = x(x - 1).$$

$$x = 0 \text{ or } x = 1.$$

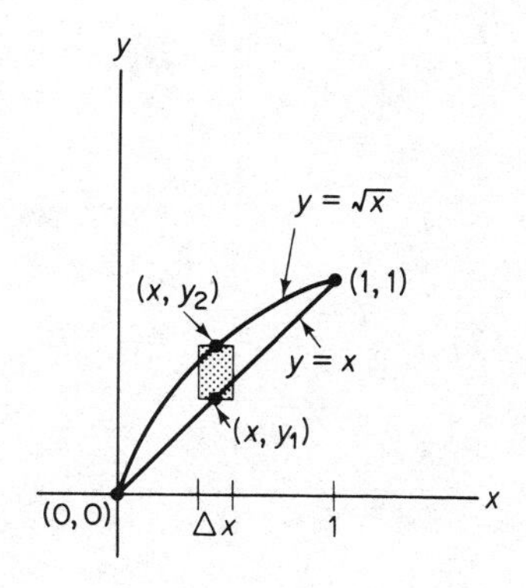

FIG. 5-26

If $x = 0$, then $y = 0$; if $x = 1$, then $y = 1$. Thus the curves intersect at (0, 0) and (1, 1). The width of the indicated element of area is Δx. The length is the y–value on the upper curve minus the y–value on the lower curve. If we distinguish between the curves by writing $y_1 = x$ and $y_2 = \sqrt{x}$, then the length of the element is

$$y_{\text{upper}} - y_{\text{lower}} = y_2 - y_1 = \sqrt{x} - x.$$

Thus the area of the element is $(\sqrt{x} - x)\,\Delta x$. Summing all such areas from $x = 0$ to $x = 1$ by the definite integral, we get the area of the entire region.

$$\sum (\sqrt{x} - x)\,\Delta x \to \int_0^1 (\sqrt{x} - x)\,dx.$$

$$\text{area} = \int_0^1 (x^{1/2} - x)\,dx = \left(\frac{x^{3/2}}{\frac{3}{2}} - \frac{x^2}{2}\right)\Bigg|_0^1$$

$$= \left(\tfrac{2}{3} - \tfrac{1}{2}\right) - (0 - 0) = \tfrac{1}{6} \text{ sq unit.}$$

It should be obvious to you that the points of intersection are important in determining the limits of integration.

Sometimes area can more easily be determined by summing areas of horizontal elements rather than vertical elements. In the following example an area will be found by both methods. In each case the element of area determines the form of the integral.

EXAMPLE 2

Find the area of the region bounded by the curve $y^2 = 4x$ and the lines $y = 3$ and $x = 0$ (the y-axis).

The region is sketched in Fig. 5-27. When the curves $y = 3$ and $y^2 = 4x$

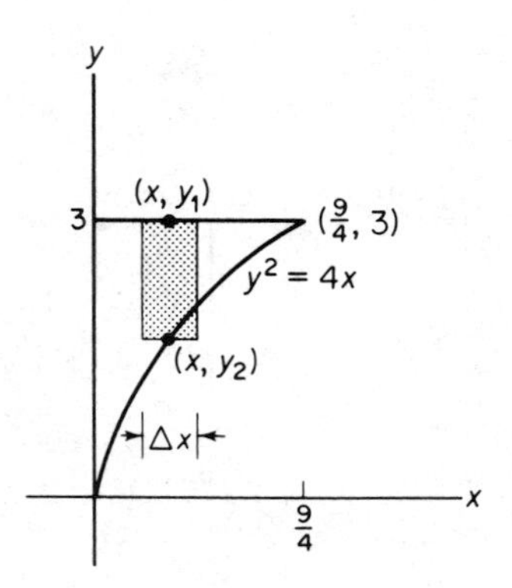

FIG. 5-27

intersect, then $9 = 4x$ and so $x = \frac{9}{4}$. Thus the point of intersection is $(\frac{9}{4}, 3)$. Since the width of the vertical strip is Δx, we integrate with respect to the variable x. Thus y_{upper} and y_{lower} must be expressed as functions of x. For the curve $y^2 = 4x$, we have $y = \pm 2\sqrt{x}$. But for the portion of this curve which bounds the region we must have $y \geqslant 0$, so we use $y = 2\sqrt{x}$. Thus the length of the strip is $y_{upper} - y_{lower} = 3 - 2\sqrt{x}$. Hence the strip has an area of $(3 - 2\sqrt{x})\,\Delta x$ and we wish to sum up all such areas from $x = 0$ to $x = \frac{9}{4}$.

$$\text{area} = \int_0^{9/4} (3 - 2\sqrt{x})\, dx = \left(3x - \frac{4x^{3/2}}{3}\right)\Bigg|_0^{9/4}$$

$$= \left[3\left(\frac{9}{4}\right) - \frac{4}{3}\left(\frac{9}{4}\right)^{3/2}\right] - [(0 - 0)]$$

$$= \frac{27}{4} - \frac{4}{3}\left[\left(\frac{9}{4}\right)^{1/2}\right]^3 = \frac{27}{4} - \frac{4}{3}\left(\frac{3}{2}\right)^3 = \frac{9}{4} \text{ sq units.}$$

Let us now approach this problem from the point of view of a **horizontal element of area** (or **horizontal strip**) as shown in Fig. 5-28. The width of the

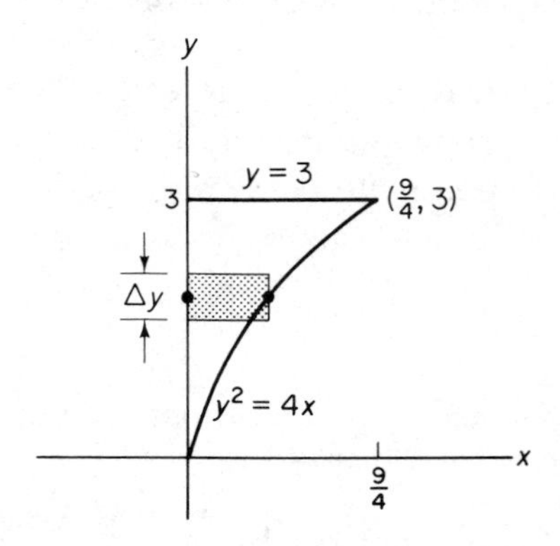

FIG. 5-28

element is Δy. The length of the element is the *x-value on the right curve minus the x-value on the left curve.* Thus the area of the element is $(x_{\text{right}} - x_{\text{left}})\,\Delta y$. We wish to sum all such areas from $y = 0$ to $y = 3$:

$$\sum (x_{\text{right}} - x_{\text{left}})\,\Delta y \to \int_0^3 (x_{\text{right}} - x_{\text{left}})\,dy.$$

Since the variable of integration is y, we must express x_{right} and x_{left} as functions of y. The right curve is $y^2 = 4x$ or, equivalently, $x = y^2/4$. The left curve is $x = 0$. Thus

$$\begin{aligned} \text{area} &= \int_0^3 (x_{\text{right}} - x_{\text{left}})\,dy \\ &= \int_0^3 \left(\frac{y^2}{4} - 0\right) dy = \frac{y^3}{12}\bigg|_0^3 = \frac{9}{4} \text{ sq units.} \end{aligned}$$

Note that for this region, horizontal strips make the definite integral easier to evaluate (and set up) than vertical strips. In any case, remember that **the limits of integration are those limits for the variable of integration.**

EXAMPLE 3

Find the area of the region bounded by the curves $y = 4x - x^2 + 8$ and $y = x^2 - 2x$.

A sketch of the region appears in Fig. 5-29. To find when the curves

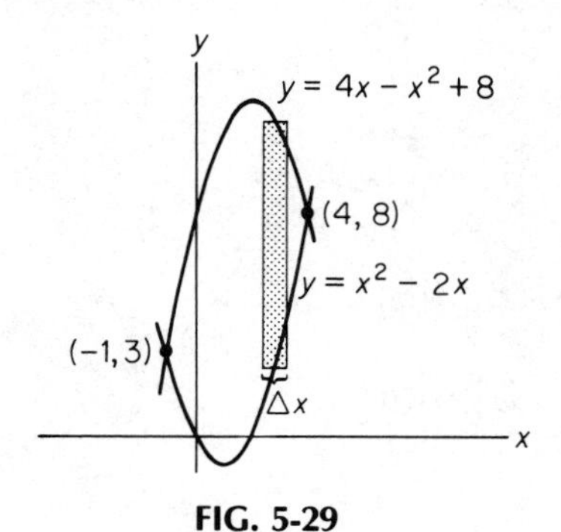

FIG. 5-29

intersect we solve

$$\begin{aligned} 4x - x^2 + 8 &= x^2 - 2x. \\ -2x^2 + 6x + 8 &= 0. \\ x^2 - 3x - 4 &= 0. \\ (x + 1)(x - 4) &= 0. \\ x = -1 \quad &\text{or} \quad x = 4. \end{aligned}$$

When $x = -1$, then $y = 3$; when $x = 4$, then $y = 8$. Thus the curves intersect at $(-1, 3)$ and $(4, 8)$. We shall use vertical strips, since they appear to present no difficulty. The area of the element is

$$(y_{\text{upper}} - y_{\text{lower}})\,\Delta x = [(4x - x^2 + 8) - (x^2 - 2x)]\,\Delta x$$
$$= (-2x^2 + 6x + 8)\,\Delta x.$$

Summing all such areas from $x = -1$ to $x = 4$, we have

$$\text{area} = \int_{-1}^{4} (-2x^2 + 6x + 8)\,dx = 41\tfrac{2}{3} \text{ sq units.}$$

EXAMPLE 4

Find the area of the region bounded by $y^2 = x$ and $x - y = 2$.

A sketch of the region appears in Fig. 5-30. The curves intersect when $y^2 = y + 2$. Thus $y^2 - y - 2 = 0$, or equivalently $(y + 1)(y - 2) = 0$, from which $y = -1$ or $y = 2$. The points of intersection are $(1, -1)$ and $(4, 2)$. Let us consider vertical elements of area [see Fig. 5-30(a)]. Solving $y^2 = x$ for y gives $y = \pm\sqrt{x}$. As seen in Fig. 5-30(a), to the *left* of $x = 1$ the upper end of the element lies on $y = \sqrt{x}$ and the lower end lies on $y = -\sqrt{x}$. To the *right* of $x = 1$, the upper curve is $y = \sqrt{x}$ and the lower curve is $x - y = 2$ (or $y = x - 2$). Thus, with vertical strips *two* integrals are needed to evaluate the area:

$$\text{area} = \int_0^1 [\sqrt{x} - (-\sqrt{x})]\,dx + \int_1^4 [\sqrt{x} - (x - 2)]\,dx.$$

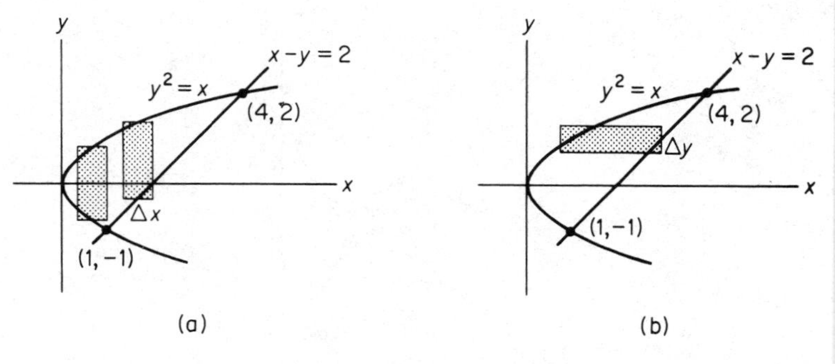

FIG. 5-30

Let us consider horizontal strips to see if we can simplify our work. In Fig. 5-30(b), the width of the strip is Δy. The rightmost curve is *always* $x - y = 2$ (or $x = y + 2$) and the leftmost curve is *always* $y^2 = x$ (or $x = y^2$). Thus the area of the horizontal strip is $[(y + 2) - y^2]\,\Delta y$ and the total area is

$$\text{area} = \int_{-1}^{2} (y + 2 - y^2)\,dy = \frac{9}{2} \text{ sq units.}$$

Clearly the use of horizontal strips is the more desirable approach for the problem.

Exercise 5-7

In Problems **1–22**, *find the area of the region bounded by the graphs of the given equations.*

1. $y = x^2, y = 2x$.

2. $y = x, y = -x + 3, y = 0$.

3. $y = x^2, x = 0, y = 4\ (x \geqslant 0)$.

4. $y = x^2, y = x$.

5. $y = x^2 + 3, y = 9$.

6. $y^2 = x, x = 2$.

7. $x = 8 + 2y, x = 0, y = -1, y = 3$.

8. $y = x - 4, y^2 = 2x$.

9. $y = 4 - x^2, y = -3x$.

10. $x = y^2 + 2, x = 6$.

11. $y^2 = x, 3x - 2y = 1$.

12. $y = x^2, y = x + 2$.

13. $2y = 4x - x^2, 2y = x - 4$.

14. $y = \sqrt{x}, y = x^2$.

15. $y^2 = x, y = x - 2$.

16. $y = 2 - x^2, y = x$.

17. $y = 8 - x^2, y = x^2, x = -1, x = 1$.

18. $y^2 = 4 - x, y = x + 2$.

19. $y = x^2, y = 2, y = 5$.

20. $y = x^3 - x$, x-axis.

21. $y = x^3, y = x$.

22. $y = x^3, y = \sqrt{x}$.

23. A *Lorentz curve* is used in studying income distributions. If x is the cumulative percentage of income recipients, ranked from poorest to richest, and y is the cumulative percentage of income, then equality of income distribution is given by the line $y = x$ in Fig. 5-31, where x and y are expressed as decimals. For example, 10 percent of the people receive 10 percent of total income, 20 percent of the people receive 20 percent of the income, etc. Suppose the actual distribution is given by the Lorentz curve defined by $y = \frac{20}{21}x^2 + \frac{1}{21}x$. Note, for example, that 30 percent of the people receive only 10 percent of total income. The degree of deviation from equality is measured by the *coefficient of inequality* **[49]** for a Lorentz curve. This coefficient is defined to be the area between the curve and the diagonal, divided by the area under the diagonal:

$$\frac{\text{area between curve and diagonal}}{\text{area under diagonal}}.$$

For example, when all incomes are equal, the coefficient of inequality is zero. Find the coefficient of inequality for the Lorentz curve defined above.

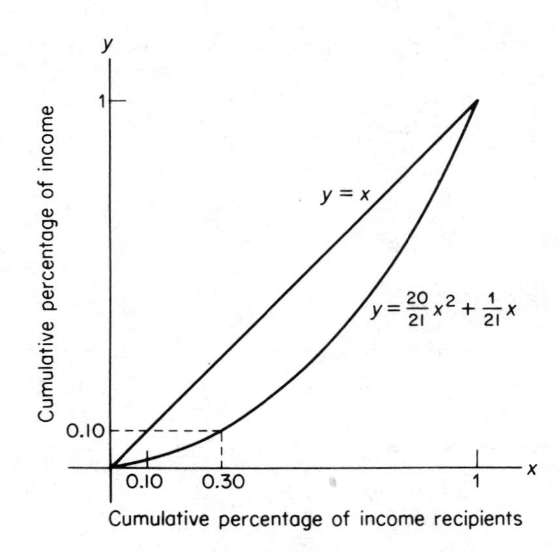

FIG. 5-31

24. Find the coefficient of inequality as in Problem 23 for the Lorentz curve defined by $y = \frac{11}{12}x^2 + \frac{1}{12}x$.

5-8 REVIEW

IMPORTANT TERMS AND SYMBOLS IN CHAPTER 5

antiderivative *(p. 211)*
indefinite integral *(p. 211)*
definite integral *(p. 240)*
integrand *(p. 212)*
limits of integration *(p. 240)*
$\int f(x)\,dx$, $\int_a^b f(x)\,dx$ *(p. 248)*
Σ *(p. 230)*
index of summation *(p. 230)*
limits of summation *(p. 230)*
Fundamental Theorem of Integral Calculus *(p. 248)*
element of area *(p. 256)*
constant of integration *(p. 212)*

REVIEW SECTION

1. If F is an antiderivative of f, then $F'(x) =$ ________.

Ans. $f(x)$.

2. $\int 5x^4\,dx =$ ________.

Ans. $x^5 + C$.

3. In $\int f(x)\,dx$, $f(x)$ is called the ________.

Ans. integrand.

4. $\int_{10}^{20} 2\,dx =$ ________.

Ans. 20.

5. True or false: $\int \sqrt{x}\,dx = \frac{3}{2}x^{3/2} + C$. ________

Ans. false.

6. $\int (2x - 5)^9\,dx =$ ________.

Ans. $\frac{1}{20}(2x - 5)^{10} + C$.

7. True or false: $\int x(x^2)\,dx = \frac{x^2}{2} \cdot \frac{x^3}{3} + C$. ________

Ans. false.

8. If $\int f(x)\,dx = F(x) + C$, then C is called the ________.

Ans. constant of integration.

9. A definite integral of a function is a (function) (number).

Ans. number.

10. $\sum_{k=0}^{2} (k + 1) =$ ________.

Ans. 6.

11. True or false: $\sum_{k=1}^{15} (k^2 + 3k + 5) = \sum_{j=1}^{15} (j^2 + 3j + 5)$. ________

Ans. true.

12. If $D_x y = 1$ and $y(0) = 1$, then $y =$ ________.

Ans. $x + 1$.

13. All antiderivatives of a given function differ at most by a ________.

Ans. constant.

14. If $F'(x) = f(x)$, then $\int_2^3 f(x)\,dx =$ ________.

Ans. $F(3) - F(2)$.

15. If $f(x) = 0$, then $\int f(x)\,dx =$ ________.

Ans. C, a constant.

REVIEW PROBLEMS

Determine the integrals in Problems **1–26.**

1. $\int (x^3 + 2x - 7)\, dx.$

2. $\int dx.$

3. $\int_0^9 (\sqrt{x} + x)\, dx.$

4. $\int \frac{5 - 3x}{2}\, dx.$

5. $\int \frac{2}{(x + 5)^3}\, dx.$

6. $\int_4^{12} (y - 8)^{501}\, dy.$

7. $\int \frac{7x}{(3x^2 - 8)^3}\, dx.$

8. $\int_0^2 \frac{2}{\sqrt{4x + 1}}\, dx.$

9. $\int_0^1 \sqrt[3]{3t + 8}\, dt.$

10. $\int \frac{4x^2 - x}{x}\, dx.$

11. $\int x^2\sqrt{3x^3 + 2}\, dx.$

12. $\int (2x^3 + x)(x^4 + x^2)^{3/4}\, dx.$

13. $\int y(y + 1)^2\, dy.$

14. $\int \frac{8x}{3\sqrt[3]{7 - 2x^2}}\, dx.$

15. $\int \left(\frac{1}{x^3} + \frac{2}{x^2}\right) dx.$

16. $\int_0^1 10^{-8}\, dx.$

17. $\int_{-2}^1 (y^4 - y + 1)\, dy.$

18. $\int_7^{70} dx.$

19. $\int_{\sqrt{3}}^2 7x\sqrt{4 - x^2}\, dx.$

20. $\int_0^1 (2x + 1)(x^2 + x)^4\, dx.$

21. $\int_0^1 \left[2x - \frac{1}{(x + 1)^{2/3}}\right] dx.$

22. $\int_2^8 (\sqrt{2x} - x + 4)\, dx.$

23. $\int \frac{\sqrt{t} - 3}{t^2}\, dt.$

24. $\int \frac{\sqrt[4]{z} - \sqrt[3]{z}}{\sqrt{z}}\, dz.$

25. $\int \frac{(.5x - .1)^4}{.4}\, dx.$

26. $\int \frac{(x^2 + 4)^2}{x^2}\, dx.$

In Problems **27–34**, *determine the area of the region bounded by the given curves, the x-axis, and the given lines.*

27. $y = x^2 - 1,\ x = 2\ (y \geqslant 0).$

28. $y = 6x - 8 - x^2.$

29. $y = \sqrt{x + 4},\ x = 0.$

30. $y = x^2 - x - 2,\ x = -2,\ x = 2.$

31. $y = 5x - x^2$.

32. $y = \sqrt[4]{x}$, $x = 1$, $x = 16$.

33. $y = \frac{1}{x^2} + 3$, $x = 1$, $x = 3$.

34. $y = x^3 - 1$, $x = -1$.

In Problems **35–40**, *find the area of the region bounded by the given curves.*

35. $y^2 = 4x$, $x = 0$, $y = 2$.

36. $y = 2x^2$, $x = 0$, $y = 2$ $(x \geqslant 0)$.

37. $y = x^2 + 4x - 5$, $y = 0$.

38. $y = 2x^2$, $y = x^2 + 9$.

39. $y = x^2 - 2x$, $y = 12 - x^2$.

40. $y = \sqrt{x}$, $x = 0$, $y = 3$.

41. If marginal revenue is given by $dr/dq = 100 - (3/2)\sqrt{2q}$, determine the corresponding demand equation.

42. If marginal cost is given by $dc/dq = q^2 + 7q + 6$, and fixed costs are 2500, determine the total cost for producing 6 units.

43. Suppose $f(x) = k + (x/10)$, where k is a constant and $0 \leqslant x \leqslant 3$. If f is a density function (see Example 4 in Sec. 5-6), find (a) the value of k, (b) $P(1 \leqslant x \leqslant 2)$, and (c) the distribution function F.

44. A manufacturer's marginal revenue function is $dr/dq = 275 - q - .3q^2$. If r is in dollars, find the increase in the manufacturer's total revenue if production is increased from 10 to 20 units.

45. A manufacturer's marginal cost function is $dc/dq = 500/\sqrt{2q + 25}$. If c is in dollars, determine the cost involved to increase production from 100 to 300 units.

46. Dizygotic twins are twins resulting from two fertilized ova. Under appropriate assumptions and over a certain time interval $[0, T_p]$, the mean (average) probability T that both twins have the same sex is given in **[42]** by

$$T = \frac{1}{T_p}\int_0^{T_p}\{[p(m)]^2 + [p(f)]^2\}\,dt,$$

where

$$p(m) = a - \frac{a-b}{T_p}t,$$

$$p(f) = 1 - p(m),$$

and both a and b are constants. (a) Show that $T = a^2 + (1-a)^2 + (a-b)(1-2a) + \frac{2}{3}(a-b)^2$. (b) Evaluate T if $a = .7$ and $b = .3$.

CHAPTER 6

Exponential and Logarithmic Functions

6-1 EXPONENTIAL FUNCTIONS

In this chapter we shall look at two special functions that have wide applications in many areas of study. The first function involves a constant raised to a variable power, such as $f(x) = 2^x$. We call this an *exponential function*.

DEFINITION. *The function f defined by*

$$y = f(x) = b^x,$$

where $b > 0$, $b \neq 1$, *and the exponent* x *is any real number, is called an* ***exponential function*** *to the base* b.

We made the restriction $b \neq 1$ to exclude the rather simple constant function $f(x) = 1^x = 1$. Since the exponent in b^x can be any real number, the question comes up as to how we define something like $b^{\sqrt{2}}$. Stated simply, we use an approximation method. First, $b^{\sqrt{2}}$ is approximately $b^{1.4} = b^{7/5} = \sqrt[5]{b^7}$, which *is* defined. Better approximations are $b^{1.41} = \sqrt[100]{b^{141}}$ and $b^{1.414}$, etc. In this way a meaning of $b^{\sqrt{2}}$ becomes clear.

In Fig. 6-1 are the graphs of the exponential functions $y = 2^x$, $y = 3^x$, and $y = (\frac{1}{2})^x = 2^{-x}$. Notice that

(1) the *domain* of an exponential function is all real numbers, and
(2) the *range* is all positive numbers.

Also, $b^0 = 1$ for every base b, as shown by the point of intersection (0, 1) of the graphs.

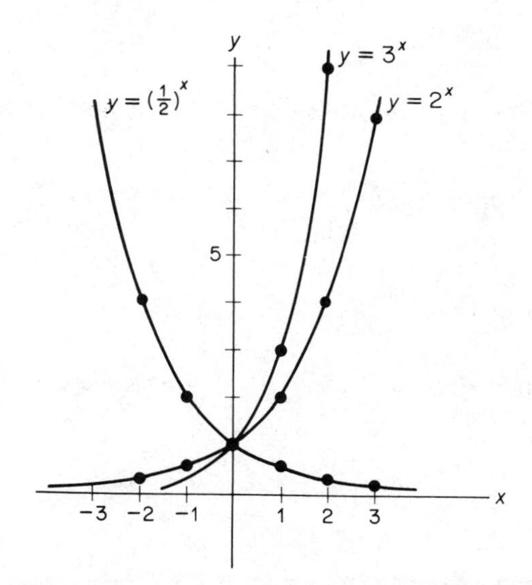

x	2^x	3^x	$(\frac{1}{2})^x$
-2	$\frac{1}{4}$	$\frac{1}{9}$	4
-1	$\frac{1}{2}$	$\frac{1}{3}$	2
0	1	1	1
1	2	3	$\frac{1}{2}$
2	4	9	$\frac{1}{4}$
3	8	27	$\frac{1}{8}$

FIG. 6-1

We also see in Fig. 6-1 that $y = b^x$ has two basic shapes, depending on whether $b > 1$ or $0 < b < 1$. If $b > 1$, then as x increases, y also increases. But y can also take on values very close to 0. In terms of limits,

$$\lim_{x\to\infty} b^x = \infty \quad \text{and} \quad \lim_{x\to-\infty} b^x = 0, \qquad \text{if } b > 1.$$

Now, suppose $0 < b < 1$ as in $y = (\frac{1}{2})^x$. Then as x increases, y decreases and takes on values close to zero. In fact,

$$\lim_{x \to \infty} b^x = 0 \quad \text{and} \quad \lim_{x \to -\infty} b^x = \infty, \qquad \text{if } 0 < b < 1.$$

The next example makes use of limits given in this paragraph.

EXAMPLE 1

Find each of the following limits.

a. $\lim_{x \to \infty} \frac{5}{2^x}$.

As $x \to \infty$, then $2^x \to \infty$ since $b = 2 > 1$. Dividing 5 by arbitrarily large numbers results in numbers arbitrarily close to 0. Thus,

$$\lim_{x \to \infty} \frac{5}{2^x} = 0.$$

b. $\lim_{t \to \infty} \left[7 + 5(.6)^{2t+1}\right]$.

As $t \to \infty$, then $2t + 1 \to \infty$. Thus $(.6)^{2t+1} \to 0$ since $b = .6$ and $0 < .6 < 1$. Hence,

$$\lim_{t \to \infty} \left[7 + 5(.6)^{2t+1}\right] = \lim_{t \to \infty} 7 + 5 \lim_{t \to \infty} (.6)^{2t+1}$$

$$= 7 + 5(0) = 7.$$

c. $\lim_{x \to -\infty} \left[\frac{1}{x} - 2(10)^{-x}\right]$.

As $x \to -\infty$, then $-x \to \infty$. Hence $(10)^{-x} \to \infty$, since $b = 10 > 1$. Also, as $x \to -\infty$, then $1/x \to 0$. Thus,

$$\lim_{x \to -\infty} \left[\frac{1}{x} - 2(10)^{-x}\right] = -\infty.$$

One of the most **useful** numbers that is used as a base in $y = b^x$ is a certain irrational number denoted by the letter e in honor of the mathematician Euler:

$$e \text{ is approximately } 2.71828\ldots.$$

In fact, e can be defined as a limit. Figure 6-2 shows the graph of $f(x) = (1 + x)^{1/x}$. As $x \to 0$, it is clear that the limit of $(1 + x)^{1/x}$ exists. This limit is defined to be the number e.

$$\lim_{x \to 0} (1 + x)^{1/x} = e.$$

x	$(1+x)^{1/x}$	x	$(1+x)^{1/x}$
0.5	2.2500	−0.5	4.0000
0.1	2.5937	−0.1	2.8680
0.01	2.7048	−0.01	2.7320
0.001	2.7169	−0.001	2.7196

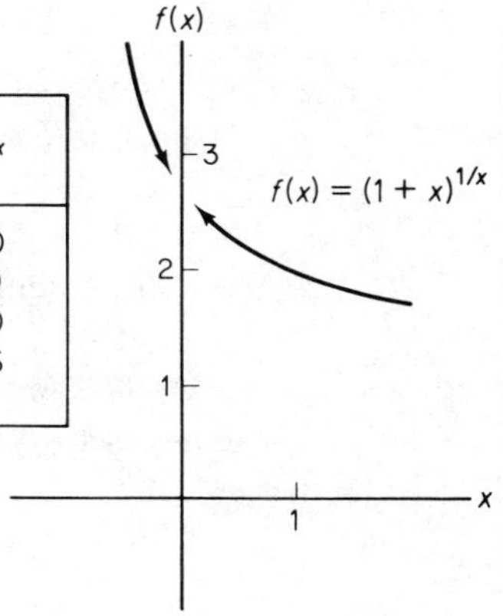

FIG. 6-2

Although e seems to be a strange number to use as a base in an exponential function, it arises quite naturally in calculus (as you will see in Sec. 6-4). It also occurs in economic analysis and problems involving natural growth (or decay), such as compound interest and population studies. A table of (approximate) values of e^x and e^{-x} is in Appendix C. The graph of $y = e^x$ is shown in Fig. 6-3.

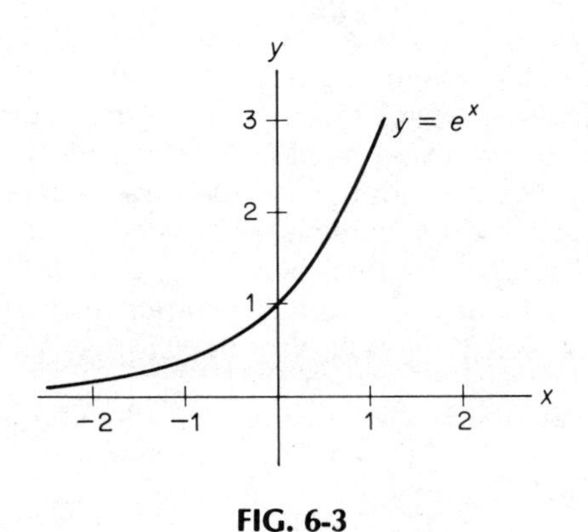

FIG. 6-3

EXAMPLE 2

The predicted population P of a city is given by

$$P = 100{,}000e^{.05t},$$

where t is the number of years after 1980. Predict the population at 2000.

We want to find P when $t = 20$.

$$P = 100{,}000e^{.05(20)} = 100{,}000e^{1} = 100{,}000e.$$

Since $e \approx 2.71828$ ($\approx$ means "is approximately"),

$$P \approx 271{,}828.$$

Many forecasts are based on population studies.

In statistics an important function that is useful as a model in describing events occurring in nature is the **Poisson distribution function:**

$$f(x) = \frac{e^{-\mu}\mu^{x}}{x!}, \qquad x = 0, 1, 2, \ldots .$$

The symbol μ (read "mu") is a Greek letter. In certain situations $f(x)$ gives the probability that exactly x events will occur in an interval of time or space. The constant μ is the mean, or average number, of occurrences in the interval. The next example illustrates the use of the Poisson distribution.

EXAMPLE 3

A hemacytometer is a counting chamber divided into squares which may be used in studying the number of microscopic structures in a liquid. In a well-known experiment **[47]**, *yeast cells were diluted and thoroughly mixed in a liquid, and the mixture was placed in a hemacytometer. With a microscope the number of yeast cells on each square were counted. The probability that there were x yeast cells on a hemacytometer square was found to fit a Poisson distribution with $\mu = 1.8$. Find the probability that there were four cells per square.*

We use the Poisson distribution function with $\mu = 1.8$ and $x = 4$.

$$f(x) = \frac{e^{-\mu}\mu^{x}}{x!}.$$

$$f(4) = \frac{e^{-1.8}(1.8)^{4}}{4!}.$$

From the table in Appendix C, $e^{-1.8} \approx .16530$ and so

$$f(4) \approx \frac{(.16530)(10.4976)}{24} \approx .072.$$

This means that in 400 squares we would *expect* $400(.072) \approx 29$ squares to contain exactly 4 cells. (In the experiment the actual number observed was 30.)

If a principal of P dollars is invested at an annual rate of $100r$ percent (for example, at 5 percent, r is .05), then at the end of one year the interest will be Pr. Thus the accumulated amount is the principal plus interest; that is, $P + Pr$, or

$$P(1 + r).$$

Under compound interest, this acts as the principal for the second year. Thus the interest at the end of the second year will be $[P(1 + r)]r$, and the accumulated amount is

$$\begin{aligned} &P(1 + r) + [P(1 + r)]r \\ &= P(1 + r)[1 + r] \qquad \text{[factoring]} \\ &= P(1 + r)^2. \end{aligned}$$

On the other hand, suppose the principal is invested at an annual rate of r compounded *twice* a year. Then in two years there are four interest periods, where the interest rate per period is $r/2$. The accumulated amount is (following the pattern above)

$$P\left(1 + \frac{r}{2}\right)^4.$$

More generally, if interest is compounded n times a year at an annual rate of r, then the rate per interest period is r/n. In t years there are nt periods and the accumulated amount A is

$$A = P\left(1 + \frac{r}{n}\right)^{nt}.$$

If $n \to \infty$, the number of interest periods increases indefinitely and the length of each period approaches 0. In this case we say that interest is being **compounded continuously**, that is, at every instant of time. The accumulated amount is

$$\lim_{n\to\infty} P\left(1 + \frac{r}{n}\right)^{nt},$$

which may be written

$$P\left[\lim_{n\to\infty}\left(1 + \frac{r}{n}\right)^{n/r}\right]^{rt}.$$

By letting $x = r/n$, then as $n \to \infty$ we have $x \to 0$. Thus the above limit within the brackets has the form $\lim_{x \to 0} (1 + x)^{1/x}$, which, as we saw before, is e. Therefore,

$$A = Pe^{rt}$$

is the accumulated amount of a principal of P dollars after t years at an annual interest rate r compounded continuously.

EXAMPLE 4

If \$100 is invested at an annual rate of 5 percent compounded continuously, find the accumulated amount at the end of (a) 1 year and (b) 5 years.

a. Here $P = 100$, $r = .05$, and $t = 1$.

$$A = Pe^{rt} = 100e^{(.05)(1)} \approx 100(1.0513) = \$105.13.$$

We can compare this value with the value after one year of a \$100 investment at an annual rate of 5 percent compounded semiannually—namely, \$105.06. The difference is not significant.

b. Here $P = 100$, $r = .05$, and $t = 5$.

$$A = 100e^{(.05)(5)} = 100e^{.25} \approx 100(1.2840) = \$128.40.$$

If we solve $A = Pe^{rt}$ for P, we get $P = A/e^{rt}$ or

$$P = Ae^{-rt}.$$

Here P is the principal that must be invested now at an annual rate of r so that at the end of t years at continuous interest the accumulated amount will be A. We call P the **present value** of A.

EXAMPLE 5

A trust fund is being set up by a single payment so that at the end of 20 years there will be \$25,000 in the fund. If interest is compounded continuously at an annual rate of 7 percent, how much money should be paid into the fund initially?

We want the present value of \$25,000 due in 20 years.

$$P = Ae^{-rt} = 25{,}000e^{-(.07)(20)}$$
$$= 25{,}000e^{-1.4} \approx 25{,}000(.24660)$$
$$= 6165.$$

The present value is \$6165.

Exercise 6-1

In Problems **1–6**, *graph each function.*

1. $y = f(x) = 4^x$.

2. $y = f(x) = (\frac{1}{3})^x$.

3. $y = f(x) = 2(4^{-x})$.

4. $y = f(x) = \frac{1}{2}(3^{x/2})$.

5. $y = f(x) = 2^x - 1$.

6. $y = f(x) = 2^{x-1}$.

In Problems **7–14**, *find the limits.*

7. $\lim_{x\to\infty} 12^x$.

8. $\lim_{x\to-\infty} (.4)^x$.

9. $\lim_{x\to\infty} (4 - 3^{-x})$.

10. $\lim_{x\to\infty} \frac{9}{7(5)^x}$.

11. $\lim_{p\to-\infty} \frac{6}{2 - (1/2)^p}$.

12. $\lim_{t\to\infty} (2^t + t^2)$.

13. $\lim_{x\to\infty} \frac{2 + 5e^{1-2x}}{8}$.

14. $\lim_{x\to\infty} \frac{e^{-x}}{e^{2x-1}}$.

In Problems **15–18**, *use the table in Appendix C to find the approximate value of each expression.*

15. $e^{1.5}$.

16. $e^{3.4}$.

17. $e^{-.4}$.

18. $e^{-3/4}$.

19. The predicted population P of a city is given by $P = 125{,}000(1.12)^{t/20}$, where t is the number of years after 1980. Predict the population in 2000.

20. For a certain city the population P grows at the rate of 2 percent per year. The formula $P = 1{,}000{,}000(1.02)^t$ gives the population t years after 1980. Find the population in (a) 1980 and (b) in 1982.

21. The probability P that a telephone operator will receive exactly x calls during a certain time interval is

$$P = \frac{e^{-3}3^x}{x!}.$$

Find the probability that exactly three calls will be received. Give the answer to four decimal places.

22. A mail order firm finds that the proportion P of small towns having exactly x persons responding to a magazine advertisement is given approximately by the formula

$$P = \frac{e^{-.5}(.5)^x}{x!}.$$

From what proportion of small towns can the firm expect exactly two people to respond? Give the answer to four decimal places.

23. At a certain time there are 100 milligrams (mg) of a radioactive substance. It decays so that after t years the number of milligrams present, A, is given by $A = 100e^{-.035t}$. How many milligrams are present after 20 years? Give your answer to the nearest milligram.

24. The demand equation for a certain product is $q = 80 - 2^p$. Sketch its graph and choose q for the horizontal axis.

25. Suppose the number x of patients admitted into a hospital emergency room during a certain hour of the day has a Poisson distribution with mean 4. Find the probability that during that hour there will be exactly two emergency patients. Give the answer to four decimal places.

26. In a classic study [29], records of 10 army corps of Prussian cavalrymen were analyzed over a period of 20 years. There were 200 readings in terms of corps-years. It was found that the probability that x cavalrymen per year per corps being kicked to death by a horse fit a Poisson distribution with $\mu = .61$. (a) Find the probability that $x = 2$. (b) To the nearest unit, how many corps-years does the Poisson distribution predict with this number of deaths?

27. In a study of the effects of food deprivation on hunger [18], an insect was fed until its appetite was completely satisfied. Then it was deprived of food for t hr (deprivation period). At the end of this period the insect was again fed until its appetite was completely satisfied. The approximate weight H (in grams) of the food that was consumed at this time was found to be a function of t, where

$$H = 1.00\left[1 - e^{-(.0464t+.0670)}\right].$$

Here H is a measure of hunger. If it is assumed that the deprivation period can increase without bound and that the insect does not die, to what value would H tend?

28. In a psychological experiment involving learning [26], subjects were asked to give particular responses after being shown certain stimuli. Each stimulus was a pair of letters, and each response was either the digit 1 or 2. After

each response the subject was told the correct answer. In this so-called *paired-associate* learning experiment, the theoretical probability P that a subject makes a correct response on the nth trial is given by

$$P = 1 - \tfrac{1}{2}(1 - c)^{n-1}, \qquad n \geqslant 1, \ \ 0 < c < 1.$$

Find P when $n = 1$, and find $\lim_{n \to \infty} P$.

29. If \$1000 is invested at an annual rate of 6 percent compounded continuously, find the accumulated amount at the end of eight years.

30. If \$100 is deposited in a savings account that earns interest at an annual rate of $5\frac{1}{2}$ percent compounded continuously, what is the value of the account at the end of two years?

31. The board of directors of a corporation agrees to redeem some of its callable preferred stock in five years. At that time \$1,000,000 will be required. If the corporation can invest money at an annual interest rate of 6 percent compounded continuously, how much should it presently invest so that the future value is sufficient to redeem the shares?

32. A trust fund is being set up by a single payment so that at the end of 30 years there will be \$50,000 in the fund. If interest is compounded continuously at an annual rate of 5 percent, how much money should be paid into the fund initially?

33. The formula $A = Pe^{-rn}$ gives the amount at the end of n years of a principal P which depreciates at a rate of r per year compounded continuously. What is the value at the end of ten years of \$60,000 of machinery which depreciates at a rate of 8 percent, compounded continuously? Give the answer to the nearest dollar.

34. An important function used in decision making is the *normal distribution density function*, which in standard form is

$$y = f(x) = \frac{1}{\sqrt{2\pi}} e^{-x^2/2}.$$

Evaluate $f(0)$, $f(-1)$, and $f(1)$ by using $\dfrac{1}{\sqrt{2\pi}} \approx .399$.

35. The demand equation for a new toy is $q = 10{,}000(.95123)^p$. It is desired to evaluate q when $p = 10$. To convert the equation into a more desirable computational form, use Appendix C to show that $q = 10{,}000e^{-.05p}$. Then evaluate and give the answer to the nearest integer. *Hint*: Find a number x such that $.95123 \approx e^{-x}$.

36. *Calculator Problem.* Evaluate $f(x) = x^{2x}$ when $x = 1, .5, .2, .1, .01, .001$, and $.0001$. From your results, draw a conclusion about $\lim_{x \to 0^+} x^{2x}$.

6-2 LOGARITHMIC FUNCTIONS

Another important type of function is a *logarithmic function*, which is related to an exponential function. Figure 6-4 shows the graph of the exponential function $s = f(t) = 2^t$. Here f sends an input number t into a *positive* output number s:

$$f : t \to s, \quad \text{where} \quad s = 2^t.$$

For example, f sends 2 into 4:

$$f : 2 \to 4.$$

Now look at the same curve in Fig. 6-5. There you can see that with each positive number s we can associate exactly one value of t.

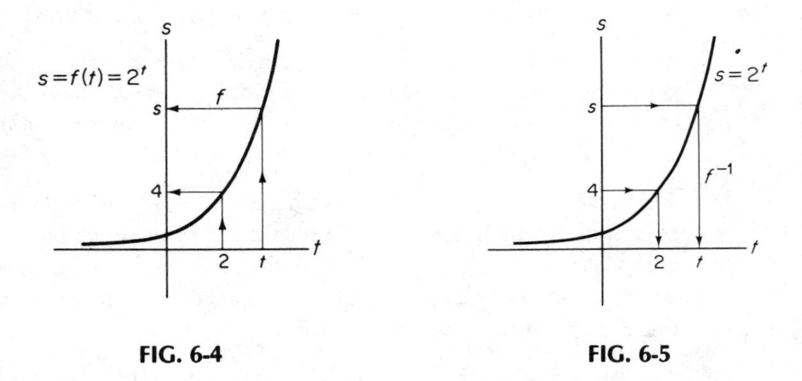

FIG. 6-4 FIG. 6-5

With $s = 4$ we associate $t = 2$. Let's think here of s as an input and t as an output. Then we have a function that sends s's into t's. We'll denote this function by the symbol f^{-1} (read "f inverse").†

$$f^{-1} : s \to t, \quad \text{where} \quad s = 2^t.$$

Thus, $f^{-1}(s) = t$.

The functions f and f^{-1} are related. In Fig. 6-6, from the arrows you can see that f^{-1} *reverses* the action of f, and vice versa. For example,

$$f \text{ sends 2 into 4, and } f^{-1} \text{ sends 4 into 2.}$$

†f^{-1} stands for a new function. It does not mean $\dfrac{1}{f}$.

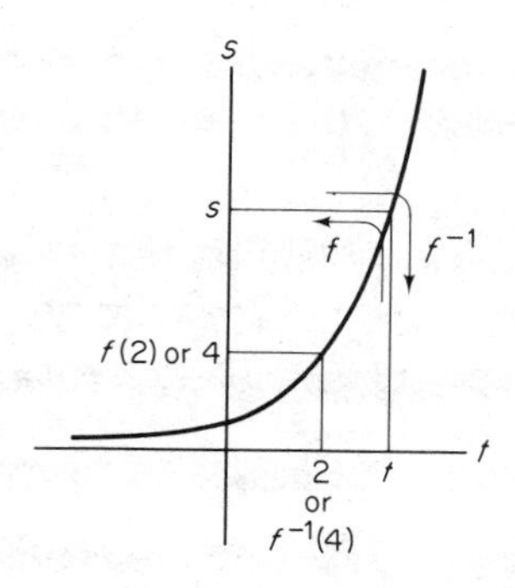

FIG. 6-6

In terms of composition,

$$f^{-1}[f(2)] = f^{-1}[4] = 2.$$

More generally,

$$f^{-1}[f(t)] = t.$$

This is what is really meant by saying f^{-1} reverses the action of f. We also have $f[f^{-1}(4)] = f[2] = 4$. Generalizing this, we have

$$f[f^{-1}(s)] = s.$$

Notice that the domain of f^{-1} is the range of f (all positive numbers), and the range of f^{-1} is the domain of f (all real numbers).

We give a special name to f^{-1}. It is called the **logarithmic function base 2**. Usually we write f^{-1} as $\log_2$ (read "log base 2"). Thus $\log_2$ is merely a symbol for a special function.

In summary,

$$\text{if } s = f(t) = 2^t, \quad \text{then} \quad f^{-1}(s) = \log_2(s) = t. \tag{1}$$

The domain of $\log_2$ is all positive numbers and the range is all reals.

Generalizing to other bases and replacing s by x and t by y in (1) gives the following definition.

DEFINITION. *The **logarithmic function base b**, denoted $\log_b$, is defined by*

$$y = \log_b x \quad \text{if and only if} \quad b^y = x.$$

The domain of $\log_b$ is all positive numbers and its range is all real numbers.

The logarithmic function reverses the action of the exponential function. Because of this we sometimes say that the logarithmic function is the *inverse* of the exponential function.

Always remember: when we say that the log base b of x is y, we mean that b raised to the y power is x.

$$\log_b x = y \quad \text{means} \quad b^y = x.$$

In this sense, *the logarithm of a number is an exponent.* For example,

$$\log_2 8 = 3 \text{ because } 2^3 = 8.$$

We say that $\log_2 8 = 3$ is the **logarithmic form** of the **exponential form** $2^3 = 8$.

EXAMPLE 1

a. Since $25 = 5^2$, then $\log_5 25 = 2$.

b. Since $10^0 = 1$, then $\log_{10} 1 = 0$.

c. If $6y = e^{2r}$, then $\log_e 6y = 2r$ and so $r = [\log_e (6y)]/2$.

d. $\log_{10} 100 = 2$ means $10^2 = 100$.

e. $\log_{64} 8 = \frac{1}{2}$ means $64^{1/2} = 8$.

f. $\log_2 \frac{1}{16} = -4$ means $2^{-4} = \frac{1}{16}$.

EXAMPLE 2

Graph the function $y = \log_2 x$.

To plot points it is more convenient to use the equivalent form $2^y = x$. If $y = 0$, then $x = 1$, giving the point (1, 0). Other points are shown in Fig. 6-7.

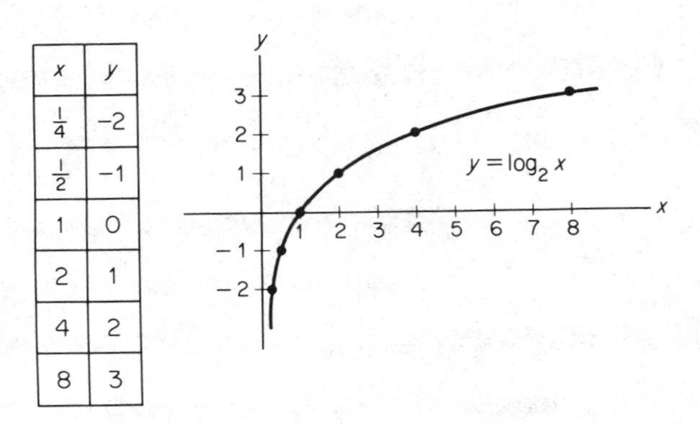

x	y
$\frac{1}{4}$	-2
$\frac{1}{2}$	-1
1	0
2	1
4	2
8	3

FIG. 6-7

Note that the domain is all positive numbers and the range is all real numbers.

Figure 6-8 shows the graph of $y = \log_e x$. Notice that it has the same shape as Fig. 6-7.

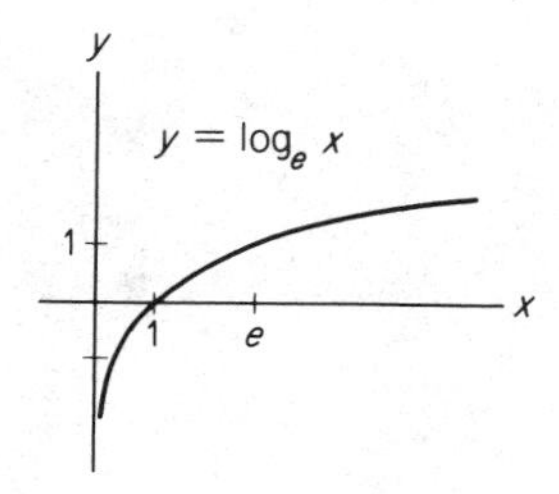

FIG. 6-8

Logarithms to the base 10 are called **common logarithms**. They were often used to simplify computations before the pocket-calculator age. The subscript 10 is generally omitted from the notation. Thus,

$$\log x \quad \text{means} \quad \log_{10} x.$$

Important in calculus are logarithms to the base e, called **natural** (or Naperian) **logarithms**. We use the symbol "ln" for such logarithms. Thus,

$$\ln x \quad \text{means} \quad \log_e x.$$

In Appendix D is a table of approximate values of natural logarithms. From there you can see that $\ln 2 \approx .69315$. This means that $e^{.69315} \approx 2$.

EXAMPLE 3

Find each of the following.

a. log 1000.

Here the base is 10. Let $\log 1000 = y$. Converting to exponential form, we have

$$10^y = 1000.$$

Clearly y must be 3. Thus $\log 1000 = 3$.

b. ln 1.

Here the base is e. Let $\ln 1 = y$. Converting to exponential form, we have

$$e^y = 1.$$

Clearly y must be 0. Thus $\ln 1 = 0$.

c. log .1.

$$\log .1 = y.$$

$$10^y = .1 = \tfrac{1}{10} = 10^{-1}.$$

Thus $y = -1$ and $\log .1 = -1$.

d. ln e.

$$\ln e = y.$$

$$e^y = e.$$

Thus $y = 1$ and $\ln e = 1$.

EXAMPLE 4

Solve each equation for x.

a. $\log_3 x = 4$.

$$\log_3 x = 4,$$

or equivalently

$$3^4 = x,$$

$$81 = x.$$

b. $x + 1 = \log_4 16$.

$$x + 1 = \log_4 16.$$

$$4^{x+1} = 16.$$

From inspection the exponent $x + 1$ must be 2 and so $x = 1$.

c. $\log_x 49 = 2$.

$$\log_x 49 = 2.$$

$$x^2 = 49.$$

$$x = 7.$$

We rejected $x = -7$ in solving $x^2 = 49$, since a negative number cannot be a base of a logarithmic function.

d. $12 = 5 + 3(4)^{x-1}$.

$$\begin{aligned} 12 &= 5 + 3(4)^{x-1} \\ 7 &= 3(4)^{x-1}, \\ \tfrac{7}{3} &= 4^{x-1}, \\ x - 1 &= \log_4 \tfrac{7}{3}, \\ x &= 1 + \log_4 \tfrac{7}{3}. \end{aligned}$$

EXAMPLE 5

If interest is compounded continuously at an annual rate r, how long would it take for a principal of P to double?

The accumulated amount A after t years at an annual rate r is $A = Pe^{rt}$. When P doubles, then $A = 2P$. Thus $2P = Pe^{rt}$ and so $2 = e^{rt}$. This means that $rt = \ln 2$ or

$$t = \frac{\ln 2}{r} \approx \frac{.69315}{r}.$$

For example, if $r = .05$, then money would double itself in $.69315/.05 \approx 13.9$ years.

Exercise 6-2

In Problems **1–12**, *express each logarithmic form exponentially and each exponential form logarithmically.*

1. $16^{1/2} = 4$.

2. $2 = \log_{12} 144$.

3. $10^4 = 10{,}000$.

4. $\log_{1/2} 4 = -2$.

5. $\log_2 64 = 6$.

6. $8^{2/3} = 4$.

7. $\log_2 x = 14$.

8. $10^{.48302} = 3.041$.

9. $e^2 = 7.3891$.

10. $e^{.33647} = 1.4$.

11. $\ln 3 = 1.09861$.

12. $\log 5 = .6990$.

In Problems **13** *and* **14**, *graph the functions.*

13. $y = f(x) = \log_3 x$.

14. $y = f(x) = \log_{1/2} x$.

In Problems **15–44**, *find x.*

15. $\log_2 x = 4$.

16. $\log_3 x = 2$.

17. $\log_5 x = 3$.

18. $\log_4 x = 0$.

19. $\log x = -1$.

20. $\ln x = 1$.

21. $\ln x = 2$.

22. $\log_x 100 = 2$.

23. $\log_x 8 = 3$.

24. $\log_x 3 = \frac{1}{2}$.

25. $\log_x \frac{1}{6} = -1$.

26. $\log_x y = 1$.

27. $\log_4 16 = x$.

28. $\log_3 1 = x$.

29. $\log 10{,}000 = x$.

30. $\log_2 \frac{1}{16} = x$.

31. $\log_{25} 5 = x$.

32. $\log_9 9 = x$.

33. $\log_3 x = -4$.

34. $\log_x (2x - 3) = 1$.

35. $\log_x (6 - x) = 2$.

36. $\log_8 64 = x - 1$.

37. $2^x = 5$.

38. $4^{x+3} = 7$.

39. $e^{3x} = 2$.

40. $\dfrac{8}{3^x} = 4$.

41. $5(3^x - 6) = 10$.

42. $.1e^{.1x} = .5$.

43. $e^{2x-5} + 1 = 4$.

44. $3e^{2x} - 1 = \frac{1}{2}$.

In Problems **45–48**, *use the tables in Appendix D to find the approximate value of each expression.*

45. $\ln 5$.

46. $\ln 3.12$.

47. $\ln 7.39$.

48. $\ln 9.98$.

49. The cost c for a firm producing q units of a product is given by the cost function $c = (2q \ln q) + 20$. Evaluate the cost when $q = 6$. Give your answer to two decimal places.

50. If interest is compounded continuously, at what annual percentage rate will a principal of P double in ten years? Give your answer to the nearest percent.

51. If interest is compounded continuously at an annual rate of .05, how many years would it take for a principal P to triple? Give your answer to the nearest year.

52. For a certain population of cells, the number N of cells at time t is given by $N = N_0(2^{t/k})$, where N_0 is the number of cells at $t = 0$ and k is a positive constant. (a) Find N when $t = k$. (b) What is the significance of k? (c) Show that the time it takes to have population N_1 can be written $t = k \log_2 (N_1/N_0)$.

53. The magnitude M of an earthquake and its energy E are related in [5] by the equation

$$1.5M = \log\left(\frac{E}{2.5 \times 10^{11}}\right).$$

Here M is given in terms of Richter's preferred scale of 1958 and E is in ergs. Solve the equation for E.

54. In a discussion of market penetration by new products, Hurter and Rubenstein [22] refer to the function

$$F(t) = \frac{q - pe^{-(t+C)(p+q)}}{q[1 + e^{-(t+C)(p+q)}]},$$

where p, q, and C are constants. They claim that if $F(0) = 0$, then

$$C = -\frac{1}{p+q}\ln\frac{q}{p}.$$

Show that their claim is true.

55. Suppose the daily output of q units of a new product on the tth day of a production run is given by $q = 500(1 - e^{-.2t})$. Such an equation is called a *learning equation* and indicates that as time progresses, output per day will increase. This may be due to the gain of the workers' proficiency at their jobs. Determine to the nearest complete unit the output on the (a) first day, and (b) tenth day after the start of a production run. (c) After how many days will a daily production run of $q = 400$ units be reached? Assume $\ln 0.2 \approx -1.6$.

56. *Calculator Problem.* Evaluate $f(x) = x \ln x$ when $x = 1, .5, .2, .1, .01, .001$, and $.0001$. From your results draw a conclusion about $\lim_{x\to 0^+} x \ln x$.

6-3 PROPERTIES OF LOGARITHMS

Some basic properties of logarithms deserve mention.

PROPERTY 1. $\log_b(mn) = \log_b m + \log_b n$.

In short, the logarithm of a product is a sum of logarithms.

PROOF. Let $x = \log_b m$ and $y = \log_b n$. Then $b^x = m$ and $b^y = n$. Thus,

$$mn = b^x b^y = b^{x+y}.$$

Since $mn = b^{x+y}$, then $\log_b(mn) = x + y$. Hence,

$$\log_b(mn) = \log_b m + \log_b n.$$

We shall not prove the next two properties, since their proofs are similar to that of Property 1.

PROPERTY 2. $\log_b \frac{m}{n} = \log_b m - \log_b n.$

In short, the logarithm of a quotient is a difference of logarithms.

PROPERTY 3. $\log_b m^n = n \log_b m.$

In short, the logarithm of a power of a number is the exponent times a logarithm.

Table 6-1 gives values of a few common logarithms. Most entries are approximate. Notice that $\log 4 \approx .6021$, which means $10^{.6021} \approx 4$. We shall use this table in some of the examples and exercises that follow.

TABLE 6-1 Common Logarithms

x	$\log x$	x	$\log x$
2	.3010	7	.8451
3	.4771	8	.9031
4	.6021	9	.9542
5	.6990	10	1.0000
6	.7782	e	.4343

EXAMPLE 1

a. *Find* log 56.

Log 56 is not in Table 6-1. However, we can write 56 as the product $8 \cdot 7$. Thus by Property 1,

$$\log 56 = \log (8 \cdot 7) = \log 8 + \log 7 \approx .9031 + .8451 = 1.7482.$$

b. *Find* $\log \frac{9}{2}$.

By Property 2,

$$\log \tfrac{9}{2} = \log 9 - \log 2 \approx .9542 - .3010 = .6532.$$

c. *Find* log 64.

Since $64 = 8^2$, then by Property 3,

$$\log 64 = \log 8^2 = 2 \log 8 \approx 2(.9031) = 1.8062.$$

d. *Find* $\log \sqrt{5}$.

$$\log\sqrt{5} = \log 5^{1/2} = \tfrac{1}{2}\log 5 \approx \tfrac{1}{2}(.6990) = .3495.$$

e. *Find* $\log \frac{15}{7}$.

$$\log \frac{15}{7} = \log \frac{3 \cdot 5}{7} = \log (3 \cdot 5) - \log 7$$
$$= \log 3 + \log 5 - \log 7$$
$$\approx .4771 + .6990 - .8451 = .3310.$$

EXAMPLE 2

a. *Simplify* $\log_3 \frac{1}{x^2}$.

$$\log_3 \frac{1}{x^2} = \log_3 x^{-2} = -2 \log_3 x \qquad \text{[Property 3].}$$

b. *Write* $-\log\frac{x}{2}$ *without using a minus sign.*

$$-\log\frac{x}{2} = (-1)\log\frac{x}{2}$$
$$= \log\left(\frac{x}{2}\right)^{-1} \qquad \text{[Property 3]}$$
$$= \log\frac{2}{x}.$$

In general, $-\log_b\frac{m}{n} = \log_b\frac{n}{m}$.

EXAMPLE 3

a. *Write* $\log_4 x - \log_4 (x + 3)$ *as a single logarithm.*

$$\log_4 x - \log_4 (x + 3) = \log_4 \frac{x}{x + 3} \qquad \text{[Property 2].}$$

b. *Write* $3 \log_2 10 + \log_2 15$ *as a single logarithm.*

$$3 \log_2 10 + \log_2 15 = \log_2 (10^3) + \log_2 15 \qquad \text{[Property 3]}$$
$$= \log_2 [(10^3)15] \qquad \text{[Property 1]}$$
$$= \log_2 15{,}000.$$

c. *Write* $\ln \sqrt[3]{\frac{x^5(x-2)^8}{x-3}}$ *in terms of* $\ln x$, $\ln(x-2)$, *and* $\ln(x-3)$.

$$
\begin{aligned}
\ln \sqrt[3]{\frac{x^5(x-2)^8}{x-3}} &= \ln\left[\frac{x^5(x-2)^8}{x-3}\right]^{1/3} = \frac{1}{3}\ln\frac{x^5(x-2)^8}{x-3} \\
&= \frac{1}{3}\{\ln [x^5(x-2)^8] - \ln (x-3)\} \\
&= \frac{1}{3}[\ln x^5 + \ln (x-2)^8 - \ln (x-3)] \\
&= \frac{1}{3}[5 \ln x + 8 \ln (x-2) - \ln (x-3)].
\end{aligned}
$$

Since $b^0 = 1$ and $b^1 = b$, then by converting to logarithmic forms we have the following properties:

PROPERTY 4. $\log_b 1 = 0.$

PROPERTY 5. $\log_b b = 1.$

By Property 3, $\log_b b^n = n \log_b b$. But, by Property 5, $\log_b b = 1$. Thus we have the next property.

PROPERTY 6. $\log_b b^n = n.$

EXAMPLE 4

a. *Find* $\ln e$.

$$\ln e = \log_e e = 1 \qquad \text{[Property 5]}.$$

b. *Find* $\log 10^c$.

$$\log 10^c = \log_{10} 10^c = c \qquad \text{[Property 6]}.$$

c. *Find* $\log \frac{200}{21}$.

$$
\begin{aligned}
\log \tfrac{200}{21} &= \log 200 - \log 21 = \log (2 \cdot 100) - \log (7 \cdot 3) \\
&= \log 2 + \log 100 - (\log 7 + \log 3) \\
&\approx .3010 + 2 - (.8451 + .4771) \quad [\text{since } \log 100 = \log 10^2 = 2] \\
&= .9788.
\end{aligned}
$$

d. *Find* $\log_7 \sqrt[9]{7^8}$.

$$\log_7 \sqrt[9]{7^8} = \log_7 7^{8/9} = \tfrac{8}{9}.$$

e. *Find* $\log_3 (\frac{27}{81})$.

$$\log_3\left(\frac{27}{81}\right) = \log_3\left(\frac{3^3}{3^4}\right) = \log_3 (3^{-1}) = -1.$$

f. *Find* $\ln e + \log \frac{1}{10}$.

$$\ln e + \log \tfrac{1}{10} = \ln e + \log 10^{-1} = \log_e e + \log_{10} 10^{-1}$$
$$= 1 + (-1) = 0.$$

For many functions f, if $f(m) = f(n)$, this does not imply that $m = n$. For example, if $f(x) = x^2$ and $m = 2$ and $n = -2$, then $f(m) = f(n)$, but $m \neq n$. This is not the case for the logarithmic function. Notice in Fig. 6-9 that if x_1 and x_2 are different, then their logarithms are

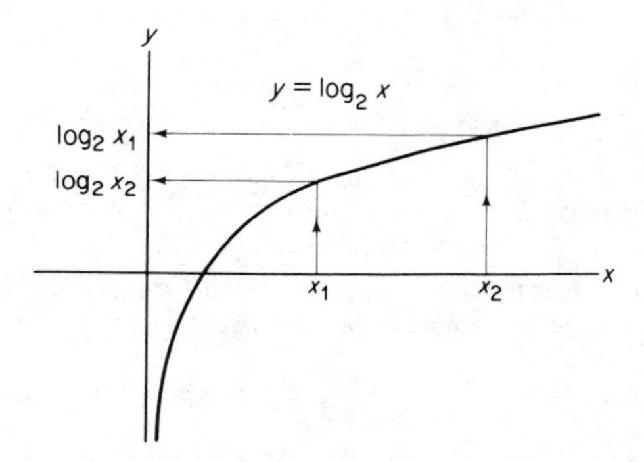

FIG. 6-9

different. This means that if $\log_2 m = \log_2 n$, then $m = n$. Generalizing to base b, we have the following property:

PROPERTY 7. *If* $\log_b m = \log_b n$, *then* $m = n$.

There is a similar property for exponentials:

PROPERTY 8. *If* $b^m = b^n$, *then* $m = n$.

EXAMPLE 5

a. *Solve* $\log_b x + \log_b (2x) = \log_b 100$ *for* x.

$$\log_b x + \log_b (2x) = \log_b 100.$$
$$\log_b [(x)(2x)] = \log_b 100.$$
$$\log_b (2x^2) = \log_b 100.$$

By Property 7,

$$2x^2 = 100.$$
$$x^2 = 50.$$
$$x = 5\sqrt{2}.$$

(Why do we ignore $x = -5\sqrt{2}$?)

b. *Find x if* $(25)^{x+2} = 5^{3x-4}$.

Since $25 = 5^2$, we can express both sides of the equation as powers of 5.

$$(25)^{x+2} = 5^{3x-4}.$$
$$(5^2)^{x+2} = 5^{3x-4}.$$
$$5^{2x+4} = 5^{3x-4}.$$
$$2x + 4 = 3x - 4 \quad \text{[Property 8]}.$$
$$x = 8.$$

EXAMPLE 6

An experiment was conducted with a particular type of small animal **[39]**. *The logarithm of the amount of oxygen consumed per hour was determined for a number of the animals and was plotted against the logarithms of the weights of the animals. It was found that*

$$\log y = \log 5.934 + .885 \log x,$$

where y was the number of microliters of oxygen consumed per hour, and x was the weight of the animal (in grams). Solve for y.

$$\log y = \log 5.934 + .885 \log x$$
$$= \log 5.934 + \log x^{.885}.$$
$$\log y = \log (5.934x^{.885}).$$

By Property 7,

$$y = 5.934x^{.885}.$$

PROPERTY 9. $b^{\log_b x} = x$, *and in particular* $10^{\log x} = x$ *and* $e^{\ln x} = x$.

PROOF. Let $t = b^{\log_b x}$. Writing this in logarithmic form, we then have $\log_b t = \log_b x$. By Property 7, $t = x$ and so $x = b^{\log_b x}$.

EXAMPLE 7

a. *Find* $2^{\log_2 6}$.

$$2^{\log_2 6} = 6 \quad \text{[Property 9]}.$$

b. *Solve* $10^{\log x^2} = 25$ *for x.*

$$10^{\log x^2} = 25.$$
$$x^2 = 25.$$
$$x = \pm 5.$$

c. *Evaluate* $e^{(\ln 3 + 2 \ln 4)}$.

$$e^{(\ln 3+2 \ln 4)} = e^{(\ln 3+\ln 4^2)} = e^{\ln(3\cdot 4^2)} = 3 \cdot 4^2 = 48.$$

EXAMPLE 8

Find $\log_5 2$.

Let $x = \log_5 2$. Then $5^x = 2$, and by taking common logs of both sides we get

$$\log 5^x = \log 2.$$
$$x \log 5 = \log 2.$$
$$x = \frac{\log 2}{\log 5} \approx \frac{.3010}{.6990} \approx .4306.$$

If we had taken natural logs of both sides, the result would be $x = (\ln 2)/(\ln 5) \approx .69315/1.60944 \approx .43068$. This differs from our previous result because of the accuracy of the tables involved.

Generalizing the method used in Example 8, we have

$$\log_b N = \frac{\log N}{\log b}$$

and

$$\log_b N = \frac{\log_a N}{\log_a b}, \tag{1}$$

which are called **change of base formulas.** With these we can convert logarithms from one base to another.

EXAMPLE 9

Write $\log x$ *in terms of natural logarithms.*

In Eq. (1), let $b = 10$, $N = x$, and $a = e$. Then

$$\log_{10} x = \frac{\log_e x}{\log_e 10},$$

$$\text{or} \quad \log x = \frac{\ln x}{\ln 10}.$$

EXAMPLE 10

A demand equation for a product is defined by $p = 12^{1-.1q}$. Use common logarithms to express q in terms of p.

Taking common logs of both sides of $p = 12^{1-.1q}$ gives

$$\log p = \log (12^{1-.1q}).$$

$$\log p = (1 - .1q) \log 12.$$

$$\frac{\log p}{\log 12} = 1 - .1q.$$

$$.1q = 1 - \frac{\log p}{\log 12}.$$

$$q = 10\left(1 - \frac{\log p}{\log 3 + \log 4}\right).$$

$$q = 10\left(1 - \frac{\log p}{1.0792}\right).$$

Exercise 6-3

In Problems **1–18**, *find the given values.*

1. $\log 35$.

2. $\log 12$.

3. $\log \frac{9}{4}$.

4. $\log \frac{7}{10}$.

5. $\log 25$.

6. $\log .0001$.

7. $\log 2000$.

8. $\log 900$.

9. $\log_7 7^{48}$.

10. $\log_5 (5\sqrt{5})^5$.

11. $\log_2 (2^6/2^{10})$.

12. $\log_7 4$.

13. $\log_2 3$.

14. $\ln e$.

15. $\log_3 \sqrt[3]{3}$.

16. $\log_2 4$.

17. $\log 10 + \ln e^3$.

18. $\log 10^e$.

In Problems **19–26**, *express each of the given forms as a single logarithm.*

19. $\log 7 + \log 4$.

20. $\log_3 10 - \log_3 5$.

21. $\log_2 (2x) - \log_2 (x + 1)$.

22. $2 \log x - \frac{1}{2}\log (x - 2)$.

23. $9 \log 7 + 5 \log 23$.

24. $3(\log x + \log y - 2 \log z)$.

25. $2 + 10 \log 1.05$.

26. $\frac{1}{2}(\log 215 + 8 \log 6 - 3 \log 121)$.

In Problems **27–32**, *write each expression in terms of* $\log x$, $\log(x+2)$, *and* $\log(x-3)$.

27. $\log[x(x+2)(x-3)]$.

28. $\log \dfrac{x^2(x+2)}{x-3}$.

29. $\log \dfrac{\sqrt{x}}{(x+2)(x-3)^2}$.

30. $\log\left[(x-3)\sqrt{x(x+2)}\right]$.

31. $\log \sqrt{\dfrac{x^2(x-3)^3}{x+2}}$.

32. $\log \dfrac{1}{x(x-3)^2(x+2)^3}$.

In Problems **33–46**, *find* x.

33. $e^{2x} \cdot e^{5x} = e^{14}$.

34. $(e^{5x+1})^2 = e$.

35. $(16)^{3x} = 2$.

36. $(27)^{2x+1} = \frac{1}{3}$.

37. $e^{\ln(2x)} = 5$.

38. $4^{\log_4 x + \log_4 2} = 3$.

39. $10^{\log x^2} = 4$.

40. $e^{3 \ln x} = 8$.

41. $\log(2x+1) = \log(x+6)$.

42. $\log x + \log 3 = \log 5$.

43. $\log x - \log(x-1) = \log 4$.

44. $\log_2 x + 3\log_2 2 = \log_2(2/x)$.

45. $\log(x+2)^2 = 2$, where $x > 0$.

46. $\ln x = \ln(3x+1) + 1$.

47. A manufacturer has determined that his supply equation is $p = \log[10 + (q/2)]$, where q is the number of units he will supply at a price p per unit. At what price will he supply (a) 1980 units; (b) 11,980 units?

48. Suppose (x, y) lies on the graph of $y = ce^{-kx}$, where c and k are constants. For what value of L does $(x+L, y/e)$ lie on the graph? (Note that if x increases by L, then y reduces to a fraction of $1/e$ of its original value.)

49. On the surface of a glass slide is a grid that divides the surface into 225 equal squares. Suppose a blood sample containing N red cells is spread on the slide and the cells are randomly distributed. Then the number of squares containing no cells is (approximately) given by $225e^{-N/225}$. If 100 of the squares contain no cells, estimate the number of cells the blood sample contained.

50. In a discussion **[48]** of the rate of cooling of isolated portions of the body when they are exposed to low temperatures, the following equation occurs:

$$T_t - T_e = (T_t - T_e)_o e^{-at},$$

where T_t is temperature of the portion at time t, T_e is environmental temperature, the subscript o refers to initial temperature difference, and a is a constant. Show that

$$a = \frac{1}{t}\ln\frac{(T_t - T_e)_o}{T_t - T_e}.$$

51. In an article concerning predators and prey, Holling [19] refers to an equation of the form $y = K(1 - e^{-ax})$, where x is prey density, y is the number of prey attacked, and K and a are constants. Verify his claim that

$$\ln\frac{K}{K - y} = ax.$$

52. In statistics the sample regression equation $y = ab^x$ is reduced to a linear form by taking logarithms of both sides. Find $\log y$.

53. According to Richter [41], the magnitude M of an earthquake occurring 100 km from a certain type of seismometer is given by $M = \log(A) + 3$, where A is the recorded trace amplitude (in millimeters) of the quake. (a) Find the magnitude of an earthquake that records a trace amplitude of 1 mm. (b) If a particular earthquake has amplitude A_1 and magnitude M_1, determine the magnitude of a quake with amplitude $100A_1$. Express your answer to (b) in terms of M_1.

54. In an article, Taagepera and Hayes [51] refer to an equation of the form

$$\log T = 1.7 + .2068 \log P - .1334 \log^2 P.$$

Here T is the percent of a country's gross national product (GNP) that corresponds to foreign trade (exports plus imports), and P is the country's population (in units of 100,000). Verify the claim that

$$T = 50P^{(.2068 - .1334 \log P)}.$$

You may assume that $\log 50 \approx 1.7$.

55. In a study of rooted plants in a certain geographic region [39], it was determined that on plots of size A (in square meters), the average number of species that occurred was S. When $\log S$ was graphed as a function of $\log A$, the result was a straight line given by

$$\log S = \log 12.4 + .26 \log A.$$

Solve for S.

56. Express $\log(x + 8)$ in terms of natural logarithms.

57. The demand equation for a certain product is $q = 80 - 2^p$. Solve for p and express your answer in terms of common logarithms as in Example 10. Evaluate p to two decimal places when $q = 60$.

58. The vice president of a company believes that for his company's product, the number of units q sold per year after t years from the date he assumed his position is given by $q = 1000(\frac{1}{2})^{.8^t}$. Such an equation is called a *Gompertz equation* and describes natural growth in many areas of study. Solve this equation for t in the same manner as in Example 10 and show that

$$t = \frac{\log\left(\dfrac{3 - \log q}{\log 2}\right)}{(3 \log 2) - 1}.$$

6-4 DERIVATIVES OF LOGARITHMIC FUNCTIONS

We now consider the derivative of the logarithmic function $y = f(x) = \ln x$, where x is positive. By the definition of the derivative,

$$\frac{d}{dx}(\ln x) = \lim_{h\to 0} \frac{f(x+h) - f(x)}{h} = \lim_{h\to 0} \frac{\ln(x+h) - \ln x}{h}.$$

Since the difference of logarithms is a logarithm of a quotient, we can write

$$\frac{d}{dx}(\ln x) = \lim_{h\to 0} \frac{\ln\left(\dfrac{x+h}{x}\right)}{h}$$

$$= \lim_{h\to 0}\left[\frac{1}{h}\ln\left(\frac{x+h}{x}\right)\right] = \lim_{h\to 0}\left[\frac{1}{h}\ln\left(1 + \frac{h}{x}\right)\right].$$

Writing $\dfrac{1}{h}$ as $\dfrac{1}{x}\cdot\dfrac{x}{h}$, we have

$$\frac{d}{dx}(\ln x) = \lim_{h\to 0}\left[\frac{1}{x}\cdot\frac{x}{h}\ln\left(1 + \frac{h}{x}\right)\right]$$

$$= \lim_{h\to 0}\left[\frac{1}{x}\ln\left(1 + \frac{h}{x}\right)^{x/h}\right] \qquad (\text{since } n \ln r = \ln r^n)$$

$$= \frac{1}{x}\cdot \lim_{h\to 0}\left[\ln\left(1 + \frac{h}{x}\right)^{x/h}\right].$$

It can be shown that we can write this as

$$\frac{d}{dx}(\ln x) = \frac{1}{x}\ln\left[\lim_{h\to 0}\left(1 + \frac{h}{x}\right)^{x/h}\right]. \tag{1}$$

To evaluate $\lim_{h\to 0}\left(1 + \frac{h}{x}\right)^{x/h}$, first note that as $h \to 0$, then $\frac{h}{x} \to 0$. Thus, if we replace $\frac{h}{x}$ by k, the limit has the form

$$\lim_{k\to 0}(1 + k)^{1/k}.$$

As mentioned in Sec. 6-1, this limit is e, the base of natural logarithms. Thus Eq. (1) becomes

$$\frac{d}{dx}(\ln x) = \frac{1}{x}\ln e = \frac{1}{x}(1) = \frac{1}{x}.$$

Hence,

$$\boxed{\frac{d}{dx}(\ln x) = \frac{1}{x}.} \tag{2}$$

EXAMPLE 1

If $y = x \ln x$, then by the product rule and Eq. (2),

$$\begin{aligned} y' &= x\frac{d}{dx}(\ln x) + (\ln x)\frac{d}{dx}(x) \\ &= x\left(\frac{1}{x}\right) + (\ln x)(1) = 1 + \ln x. \end{aligned}$$

We now extend Eq. (2) to cover a broader class of functions. Let

$$y = \ln u, \quad \text{where} \quad u = f(x)$$

and u is positive and differentiable. By the chain rule,

$$\frac{d}{dx}(\ln u) = \frac{dy}{du}\cdot\frac{du}{dx} = \frac{d}{du}(\ln u)\cdot\frac{du}{dx} = \frac{1}{u}\cdot\frac{du}{dx}.$$

Thus,

$$\boxed{\frac{d}{dx}(\ln u) = \frac{1}{u}\cdot\frac{du}{dx}.} \tag{3}$$

EXAMPLE 2

Differentiate each of the following.

a. $y = \ln (x^2 + 1)$.

This function has the form $\ln u$ with $u = x^2 + 1$. Using Eq. (3) we have

$$\frac{dy}{dx} = \frac{1}{x^2 + 1} \frac{d}{dx}(x^2 + 1) = \frac{1}{x^2 + 1}(2x) = \frac{2x}{x^2 + 1}.$$

b. $y = x^2 \ln (4x + 2)$.

Using the product rule and then Eq. (3) with $u = 4x + 2$, we obtain

$$\begin{aligned} D_x y &= x^2 D_x[\ln (4x + 2)] + [\ln (4x + 2)] D_x(x^2) \\ &= x^2\left(\frac{1}{4x + 2}\right)(4) + [\ln (4x + 2)](2x) \\ &= \frac{4x^2}{4x + 2} + 2x \ln (4x + 2). \end{aligned}$$

c. $y = \ln (\ln x)$.

This has the form $y = \ln u$ where $u = \ln x$. Using Eqs. (3) and (2) we obtain

$$y' = \frac{1}{\ln x} \frac{d}{dx}(\ln x) = \frac{1}{\ln x}\left(\frac{1}{x}\right) = \frac{1}{x \ln x}.$$

To differentiate some functions involving logarithms, such as $y = \ln (2x + 5)^3$, it may be easier to simplify the function by using properties of logarithms *before* the differentiation. Example 3 will illustrate this.

EXAMPLE 3

Differentiate each of the following.

a. $y = \ln (2x + 5)^3$.

First we simplify the right side by using properties of logarithms.

$$y = \ln (2x + 5)^3 = 3 \ln (2x + 5).$$

$$\frac{dy}{dx} = 3\left(\frac{1}{2x + 5}\right)(2) = \frac{6}{2x + 5}.$$

If the simplification were not performed first,

$$\frac{dy}{dx} = \frac{1}{(2x+5)^3} D_x[(2x+5)^3]$$

$$= \frac{1}{(2x+5)^3}(3)(2x+5)^2(2) = \frac{6}{2x+5}.$$

b. $f(p) = \ln[(p+1)^2(p+2)^3(p+3)^4]$.

We simplify the right side and then differentiate.

$$f(p) = 2\ln(p+1) + 3\ln(p+2) + 4\ln(p+3).$$

$$f'(p) = 2\left(\frac{1}{p+1}\right)(1) + 3\left(\frac{1}{p+2}\right)(1) + 4\left(\frac{1}{p+3}\right)(1)$$

$$= \frac{2}{p+1} + \frac{3}{p+2} + \frac{4}{p+3}.$$

c. $f(w) = \ln\sqrt{\dfrac{1+w^2}{w^2-1}}$.

Again, using properties of logarithms will simplify our work.

$$f(w) = \frac{1}{2}[\ln(1+w^2) - \ln(w^2-1)].$$

$$f'(w) = \frac{1}{2}\left[\frac{1}{1+w^2}(2w) - \frac{1}{w^2-1}(2w)\right]$$

$$= \frac{w}{1+w^2} - \frac{w}{w^2-1} = -\frac{2w}{w^4-1}.$$

d. $f(x) = \ln^3[(2x+1)^4]$.

The exponent 3 refers to the cubing of $\ln(2x+1)^4$. That is,

$$\ln^3[(2x+1)^4] \quad \text{means} \quad [\ln(2x+1)^4]^3.$$

Thus,

$$f(x) = \ln^3[(2x+1)^4] = [\ln(2x+1)^4]^3 = [4\ln(2x+1)]^3.$$

By the power rule,

$$f'(x) = 3[4\ln(2x+1)]^2 D_x[4\ln(2x+1)]$$

$$= 3[4\ln(2x+1)]^2\left[4\left(\frac{1}{2x+1}\right)(2)\right]$$

$$= \frac{24}{2x+1}[\ln(2x+1)^4]^2$$

$$= \frac{24}{2x+1}\ln^2[(2x+1)^4].$$

EXAMPLE 4

Approximate ln (1.06) *by using differentials.*

Let $y = f(x) = \ln x$. Letting $x = 1$ and $dx = .06$, we have

$$f(x + dx) \approx f(x) + dy.$$

$$\ln (x + dx) \approx \ln x + \frac{1}{x} dx.$$

$$\ln (1 + .06) \approx \ln (1) + \frac{1}{1}(.06).$$

$$\ln (1.06) \approx 0 + .06 = .06.$$

The actual value of ln (1.06) to five decimal places is .05827.

We can generalize Eq. (3) to any base b. Since $\ln u = \dfrac{\log_b u}{\log_b e}$ (from Sec. 6-3), then $\log_b u = (\log_b e) \ln u$. Hence,

$$\begin{aligned} \frac{d}{dx}(\log_b u) &= \frac{d}{dx}[(\log_b e) \ln u] \\ &= (\log_b e)\frac{d}{dx}(\ln u) \\ &= (\log_b e)\left(\frac{1}{u}\frac{du}{dx}\right). \end{aligned}$$

Thus,

$$\boxed{\frac{d}{dx}(\log_b u) = \frac{1}{u}(\log_b e)\frac{du}{dx}.} \tag{4}$$

Since the use of natural logarithms gives a value of 1 to the factor $\log_b e$ in Eq. (4), natural logarithms are used extensively in calculus.

EXAMPLE 5

If $y = \log (2x + 1)$, *find the rate of change of* y *with respect to* x.

We want to find dy/dx. By Eq. (4) with $u = 2x + 1$ and $b = 10$,

$$\frac{dy}{dx} = \frac{1}{2x + 1}(\log e)D_x(2x + 1) = \frac{1}{2x + 1}(\log e)(2) = \frac{2 \log e}{2x + 1}.$$

EXAMPLE 6

If $q + p = \ln q + \ln p$, *find the rate of change of* q *with respect to* p.

We assume that q is a function of p and implicitly differentiate both sides with respect to p.

$$D_p(q) + D_p(p) = D_p(\ln q) + D_p(\ln p).$$

$$\frac{dq}{dp} + 1 = \frac{1}{q}\frac{dq}{dp} + \frac{1}{p}.$$

$$\frac{dq}{dp}\left(1 - \frac{1}{q}\right) = \frac{1}{p} - 1.$$

$$\frac{dq}{dp}\left(\frac{q-1}{q}\right) = \frac{1}{p} - 1.$$

$$\frac{dq}{dp} = \left(\frac{1}{p} - 1\right)\frac{q}{q-1}.$$

Exercise 6-4

In Problems **1–34**, *differentiate the functions. If possible, first use properties of logarithms to simplify the given function.*

1. $y = \ln(3x - 4)$.

2. $y = \ln(5x - 6)$.

3. $y = \ln x^2$.

4. $y = \ln(ax^2 + b)$.

5. $y = \ln(1 - x^2)$.

6. $y = \ln(-x^2 + 6x)$.

7. $f(p) = \ln(2p^3 + 3p)$.

8. $f(r) = \ln(2r^4 - 3r^2 + 2r + 1)$.

9. $f(t) = t \ln t$.

10. $y = x^2 \ln x$.

11. $y = \log_3(2x - 1)$.

12. $f(w) = \log(w^2 + w)$.

13. $f(z) = \dfrac{\ln z}{z}$.

14. $y = \dfrac{x^2 - 1}{\ln x}$.

15. $y = \ln(x^2 + 4x + 5)^3$.

16. $y = \ln x^{100}$.

17. $y = \ln\sqrt{1 + x^2}$.

18. $f(s) = \ln\left(\dfrac{s^2}{1 + s^2}\right)$.

19. $f(l) = \ln\left(\dfrac{1 + l}{1 - l}\right)$.

20. $y = \ln\left(\dfrac{2x + 3}{3x - 4}\right)$.

21. $y = \ln\sqrt[4]{\dfrac{1 + x^2}{1 - x^2}}$.

22. $y = \ln\sqrt{\dfrac{x^4 - 1}{x^4 + 1}}$.

23. $y = \ln\left[(x^2 + 2)^2(x^3 + x - 1)\right]$.

24. $y = \ln\left[(5x + 2)^4(8x - 3)^6\right]$.

25. $y = (x^2 + 1)\ln(2x + 1)$.

26. $y = (ax + b)\ln(ax)$.

27. $y = \ln x^3 + \ln^3 x$.

28. $y = x^{\ln 3}$.

29. $y = \ln^4 (ax)$.

30. $y = \ln^2 (2x + 3)$.

31. $y = x \ln\sqrt{x - 1}$.

32. $y = \ln (x^2\sqrt{3x - 2})$.

33. $y = \sqrt{4 + \ln x}$.

34. $y = \ln (x + \sqrt{1 + x^2})$.

35. If $y = x^2 \ln x$, find y''.

36. If $y = \ln (x^2 + x)$, find y''.

37. If $y = \ln (2x - 1)$, find the rate of change of y''.

38. If $y = \ln^2 x$, find the relative rate of change of y.

39. Find an equation of the tangent line to the curve $y = \ln (x^2 - 2x - 2)$ when $x = 3$.

40. Find the slope of the curve $y = \dfrac{x}{\ln x}$ when $x = 2$.

41. If $y = 2x \ln x$, find dy.

42. If $y = \ln\sqrt{x^4 + 1}$, find dy.

In Problems **43** *and* **44**, *determine when the function is increasing or decreasing and locate all relative extrema.*

43. $y = f(x) = x \ln x$.

44. $y = f(x) = x^2 - 2 \ln x$.

In Problems **45** *and* **46**, *approximate each expression by using differentials.*

45. $\ln .97$.

46. $\ln 1.01$.

In Problems **47** *and* **48**, *find* dy/dx *by implicit differentiation.*

47. $y^2 + y = \ln x$.

48. $\ln (xy) + x = 4$.

49. Find the marginal revenue for a product if its demand function is $p = 25/\ln (q + 2)$.

50. A total cost function is given by $c = 25 \ln (q + 1) + 12$. Find the marginal cost when $q = 6$.

51. The magnitude M of an earthquake and its energy E are related by the equation [5]

$$1.5M = \log\left(\frac{E}{2.5 \times 10^{11}}\right).$$

Here M is given in terms of Richter's preferred scale of 1958 and E is in ergs. Determine the rate of change of energy with respect to magnitude.

52. According to Richter [41], the number N of earthquakes of magnitude M or greater per unit of time is given by $\log N = A - bM$, where A and b are constants. He claims that

$$\log\left(-\frac{dN}{dM}\right) = A + \log\left(\frac{b}{q}\right) - bM,$$

where $q = \log e$. Verify this statement.

53. In an experiment (adapted from [39]) on dispersal of a particular insect, a large number of insects are placed at a release point in an open field. Surrounding this point are traps that are placed in a concentric circular arrangement at a distance of 1 m, 2 m, 3 m, etc. from the release point. Twenty-four hours after the insects are released, the number of insects in each trap are counted. It is determined that at a distance of r meters from the release point, the average number of insects contained in a trap is n, where

$$n = f(r) = .1 \ln(r) + \frac{7}{r} - .8, \qquad 1 \leqslant r \leqslant 10.$$

(a) Show that the graph of f is always falling and concave up. (b) Sketch the graph of f. (c) When $r = 5$, at what rate is the average number of insects in a trap decreasing with respect to distance?

54. Show that the relative rate of change of $y = f(x)$ with respect to x is equal to the derivative of $y = \ln[f(x)]$.

6-5 DERIVATIVES OF EXPONENTIAL FUNCTIONS

We now turn to the derivative of the exponential function $y = e^u$, where u is a differentiable function of x. Since $y = e^u$, then in logarithmic form we have

$$u = \ln y.$$

Differentiating implicitly with respect to x, we have

$$\frac{du}{dx} = \frac{1}{y}\frac{dy}{dx}.$$

Solving for dy/dx and replacing y by e^u give

$$\frac{dy}{dx} = y\frac{du}{dx} = e^u\frac{du}{dx}.$$

Thus,

$$\boxed{\frac{d}{dx}(e^u) = e^u \frac{du}{dx}.} \tag{1}$$

In particular, if $u = x$, then $du/dx = 1$ and

$$\boxed{\frac{d}{dx}(e^x) = e^x.} \tag{2}$$

PITFALL. *Do **not** use the power rule to find $D_x(e^x)$. That is,*

$$D_x(e^x) \neq xe^{x-1}.$$

EXAMPLE 1

a. *Find* $\dfrac{d}{dx}(e^{x^3+3x})$.

The function has the form e^u with $u = x^3 + 3x$. Using Eq. (1), we obtain

$$\frac{d}{dx}(e^{x^3+3x}) = e^{x^3+3x}D_x(x^3 + 3x) = e^{x^3+3x}(3x^2 + 3)$$
$$= 3(x^2 + 1)e^{x^3+3x}.$$

b. *If* $y = \dfrac{x}{e^x}$, *find* y'.

We *first* use the quotient rule and then Eq. (2).

$$\frac{dy}{dx} = \frac{e^xD_x(x) - xD_x(e^x)}{(e^x)^2} = \frac{e^x(1) - x(e^x)}{(e^x)^2} = \frac{e^x(1 - x)}{e^{2x}} = \frac{1 - x}{e^x}.$$

c. *If* $f(w) = w^4e^{2w}$, *find* $f'(w)$.

We *first* use the product rule and then Eq. (1) with $u = 2w$.

$$f'(w) = w^4D_w(e^{2w}) + e^{2w}D_w(w^4)$$
$$= w^4(e^{2w})(2) + e^{2w}(4w^3) = 2e^{2w}w^3(w + 2).$$

d. *Find* $D_x[e^{x+1} \ln (x^2 + 1)]$.

By the product rule,

$$D_x[e^{x+1} \ln (x^2 + 1)] = e^{x+1}D_x[\ln (x^2 + 1)] + [\ln (x^2 + 1)]D_x(e^{x+1})$$
$$= e^{x+1}\left(\frac{1}{x^2 + 1}\right)(2x) + [\ln (x^2 + 1)]e^{x+1}(1)$$
$$= e^{x+1}\left[\frac{2x}{x^2 + 1} + \ln (x^2 + 1)\right].$$

e. *If* $y = e^2 + e^x + \ln 3$, *find* y'.

Since e^2 and $\ln 3$ are constants,

$$y' = 0 + e^x + 0 = e^x.$$

We can generalize Eq. (1) by considering the derivative of a^u where $a > 0$, $a \neq 1$, and u is a differentiable function of x. First we shall write a^u as an exponential function with base e. By Property 9 of Sec. 6-3, we have $a = e^{\ln a}$. Thus,

$$\begin{aligned} D_x(a^u) &= D_x\left[(e^{\ln a})^u\right] = D_x(e^{u \ln a}) \\ &= e^{u \ln a} D_x(u \ln a) && [\text{by Eq. } (1)] \\ &= e^{u \ln a}\left(\frac{du}{dx}\right) \ln a && [\ln a \text{ is constant}] \\ &= a^u (\ln a) \frac{du}{dx} && [\text{since } e^{u \ln a} = a^u]. \end{aligned}$$

Thus,

$$\boxed{\frac{d}{dx}(a^u) = a^u (\ln a) \frac{du}{dx}.} \qquad (3)$$

EXAMPLE 2

For a certain population of cells, the number N of cells at time t is given by

$$N(t) = N_0(2^{t/k}),$$

where N_0 *is the number of cells at* $t = 0$ *and k is a positive constant. Find the relative rate of change of* $N(t)$.

The relative rate of change is $N'(t)/N(t)$. To find $N'(t)$ we use Eq. (3) with $a = 2$ and $u = t/k$. Thus,

$$N'(t) = N_0 2^{t/k} (\ln 2) D_t\left(\frac{t}{k}\right) = N_0 2^{t/k} (\ln 2)\left(\frac{1}{k}\right).$$

Thus,

$$\frac{N'(t)}{N(t)} = \frac{N_0 2^{t/k} (\ln 2)\left(\frac{1}{k}\right)}{N_0 2^{t/k}} = \frac{\ln 2}{k}.$$

EXAMPLE 3

If $y = x^{100} + 100^x$, find $D_x y$.

Note that this function involves a variable to a constant power and a constant raised to a variable power. Do not confuse these forms!

$$D_x y = 100x^{99} + 100^x(\ln 100)\, D_x(x) = 100x^{99} + 100^x \ln 100.$$

EXAMPLE 4

An important function in the social sciences is the ***normal density function***

$$y = f(x) = \frac{1}{\sigma\sqrt{2\pi}}\, e^{-(1/2)[(x-\mu)/\sigma]^2},$$

where σ (a Greek letter read "sigma") and μ (a Greek letter read "mu") are constants. Its graph, called the normal curve, is "bell-shaped" (see Fig. 6-10).

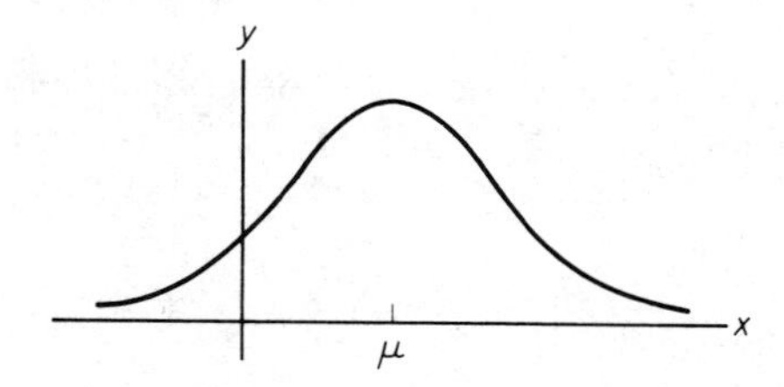

FIG. 6-10

Determine the rate of change of y with respect to x when $x = \mu$.

$$\frac{dy}{dx} = \frac{1}{\sigma\sqrt{2\pi}}\left[e^{-(1/2)[(x-\mu)/\sigma]^2}\right]\left[-\frac{1}{2}(2)\left(\frac{x-\mu}{\sigma}\right)\left(\frac{1}{\sigma}\right)\right].$$

Evaluating dy/dx when $x = \mu$, we obtain

$$\left.\frac{dy}{dx}\right|_{x=\mu} = 0.$$

EXAMPLE 5

If $e^{xy} = x + y$, find y' by implicit differentiation.

Differentiating both sides with respect to x and treating y as a function of x, we obtain

$$D_x(e^{xy}) = D_x(x) + D_x(y),$$
$$e^{xy}D_x(xy) = 1 + D_x(y),$$
$$e^{xy}(xD_xy + yD_xx) = 1 + D_x(y) \qquad \text{[product rule]},$$
$$e^{xy}(xy' + y) = 1 + y',$$
$$y'(xe^{xy} - 1) = 1 - ye^{xy},$$
$$y' = \frac{1 - ye^{xy}}{xe^{xy} - 1}.$$

EXAMPLE 6

Test $y = f(x) = x^2e^x$ for relative extrema.

By the product rule,

$$f'(x) = x^2e^x + e^x(2x) = xe^x(x + 2).$$

Since e^x is always positive, the critical values of f are $x = 0, -2$. From the signs of $f'(x)$ given in Fig. 6-11, we conclude that there is a relative maximum when $x = -2$ and a relative minimum when $x = 0$.

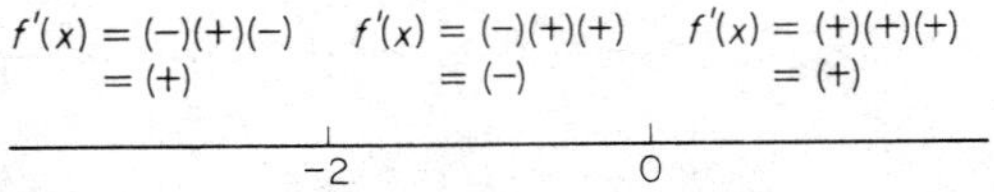

$f'(x) = (-)(+)(-) = (+)$	$f'(x) = (-)(+)(+) = (-)$	$f'(x) = (+)(+)(+) = (+)$
	-2	0

FIG. 6-11

Exercise 6-5

In Problems **1–32**, *differentiate the functions.*

1. $y = e^{x^2+1}$.

2. $y = e^{2x^2+5}$.

3. $y = e^{3-5x}$.

4. $f(q) = e^{-q^3+6q-1}$.

5. $f(r) = e^{3r^2+4r+4}$.

6. $y = e^{9x^2+5x^3-6}$.

7. $y = xe^x$.

8. $y = x^2e^{-x}$.

9. $y = x^2e^{-x^2}$.

10. $y = xe^{2x}$.

11. $y = \dfrac{e^x + e^{-x}}{2}$.

12. $y = \dfrac{e^x - e^{-x}}{2}$.

13. $y = 4^{3x^2}$.

14. $y = 4^{3x+1}$.

15. $f(w) = \frac{e^{2w}}{w^2}$.

16. $y = 2^x x^2$.

17. $y = e^{1+\sqrt{x}}$.

18. $y = e^{x-\sqrt{x}}$.

19. $y = \frac{e^x - 1}{e^x + 1}$.

20. $f(z) = e^{1/z}$.

21. $y = (e^{3x} + 1)^4$.

22. $y = e^{\ln 4}$.

23. $y = e^{e^x}$.

24. $y = e^{2x}(x + 1)$.

25. $y = e^{\ln x}$.

26. $y = e^{\ln(x^2+1)}$.

27. $y = \frac{e^x}{\ln x}$.

28. $y = e^{-x} \ln x$.

29. $y = (\log 2)^x$.

30. $y = \ln e^{4x+1}$.

31. $y = x^3 - 3^x$.

32. $y = \sqrt{e^x + \ln x}$.

33. Find an equation of the tangent line to the graph of $y = e^x$ when $x = 2$.

34. Find the slope of the tangent line to the graph of $y = 2e^{-4x^2}$ when $x = 0$.

In Problems **35–38**, *find the indicated derivatives.*

35. $y = x^3 + e^x$; $y^{(4)}$.

36. $y = e^{-4x^2}$; y''.

37. $f(z) = z^2 e^z$; $f''(z)$.

38. $y = \frac{x}{e^x}$; $\frac{d^2y}{dx^2}$.

39. If $y = e^{x^3+5}$, find dy.

40. If $u = (4x + 3)e^{2x^2+3}$, find du.

41. Use differentials to approximate $e^{.01}$.

42. Use differentials to approximate $e^{-.01}$.

In Problems **43** *and* **44**, *determine when the function is increasing or decreasing and locate all relative extrema.*

43. $y = xe^x$.

44. $y = e^{-x^2}$.

In Problems **45** *and* **46**, *find* dy/dx *by implicit differentiation.*

45. $xe^y + y = 4$.

46. $y \ln x = xe^y$.

For each of the demand equations in Problems **47** *and* **48**, *find the rate of change of price* p *with respect to quantity* q. *What is the rate of change for the indicated value of* q?

47. $p = 15e^{-.001q}$; $q = 500$.

48. $p = 8e^{-3q/800}$; $q = 400$.

49. For a manufacturer's product, the demand function is $q = 10{,}000e^{-.02p}$. Find the value of p for which maximum revenue is obtained.

50. The population P of a city t years from now is given by $P = 20{,}000e^{.03t}$. Show that $dP/dt = kP$ where k is a constant. This means that the rate of change of population at any time is proportional to the population at that time.

In Problems 51 *and* 52, $\bar{c}$ *is the average cost of producing* q *units of a product. Find the marginal cost function and the marginal cost for the given values of* q.

51. $\bar{c} = (7000e^{q/700})/q;\quad q = 350,\ q = 700.$

52. $\bar{c} = \dfrac{850}{q} + 4000\dfrac{e^{(2q+6)/800}}{q};\quad q = 97,\ q = 197.$

53. For a firm the daily output q on the t-th day of a production run is given by $q = 500(1 - e^{-.2t})$. Find the rate of change of output q with respect to t on the tenth day.

54. After t years, the value A of a principal of P dollars which is invested at a nominal rate of r per year and compounded continuously is given by $A = Pe^{rt}$. Show that the relative rate of change of A with respect to t is r.

55. Sketch the graph of the normal density function

$$f(x) = \frac{1}{\sqrt{2\pi}}e^{-x^2/2}.$$

Include relative extrema and inflection points.

56. In an article concerning predators and prey, Holling **[19]** refers to an equation of the form

$$y = K(1 - e^{-ax}),$$

where x is the prey density, y is the number of prey attacked, and K and a are constants. Verify his statement that

$$\frac{dy}{dx} = a(K - y).$$

57. According to Richter **[41]**, the number N of earthquakes of magnitude M or greater per unit of time is given by $N = 10^A 10^{-bM}$, where A and b are constants. Find dN/dM.

58. In a study of the effects of food deprivation on hunger **[18]**, an insect was fed until its appetite was completely satisfied. Then it was deprived food for t hours (deprivation period). At the end of this period, the insect was refed until its appetite was again completely satisfied. The weight H (in grams) of the food that was consumed at this time was statistically found to be a function of t, where

$$H = 1.00\left[1 - e^{-(.0464t + .0670)}\right].$$

Here H is a measure of hunger. Show that H is increasing with respect to t, but at a decreasing rate.

59. Short-term retention was studied by Peterson and Peterson **[38]**. They analyzed a procedure in which an experimenter verbally gave a subject a three-letter consonant syllable, such as CHJ, followed by a three-digit number, such as 309. The subject then repeated the number and counted backwards by 3's, such as 309, 306, 303, ... After a period of time the subject was signaled by a light to recite the three-letter consonant syllable. The time interval between the experimenter's completion of the last consonant to the onset of the light was called the *recall interval*. The time between the onset of the light and the completion of a response was referred to as *latency*. After many trials, it was determined that for a recall interval of t seconds, the approximate proportion of correct recalls with latency below 2.83 sec was p, where

$$p = .89\left[.01 + .99(.85)^t\right].$$

(a) Find dp/dt and interpret your result.

(b) If you have a calculator, evaluate dp/dt when $t = 2$. Give your answer to two decimal places.

60. When a deep-sea diver undergoes decompression, or a pilot climbs to a high altitude, nitrogen may bubble out of the blood, causing what is commonly called the *bends*. Suppose the percentage P of people who suffer effects of the bends at an altitude of h thousand feet is given by (adapted from **[13]**)

$$P = \frac{100}{1 + 100{,}000e^{-.36h}}.$$

(a) Determine $\lim_{h\to\infty} P$. What interpretation can you attach to this limit? (b) Find dP/dh. (c) Is P an increasing function of h?

61. Hurter and Rubenstein **[22]** discuss the diffusion of a new process into a market. They refer to an equation of the form

$$Y = k\alpha^{\beta^t},$$

where Y is the cumulative level of diffusion of the new process at time t, and k, α, and β are positive constants. Verify their claim that

$$\frac{dY}{dt} = k\alpha^{\beta^t}(\beta^t \ln \alpha) \ln \beta.$$

62. Population growth is sometimes described by the S-shaped Gompertz curve given by

$$N(t) = ae^{-be^{-ct}},$$

where $N(t)$ is the number of individuals in the population at time t, and a, b, and c are constants. Suppose $b = 3$ and $c = .05$.
(a) Find $N(0)$. (b) Find $N\left(\frac{\ln b}{c}\right)$. (c) Find dN/dt. (d) Show that $N''(t) = 0$ when $t = (\ln 3)/.05$. It is at this value of t that there is maximum rate of growth. (e) Find $\lim_{t\to\infty} N(t)$. (f) Show that $N(t)$ has the form $N(t) = AB^{C^t}$, where A, B, and C are constants.

63. Suppose a tracer, such as a colored dye, is injected instantly into the heart at time $t = 0$ and mixes uniformly with blood inside the heart. Let the initial concentration of the tracer in the heart be C_0 and assume the heart has constant volume V. As fresh blood flows into the heart, assume that the diluted mixture of blood and tracer flows out at the constant positive rate r. Then the concentration $C(t)$ of the tracer in the heart at time t is given by

$$C(t) = C_0e^{-(r/V)t}.$$

(a) Find $\lim_{t\to\infty} C(t)$. (b) Show that $dC/dt = (-r/V)C(t)$.

64. In Problem 63, suppose the tracer is injected at a constant rate R. Then the concentration at time t is

$$C(t) = \frac{R}{r}\left[1 - e^{-(r/V)t}\right].$$

(a) Find $C(0)$. (b) Find $\lim_{t\to\infty} C(t)$. (c) Show that

$$\frac{dC}{dt} = \frac{R}{V} - \frac{r}{V}C(t).$$

65. Several models have been used to analyze the length of stay in a hospital. For a particular group of schizophrenics, one such model is (adapted from [9])

$$f(t) = 1 - e^{-.008t},$$

where $f(t)$ is the proportion of the group that was discharged at the end of t days of hospitalization. (a) Find $\lim_{t\to\infty} f(t)$ and interpret your result. (b) Find the rate of discharge (proportion discharged per day) at the end of 100 days. Give your answer to four decimal places.

66. In a psychological experiment involving conditioned response (adapted from [20]), subjects listened to four tones, denoted 0, 1, 2, and 3. Initially, the subjects were conditioned to tone 0 by receiving a shock whenever this tone was heard. Later, when each of the four tones (stimuli) were heard without shocks, the subjects' responses were recorded by means of a tracking device that measures galvanic skin reaction. The average response to each stimulus (without shock) was determined, and the results were plotted on a coordinate plane where the x- and y-axes represent the stimuli (0, 1, 2, 3) and the average galvanic responses, respectively. It was determined that the points fit a curve that is approximated by the graph of $y = 12.5 + 5.8(.42)^x$. Show that this function is decreasing and concave up.

6-6 INTEGRALS INVOLVING EXPONENTIAL AND LOGARITHMIC FUNCTIONS

We now turn our attention to integrating an exponential function. If $u = u(x)$ is a differentiable function, then $d(e^u) = D_u(e^u)\,du = e^u\,du$. Thus,

$$\int e^u\,du = e^u + C. \tag{1}$$

EXAMPLE 1

Find the following integrals.

a. $\int e^x\,dx$.

If $u = x$, then $du = dx$ and by Eq. (1) we have

$$\int e^x\,dx = e^x + C.$$

b. $\int 2xe^{x^2}\,dx$.

Let $u = x^2$. Then $du = 2x\,dx$ and by Eq. (1),

$$\int 2xe^{x^2}\,dx = \int e^{x^2}[2x\,dx] = \int e^u\,du$$
$$= e^u + C = e^{x^2} + C.$$

c. $\int (x^2 + 1)e^{x^3+3x}\,dx$.

If $u = x^3 + 3x$, then $du = (3x^2 + 3)\,dx = 3(x^2 + 1)\,dx$. If the integrand contained a factor of 3, the integral would have the form $\int e^u\,du$. Thus we write

$$\begin{aligned}\int (x^2 + 1)e^{x^3+3x}\,dx &= \tfrac{1}{3}\int e^{x^3+3x}[3(x^2+1)\,dx] \\ &= \tfrac{1}{3}\int e^u\,du = \tfrac{1}{3}e^u + C \\ &= \tfrac{1}{3}e^{x^3+3x} + C.\end{aligned}$$

d. $\int_0^1 e^{3t}\,dt.$

$$\begin{aligned}\int_0^1 e^{3t}\,dt &= \tfrac{1}{3}\int_0^1 e^{3t}[3\,dt] \\ &= \left(\tfrac{1}{3}\right)e^{3t}\Big|_0^1 = \tfrac{1}{3}(e^3 - e^0) = \tfrac{1}{3}(e^3 - 1).\end{aligned}$$

PITFALL. *Do not apply the formula for* $\int u^n\,du$ *to* $\int e^u\,du$. *For example,*

$$\int e^x\,dx \neq \frac{e^{x+1}}{x+1} + C.$$

Integrals of the form $\int a^u\,du$ can be handled by writing a as $e^{\ln a}$ (see Property 9 in Sec. 6-3) and then using Eq. (1), as the next example shows.

EXAMPLE 2

Determine $\int 2^{3-x}\,dx.$

Since $2 = e^{\ln 2}$,

$$\int 2^{3-x}\,dx = \int (e^{\ln 2})^{3-x}\,dx = \int e^{(\ln 2)(3-x)}\,dx.$$

If we let $u = (\ln 2)(3 - x)$, then $du = (-\ln 2)\,dx$. Thus,

$$\begin{aligned}\int e^{(\ln 2)(3-x)}\,dx &= -\frac{1}{\ln 2}\int e^{(\ln 2)(3-x)}[(-\ln 2)\,dx] \\ &= -\frac{1}{\ln 2}\int e^u\,du = -\frac{1}{\ln 2}e^u + C \\ &= -\frac{1}{\ln 2}e^{(\ln 2)(3-x)} + C.\end{aligned}$$

Since $e^{(\ln 2)(3-x)} = 2^{3-x}$, we have

$$\int 2^{3-x}\, dx = -\frac{2^{3-x}}{\ln 2} + C.$$

As you know, the formula $\int u^n\, du = u^{n+1}/(n+1) + C$ assumes that $n \neq -1$. To find $\int u^{-1}\, du = \int \frac{1}{u}\, du$, we first observe that

$$d(\ln u) = \frac{d}{du}(\ln u)\, du = \frac{1}{u}\, du.$$

It would seem that $\int \frac{1}{u}\, du = \ln u + C$. However, the logarithm of a number u is defined if and only if u is positive. If $u < 0$, then $\ln u$ is not defined. Thus, $\int \frac{1}{u}\, du = \ln u + C$ as long as $u > 0$. On the other hand, if $u < 0$, then $-u > 0$ and $\ln(-u)$ *is* defined. Moreover,

$$d[\ln(-u)] = \frac{d}{du}[\ln(-u)]\, du = \frac{1}{-u}(-1)\, du = \frac{1}{u}\, du.$$

In this case ($u < 0$), $\int \frac{1}{u}\, du = \ln(-u) + C$. In summary, if $u > 0$, then $\int \frac{1}{u}\, du = \ln u + C$; if $u < 0$, then $\int \frac{1}{u}\, du = \ln(-u) + C$. Combining these cases, we write

$$\boxed{\int \frac{1}{u}\, du = \ln |u| + C, \quad \text{if } u \neq 0.} \qquad (2)$$

EXAMPLE 3

Find the following integrals.

a. $\int \frac{7}{x}\, dx.$

$$\int \frac{7}{x}\, dx = 7\int \frac{1}{x}\, dx.$$

If $u = x$, then $du = dx$ and by Eq. (2) we have

$$\int \frac{7}{x}\, dx = 7 \ln |x| + C.$$

Using properties of logarithms, we can write this answer another way:

$$\int \frac{7}{x}\,dx = \ln |x^7| + C.$$

b. $\displaystyle\int \frac{2x}{x^2 + 5}\,dx.$

Let $u = x^2 + 5$. Then $du = 2x\,dx$. Thus,

$$\int \frac{2x}{x^2 + 5}\,dx = \int \frac{1}{x^2 + 5}[2x\,dx] = \int \frac{1}{u}\,du$$
$$= \ln |u| + C = \ln |x^2 + 5| + C.$$

Since $x^2 + 5$ is always positive, we can omit the absolute–value bars and write

$$\int \frac{2x}{x^2 + 5}\,dx = \ln (x^2 + 5) + C.$$

c. $\displaystyle\int \frac{2x^3 + 3x}{x^4 + 3x^2 + 7}\,dx.$

If $u = x^4 + 3x^2 + 7$, then $du = (4x^3 + 6x)\,dx$ which is 2 times the given numerator. Thus we insert a factor 2 and adjust for it with a factor of $\frac{1}{2}$.

$$\int \frac{2x^3 + 3x}{x^4 + 3x^2 + 7}\,dx = \frac{1}{2}\int \frac{2(2x^3 + 3x)}{x^4 + 3x^2 + 7}\,dx$$
$$= \frac{1}{2}\int \frac{1}{x^4 + 3x^2 + 7}[(4x^3 + 6x)\,dx]$$
$$= \frac{1}{2}\int \frac{1}{u}\,du = \frac{1}{2}\ln |u| + C$$
$$= \frac{1}{2}\ln |x^4 + 3x^2 + 7| + C$$
$$= \ln \sqrt{x^4 + 3x^2 + 7} + C.$$

EXAMPLE 4

Find the following integrals.

a. $\displaystyle\int \left[\frac{1}{(1 - w)^2} + \frac{1}{w - 1}\right] dw.$

$$\int \left[\frac{1}{(1 - w)^2} + \frac{1}{w - 1}\right] dw = \int (1 - w)^{-2}\,dw + \int \frac{1}{w - 1}\,dw$$
$$= -1\int (1 - w)^{-2}[-dw] + \int \frac{1}{w - 1}\,dw.$$

On the last line the first integral has the form $\int u^{-2}\,du$, and the second has the form $\int \frac{1}{v}\,dv$. Thus,

$$\int\left[\frac{1}{(1-w)^2}+\frac{1}{w-1}\right]dw = -\frac{(1-w)^{-1}}{-1}+\ln|w-1|+C$$
$$=\frac{1}{1-w}+\ln|w-1|+C.$$

b. $\int_1^e \frac{1}{y}\,dy.$

$$\int_1^e \frac{1}{y}\,dy = \ln|y|\Big|_1^e = \ln e - \ln 1 = 1 - 0 = 1.$$

c. $\int \frac{1}{x\ln x}\,dx.$

We can write the integral as

$$\int \frac{1}{\ln x}\left(\frac{1}{x}\,dx\right).$$

If $u = \ln x$, then $du = \frac{1}{x}\,dx$ and

$$\int \frac{1}{x\ln x}\,dx = \int \frac{1}{\ln x}\left(\frac{1}{x}\,dx\right) = \int \frac{1}{u}\,du$$
$$= \ln|u| + C = \ln|\ln x| + C.$$

d. $\int \frac{5}{w(\ln w)^{3/2}}\,dw.$

If $u = \ln w$, then $du = \frac{1}{w}\,dw$ and

$$\int \frac{5}{w(\ln w)^{3/2}}\,dw = 5\int (\ln w)^{-3/2}\left[\frac{1}{w}\,dw\right]$$
$$= 5\int u^{-3/2}\,du = 5\cdot\frac{u^{-1/2}}{-\frac{1}{2}}+C$$
$$= \frac{-10}{u^{1/2}}+C = -\frac{10}{(\ln w)^{1/2}}+C.$$

When one is integrating fractions, sometimes a preliminary division is needed to get familiar integration forms, as the next example shows.

EXAMPLE 5

a. $\int \frac{x^3 + x - 1}{x^2}\,dx.$

A familiar integration form is not apparent. However, we can write the integrand as the sum and difference of three fractions by dividing each term in the numerator by the denominator.

$$\int \frac{x^3 + x - 1}{x^2}\,dx = \int \left[\frac{x^3}{x^2} + \frac{x}{x^2} - \frac{1}{x^2}\right] dx = \int \left[x + \frac{1}{x} - \frac{1}{x^2}\right] dx$$

$$= \frac{x^2}{2} + \ln|x| - \int x^{-2}\,dx = \frac{x^2}{2} + \ln|x| + \frac{1}{x} + C.$$

b. $\int \frac{2x^3 + 3x^2 + x + 1}{2x + 1}\,dx.$

Here the integrand is a quotient of polynomials in which the degree of the numerator is greater than or equal to that of the denominator, and the denominator has more than one term. In such a situation, in order to integrate we first use long division until the degree of the remainder is less than that of the divisor.

$$\int \frac{2x^3 + 3x^2 + x + 1}{2x + 1}\,dx = \int \left(x^2 + x + \frac{1}{2x + 1}\right) dx$$

$$= \frac{x^3}{3} + \frac{x^2}{2} + \int \frac{1}{2x + 1}\,dx$$

$$= \frac{x^3}{3} + \frac{x^2}{2} + \frac{1}{2}\int \frac{1}{2x + 1}[2\,dx]$$

$$= \frac{x^3}{3} + \frac{x^2}{2} + \frac{1}{2}\ln|2x + 1| + C.$$

Exercise 6-6

In Problems **1–68**, *determine the integrals.*

1. $\int 3e^{3x}\,dx.$

2. $\int 2e^{2t+5}\,dt.$

3. $\int (2t + 1)e^{t^2+t}\,dt.$

4. $\int -3w^2e^{-w^3}\,dw.$

5. $\int \frac{1}{x + 5}\,dx.$

6. $\int \frac{2x + 1}{x + x^2}\,dx.$

7. $\int \frac{3x^2 + 4x^3}{x^3 + x^4}\,dx.$

8. $\int \frac{3x^2 - 2x}{1 - x^2 + x^3}\,dx.$

9. $\int xe^{5x^2}\,dx.$

10. $\int x^3e^{4x^4}\,dx.$

11. $\int 6e^{-2x}\,dx.$

12. $\int x^4e^{-6x^5}\,dx.$

13. $\int \frac{6z}{(z^2-6)^5}\,dz.$

14. $\int \frac{1}{(8y-3)^3}\,dy.$

15. $\int \frac{4}{x}\,dx.$

16. $\int \frac{3}{1+2y}\,dy.$

17. $\int \frac{s^2}{s^3+5}\,ds.$

18. $\int \frac{2x^2}{3-4x^3}\,dx.$

19. $\int \frac{7}{5-3x}\,dx.$

20. $\int \frac{7t}{5t^2-6}\,dt.$

21. $\int_0^{e-1} \frac{1}{x+1}\,dx.$

22. $\int_1^8 \frac{4}{y}\,dy.$

23. $\int 2y^3e^{y^4+1}\,dy.$

24. $\int \frac{7}{3-2x}\,dx.$

25. $\int v^2e^{-2v^3+1}\,dv.$

26. $\int 2ye^{3y^2}\,dy.$

27. $\int (e^{-5x}+2e^x)\,dx.$

28. $\int (e^x-e^{-x}+e^{2x})\,dx.$

29. $\int_0^2 x^2e^{x^3}\,dx.$

30. $\int_0^1 (e^x-e^{-2x})\,dx.$

31. $\int \frac{x^2+2}{x^3+6x}\,dx.$

32. $\int \frac{3}{7}(v-2)e^{2-4v+v^2}\,dv.$

33. $\int \frac{16s-4}{3-2s+4s^2}\,ds.$

34. $\int (e^{3.1})^2\,dx.$

35. $\int_1^e (x^{-1}+x^{-2}-x^{-3})\,dx.$

36. $\int_1^3 (x+1)e^{x^2+2x}\,dx.$

37. $\int_0^1 \frac{2x^3+x}{x^2+x^4+1}\,dx.$

38. $\int_3^4 \frac{e^{\ln x}}{x}\,dx.$

39. $\int (e^x-e^{-x})^2\,dx.$

40. $\int x(2x^2+1)^{-1}\,dx.$

41. $\int \left[\frac{x}{x^2+1}+\frac{x^5}{(x^6+1)^2}\right]dx.$

42. $\int \left[x(x^2-16)^2-\frac{1}{2x+5}\right]dx.$

43. $\int \left[\sqrt{2x+3}-\frac{x}{x^2+3}\right]dx.$

44. $\int \left[\frac{1}{x-1}+\frac{1}{(x-1)^2}\right]dx.$

45. $\int \frac{e^{\sqrt{x}}}{\sqrt{x}}\,dx.$

46. $\int 3^x\,dx.$

47. $\int \frac{3x^3 + x^2 - x}{x^2}\,dx.$

48. $\int \left(e^x + x^e + ex + \frac{e}{x}\right)dx.$

49. $\int 4^{7x}\,dx.$

50. $\int (e^{4-3x})^2\,dx.$

51. $\int 2x(7 - e^{x^2/4})\,dx.$

52. $\int \frac{2x^4 - 6x^3 + x - 2}{x - 2}\,dx.$

53. $\int \frac{3e^{2x}}{e^{2x} + 1}\,dx.$

54. $\int \frac{3e^s}{6 + 5e^s}\,ds.$

55. $\int \frac{\ln x}{x}\,dx.$

56. $\int \frac{8x^3 - 6x^2 - ex^4}{3x^3}\,dx.$

57. $\int \frac{\ln^2 (r + 1)}{r + 1}\,dr.$

58. $\int \frac{x + 3}{x + 6}\,dx.$

59. $\int x\sqrt{e^{x^2+3}}\,dx.$

60. $\int (x^{e^2} + 2x)\,dx.$

61. $\int \frac{1}{(x + 3)\ln (x + 3)}\,dx.$

62. $\int \frac{x - x^{-2}}{x^2 + 2x^{-1}}\,dx.$

63. $\int \frac{e^{7/x}}{x^2}\,dx.$

64. $\int \frac{e^x + e^{-x}}{e^x - e^{-x}}\,dx.$

65. $\int \frac{6x^2 - 11x + 5}{3x - 1}\,dx.$

66. $\int \frac{(2x - 1)(x + 3)}{x - 5}\,dx.$

67. $\int \frac{x}{x - 1}\,dx.$

68. $\int \frac{x}{(x^2 + 1)\ln (x^2 + 1)}\,dx.$

In Problems 69 *and* 70, $\frac{dc}{dq}$ *is a marginal cost function. Find the total cost function if fixed costs in each case are* 2000.

69. $\frac{dc}{dq} = \frac{20}{q + 5}.$

70. $\frac{dc}{dq} = 2e^{.001q}.$

In Problems 71–74, *find the area of the region bounded by the given curves, the x-axis, and the given lines. In each case, sketch the region.*

71. $y = e^x,\ x = 0,\ x = 2.$

72. $y = \frac{1}{x},\ x = 1,\ x = e^2.$

73. $y = x + \frac{2}{x},\ x = 1,\ x = 2.$

74. $y = e^{-x},\ x = 0,\ x = 1.$

In Problems 75 *and* 76, *find* y *subject to the given conditions.*

75. $y' = \dfrac{x}{x^2 + 4}$; $y(1) = 0$.

76. $y'' = 2e^x$; $y'(2) = 0$ and $y(0) = 3$.

77. The present value (in dollars) of a continuous flow of income of \$2000 a year for 5 years at 6 percent compounded continuously is given by

$$\int_0^5 2000e^{-.06t}\,dt.$$

Evaluate the present value to the nearest dollar.

78. The total expenditures (in dollars) of a business over the next 5 years is given by

$$\int_0^5 4000e^{.05t}\,dt.$$

Evaluate the expenditures.

79. In biology, problems frequently arise involving transfer of a substance between compartments. An example would be transfer from the bloodstream to tissue. Evaluate the following integral which occurs in [45] in a two-compartment diffusion problem:

$$\int_0^t (e^{-a\tau} - e^{-b\tau})\,d\tau,$$

where τ (read "tau") is a Greek letter and a and b are constants.

80. If c_0 is the yearly consumption of a mineral at time $t = 0$, then under continuous consumption the total amount of the mineral used in the interval $[0, t_1]$ is

$$\int_0^{t_1} c_0 e^{kt}\,dt,$$

where k is the consumption rate. For the rare-earth mineral *junko* it has been determined that $c_0 = 3000$ units and $k = .05$. Evaluate the above integral for these data.

81. In a discussion of diffusion of oxygen from capillaries [45], concentric cylinders of radius r are used as a model for a capillary. The concentration C of oxygen in the capillary is given by

$$C = \int \left(\frac{Rr}{2K} + \frac{B_1}{r} \right) dr,$$

where R is the constant rate at which oxygen diffuses from the capillary, and K and B_1 are constants. Find C. (Write the constant of integration as B_2).

82. In a discussion of gene mutation **[33]**, the following equation occurs:

$$\int_{q_0}^{q_n} \frac{dq}{q - \hat{q}} = -(u + v)\int_0^n dt,$$

where u and v are gene mutation rates, the q's are gene frequencies, and n is the number of generations. Assume that all letters represent constants except q and t. Integrate both sides and then use your result to show that

$$n = \frac{1}{u + v}\ln\left|\frac{q_0 - \hat{q}}{q_n - \hat{q}}\right|.$$

83. In statistics the function $f(x) = k/x$ is a density function for $e \leqslant x \leqslant e^2$ if $\int_e^{e^2} f(x) = 1$.

(a) Find k so that f is a density function.

(b) The probability that $3 \leqslant x \leqslant 5$ is given by $\int_3^5 f(x)\,dx$. Evaluate this integral by using the value of k found in part a.

(c) The distribution function for f is given by $F(t) = \int_e^t f(x)\,dx$, where $e \leqslant t \leqslant e^2$. Find $F(t)$ by using the value of k found in part a.

84. In a discussion of inventory, Barbosa and Friedman **[3]** refer to the function

$$g(x) = \frac{1}{k}\int_1^{1/x} ku^r\,du,$$

where k and r are constants, $k > 0$ and $r > -2$, and $x > 0$. They claim that

$$g'(x) = -\frac{1}{x^{r+2}}.$$

Show that their claim is indeed true. (*Hint*: Consider two cases: when $r \neq -1$, and when $r = -1$.)

85. Taagepera **[50]** considers a "one-dimensional" country of length $2R$ (see Fig. 6-12). Under certain conditions the amount E of the country's exports

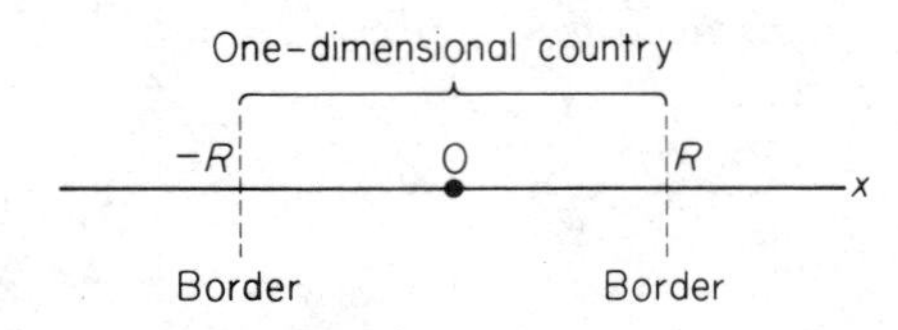

FIG. 6-12

is given by

$$E = \int_{-R}^{R} \frac{i}{2}\left[e^{-k(R-x)} + e^{-k(R+x)}\right] dx,$$

where i and k are constants ($k \neq 0$). Evaluate E.

6-7 LOGARITHMIC DIFFERENTIATION

There is a technique which may be used to simplify the differentiation of $y = f(x)$ when $f(x)$ involves products, quotients, or powers. We first take the natural logarithm of both sides of $y = f(x)$. After simplifying by using properties of logarithms, we then differentiate both sides with respect to x. The next example illustrates this method of **logarithmic differentiation.**

EXAMPLE 1

Find y' if $y = \sqrt[4]{\dfrac{(x-1)(x^2+2)^2}{(x+3)(x-4)^3}}$

This function has the form $y = u^{1/4}$, and so the power rule for differentiation may be used. However, finding du/dx would involve considerable work, since we would get quite involved with the quotient and product rules. Logarithmic differentiation makes the work less tedious. First we take the natural logarithm of both sides of the original equation:

$$\ln y = \ln \sqrt[4]{\frac{(x-1)(x^2+2)^2}{(x+3)(x-4)^3}}.$$

Using properties of logarithms to simplify the right side, we have

$$\ln y = \frac{1}{4} \ln \frac{(x-1)(x^2+2)^2}{(x+3)(x-4)^3}.$$

$$\ln y = \frac{1}{4}[\ln (x-1) + 2 \ln (x^2+2) - \ln (x+3) - 3 \ln (x-4)].$$

Differentiating both sides with respect to x gives

$$\left(\frac{1}{y}\right)y' = \frac{1}{4}\left[\frac{1}{x-1} + (2)\cdot\frac{1}{x^2+2}(2x) - \frac{1}{x+3} - (3)\cdot\frac{1}{x-4}\right].$$

$$\frac{y'}{y} = \frac{1}{4}\left[\frac{1}{x-1} + \frac{4x}{x^2+2} - \frac{1}{x+3} - \frac{3}{x-4}\right].$$

Solving for y' yields

$$y' = \frac{y}{4}\left[\frac{1}{x-1} + \frac{4x}{x^2+2} - \frac{1}{x+3} - \frac{3}{x-4}\right],$$

where y is given in the original equation.

Logarithmic differentiation can also be used to differentiate a function of the form $y = u^v$, where both u and v are differentiable functions of x.

EXAMPLE 2

Find y' for each of the following.

a. $y = x^x$.

This has the form $y = u^v$, where u and v are functions of x. Taking the natural logarithm of each side, we obtain

$$\ln y = \ln x^x.$$

Using properties of logarithms, we have

$$\ln y = x \ln x.$$

Differentiating both sides with respect to x gives

$$\left(\frac{1}{y}\right)y' = x\left(\frac{1}{x}\right) + (\ln x)(1).$$

$$\frac{y'}{y} = 1 + \ln x.$$

Solving for y', we obtain

$$y' = y(1 + \ln x).$$

Since $y = x^x$ (as originally given), by substitution we can write the answer in terms of x only:

$$y' = x^x(1 + \ln x).$$

b. $y = x^{e^{-x^2}}$

This has the form $y = u^v$ where $v = e^{-x^2}$. Using logarithmic differentiation, we have

$$\ln y = \ln x^{e^{-x^2}} = e^{-x^2} \ln x.$$

$$\left(\frac{1}{y}\right)y' = e^{-x^2}\left(\frac{1}{x}\right) + (\ln x)[(e^{-x^2})(-2x)].$$

$$\frac{y'}{y} = \frac{e^{-x^2}}{x} - 2xe^{-x^2} \ln x = e^{-x^2}\left(\frac{1}{x} - 2x \ln x\right).$$

$$y' = ye^{-x^2}\left(\frac{1}{x} - 2x \ln x\right)$$

$$= x^{e^{-x^2}}e^{-x^2}\left(\frac{1}{x} - 2x \ln x\right) \quad \text{[by substitution]}.$$

PITFALL. *When using properties of logarithms, you must at all times be able to justify your steps. Thus, if* $y = \ln(x + y)$, *then* $\ln y \neq \ln(x + y)$. *Similarly, if* $y = x^x + x^5$, *then* $\ln y \neq \ln x^x + \ln x^5$.

Be sure you know how to differentiate each of the following forms:

$$y = \begin{cases} [f(x)]^n, & \text{(a)} \\ a^{f(x)}, & \text{(b)} \\ [f(x)]^{g(x)}. & \text{(c)} \end{cases}$$

For type (a), use the power rule; for type (b), use the differentiation formula for exponential functions; for type (c), use logarithmic differentiation. It would be nonsense to write $D_x(x^x) = x \cdot x^{x-1}$. Be sure that all steps are justified by the basic concepts that have been developed.

Exercise 6-7

In Problems **1–20**, *find* y' *by using logarithmic differentiation.*

1. $y = (x + 1)^2(x - 1)(x^2 + 3)$.

2. $y = (3x + 4)(8x - 1)^2(3x^2 + 1)^4$.

3. $y = (3x^3 - 1)^2(2x + 5)^3$.

4. $y = (3x + 1)\sqrt{8x - 1}$.

5. $y = \sqrt{x + 1}\sqrt{x^2 - 2}\sqrt{x + 4}$.

6. $y = (x + 2)\sqrt{x^2 + 9}\sqrt[3]{2x + 1}$.

7. $y = \dfrac{(2x^2 + 2)^2}{(x + 1)^2(3x + 2)}$.

8. $y = \sqrt{\dfrac{(x - 1)(x + 1)}{3x - 4}}$.

9. $y = \dfrac{(8x + 3)^{1/2}(x^2 + 2)^{1/3}}{(1 + 2x)^{1/4}}$.

10. $y = \dfrac{x(1 + x^2)^2}{\sqrt{2 + x^2}}$.

11. $y = \dfrac{\sqrt{1 - x^2}}{1 - 2x}$.

12. $y = \sqrt{\dfrac{x^2 + 5}{x + 9}}$.

13. $y = x^{2x+1}$.

14. $y = x^{\sqrt{x}}$.

15. $y = x^{1/x}$.

16. $y = \left(\dfrac{2}{x}\right)^x$.

17. $y = x^{x^2}$.

18. $y = x^{e^x}$.

19. $y = e^x x^{3x}$.

20. $y = (\ln x)^{e^x}$.

21. Without using logarithmic differentiation, find the derivative of $y = x^x$. *Hint:* First show that $y = x^x = e^{x \ln x}$.

6-8 REVIEW

IMPORTANT TERMS AND SYMBOLS IN CHAPTER 6

exponential function *(p. 272)*

e *(p. 274)*

interest compounded continuously *(p. 277)*

present value *(p. 278)*

logarithmic function *(p. 283)*

$\log_b x$ *(p. 283)*

common logarithm *(p. 285)*

$\log x$ *(p. 285)*

natural logarithm *(p. 285)*

$\ln x$ *(p. 285)*

logarithmic differentiation *(p. 325)*

REVIEW SECTION

1. The domain of the exponential function $f(x) = b^x$ is ___(a)___ and its range is ___(b)___.

Ans. (a) all real numbers; (b) all positive numbers.

2. The domain of the logarithmic function $g(x) = \log_b x$ is ___(a)___ and its range is ___(b)___.

Ans. (a) all positive numbers; (b) all real numbers.

3. The graph in Fig. 6-13 is typical of a(n) (exponential)(logarithmic) function.

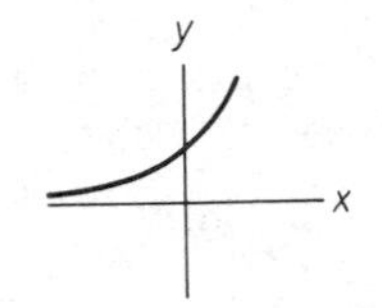

FIG. 6-13

Ans. exponential.

4. $10^{\log 4} =$ ________.

Ans. 4.

5. If $\log_2 (x + 1) = \log_2 4$, then $x =$ ________.

Ans. 3.

6. $\lim_{x\to\infty} (\frac{1}{2})^x =$ ________.

Ans. 0.

7. If $\log x = 1.2222$, then $\log\sqrt{x} =$ ________.

Ans. .6111.

8. $e^{\ln x} =$ ________.

Ans. x.

9. $\log 10^{5x} =$ ________.

Ans. $5x$.

10. $\ln \frac{x^2y^3}{z^4} =$ __(a)__ $\ln x +$ __(b)__ $\ln y -$ __(c)__ $\ln z$.

Ans. (a) 2; (b) 3; (c) 4.

11. The graphs of $y = e^{x+2}$ and $y = e^2e^x$ (are)(are not) identical.

Ans. are.

12. The graph of $y = e^x$ has a __(a) (horizontal)(vertical)__ but no __(b) (horizontal)(vertical)__ asymptote.

Ans. (a) horizontal; (b) vertical.

13. If $y = e^{8x^2+3}$, then $y' =$ ________.

Ans. $16xe^{8x^2+3}$.

14. If $y = \ln (8x^2 + 3)$, then $dy/dx =$ ________.

Ans. $16x/(8x^2 + 3)$.

15. If $y = e^x$, then $d^4y/dx^4 =$ ________.

Ans. e^x.

16. The slope of the curve $y = x + e^x$ when $x = 0$ is ________.

Ans. 2.

17. True or false: $\int x^{-1}\, dx = \frac{x^{-1+1}}{-1+1} + C$, __(a)__; $\int e^x\, dx = \frac{e^{x+1}}{x+1} + C$, __(b)__.

Ans. (a) false; (b) false.

18. $\int e^{2x+1}\, dx =$ ________.

Ans. $\frac{1}{2}e^{2x+1} + C$.

19. $\int \frac{2}{x+5}\, dx =$ ________.

Ans. $2 \ln |x + 5| + C$.

20. $\int \frac{x+1}{x}\,dx =$ ________.

Ans. $x + \ln|x| + C$.

REVIEW PROBLEMS

1. Convert $3^4 = 81$ to logarithmic form.

2. Convert $\log_5 \frac{1}{5} = -1$ to exponential form.

In Problems **3** *and* **4**, *find the limits.*

3. $\lim_{x\to\infty} \frac{7 + 3e^{-x}}{5}$.

4. $\lim_{x\to-\infty} \frac{10}{4 + 2^{x+1}}$.

In Problems **5–12** *find x.*

5. $\log_5 125 = x$.

6. $\log_x \frac{1}{8} = -3$.

7. $\log x = -2$.

8. $\ln \frac{1}{e} = x$.

9. $\log_x (2x + 3) = 2$.

10. $\log (4x + 1) = \log (x + 2)$.

11. $e^{\ln(x+4)} = 7$.

12. $\log x + \log 2 = 1$.

13. Find the value of $\log 2500$.

14. Find the value of $\log_3 4$.

15. If $\log 3 = x$ and $\log 4 = y$, express $\log (16\sqrt{3})$ in terms of x and y.

16. Express

$$\log \frac{x^2\sqrt{x+1}}{\sqrt[3]{x^2+2}}$$

in terms of $\log x$, $\log (x + 1)$, and $\log (x^2 + 2)$.

17. Simplify $e^{\ln x} + \ln e^x + \ln 1$.

18. Simplify $\log 10^2 + \log 1000 - 5$.

19. If $\ln y = x^2 + 2$, find y.

20. Sketch the graphs of $y = 3^x$ and $y = \log_3 x$.

In Problems **21–48**, *differentiate.*

21. $y = 2e^x + e^2 + e^{x^2}$.

22. $f(w) = we^w + w^2$.

23. $f(r) = \ln (r^2 + 5r)$.

24. $y = e^{\ln x}$.

25. $y = e^{x^2+4x+5}$.

26. $f(t) = \log_6 \sqrt{t^2+1}$.

27. $y = e^x(x^2+2)$.

28. $y = 2^{7x^2}$.

29. $y = \sqrt{(x-6)(x+5)(9-x)}$.

30. $f(t) = e^{\sqrt{t}}$.

31. $y = \dfrac{\ln x}{e^x}$.

32. $y = \dfrac{e^x + e^{-x}}{x^2}$.

33. $f(q) = \ln [(q+1)^2(q+2)^3]$.

34. $y = (x-6)^4(x+4)^3(6-x)^2$.

35. $y = 10^{2-7x}$.

36. $y = (e+e^2)^0$.

37. $y = \dfrac{4e^{3x}}{xe^{x-1}}$.

38. $y = \dfrac{e^x}{\ln x}$.

39. $y = \log_2 (8x+5)^2$.

40. $y = \ln (1/x)$.

41. $f(l) = \ln (1 + l + l^2 + l^3)$.

42. $y = x^{x^3}$.

43. $y = (x+1)^{x+1}$.

44. $y = \dfrac{1+e^x}{1-e^x}$.

45. $f(t) = \ln (t^2\sqrt{1-t})$.

46. $y = (x+2)^{\ln x}$.

47. $y = \dfrac{(x^2+2)^{3/2}(x^2+9)^{4/9}}{(x^3+6x)^{4/11}}$.

48. $y = \dfrac{\ln x}{\sqrt{x}}$.

In Problems 49–52, *find the indicated derivative at the given point. It is not necessary to simplify the derivative before substituting the coordinates.*

49. $y = e^{x^2-4}, y'', (2, 1)$.

50. $y = x^2e^x, y''', (1, e)$.

51. $y = \ln (2x), y''', (1, \ln 2)$.

52. $y = x \ln x, y'', (1, 0)$.

In Problems 53 *and* 54, *find an equation of the tangent line to the curve at the point corresponding to the given value of x.*

53. $y = e^x, x = \ln 2$.

54. $y = x + x^2 \ln x, x = 1$.

In Problems 55 *and* 56, *find the rate of change of y with respect to x at the given point.*

55. $\ln (3y) = x, (0, \frac{1}{3})$.

56. $e^{x-y} = y, (1, 1)$.

In Problems 57 *and* 58, *use differentials to approximate the given values.*

57. $e^{-.2}$.

58. $\ln (1.1)$.

In Problems 59 *and* 60, *find intervals on which the function is increasing, decreasing, concave up, concave down; find relative extrema and points of inflection. Sketch the graphs of the functions.*

59. $f(x) = \dfrac{e^x + e^{-x}}{2}$.

60. $f(x) = 1 + \ln (x^2)$.

In Problems **61–78**, *determine the integrals.*

61. $\int 4e^{3x+5}\, dx$.

62. $\int_0^2 xe^{4-x^2}\, dx$.

63. $\int \dfrac{1}{2(x+1)}\, dx$.

64. $\int \dfrac{2}{5-3x}\, dx$.

65. $\int \dfrac{6x^2 - 12}{x^3 - 6x + 1}\, dx$.

66. $\int \dfrac{2t}{5t^2 + 1}\, dt$.

67. $\int_1^2 \dfrac{t^2}{2 + t^3}\, dt$.

68. $\int \dfrac{1}{e^{-3x}}\, dx$.

69. $\int_1^e \dfrac{p + 8}{p}\, dp$.

70. $\int_0^1 \dfrac{e^{2x}}{1 + e^{2x}}\, dx$.

71. $\int (e^{2y} - e^{-2y})\, dy$.

72. $\int x(5x^2 + 3)^{-1}\, dx$.

73. $\int \left(\dfrac{1}{x} + \dfrac{2}{x^2}\right) dx$.

74. $\int \dfrac{z^2}{z - 1}\, dz$.

75. $\int 5^{2x}\, dx$.

76. $\int \ln (e^{4x})\, dx$.

77. $\int_{-1}^0 \dfrac{x^2 + 4x - 1}{x + 2}\, dx$.

78. $\int \dfrac{1}{\sqrt{x}\,(1 + \sqrt{x}\,)}\, dx$.

In Problems **79** *and* **80**, *find the area of the region bounded by the given curves, the x-axis, and the given lines. Assume that the answer is in square units.*

79. $y = \dfrac{1}{x} + 3$, $x = 1$, $x = 3$.

80. $y = 4e^{2x}$, $x = 0$, $x = 3$.

81. For a new product the yearly number of thousand packages sold, y, after t years from its introduction is predicted to be

$$y = f(t) = 150 - 76e^{-t}.$$

Show that $y = 150$ is a horizontal asymptote of the graph. This shows that after the product is established with consumers, the market tends to be constant.

82. Show that the graphs of $y = 6 - 3e^{-x}$ and $y = 6 + 3e^{-x}$ are both asymptotic to the same line. What is the equation of this line?

83. Due to ineffective advertising, the Kleer-Kut Razor Company finds its annual revenues have been cut sharply. Moreover, the annual revenue R at the end of t years of business satisfies the equation $R = 200{,}000e^{-.2t}$. Find the annual revenue at the end of 2 years; at the end of 3 years.

84. In a study of water mites infesting a group of 589 adult flies [47], a Poisson distribution was fitted to the data. If x is the number of mites per fly and $\mu = .44$, find the probability of 3 mites per fly. (See Sec. 6-1 for a discussion of the Poisson distribution.)

85. For a certain population of cells, the number N of cells at time t is given by $N = N_0(2^{t/k})$, where N_0 is the number of cells at $t = 0$ and k is a positive constant. Show that

$$k = \frac{t \ln 2}{\ln (N/N_0)}.$$

86. If $\bar{c} = (500/q)e^{q/300}$ is an average cost function, find the marginal cost function.

87. The demand function for a manufacturer's product is given by $p = 100e^{-.1q}$, where p is the price per unit at which q units are demanded. For what value of q does the manufacturer maximize total revenue?

88. For a group of hospitalized individuals, suppose the discharge rate is given by $f(t) = .008e^{-.008t}$, where $f(t)$ is the proportion discharged per day at the end of t days of hospitalization. What proportion of the group is discharged at the end of 100 days?

89. Several models have been used to analyze the length of stay in a hospital. For a particular group of schizophrenics, one such model is (adapted from [9]) $f(t) = 1 - (.8e^{-.01t} + .2e^{-.0002t})$, where $f(t)$ is the proportion of the group that was discharged at the end of t days of hospitalization. Determine the discharge rate (proportion discharged per day) at the end of t days.

90. Another model for the group of schizophrenics of Problem 89 is

$$f(t) = 1 - e^{-.01t^{0.9}}.$$

Show that the rate of discharge is decreasing for $t > 0$.

91. New products or technologies often tend to replace old ones. For example, today most commercial airlines use jet engines rather than prop engines. In discussing the forecasting of technological substitution, Hurter and Rubenstein [22] refer to the equation

$$\ln \frac{f(t)}{1 - f(t)} + \sigma \frac{1}{1 - f(t)} = C_1 + C_2 t,$$

where $f(t)$ is the market share of the substitute over time t, and C_1, C_2, and σ (a Greek letter read "sigma") are constants. Verify their claim that the rate of substitution is

$$f'(t) = \frac{C_2 f(t)[1 - f(t)]^2}{\sigma f(t) + [1 - f(t)]}.$$

92. The "logit" function is employed in the study of the behavioral sciences. It is defined as follows:

$$\text{logit}\,(x) = \frac{1}{2}\ln\frac{1-x}{x}, \quad \text{where } 0 < x < 1.$$

If logit $(y) = a$ logit $(x) + b$, show that

$$y = \frac{1}{1 + e^{2b}\left(\frac{1-x}{x}\right)^a}.$$

CHAPTER 7

Methods and Applications of Integration

7-1 INTEGRATION BY PARTS[†]

Many integrals cannot be found by our previous methods. However, there are ways of changing certain integrals to forms that are easier to integrate. Of these methods, we shall discuss *integration by parts*.

If u and v are differentiable functions of x, by the product rule we have

$$(uv)' = uv' + vu'.$$

By rearranging,

$$uv' = (uv)' - vu'.$$

[†]May be omitted without loss of continuity.

Integrating both sides with respect to x, we get

$$\int uv'\,dx = \int (uv)'\,dx - \int vu'\,dx. \tag{1}$$

For $\int (uv)'\,dx$, we must find a function whose derivative with respect to x is $(uv)'$. Clearly uv is such a function. Hence $\int (uv)'\,dx = uv + C_1$ and Eq. (1) becomes

$$\int uv'\,dx = uv + C_1 - \int vu'\,dx.$$

Incorporating C_1 into the constant of integration for $\int vu'\,dx$ and replacing $v'\,dx$ by dv and $u'\,dx$ by du, we have the *integration by parts formula*:

INTEGRATION BY PARTS FORMULA

$$\int u\,dv = uv - \int v\,du. \tag{2}$$

This formula expresses a given integral, $\int u\,dv$, in terms of another integral, $\int v\,du$, which may be easier to find.

To apply the formula to $\int f(x)\,dx$, we must write $f(x)\,dx$ as the product of two factors (or *parts*) by choosing a function u and a differential dv such that $f(x)\,dx = u\,dv$. For the formula to be useful, we must be able to integrate the part chosen for dv. To illustrate, consider

$$\int xe^x\,dx.$$

This integral cannot be determined by previous integration formulas. We can write $xe^x\,dx$ in the form $u\,dv$ by letting

$$u = x \quad \text{and} \quad dv = e^x\,dx.$$

To apply the integration by parts formula, we must find du and v:

$$du = dx \quad \text{and} \quad v = \int e^x\,dx = e^x + C_1.$$

Thus,

$$\begin{aligned}\int \underbrace{x}_{u}\,\underbrace{e^x\,dx}_{dv} &= uv - \int v\,du \\ &= x(e^x + C_1) - \int (e^x + C_1)\,dx \\ &= xe^x + C_1x - e^x - C_1x + C \\ &= xe^x - e^x + C = e^x(x-1) + C.\end{aligned}$$

The first constant C_1 does not appear in the final answer. This is a characteristic of integration by parts and from now on this constant will not be written when finding v from dv.

When you are using the integration by parts formula, sometimes the "best choice" for u and dv may not be obvious. In some cases one choice may be as good as another; in other cases only one choice may be suitable. Insight into making a good choice (if any exists) will come only with practice and, of course, trial and error.

EXAMPLE 1

Find $\int \frac{\ln x}{\sqrt{x}}\,dx$ by integration by parts.

We try

$$u = \ln x \quad \text{and} \quad dv = \frac{1}{\sqrt{x}}\,dx.$$

Then $$du = \frac{1}{x}\,dx \quad \text{and} \quad v = \int x^{-1/2}\,dx = 2x^{1/2}.$$

Thus, $$\int \underbrace{\ln x}_{u}\underbrace{\left(\frac{1}{\sqrt{x}}\,dx\right)}_{dv} = uv - \int v\,du$$

$$= (\ln x)(2\sqrt{x}\,) - \int (2x^{1/2})\left(\frac{1}{x}\,dx\right)$$

$$= 2\sqrt{x}\ln x - 2\int x^{-1/2}\,dx$$

$$= 2\sqrt{x}\ln x - 2(2\sqrt{x}\,) + C$$

$$= 2\sqrt{x}\,(\ln x - 2) + C.$$

EXAMPLE 2

Evaluate $\int_1^2 x \ln x\,dx$.

Let $u = x$ and $dv = \ln x\,dx$. Then $du = dx$, but $v = \int \ln x\,dx$ is not apparent by inspection. Let us try

$$u = \ln x \quad \text{and} \quad dv = x\,dx.$$

Then $$du = \frac{1}{x}\,dx \quad \text{and} \quad v = \int x\,dx = \frac{x^2}{2}.$$

Thus, $$\int_1^2 x \ln x\, dx = (\ln x)\left(\frac{x^2}{2}\right)\Bigg|_1^2 - \int_1^2 \left(\frac{x^2}{2}\right)\frac{1}{x}\, dx$$
$$= (\ln x)\left(\frac{x^2}{2}\right)\Bigg|_1^2 - \frac{1}{2}\int_1^2 x\, dx$$
$$= \frac{x^2 \ln x}{2}\Bigg|_1^2 - \frac{1}{2}\left(\frac{x^2}{2}\right)\Bigg|_1^2$$
$$= (2 \ln 2 - 0) - \left(1 - \tfrac{1}{4}\right) = 2 \ln 2 - \tfrac{3}{4}.$$

EXAMPLE 3

Determine $\int \ln y\, dy$.

Let $u = \ln y$ and $dv = dy$. Then $du = (1/y)\, dy$ and $v = y$.

$$\int \ln y\, dy = (\ln y)(y) - \int y\left(\frac{1}{y}\, dy\right)$$
$$= y \ln y - \int dy = y \ln y - y + C$$
$$= y(\ln y - 1) + C.$$

EXAMPLE 4

Determine $\int xe^{x^2}\, dx$.

PITFALL. *Do not forget about basic integration forms. Integration by parts is not needed here!*

$$\int xe^{x^2}\, dx = \tfrac{1}{2}\int e^{x^2}[2x\, dx]$$
$$= \tfrac{1}{2}\int e^u\, du \quad \text{where } u = x^2$$
$$= \tfrac{1}{2}e^u + C = \tfrac{1}{2}e^{x^2} + C.$$

Sometimes integration by parts must be used more than once, as shown in the following example.

EXAMPLE 5

Determine $\int x^2 e^{2x+1}\, dx$.

Let $u = x^2$ and $dv = e^{2x+1}\,dx$. Then $du = 2x\,dx$ and $v = e^{2x+1}/2$.

$$\int x^2 e^{2x+1}\,dx = \frac{x^2 e^{2x+1}}{2} - \int \frac{e^{2x+1}}{2}(2x\,dx)$$
$$= \frac{x^2 e^{2x+1}}{2} - \int x e^{2x+1}\,dx.$$

To find $\int xe^{2x+1}\,dx$, we shall again use integration by parts. Here, let $u = x$ and $dv = e^{2x+1}\,dx$. Then $du = dx$ and $v = e^{2x+1}/2$.

$$\int x e^{2x+1}\,dx = \frac{xe^{2x+1}}{2} - \int \frac{e^{2x+1}}{2}\,dx$$
$$= \frac{xe^{2x+1}}{2} - \frac{e^{2x+1}}{4} + C_1.$$

Thus,

$$\int x^2 e^{2x+1}\,dx = \frac{x^2 e^{2x+1}}{2} - \frac{xe^{2x+1}}{2} + \frac{e^{2x+1}}{4} + C \qquad (\text{where } C = -C_1)$$
$$= \frac{e^{2x+1}}{2}\left(x^2 - x + \frac{1}{2}\right) + C.$$

Exercise 7-1

In Problems **1–20**, *find the integrals.*

1. $\int xe^{-x}\,dx.$

2. $\int xe^{2x}\,dx.$

3. $\int y^3 \ln y\,dy.$

4. $\int x^2 \ln x\,dx.$

5. $\int \ln(4x)\,dx.$

6. $\int \frac{t}{e^t}\,dt.$

7. $\int x\sqrt{x+1}\,dx.$

8. $\int \frac{x}{\sqrt{1+4x}}\,dx.$

9. $\int \sqrt{x}\ln x\,dx.$

10. $\int \frac{\ln(x+1)}{2(x+1)}\,dx.$

11. $\int_1^2 xe^{2x}\,dx.$

12. $\int_0^1 xe^{-x}\,dx.$

13. $\int_0^1 xe^{-x^2}\,dx.$

14. $\int \frac{x^3}{\sqrt{4-x^2}}\,dx.$

15. $\int_1^2 \frac{x}{\sqrt{4-x}}\,dx.$

16. $\int (\ln x)^2\,dx.$

17. $\int x^2 e^x\,dx.$

18. $\int x^2 e^{-2x}\,dx.$

19. $\int (x - e^{-x})^2\,dx.$

20. $\int x^3 e^{x^2}\,dx.$

21. Find the area of the region bounded by the x-axis, the curve $y = \ln x$ and the line $x = e^3$.

22. Find the area of the region bounded by the x-axis and the curve $y = xe^{-x}$ between $x = 0$ and $x = 4$.

7-2 INTEGRATION USING TABLES

Certain forms of integrals which occur frequently may be found in standard tables of integration formulas†. A short table appears in Appendix E. In this section its use will be illustrated.

No table of integrals is exhaustive. We may still have to appeal to prior techniques of integration when a suitable form is not listed in the tables. Moreover, a given integral may have to be replaced by an equivalent form before it will fit a formula in the table. The equivalent form must match the formula *exactly*. Consequently, the steps that you perform should *not* be done mentally. *Write them down!* Failure to do this can easily lead to incorrect results. Before proceeding with the exercises, be sure you understand the illustrative examples *thoroughly*.

In the following examples the formula numbers refer to the Table of Selected Integrals given in Appendix E.

EXAMPLE 1

Find $\int \frac{x\,dx}{(2+3x)^2}$.

Scanning the tables, we identify the integrand with formula 7:

$$\int \frac{u\,du}{(a+bu)^2} = \frac{1}{b^2}\left(\ln|a+bu| + \frac{a}{a+bu}\right) + C.$$

†See, for example, S. M. Selby, *Standard Mathematical Tables*, 22nd ed. Cleveland: Chemical Rubber Co., 1974.

For the given integrand, let $u = x$, $a = 2$, and $b = 3$. Then $du = dx$ and we have

$$\int \frac{x\,dx}{(2 + 3x)^2} = \int \frac{u\,du}{(a + bu)^2}.$$

By the formula,

$$\int \frac{x\,dx}{(2 + 3x)^2} = \int \frac{u\,du}{(a + bu)^2} = \frac{1}{b^2}\left(\ln|a + bu| + \frac{a}{a + bu}\right) + C.$$

Returning to the variable x and replacing a by 2 and b by 3, we obtain

$$\int \frac{x\,dx}{(2 + 3x)^2} = \frac{1}{9}\left(\ln|2 + 3x| + \frac{2}{2 + 3x}\right) + C.$$

EXAMPLE 2

Find $\int x^2\sqrt{x^2 - 1}\,dx$.

This integral is identified with formula 24:

$$\int u^2\sqrt{u^2 \pm a^2}\,du = \frac{u}{8}(2u^2 \pm a^2)\sqrt{u^2 \pm a^2} - \frac{a^4}{8}\ln\left|u + \sqrt{u^2 \pm a^2}\right| + C.$$

In applying this formula, if the bottommost sign in the dual symbol "$\pm$" on the left side is used, the bottommost sign in the dual symbols on the right must also be used. Letting $u = x$ and $a = 1$, then we have $du = dx$. Thus,

$$\begin{aligned}\int x^2\sqrt{x^2 - 1}\,dx &= \int u^2\sqrt{u^2 - a^2}\,du\\ &= \frac{u}{8}(2u^2 - a^2)\sqrt{u^2 - a^2} - \frac{a^4}{8}\ln\left|u + \sqrt{u^2 - a^2}\right| + C.\end{aligned}$$

Since $u = x$ and $a = 1$,

$$\int x^2\sqrt{x^2 - 1}\,dx = \frac{x}{8}(2x^2 - 1)\sqrt{x^2 - 1} - \frac{1}{8}\ln\left|x + \sqrt{x^2 - 1}\right| + C.$$

EXAMPLE 3

Find $\int \frac{dx}{x\sqrt{16x^2 + 3}}$.

The integrand can be identified with formula 28:

$$\int \frac{du}{u\sqrt{u^2 + a^2}} = \frac{1}{a}\ln\left|\frac{\sqrt{u^2 + a^2} - a}{u}\right| + C.$$

If we let $u = 4x$ and $a = \sqrt{3}$, then $du = 4\,dx$. Thus (watch closely)

$$\int \frac{dx}{x\sqrt{16x^2 + 3}} = \int \frac{(4\,dx)}{(4x)\sqrt{(4x)^2 + (\sqrt{3})^2}} = \int \frac{du}{u\sqrt{u^2 + a^2}}.$$

By formula 28, the latter integral is

$$\frac{1}{a}\ln\left|\frac{\sqrt{u^2 + a^2} - a}{u}\right| + C.$$

Hence, replacing u by $4x$ and a by $\sqrt{3}$, we have

$$\int \frac{dx}{x\sqrt{16x^2 + 3}} = \frac{1}{\sqrt{3}}\ln\left|\frac{\sqrt{16x^2 + 3} - \sqrt{3}}{4x}\right| + C.$$

EXAMPLE 4

Find $\int \frac{dx}{x^2(2 - 3x^2)^{1/2}}$.

The integrand is identified with formula 21:

$$\int \frac{du}{u^2\sqrt{a^2 - u^2}} = -\frac{\sqrt{a^2 - u^2}}{a^2u} + C.$$

Letting $u = \sqrt{3}\,x$ and $a^2 = 2$, then we have $du = \sqrt{3}\,dx$. Hence

$$\int \frac{dx}{x^2(2 - 3x^2)^{1/2}} = \sqrt{3}\int \frac{(\sqrt{3}\,dx)}{(\sqrt{3}\,x)^2(2 - 3x^2)^{1/2}} = \sqrt{3}\int \frac{du}{u^2(a^2 - u^2)^{1/2}}$$

$$= \sqrt{3}\left[-\frac{\sqrt{a^2 - u^2}}{a^2u}\right] + C = \sqrt{3}\left[-\frac{\sqrt{2 - 3x^2}}{2(\sqrt{3}\,x)}\right] + C$$

$$= -\frac{\sqrt{2 - 3x^2}}{2x} + C.$$

EXAMPLE 5

Find $\int 7x^2 \ln(4x)\,dx$.

This is similar to formula 42 where $n = 2$;

$$\int u^n \ln u\,du = \frac{u^{n+1}\ln u}{n + 1} - \frac{u^{n+1}}{(n + 1)^2} + C.$$

If we let $u = 4x$, then $du = 4\,dx$. Hence

$$\int 7x^2 \ln (4x)\, dx = \frac{7}{64}\int (4x)^2 \ln (4x)(4\, dx)$$

$$= \frac{7}{64}\int u^2 \ln u\, du = \frac{7}{64}\left(\frac{u^3 \ln u}{3} - \frac{u^3}{9}\right) + C$$

$$= \frac{7}{64}\left[\frac{(4x)^3 \ln (4x)}{3} - \frac{(4x)^3}{9}\right] + C$$

$$= 7x^3\left[\frac{\ln (4x)}{3} - \frac{1}{9}\right] + C.$$

EXAMPLE 6

Find $\int \frac{e^{2x}\, dx}{7 + e^{2x}}$.

At first glance we do not identify the integrand with any form in the table. Perhaps rewriting the integral will help. Let $u = 7 + e^{2x}$; then $du = 2e^{2x}\, dx$.

$$\int \frac{e^{2x}\, dx}{7 + e^{2x}} = \frac{1}{2}\int \frac{(2e^{2x}\, dx)}{7 + e^{2x}} = \frac{1}{2}\int \frac{du}{u} = \frac{1}{2} \ln |u| + C$$

$$= \frac{1}{2}\ln |7 + e^{2x}| + C = \frac{1}{2}\ln (7 + e^{2x}) + C.$$

Thus we had only to use our knowledge of basic integration forms. Actually, this form appears as formula 2 in the tables.

EXAMPLE 7

Evaluate $\int_1^4 \frac{dx}{(4x^2 + 2)^{3/2}}$.

We shall use formula 32 to get the indefinite integral first:

$$\int \frac{du}{(u^2 \pm a^2)^{3/2}} = \frac{\pm u}{a^2\sqrt{u^2 \pm a^2}} + C.$$

Letting $u = 2x$ and $a^2 = 2$, then we have $du = 2\, dx$. Thus,

$$\int \frac{dx}{(4x^2 + 2)^{3/2}} = \frac{1}{2}\int \frac{(2\, dx)}{[(2x)^2 + 2]^{3/2}} = \frac{1}{2}\int \frac{du}{(u^2 + 2)^{3/2}}$$

$$= \frac{1}{2}\left[\frac{u}{2\sqrt{u^2 + 2}}\right] + C.$$

Instead of substituting back to x and evaluating from $x = 1$ to $x = 4$, we can determine the corresponding limits of integration with respect to u. Since

$u = 2x$, then when $x = 1$ we have $u = 2$; when $x = 4$ we have $u = 8$. Thus,

$$\int_1^4 \frac{dx}{(4x^2+2)^{3/2}} = \frac{1}{2}\int_2^8 \frac{du}{(u^2+2)^{3/2}}$$

$$= \frac{1}{2}\left(\frac{u}{2\sqrt{u^2+2}}\right)\Bigg|_2^8 = \frac{2}{\sqrt{66}} - \frac{1}{2\sqrt{6}}.$$

PITFALL. *In Example 7, when changing the variable of integration x to the variable of integration u, be certain to change the limits of integration so that they agree with u. That is,*

$$\int_1^4 \frac{dx}{(4x^2+2)^{3/2}} \neq \frac{1}{2}\int_1^4 \frac{du}{(u^2+2)^{3/2}}.$$

Suppose that you must pay out $100 at the end of each year for the next two years. Any series of payments over a period of time, such as this, is called an *annuity*. If money is worth 5 percent and you were to pay off the debt immediately, you would *not* pay $200. Instead, you would pay the present value of the $100 due at the end of the first year, plus the present value of the $100 due at the end of the second year. (Present value is discussed in Sec. 6-1.) The sum of these present values is called the present value of the annuity. We shall now consider the present value of payments made continuously over the time interval from $t = 0$ to $t = T$, t in years, at continuous interest at an annual rate of r.

Suppose a payment is made at time t such that on an annual basis this payment is $f(t)$. Then over a small interval $\Delta t = [t_i, t_{i+1}]$ the total amount of all payments is approximately $f(t_i)\,\Delta t$. [For example, if $f(t) = 2000$ and Δt were one day, then the total amount of the payments would be $2000(\frac{1}{365})$.] The present value of these payments is approximately $e^{-rt_i}f(t_i)\,\Delta t$. Over the interval $[0, T]$, the total of all such present values is

$$\Sigma\, e^{-rt_i}f(t_i)\,\Delta t.$$

This sum approximates the present value P of the annuity. The smaller Δt is, the better is the approximation. That is, as $\Delta t \to 0$ the limit of the sum *is* the present value. However, this limit is also a definite integral. That is,

$$\boxed{P = \int_0^T f(t)e^{-rt}\,dt} \qquad (1)$$

is the **present value of a continuous annuity** at an annual rate r (compounded continuously) for T years if a payment at time t is at the rate of $f(t)$ per year. Sometimes we speak of this integral as the **present value of a continuous income stream.**

We can also look at the *future* value of an annuity rather than its present value. If a payment is made at time t, then it has a certain value at the *end* of the period of the annuity, that is, $T - t$ years later. This value is

$$\begin{pmatrix}\text{amount of} \\ \text{payment}\end{pmatrix} + \begin{pmatrix}\text{interest on this} \\ \text{payment for } T - t \text{ years}\end{pmatrix}.$$

If A is the total of such values for all payments, then A is called the **accumulated amount of a continuous annuity** and is given by

$$\boxed{A = \int_0^T f(t)e^{r(T-t)}\, dt.}$$

EXAMPLE 8

Find the present value (to the nearest dollar) of a continuous annuity at an annual rate of 4 percent for 10 years if the payment at time t is at the rate of t^2 dollars per year.

The present value is given by

$$P = \int_0^T f(t)e^{-rt}\, dt = \int_0^{10} t^2 e^{-.04t}\, dt.$$

We shall use formula 39,

$$\int u^n e^{au}\, du = \frac{u^n e^{au}}{a} - \frac{n}{a}\int u^{n-1}e^{au}\, du,$$

called a *reduction formula* since it reduces an integral into an expression that involves an integral which is easier to determine. If $u = t$, $n = 2$, and $a = -.04$, then $du = dt$ and we have

$$P = \left.\frac{t^2 e^{-.04t}}{-.04}\right|_0^{10} - \frac{2}{-.04}\int_0^{10} te^{-.04t}\, dt.$$

In the new integral the exponent of t has been reduced to 1. By applying formula 38,

$$\int ue^{au}\, du = \frac{e^{au}}{a^2}(au - 1) + C,$$

to this integral where $u = t$, $du = dt$, and $a = -.04$, we then have

$$P = \int_0^{10} t^2 e^{-.04t}\, dt$$

$$= \frac{t^2 e^{-.04t}}{-.04}\Bigg|_0^{10} - \frac{2}{-.04}\left[\frac{e^{-.04t}}{(-.04)^2}(-.04t - 1)\right]\Bigg|_0^{10}$$

$$= \frac{100e^{-.4}}{-.04} + \frac{2}{.04}\left[\frac{e^{-.4}}{(-.04)^2}(-.4 - 1) - \frac{1}{(-.04)^2}(-1)\right]$$

$$= -2500e^{-.4} + 50[-875e^{-.4} + 625]$$

$$= -46{,}250e^{-.4} + 31{,}250 \approx -46{,}250(.67032) + 31{,}250$$

$$\approx 248.$$

The present value is \$248.

Equation (1) may also be used for the present value of future profits of a business. In this situation, $f(t)$ would be the annual rate of profit at time t.

Exercise 7-2

In Problems **1–34**, *find the integrals by using the tables in Appendix E.*

1. $\displaystyle\int \frac{dx}{x(6 + 7x)}$.

2. $\displaystyle\int \frac{x^2\, dx}{(1 + 2x)^2}$.

3. $\displaystyle\int \frac{dx}{x\sqrt{x^2 + 9}}$.

4. $\displaystyle\int \frac{dx}{(x^2 + 7)^{3/2}}$.

5. $\displaystyle\int \frac{x\, dx}{(2 + 3x)(4 + 5x)}$.

6. $\displaystyle\int 2^{5x}\, dx$.

7. $\displaystyle\int \frac{dx}{4 + 3e^{2x}}$.

8. $\displaystyle\int x^2\sqrt{1 + x}\, dx$.

9. $\displaystyle\int \frac{2\, dx}{x(1 + x)^2}$.

10. $\displaystyle\int \frac{dx}{x\sqrt{5 - 11x^2}}$.

11. $\displaystyle\int_0^1 \frac{x\, dx}{2 + x}$.

12. $\displaystyle\int \frac{x^2\, dx}{2 + 5x}$.

13. $\displaystyle\int \sqrt{x^2 - 3}\, dx$.

14. $\displaystyle\int \frac{dx}{(4 + 3x)(4x + 3)}$.

15. $\int_0^{1/12} xe^{12x}\,dx.$

16. $\int \sqrt{\frac{2+3x}{5+3x}}\,dx.$

17. $\int x^2e^x\,dx.$

18. $\int_1^2 \frac{dx}{x^2(1+x)}.$

19. $\int \frac{\sqrt{4x^2+1}}{x^2}\,dx.$

20. $\int \frac{dx}{x\sqrt{2-x}}.$

21. $\int \frac{x\,dx}{(1+3x)^2}.$

22. $\int \frac{dx}{\sqrt{(1+2x)(3+2x)}}.$

23. $\int \frac{dx}{7-5x^2}.$

24. $\int x^2\sqrt{2x^2-9}\,dx.$

25. $\int x^5 \ln(3x)\,dx.$

26. $\int \frac{dx}{x^2(1+x)^2}.$

27. $\int 2x\sqrt{1+3x}\,dx.$

28. $\int x^2 \ln x\,dx.$

29. $\int \frac{dx}{\sqrt{4x^2-13}}.$

30. $\int \frac{dx}{x\ln(2x)}.$

31. $\int x\ln(2x)\,dx.$

32. $\int \frac{\sqrt{2-3x^2}}{x}\,dx.$

33. $\int \frac{dx}{x^2\sqrt{9-4x^2}}.$

34. $\int_0^1 \frac{x^3\,dx}{1+x^4}.$

In Problems **35–52**, *find the integrals by any method.*

35. $\int \frac{x\,dx}{x^2+1}.$

36. $\int \sqrt{x}\,e^{x^{3/2}}\,dx.$

37. $\int x\sqrt{2x^2+1}\,dx.$

38. $\int \frac{4x^2-\sqrt{x}}{x}\,dx.$

39. $\int \frac{dx}{x^2-5x+6}.$

40. $\int \frac{e^{2x}}{\sqrt{e^{2x}+3}}\,dx.$

41. $\int x^3 \ln x\,dx.$

42. $\int_0^3 xe^{-x}\,dx.$

43. $\int xe^{2x}\,dx.$

44. $\int_1^2 x^2\sqrt{3+2x}\,dx.$

45. $\int \ln^2 x\,dx.$

46. $\int_1^e \ln x\,dx.$

47. $\int_1^2 \frac{x\,dx}{\sqrt{4-x}}$.

48. $\int_1^2 x\sqrt{1+2x}\,dx$.

49. $\int_0^1 \frac{2x\,dx}{\sqrt{8-x^2}}$.

50. $\int_0^{\ln 2} x^3 e^{2x}\,dx$.

51. $\int_1^2 x \ln(2x)\,dx$.

52. $\int_1^2 dx$.

53. In a discussion [33] about gene frequency, the following integral occurs:

$$\int_{q_0}^{q_n} \frac{dq}{q(1-q)},$$

where the q's represent gene frequencies. Evaluate this integral.

54. Under certain conditions, the number n of generations required to change the frequency of a gene from .3 to .1 is given in [56] by

$$n = -\frac{1}{.4}\int_{.3}^{.1} \frac{dq}{q^2(1-q)}.$$

Find n (to the nearest integer).

55. Find the present value, to the nearest dollar, of a continuous annuity at an annual rate of r for T years if the payment at time t is at the annual rate of $f(t)$ dollars given that

(a) $r = .06$, $T = 10$, $f(t) = 5000$,

(b) $r = .05$, $T = 8$, $f(t) = 200t$.

56. If $f(t) = k$, where k is a positive constant, show that the value of the integral in Eq. (1) of this section is

$$k\left(\frac{1-e^{-rT}}{r}\right).$$

57. Find the accumulated amount, to the nearest dollar, of a continuous annuity at an annual rate of r for T years if the payment at time t is at the annual rate of $f(t)$ dollars given that

(a) $r = .06$, $T = 10$, $f(t) = 400$,

(b) $r = .04$, $T = 5$, $f(t) = 40t$.

58. Over the next five years the profits of a business at time t are estimated to be $20{,}000t$ dollars per year. The business is to be sold at a price equal to the present value of these future profits. If interest is compounded continuously at the annual rate of 6 percent, at what price should the business be sold?

7-3 AVERAGE VALUE OF A FUNCTION

If we are given the three numbers 1, 2, and 9, then their average value, or *mean*, is their sum divided by 3. Calling this average $\bar{y}$, we have

$$\bar{y} = \frac{1 + 2 + 9}{3} = 4.$$

Similarly, suppose we are given a function f defined on the interval $[a, b]$ and the points $x_1, x_2, \ldots, x_n$ are in the interval. Then the average value of the n corresponding functional values $f(x_1)$, $f(x_2), \ldots, f(x_n)$ is

$$\bar{y} = \frac{f(x_1) + f(x_2) + \cdots + f(x_n)}{n} = \frac{\sum_{k=1}^{n} f(x_k)}{n}. \qquad (1)$$

We can go a step further. Let's divide the interval $[a, b]$ into n subintervals of equal length. We shall choose x_1 to be in the first subinterval, x_2 to be in the second, etc. Since $[a, b]$ has length $b - a$, each subinterval has length $\frac{b - a}{n}$, which we shall call Δx. Thus, (1) can be written

$$\bar{y} = \frac{\sum_{k=1}^{n} f(x_k)\left(\frac{\Delta x}{\Delta x}\right)}{n} = \frac{1}{\Delta x}\left[\frac{\sum_{k=1}^{n} f(x_k)\,\Delta x}{n}\right] = \frac{1}{n\,\Delta x}\sum_{k=1}^{n} f(x_k)\,\Delta x. \qquad (2)$$

Since $\Delta x = \frac{b - a}{n}$, the expression $n\,\Delta x$ in (2) can be replaced by $b - a$. Moreover, as $n \to \infty$ the number of functional values used in computing $\bar{y}$ increases, and we get the so-called *average value of the function* f, *denoted* $\bar{f}$:

$$\bar{f} = \lim_{n\to\infty} \frac{1}{b - a}\sum_{k=1}^{n} f(x_k)\,\Delta x = \frac{1}{b - a}\lim_{n\to\infty}\sum_{k=1}^{n} f(x_k)\,\Delta x.$$

But the limit on the right is just the definite integral $\int_a^b f(x)\,dx$.

DEFINITION. *The* ***average*** *(or* ***mean****)* ***value of a function*** $y = f(x)$ *over the interval* $[a, b]$ *is denoted* $\bar{f}$ *(or* $\bar{y}$ *) and is given by*

$$\bar{f} = \frac{1}{b - a}\int_a^b f(x)\,dx.$$

EXAMPLE 1

Find the average value of the function $f(x) = x^2$ *over the interval* [1, 2].

$$\bar{f} = \frac{1}{b-a}\int_a^b f(x)\,dx$$

$$= \frac{1}{2-1}\int_1^2 x^2\,dx = \frac{x^3}{3}\bigg|_1^2 = \frac{7}{3}.$$

In Example 1 we found that the average value of $y = f(x) = x^2$ over [1, 2] was $\frac{7}{3}$. We can interpret this value geometrically. Since

$$\frac{1}{2-1}\int_1^2 x^2\,dx = \frac{7}{3},$$

then

$$\int_1^2 x^2\,dx = \frac{7}{3}(2-1).$$

However, this integral gives the area of the region bounded by $f(x) = x^2$ and the x-axis from $x = 1$ to $x = 2$. See Fig. 7-1. This area also equals

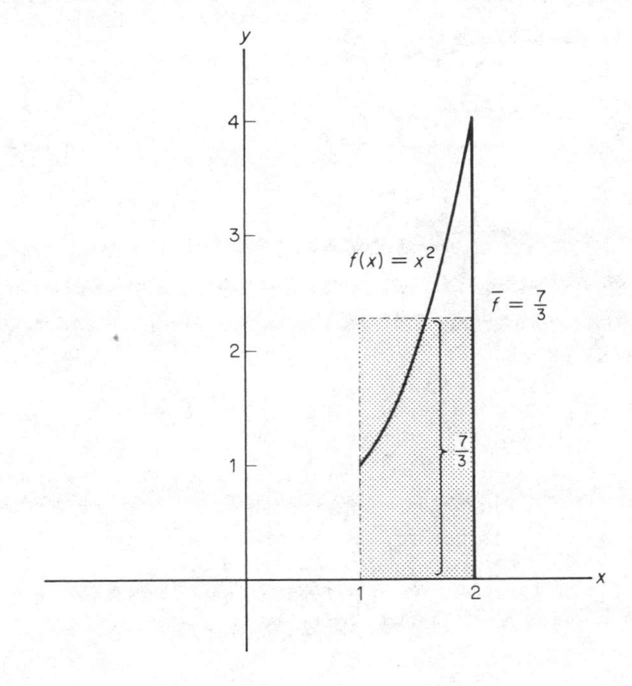

FIG. 7-1

$(\frac{7}{3})(2-1)$, which is the area of a rectangle with height $\bar{f} = \frac{7}{3}$ and width $b - a = 2 - 1 = 1$.

EXAMPLE 2

Suppose the flow of blood at time t in a system is given by

$$F(t) = \frac{F_1}{(1+\alpha t)^2}, \qquad 0 \leq t \leq T,$$

where F_1 and α (a Greek letter read "alpha") are constant **[45]**. *Find the average flow $\bar{F}$ on the interval $[0, T]$.*

$$\bar{F} = \frac{1}{T-0}\int_0^T F(t)\,dt$$

$$= \frac{1}{T}\int_0^T \frac{F_1}{(1+\alpha t)^2}\,dt = \frac{F_1}{\alpha T}\int_0^T (1+\alpha t)^{-2}(\alpha\,dt)$$

$$= \frac{F_1}{\alpha T}\left(-\frac{1}{1+\alpha t}\right)\Bigg|_0^T = \frac{F_1}{\alpha T}\left[-\frac{1}{1+\alpha T}+1\right]$$

$$= \frac{F_1}{\alpha T}\left[\frac{-1+1+\alpha T}{1+\alpha T}\right] = \frac{F_1}{\alpha T}\left[\frac{\alpha T}{1+\alpha T}\right] = \frac{F_1}{1+\alpha T}.$$

Exercise 7-3

In Problems **1–8**, *find the average value of the function over the given interval.*

1. $f(x) = x^2$; $[0, 4]$.

2. $f(x) = 3x - 1$; $[1, 2]$.

3. $f(x) = 2 - 3x^2$; $[-1, 2]$.

4. $f(x) = x^2 + x + 1$; $[1, 3]$.

5. $f(t) = 4t^3$; $[-2, 2]$.

6. $f(t) = t\sqrt{t^2+9}$; $[0, 4]$.

7. $f(x) = \sqrt{x}$; $[1, 9]$.

8. $f(x) = \frac{1}{x}$; $[2, 4]$.

9. The profit P of a business is given by

$$P = P(q) = 396q - 2.1q^2 - 400,$$

where q is the number of units of the product sold. Find the average profit on the interval from $q = 0$ to $q = 100$.

10. Suppose the cost c of producing q units of a product is given by

$$c = 4000 + 10q + .1q^2.$$

Find the average cost on the interval from $q = 100$ to $q = 500$.

11. The value A (in dollars) of an investment of \$3000 at 5 percent compounded continuously for t years is given by $A = 3000e^{.05t}$. Find the average value of a two-year investment.

12. Suppose that colored dye is injected into the bloodstream at a constant rate R (see [45]). Let $C(t)$ be the concentration at time t of dye at a location distant (distal) from the point of injection, where

$$C(t) = \frac{R}{F(t)}$$

and $F(t)$ is given in Example 2. Show that the average concentration $\bar{C}$ on $[0, T]$ is

$$\bar{C} = \frac{R\left(1 + \alpha T + \frac{1}{3}\alpha^2 T^2\right)}{F_1}.$$

7-4 APPROXIMATE INTEGRATION

When using the Fundamental Theorem to evaluate $\int_a^b f(x)\,dx$, it may be extremely difficult, or perhaps impossible, to find an elementary antiderivative of f. Fortunately there are numerical methods, involving arithmetic, which can be used to estimate a definite integral. These methods use values of $f(x)$ at various points and are especially suitable for electronic computers or hand calculators. We shall consider two methods: the *trapezoidal rule* and *Simpson's rule*. In both cases we assume that f is continuous on $[a, b]$.

In developing the trapezoidal rule, for convenience we shall assume $f(x) \geqslant 0$ on $[a, b]$ so that we can think in terms of area. Basically, this rule involves approximating the graph of f by straight line segments.

In Fig. 7-2, the interval $[a, b]$ is divided into n subintervals of equal length by the points $a = x_0, x_1, x_2, \ldots,$ and $x_n = b$. Since the length of $[a, b]$ is $b - a$, then the length of each subinterval is $(b - a)/n$, which we shall call h. Clearly, $x_1 = a + h$, $x_2 = a + 2h, \ldots, x_n = a + nh = b$. With each subinterval we can associate a trapezoid (a four-sided figure with two parallel sides). The area of the region bounded by the curve, the x-axis, and the lines $x = a$ and $x = b$ is approximated by the sum of the areas of the trapezoids determined by the subintervals.

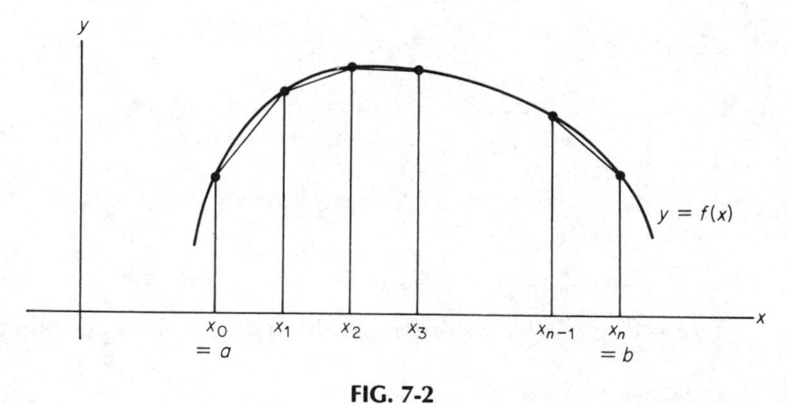

FIG. 7-2

Consider the first trapezoid, which is redrawn in Fig. 7-3. Since the area of a trapezoid is equal to one-half the base times the sum of the

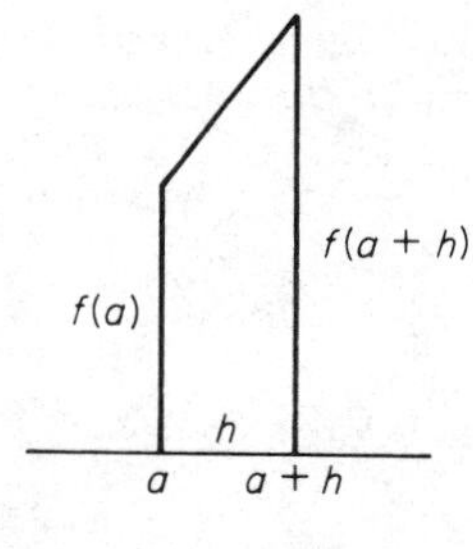

FIG. 7-3

parallel sides, this trapezoid has area

$$\tfrac{1}{2}h[f(a) + f(a + h)].$$

Similarly, the second trapezoid has area

$$\tfrac{1}{2}h[f(a + h) + f(a + 2h)].$$

For n trapezoids the area A under the curve is approximated by

$$A \approx \tfrac{1}{2}h[f(a) + f(a + h)] + \tfrac{1}{2}h[f(a + h) + f(a + 2h)] +$$

$$\tfrac{1}{2}h[f(a + 2h) + f(a + 3h)] + \ldots + \tfrac{1}{2}h[f(a + (n - 1)h) + f(b)].$$

Since $A = \int_a^b f(x)\,dx$, by simplifying the above we have

THE TRAPEZOIDAL RULE

$$\int_a^b f(x)\,dx \approx \frac{h}{2}\{f(a) + 2f(a+h) + 2f(a+2h) + \dots + 2f[a+(n-1)h] + f(b)\}.$$

Usually, the more subintervals the better is the approximation.

EXAMPLE 1

Use the trapezoidal rule to estimate the value of

$$\int_0^1 \frac{1}{1+x^2}\,dx$$

by using $n = 5$. Compute each term to four decimal places and round off the answer to three decimal places.

Here $f(x) = 1/(1+x^2)$, $n = 5$, $a = 0$, and $b = 1$. Thus,

$$h = \frac{b-a}{n} = \frac{1-0}{5} = \frac{1}{5} = .2.$$

The terms to be added are

$$\begin{aligned} f(a) &= f(0) &&= 1.0000 \\ 2f(a+h) &= 2f(.2) &&= 1.9231 \\ 2f(a+2h) &= 2f(.4) &&= 1.7241 \\ 2f(a+3h) &= 2f(.6) &&= 1.4706 \\ 2f(a+4h) &= 2f(.8) &&= 1.2195 \\ f(b) &= f(1) &&= \underline{\ \ .5000} \\ & && \ 7.8373 = \text{Sum}. \end{aligned}$$

Thus our estimate for the integral is

$$\int_0^1 \frac{1}{1+x^2}\,dx \approx \frac{.2}{2}(7.8373) \approx .784.$$

The actual value is approximately .785.

Another method for estimating $\int_a^b f(x)\,dx$ is given by Simpson's rule, which involves approximating the graph of f by parabolic segments. We shall omit the derivation.

SIMPSON'S RULE (n even)

$$\int_a^b f(x)\,dx \approx \frac{h}{3}\{f(a) + 4f(a+h) + 2f(a+2h) + \dots + 4f[a+(n-1)h] + f(b)\}.$$

Note that n must be even and that the pattern of coefficients inside the braces is 1, 4, 2, 4, 2, . . . , 2, 4, 1. Let us use this rule for the integral in Example 1.

EXAMPLE 2

Use Simpson's rule to estimate the value of $\int_0^1 \frac{1}{1+x^2}\,dx$ *by using* $n = 4$. *Compute each term to four decimal places and round off the answer to three decimal places.*

Here $f(x) = 1/(1 + x^2)$, $n = 4$, $a = 0$, and $b = 1$. Thus $h = (b - a)/n = 1/4 = .25$. The terms to be added are

$$\begin{aligned} f(a) &= f(0) = 1.0000 \\ 4f(a+h) &= 4f(.25) = 3.7647 \\ 2f(a+2h) &= 2f(.5) = 1.6000 \\ 4f(a+3h) &= 4f(.75) = 2.5600 \\ f(b) &= f(1) = \underline{.5000} \\ & \quad 9.4247 = \text{Sum}. \end{aligned}$$

Thus, by Simpson's rule,

$$\int_0^1 \frac{1}{1+x^2}\,dx \approx \frac{.25}{3}(9.4247) \approx .785.$$

This is a better approximation than we obtained by using the trapezoidal rule.

EXAMPLE 3

A function often used in demography (the study of births, marriages, mortality, etc. in a population) is the *life table function,* denoted $l = l(x)$. In a population having 100,000 births in any year of time, $l(x)$ represents the number of persons who reach the age of x in any year of time. If $l(20) = 95{,}961$, then the number of persons who attain age 20 in any year of time is 95,961. Suppose that the function l applies to all people born over an extended period of time. It can be shown that at any time, the expected number of persons in the population between the exact ages of x and $x + n$

inclusive is given by

$$\int_x^{x+n} l(t)\, dt.$$

In Table 7-1† are values of $l(x)$ for males and females in a certain population. Approximate the number of women in the 20–35 age group by using the trapezoidal rule with $n = 3$.

TABLE 7-1 Life Table

AGE, x	MALES $l(x)$	FEMALES $l(x)$
0	100,000	100,000
5	97,158	97,791
10	96,921	97,618
15	96,672	97,473
20	95,961	97,188
25	95,000	96,839
30	94,097	96,429
35	93,067	95,844
40	91,628	94,961
45	89,489	93,667
50	86,195	91,726
55	81,154	88,935
60	73,830	84,971
65	64,108	79,455
70	52,007	71,196
75	38,044	59,946
80	24,900	45,662

We want to evaluate

$$\int_{20}^{35} l(t)\, dt.$$

We have $h = \dfrac{b-a}{n} = \dfrac{35-20}{3} = 5$. The terms to be added are

$$\begin{aligned} l(20) &= 97{,}188 \\ 2l(25) = 2(96{,}839) &= 193{,}678 \\ 2l(30) = 2(96{,}429) &= 192{,}858 \\ l(35) &= \underline{\ 95{,}844} \\ & \quad 579{,}568 = \text{Sum}. \end{aligned}$$

†For United States, 1967. Adapted from *Population: Facts and Methods of Demography* by Nathan Keyfitz and William Flieger, W. H. Freeman and Company. Copyright © 1971.

Thus, $$\int_{20}^{35} l(t)\,dt \approx \frac{5}{2}(579{,}568) = 1{,}448{,}920.$$

There are formulas which are used to determine the accuracy of answers obtained by using the trapezoidal or Simpson's rule. They may be found in standard texts on numerical analysis.

Exercise 7-4

In each problem, compute each term to four decimal places and round off the answer to three decimal places.

In Problems **1–6**, *use the trapezoidal rule or Simpson's rule (as indicated) and the given value of n to estimate the integral. In Problems* **1–4**, *also find the answer by antidifferentiation (the Fundamental Theorem of Integral Calculus).*

1. $\int_0^1 x^2\,dx$; trapezoidal rule, $n = 5$.

2. $\int_0^1 x^2\,dx$; Simpson's rule, $n = 4$.

3. $\int_1^4 \frac{dx}{x}$; Simpson's rule, $n = 6$.

4. $\int_1^4 \frac{dx}{x}$; trapezoidal rule, $n = 6$.

5. $\int_0^2 \frac{x\,dx}{x+1}$; trapezoidal rule, $n = 4$.

6. $\int_2^4 \frac{dx}{x+x^2}$; Simpson's rule, $n = 4$.

In Problems **7** *and* **8**, *use the life table (Table* 7-1*) in Example* 3 *to estimate the given integrals by the trapezoidal rule.*

7. $\int_{15}^{40} l(t)\,dt$, males, $n = 5$.

8. $\int_{35}^{55} l(t)\,dt$, females, $n = 4$.

In Problems **9** *and* **10**, *suppose the graph of a continuous function f, where $f(x) \geqslant 0$, contains the given points. Use Simpson's rule and all of the points to approximate the area between the graph and the x-axis on the given interval.*

9. (1, .4), (2, .6), (3, 1.2), (4, .8), (5, .5); [1, 5].

10. (2, 0), (2.5, 3.6), (3, 10), (3.5, 19.9), (4, 34); [2, 4].

Problems **11** *and* **12** *are designed for students with calculators having the square-root function. Estimate the given integrals.*

11. $\int_0^1 \sqrt{1 - x^2}\, dx$; Simpson's rule, $n = 4$.

12. $\int_4^6 \frac{1}{\sqrt{1 + x}}\, dx$; Simpson's rule, $n = 4$. Also find the answer by antidifferentiation.

7-5 IMPROPER INTEGRALS[†]

Any integral of the form

$$\int_a^\infty f(x)\, dx, \tag{1}$$

$$\int_{-\infty}^b f(x)\, dx, \tag{2}$$

or

$$\int_{-\infty}^\infty f(x)\, dx \tag{3}$$

is called an **improper integral**. In each case the interval over which the integral is evaluated has infinite length.

We define (1) as follows:

$$\int_a^\infty f(x)\, dx = \lim_{r \to \infty} \int_a^r f(x)\, dx.$$

When this limit exists, $\int_a^\infty f(x)\, dx$ is said to be *convergent* or to *converge to that limit*. When the limit does not exist, the integral is said to be *divergent*.

We can give a geometric interpretation of this improper integral for the case where f is nonnegative for $a \leqslant x < \infty$. See Fig. 7-4. The integral $\int_a^r f(x)\, dx$ is the area under the curve and above the x-axis from $x = a$ to $x = r$. As $r \to \infty$, we may think of $\int_a^r f(x)\, dx$ as the area of the unbounded region that is shaded in Fig. 7-4. If $\int_a^\infty f(x)\, dx$ converges, then the unbounded region is considered to have a finite area, and this

[†]May be omitted without loss of continuity.

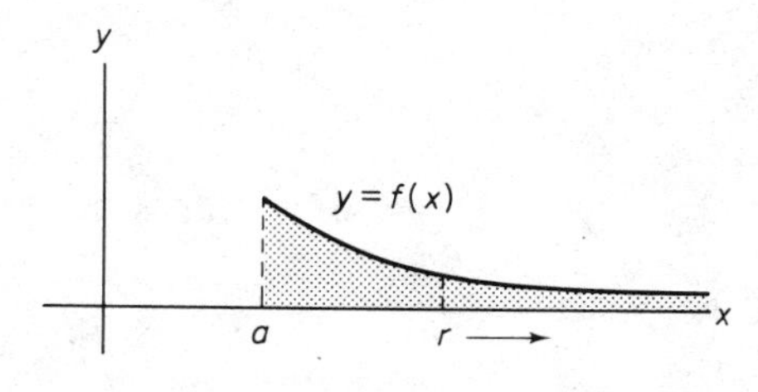

FIG. 7-4

area is represented by $\int_a^{\infty} f(x)\,dx$. If the improper integral is divergent, then the region does not have a finite area.

The improper integral in (2) is defined as

$$\int_{-\infty}^{b} f(x)\,dx = \lim_{r\to-\infty} \int_r^b f(x)\,dx.$$

If this limit exists, $\int_{-\infty}^{b} f(x)\,dx$ is said to be convergent. Otherwise it is divergent.

EXAMPLE 1

Determine whether the following improper integrals are convergent or divergent. If convergent, determine the value of the integral.

a. $\int_1^{\infty} \frac{1}{x^3}\,dx.$

$$\int_1^{\infty} \frac{1}{x^3}\,dx = \lim_{r\to\infty} \int_1^r x^{-3}\,dx = \lim_{r\to\infty} -\frac{x^{-2}}{2}\Bigg|_1^r$$

$$= \lim_{r\to\infty}\left[-\frac{1}{2r^2} + \frac{1}{2}\right] = -0 + \frac{1}{2} = \frac{1}{2}.$$

Therefore, $\int_1^{\infty} \frac{1}{x^3}\,dx$ converges to $\frac{1}{2}$.

b. $\int_{-\infty}^{0} e^x\,dx.$

$$\int_{-\infty}^{0} e^x\,dx = \lim_{r\to-\infty} \int_r^0 e^x\,dx = \lim_{r\to-\infty} e^x\Big|_r^0$$

$$= \lim_{r\to-\infty} (1 - e^r) = 1 - 0 = 1.$$

Therefore, $\int_{-\infty}^{0} e^x\,dx$ converges to 1.

c. $\int_1^\infty \frac{1}{\sqrt{x}}\, dx.$

$$\int_1^\infty \frac{1}{\sqrt{x}}\, dx = \lim_{r\to\infty} \int_1^r x^{-1/2}\, dx = \lim_{r\to\infty} 2x^{1/2}\Big|_1^r$$

$$= \lim_{r\to\infty} 2(\sqrt{r} - 1) = \infty.$$

Therefore, the improper integral diverges.

The improper integral $\int_{-\infty}^{\infty} f(x)\, dx$ is defined in terms of improper integrals of the forms (1) and (2):

$$\int_{-\infty}^{\infty} f(x)\, dx = \int_{-\infty}^{0} f(x)\, dx + \int_0^\infty f(x)\, dx. \tag{4}$$

If *both* integrals on the right side of (4) are convergent, then $\int_{-\infty}^{\infty} f(x)\, dx$ is said to be convergent; otherwise, it is divergent.

EXAMPLE 2

Determine whether $\int_{-\infty}^{\infty} e^x\, dx$ *is convergent or divergent.*

$$\int_{-\infty}^{\infty} e^x\, dx = \int_{-\infty}^{0} e^x\, dx + \int_0^\infty e^x\, dx.$$

By Example 1(b), $\int_{-\infty}^{0} e^x\, dx = 1$. On the other hand,

$$\int_0^\infty e^x\, dx = \lim_{r\to\infty} \int_0^r e^x\, dx = \lim_{r\to\infty} e^x\Big|_0^r$$

$$= \lim_{r\to\infty} (e^r - 1) = \infty.$$

Since $\int_0^\infty e^x\, dx$ is divergent, $\int_{-\infty}^{\infty} e^x\, dx$ is also divergent.

EXAMPLE 3

In statistics, a function f is called a density function if $f(x) \geqslant 0$ *and*

$$\int_{-\infty}^{\infty} f(x)\, dx = 1.$$

Suppose

$$f(x) = \begin{cases} ke^{-x}, & \text{for } x \geq 0, \\ 0, & \text{elsewhere} \end{cases}$$

is a density function. Find k.

We can write $\int_{-\infty}^{\infty} f(x)\,dx = 1$ as

$$\int_{-\infty}^{\infty} f(x)\,dx = \int_{-\infty}^{0} f(x)\,dx + \int_{0}^{\infty} f(x)\,dx = 1.$$

Since $f(x) = 0$ for $x < 0$, $\int_{-\infty}^{0} f(x)\,dx = 0$. Thus,

$$\int_{0}^{\infty} ke^{-x}\,dx = 1,$$

$$\lim_{r\to\infty} \int_{0}^{r} ke^{-x}\,dx = 1,$$

$$\lim_{r\to\infty} -ke^{-x}\Big|_{0}^{r} = 1,$$

$$\lim_{r\to\infty} (-ke^{-r} + k) = 1,$$

$$0 + k = 1,$$

$$k = 1.$$

Exercise 7-5

In Problems **1–12**, *determine the integrals, if they exist. Indicate those which are divergent.*

1. $\int_{3}^{\infty} \frac{1}{x^2}\,dx.$ **2.** $\int_{2}^{\infty} \frac{1}{(2x-1)^3}\,dx.$ **3.** $\int_{1}^{\infty} \frac{1}{x}\,dx.$

4. $\int_{1}^{\infty} \frac{1}{\sqrt[3]{x+1}}\,dx.$ **5.** $\int_{1}^{\infty} e^{-x}\,dx.$ **6.** $\int_{0}^{\infty} (5 + e^{-x})\,dx.$

7. $\int_{1}^{\infty} \frac{1}{\sqrt{x}}\,dx.$ **8.** $\int_{4}^{\infty} \frac{x\,dx}{\sqrt{(x^2+9)^3}}.$ **9.** $\int_{-\infty}^{-2} \frac{1}{(x+1)^3}\,dx.$

10. $\int_{-\infty}^{3} \frac{1}{\sqrt{7-x}}\,dx.$ **11.** $\int_{-\infty}^{\infty} xe^{-x^2}\,dx.$ **12.** $\int_{-\infty}^{\infty} (5 - 3x)\,dx.$

In Problems **13** *and* **14**, *determine the integrals, if they exist. Indicate those which are divergent.*

13. $\int_{1}^{\infty} xe^{x}\,dx.$ **14.** $\int_{1}^{\infty} x \ln x\,dx.$

15. The density function of the life in hours, x, of an electronic component in a calculator is given by

$$f(x) = \begin{cases} \dfrac{k}{x^2}, & \text{for } x \geqslant 800, \\ 0, & \text{for } x < 800. \end{cases}$$

(a) Find k so that $\int_{800}^{\infty} f(x)\,dx = 1$. (b) The probability that the component will last at least 1200 hours is given by $\int_{1200}^{\infty} f(x)\,dx$. Evaluate this integral.

16. Given the density function

$$f(x) = \begin{cases} ke^{-4x}, & \text{for } x \geqslant 0, \\ 0, & \text{elsewhere,} \end{cases}$$

find k.

17. In a psychological model for signal detection [26], the probability α (a Greek letter read "alpha") of reporting a signal when no signal is present is given by

$$\alpha = \int_{x_c}^{\infty} e^{-x}\,dx, \qquad x \geqslant 0.$$

The probability β (a Greek letter read "beta") of detecting a signal when it is present is

$$\beta = \int_{x_c}^{\infty} ke^{-kx}\,dx, \qquad x \geqslant 0.$$

In both integrals x_c is a fixed criterion value. Find α and β if $k = \frac{1}{8}$.

18. For a business the present value of all future profits at an interest rate r compounded continuously is given by

$$\int_0^{\infty} p(t)e^{-rt}\,dt,$$

where $p(t)$ is the profit per year in dollars at time t. If $p(t) = 240{,}000$ and $r = .06$, evaluate the above integral.

19. In discussing entrance of a firm into an industry, Stigler [49] uses the equation

$$V = \pi_0 \int_0^{\infty} e^{\theta t} e^{-\rho t}\,dt,$$

where π_0, θ (a Greek letter read "theta"), and ρ (a Greek letter read "rho") are constants. Show that $V = \pi_0/(\rho - \theta)$ if $\theta < \rho$.

20. Find the area of the region in the first quadrant bounded by the curve $y = e^{-2x}$ and the x-axis.

21. The predicted rate of growth per year of a population of a certain small city is given by $10{,}000/(t + 2)^2$, where t is the number of years from now. In the long run (that is, as $t \to \infty$), what is the expected change in population from today's level?

7-6 DIFFERENTIAL EQUATIONS

Occasionally you may have to solve an equation that involves the derivative of an unknown function. For example,

$$y' = xy^2 \tag{1}$$

is such an equation. It is called a **differential equation**. More precisely, it is a *first order differential equation*, since it involves a derivative of the first order and none of higher order. A solution of Eq. (1) is any function $y = f(x)$ defined on an interval which satisfies Eq. (1) for all x in the interval.

To solve $y' = xy^2$ or, equivalently

$$\frac{dy}{dx} = xy^2, \tag{2}$$

we consider dy/dx as a quotient of differentials and algebraically "separate variables" by rewriting the equation so that each side contains only one variable:

$$\frac{1}{y^2}\,dy = x\,dx.$$

Integrating both sides and combining the constants of integration, we obtain

$$\int \frac{1}{y^2}\,dy = \int x\,dx,$$

$$-\frac{1}{y} = \frac{x^2}{2} + C_1,$$

$$-\frac{1}{y} = \frac{x^2 + 2C_1}{2}. \tag{3}$$

Letting $2C_1 = C$ and solving Eq. (3) for y, we have

$$y = \frac{-2}{x^2 + C}. \tag{4}$$

On any interval on which y is defined, we can verify by substitution that y is a solution to differential equation (2):

$$\frac{dy}{dx} = xy^2 \ ?$$

$$\frac{4x}{(x^2 + C)^2} = x\left[\frac{-2}{x^2 + C}\right]^2 ?$$

$$\frac{4x}{(x^2 + C)^2} = \frac{4x}{(x^2 + C)^2}.$$

Note in Eq. (4) that for *each* value of C, a different solution is obtained. We call Eq. (4) the **general solution** of the differential equation. The method that we used to get it is called **separation of variables.**

In the example above, suppose we are given the condition that $y = -\frac{2}{3}$ when $x = 1$; that is, $y(1) = -\frac{2}{3}$. Then the *particular* function that satisfies Eq. (2) can be found by substituting the values $x = 1$ and $y = -\frac{2}{3}$ into Eq. (4) and solving for C:

$$-\frac{2}{3} = -\frac{2}{1^2 + C},$$

$$C = 2.$$

Therefore, the solution of $dy/dx = xy^2$ such that $y(1) = -\frac{2}{3}$ is

$$y = -\frac{2}{x^2 + 2}. \tag{5}$$

We call Eq. (5) a **particular solution** to the differential equation.

EXAMPLE 1

Solve $y' = -\frac{y}{x}$ *if* $x, y > 0$.

Writing y' as dy/dx, separating variables, and integrating, we have

$$\frac{1}{y}\,dy = -\frac{1}{x}\,dx,$$

$$\int \frac{1}{y}\,dy = -\int \frac{1}{x}\,dx,$$

$$\ln|y| = C_1 - \ln|x|.$$

Since $x, y > 0$,

$$\ln y = C_1 - \ln x. \tag{6}$$

The constant of integration, C_1, can be any real number. Since the range of the logarithmic function is all real numbers, C_1 can be replaced by $\ln C$ where $C > 0$. Therefore, Eq. (6) becomes

$$\ln y = \ln C - \ln x,$$

$$\ln y = \ln \frac{C}{x}.$$

This can only happen if

$$y = \frac{C}{x}, \qquad C, x > 0.$$

In Sec. 6-1 interest compounded continuously was developed. Let us now take a different approach to this topic. Suppose that P dollars are invested at an annual rate r compounded n times a year. Let the function $A = A(t)$ give the total amount A present after t years from the date of the initial investment. Then the initial principal is $A(0) = P$. Furthermore, since there are n interest periods per year, each period has length $1/n$ years, which we shall denote by Δt. At the end of the first period, the accrued interest for that period is added to the principal, and the sum acts as the principal for the second period, etc. Hence, if the beginning of an interest period occurs at time t, then the increase in the amount present (that is, the interest earned) at the end of a period of Δt is $A(t + \Delta t) - A(t) = \Delta A$. Equivalently, the interest earned is principal times rate times time:

$$\Delta A = A \cdot r \cdot \Delta t.$$

Dividing both sides by Δt, we obtain

$$\frac{\Delta A}{\Delta t} = rA. \tag{7}$$

As $\Delta t \to 0$, then $n = \dfrac{1}{\Delta t} \to \infty$ and consequently interest is being *compounded continuously*; that is, the principal is subject to continuous growth at every instant. However as $\Delta t \to 0$, then $\Delta A/\Delta t \to dA/dt$ and Eq. (7) takes the form

$$\frac{dA}{dt} = rA. \tag{8}$$

This differential equation means that *when interest is compounded continuously, the rate of change of the amount of money present at time t is proportional to the amount present at time t.*

To determine the actual function A, we solve differential equation (8) by the method of separation of variables.

$$\frac{dA}{dt} = rA.$$

$$\frac{1}{A}\,dA = r\,dt.$$

$$\int \frac{1}{A}\,dA = \int r\,dt.$$

$$\ln |A| = rt + C_1.$$

Since it can be assumed that $A > 0$, then $\ln |A| = \ln A$.

$$\ln A = rt + C_1.$$

Solving for A, we have

$$A = e^{rt+C_1} = e^{C_1}e^{rt}.$$

Replacing e^{C_1} by C, we obtain

$$A = Ce^{rt}.$$

Since $A(0) = P$,

$$P = Ce^{r(0)} = C(1).$$

Hence $C = P$ and

$$A = Pe^{rt}. \tag{9}$$

Equation (9) gives the total value after t years of an initial investment of P dollars, compounded continuously at an annual rate r.

In our compound interest discussion we saw from Eq. (8) that the rate of change in the amount present was proportional to the amount present. There are many natural quantities, such as population, whose rate of growth or decay at any time is considered proportional to the amount of that quantity present.

If N denotes the amount of such a quantity at time t, then the above rate of growth means that

$$\frac{dN}{dt} = kN.$$

If we separate variables and solve for N as we did for Eq. (8), we get

$$N = N_0 e^{kt}, \tag{10}$$

where N_0 and k are constants. Due to the form of Eq. (10), we say that the quantity follows an **exponential law of growth** if k is positive and **exponential decay** if k is negative.

EXAMPLE 2

In a certain city the rate at which the population grows at any time is proportional to the size of the population. If the population was 125,000 in 1950 and 140,000 in 1970, what is the expected population in 1990?

Let N be the size of the population at time t. Since the exponential law of growth applies,

$$N = N_0 e^{kt}.$$

We must first find the constants N_0 and k. The year 1950 will correspond to $t = 0$. Thus $t = 20$ is 1970 and $t = 40$ is 1990. Now, if $t = 0$, then $N =$ 125,000. Thus,

$$N = N_0 e^{kt},$$
$$125{,}000 = N_0 e^0 = N_0.$$

Hence, $N_0 = 125{,}000$ and

$$N = 125{,}000e^{kt}.$$

But if $t = 20$, then $N = 140{,}000$. This means

$$140{,}000 = 125{,}000e^{20k}.$$

Thus,

$$e^{20k} = \frac{140{,}000}{125{,}000} = 1.12,$$
$$20k = \ln (1.12) \qquad \text{[logarithmic form]},$$
$$k = \tfrac{1}{20}\ln (1.12).$$

Therefore,

$$N = 125{,}000e^{(t/20)\ln 1.12} \tag{11}$$
$$= 125{,}000[e^{\ln 1.12}]^{t/20}.$$
$$N = 125{,}000(1.12)^{t/20}. \tag{12}$$

If $t = 40$,

$$N = 125{,}000(1.12)^2 = 156{,}800.$$

Note that we can write Eq. (11) in a form different from Eq. (12). Since $\ln 1.12 \approx .11333$, then $k \approx .11333/20 \approx .0057$. Thus,

$$N \approx 125{,}000e^{.0057t}.$$

The rate at which a radioactive element decays at any time is found to be proportional to the amount of that element present. If N is the amount of a radioactive substance at time t, we can write this as

$$\frac{dN}{dt} = -\lambda N. \tag{13}$$

The positive constant λ (a Greek letter read "lambda") is called the **decay constant**, and the minus sign indicates that N is decreasing as t increases. Thus we have exponential decay. From Eq. (10), the solution of this equation is

$$N = N_0 e^{-\lambda t}. \tag{14}$$

If $t = 0$, then $N = N_0(1) = N_0$, and so N_0 represents the amount of the radioactive substance present when $t = 0$.

The time for one-half of the substance to decay is called the **half-life** of the substance. It is the value of t when $N = N_0/2$. From Eq. (14),

$$\frac{N_0}{2} = N_0 e^{-\lambda t},$$

$$\frac{1}{2} = e^{-\lambda t}.$$

In logarithmic form we have

$$-\lambda t = \ln \tfrac{1}{2} = \ln 1 - \ln 2 = -\ln 2,$$

$$t = \frac{\ln 2}{\lambda} \approx \frac{.69315}{\lambda}. \tag{15}$$

Note that the half-life depends on λ. Figure 7-5 shows the graph of radioactive decay.

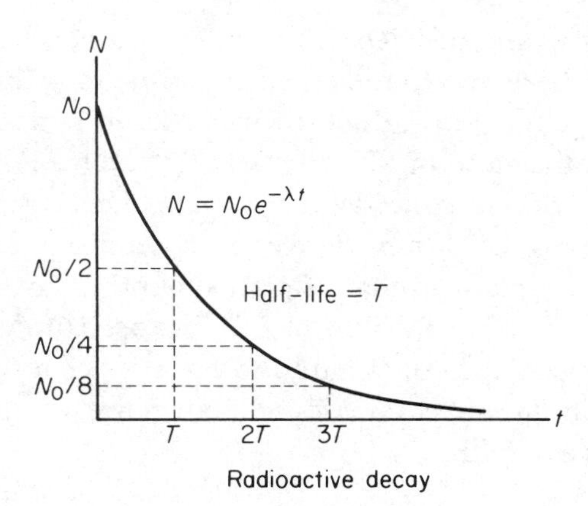

FIG. 7-5

EXAMPLE 3

If 60 *percent of a radioactive substance remains after* 50 *days, find the decay constant and the half-life of the element.*

From Eq. (14),

$$N = N_0e^{-\lambda t},$$

where N_0 is the amount of the element present at $t = 0$. Since $N = .6N_0$ when $t = 50$, we have

$$.6N_0 = N_0e^{-50\lambda},$$

$$.6 = e^{-50\lambda},$$

$$-50\lambda = \ln(.6) \qquad [\text{logarithmic form}^\dagger],$$

$$\lambda = -\tfrac{1}{50}\ln(.6) \approx -\tfrac{1}{50}(-.51083)$$

$$\approx .01022.$$

Thus, $N = N_0e^{-.01022t}$. The half-life, from Eq. (15), is approximately

$$\frac{.69315}{\lambda} \approx \frac{.69315}{.01022} \approx 67.82 \text{ days}.$$

Radioactivity is useful in dating such things as fossil plant remains and archaeological remains made from organic material. Plants and other living organisms contain a small amount of radioactive carbon 14 (C^{14}) in addition to ordinary carbon (C^{12}). The C^{12} atoms are stable, but

$^\dagger\ln(.6) = \ln(6/10) = \ln 6 - \ln 10 \approx 1.79176 - 2.30259 = -.51083.$

the C^{14} atoms are decaying exponentially. However, C^{14} is formed in the atmosphere due to the effect of cosmic rays. Eventually, this C^{14} is taken up by plants during photosynthesis and replaces what has decayed. As a result, the ratio of C^{14} atoms to C^{12} atoms is considered constant over a long period of time. When a plant dies, it stops absorbing C^{14} and the remaining C^{14} atoms decay. By comparing the proportion of C^{14} to C^{12} in a fossil plant to that of plants found today, we can estimate the age of the fossil. The half-life of C^{14} is approximately 5600 years. Thus, for example, if a fossil is found to have a C^{14} to C^{12} ratio which is half that of a similar substance found today, we would estimate the fossil to be 5600 years old.

EXAMPLE 4

A wooden tool found in a Middle East excavation site is found to have a C^{14} to C^{12} ratio which is .6 of the corresponding ratio in a present-day tree. Estimate the age of the tool.

Let N be the amount of C^{14} present in the wood t years after the tool was made. Then $N = N_0 e^{-\lambda t}$ where N_0 is the amount of C^{14} when $t = 0$. Since the C^{14} to C^{12} ratio is .6 of the corresponding ratio in a present-day tree, this means that we want to find the value of t for which $N = .6N_0$.

$$\begin{aligned} .6N_0 &= N_0 e^{-\lambda t}, \\ .6 &= e^{-\lambda t}, \\ -\lambda t &= \ln(.6), \\ t &= -\frac{1}{\lambda}\ln(.6). \end{aligned}$$

From Eq. (15), the half-life is (approximately) $.69315/\lambda$ which equals 5600, and so $\lambda \approx .69315/5600$. Thus,

$$\begin{aligned} t &\approx -\frac{1}{.69315/5600}\ln(.6) \\ &\approx -\frac{5600}{.69315}(-.51083) \\ &\approx 4100 \text{ years.} \end{aligned}$$

Exercise 7-6

In Problems **1–8**, *solve the differential equations.*

1. $y' = 2xy^2$.

2. $y' = x^3y^3$.

3. $\frac{dy}{dx} = y,\ y > 0.$

4. $\frac{dy}{dx} = \frac{x}{y}.$

5. $y' = \frac{y}{x},\ x, y > 0.$

6. $y' = e^x y^2.$

7. $\frac{dy}{dx} - x\sqrt{x^2 + 1} = 0.$

8. $\frac{dy}{dx} + xe^x = 0.$

In Problems **9–14**, *solve each of the differential equations subject to the given conditions.*

9. $y' = \frac{1}{y};\quad y > 0, y(2) = 2.$

10. $y' = e^{x-y};\quad y(0) = 0.$ *Hint:* $e^{x-y} = e^x/e^y.$

11. $e^y y' - x^2 = 0;\quad y = 0$ when $x = 0.$

12. $x^2 y' + \frac{1}{y^2} = 0;\quad y(1) = 2.$

13. $(4x^2 + 3)^2 y' - 4xy^2 = 0;\quad y(0) = \frac{3}{2}.$

14. $y' + x^2 y = 0;\quad y > 0, y = 1$ when $x = 0.$

15. In a certain town the population at any time changes at a rate proportional to the population. If the population in 1970 was 20,000 and in 1980 it was 24,000, find an equation for the population at time t, where t is the number of years past 1970. Write your answer in two forms, one involving e. You may assume $\ln 1.2 = .18$. What is the expected population in 1990?

16. The population of a town increases by natural growth at a rate which is proportional to the number N of persons present. If the population at time $t = 0$ is 10,000, find two expressions for the population N, t years later, if the population doubles in 50 years. Assume $\ln 2 = .69$. Also, find N for $t = 100$.

17. Suppose that the population of the world in 1930 was 2 billion and in 1960 it was 3 billion. If the exponential law of growth is assumed, what is the expected population in 2000? Give your answer in terms of e.

18. If exponential growth is assumed, in approximately how many years will a population triple if it doubles in 50 years? *Hint*: Let the population at $t = 0$ be N_0.

19. If 30 percent of the initial amount of a radioactive sample remains after 100 seconds, find the decay constant and the half-life of the element.

20. If 30 percent of the initial amount of a radioactive sample *has decayed* after 100 seconds, find the decay constant and the half-life of the element.

21. An Egyptian scroll was found to have a C^{14} to C^{12} ratio that is .7 of the

corresponding ratio in similar present-day material. Estimate the age of the scroll to the nearest hundred years.

22. A recently discovered archeological specimen has a C^{14} to C^{12} ratio that is .2 of the corresponding ratio found in present-day organic material. Estimate the age of the specimen to the nearest hundred years.

23. Suppose a population follows exponential growth given by $dN/dt = kN$ for $t \geqslant t_0$, and $N = N_0$ when $t = t_0$. Find N, the population size at time t.

24. Radon has a half-life of 3.82 days. (a) Find the decay constant in terms of ln 2. (b) What fraction of the original amount of it remains after $2(3.82) = 7.64$ days?

25. Radioactive isotopes are used in medical diagnoses as tracers to determine abnormalities that may exist in an organ. For example, if radioactive iodine is swallowed, after some time it is taken up by the thyroid gland. With the use of a detector, the rate at which it is taken up can be measured and a determination can be made as to whether the uptake is normal. Suppose radioactive technetium-99m, which has a half-life of 6 hours, is to be used in a brain scan two hours from now. What should be its activity now if the activity when it is used is to be 10 units? Give your answer to one decimal place. *Hint*: In Eq. (14), let $N =$ activity t hours from now, and $N_0 =$ activity now.

26. A radioactive substance that has a half-life of 8 days is to be temporarily implanted in a hospital patient until there remains three-fifths of the amount originally present. How long should the implant remain in the patient?

27. Suppose q is the amount of penicillin in the body at time t and let q_0 be the amount at $t = 0$. Assume that the rate of change of q with respect to t is proportional to q and that q decreases as t increases. Then we have $dq/dt = -kq$, where $k > 0$. Solve for q and find $\lim_{t \to \infty} q$. What percentage of the original amount present is there when $t = 2/k$?

†28. Suppose $A(t)$ is the amount of a product that is consumed at time t and A follows an exponential law of growth. If $t_1 < t_2$ and at time t_2 the amount consumed, $A(t_2)$, is double the amount consumed at time t_1, $A(t_1)$, then $t_2 - t_1$ is called a doubling period. In a discussion of exponential growth, Shonle **[43]** states that under exponential growth, "... the amount of a product consumed during one doubling period is equal to the total used for all time up to the beginning of the doubling period in question." To

†Refers to Sec. 7-5.

justify this statment, reproduce his argument as follows. The amount of the product used up to time t_1 is given by

$$\int_{-\infty}^{t_1} A_0 e^{kt}\, dt, \qquad k > 0,$$

where A_0 is the amount when $t = 0$. Show that this is equal to $(A_0/k)e^{kt_1}$. Next, the amount used during the time interval from t_1 to t_2 is

$$\int_{t_1}^{t_2} A_0 e^{kt}\, dt.$$

Show that this is equal to

$$\frac{A_0}{k} e^{kt_1}\left[e^{k(t_2 - t_1)} - 1\right]. \tag{16}$$

If the interval $[t_1, t_2]$ is a doubling period, then

$$A_0 e^{kt_2} = 2A_0 e^{kt_1}.$$

Show that this implies $e^{k(t_2 - t_1)} = 2$. Substitute this into (16); your result should be the same as the total used during all time up to t_1, namely, $(A_0/k)e^{kt_1}$.

29. In a forest natural litter occurs, such as fallen leaves and branches, dead animals, etc. **[39]**. Let $A = A(t)$ denote the amount of litter present at time t, where $A(t)$ is expressed in grams per square meter and t is in years. Suppose that there is no litter at $t = 0$. Thus, $A(0) = 0$. Assume that

(1) litter falls to the ground continuously at a constant rate of 200 grams per square meter per year, and
(2) the accumulated litter decomposes continuously at the rate of 50 percent of the amount present per year (which is $.50A$).

The difference of the two rates is the rate of change of the amount of litter present with respect to time:

$$\begin{pmatrix}\text{rate of change}\\ \text{of litter present}\end{pmatrix} = \begin{pmatrix}\text{rate of falling}\\ \text{to ground}\end{pmatrix} - \begin{pmatrix}\text{rate of}\\ \text{decomposition}\end{pmatrix}.$$

Thus,

$$\frac{dA}{dt} = 200 - .50A.$$

(a) Solve for A. (b) To the nearest gram, determine the amount of litter per square meter after one year. (c) As $t \to \infty$, to what value does A tend?

7-7 MORE APPLICATIONS OF DIFFERENTIAL EQUATIONS

Suppose that the number N of individuals in a population at time t follows an exponential law of growth. From the last section, $N = N_0e^{kt}$, where $k > 0$ and N_0 is the population when $t = 0$. This law assumes that at time t the rate of growth, dN/dt, of the population is proportional to the number of individuals in the population. That is, $dN/dt = kN$.

Under exponential growth, a population would get infinitely large as time goes on. In reality, however, when the population gets large enough, there are environmental factors that slow down the rate of growth. Examples are food supply, predators, overcrowding, etc. Since these factors cause dN/dt to eventually decrease, it is reasonable to assume that population size is limited to some maximum number M, where $0 < N < M$, and as $N \to M$, then $dN/dt \to 0$ and the population size tends to be stable.

In summary, we want a population model which has exponential growth initially but which also includes the effects of environmental resistance to large population growth. Such a model is obtained by multiplying the right side of $dN/dt = kN$ by the factor $(M - N)/M$:

$$\frac{dN}{dt} = kN\left(\frac{M - N}{M}\right).$$

Notice that if N is small, then $(M - N)/M$ is close to 1 and we have growth that is approximately exponential. As $N \to M$, then $M - N \to 0$ and $dN/dt \to 0$, as we wanted in our model. Replacing k/M by K, we have

$$\frac{dN}{dt} = KN(M - N). \tag{1}$$

This states that the rate of growth is proportional to the product of the population size and the difference between the maximum size and the population size. We can solve for N in Eq. (1) by the method of separation of variables.

$$\frac{1}{N(M - N)}\,dN = K\,dt.$$

$$\int \frac{1}{N(M - N)}\,dN = \int K\,dt. \tag{2}$$

The integral on the left side can be found by using formula (5) in the table of integrals. Thus Eq. (2) becomes

$$\frac{1}{M}\ln\left|\frac{N}{M-N}\right| = Kt + C,$$

$$\ln\left|\frac{N}{M-N}\right| = MKt + MC.$$

Since $N > 0$ and $M - N > 0$, we can write

$$\ln\frac{N}{M-N} = MKt + MC.$$

In exponential form we have

$$\frac{N}{M-N} = e^{MKt+MC} = e^{MKt}e^{MC}.$$

Replacing the positive constant e^{MC} by A gives

$$\frac{N}{M-N} = Ae^{MKt},$$

$$N = (M-N)Ae^{MKt},$$

$$N = MAe^{MKt} - NAe^{MKt},$$

$$N(Ae^{MKt} + 1) = MAe^{MKt},$$

$$N = \frac{MAe^{MKt}}{Ae^{MKt} + 1}.$$

Dividing numerator and denominator by Ae^{MKt}, we have

$$N = \frac{M}{1 + \dfrac{1}{Ae^{MKt}}} = \frac{M}{1 + \dfrac{1}{A}e^{-MKt}}.$$

Replacing $1/A$ by b and MK by c gives

$$\boxed{N = \frac{M}{1 + be^{-ct}}.} \tag{3}$$

Equation (3) is called the **logistic function** or the **Verhulst-Pearl logistic function**. Its graph, called a *logistic curve*, is S-shaped and appears in Fig. 7-6. Notice in the graph that $N = M$ is a horizontal

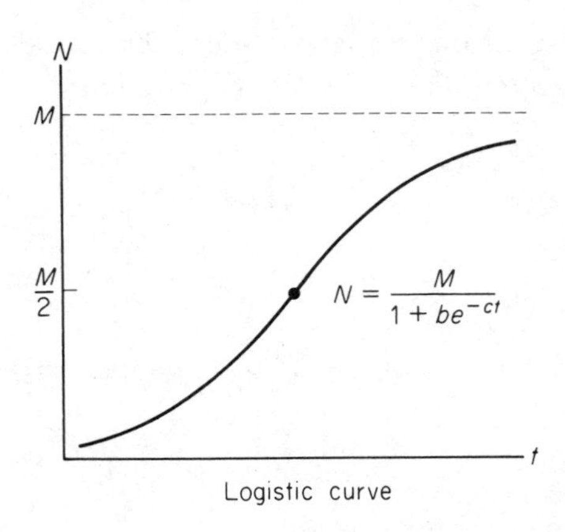

FIG. 7-6

asymptote; that is,

$$\lim_{t\to\infty} \frac{M}{1 + be^{-ct}} = \frac{M}{1 + b(0)} = M.$$

Moreover, from Eq. (1) the rate of growth is

$$KN(M - N).$$

To find when the maximum rate of growth occurs, we solve

$$\begin{aligned}\frac{d}{dN}[KN(M - N)] &= \frac{d}{dN}[K(MN - N^2)] \\ &= K[M - 2N] = 0.\end{aligned}$$

Thus $N = M/2$. The rate of growth increases until the population size is $M/2$ and decreases thereafter. The maximum rate of growth occurs when $N = M/2$ and corresponds to a point of inflection in the graph of N. To find the value of t for which this occurs, we substitute $M/2$ for N in Eq. (3) and solve for t.

$$\begin{aligned}\frac{M}{2} &= \frac{M}{1 + be^{-ct}}, \\ 1 + be^{-ct} &= 2, \\ e^{-ct} &= \frac{1}{b}, \\ e^{ct} &= b, \\ ct &= \ln b, \\ t &= \frac{\ln b}{c}.\end{aligned}$$

Thus the maximum rate of growth occurs at the point $([\ln b]/c, M/2)$. We point out that in Eq. (3), we may replace e^c by C and then the logistic function has the form

$$N = \frac{M}{1 + bC^{-t}}.$$

EXAMPLE 1

Suppose the membership in a new country club is to be a maximum of 800 persons due to limitations of the physical plant. One year ago the initial membership was 50 persons, and now there are 200. Assuming that enrollment follows a logistic function, how many members will there be three years from now?

Let N be the number of members enrolled t years after the formation of the club. Then

$$N = \frac{M}{1 + be^{-ct}}.$$

Here $M = 800$, and when $t = 0$ we have $N = 50$.

$$50 = \frac{800}{1 + b},$$

$$1 + b = \frac{800}{50} = 16,$$

$$b = 15.$$

Thus,

$$N = \frac{800}{1 + 15e^{-ct}}. \qquad (4)$$

When $t = 1$, then $N = 200$.

$$200 = \frac{800}{1 + 15e^{-c}},$$

$$1 + 15e^{-c} = \tfrac{800}{200} = 4,$$

$$e^{-c} = \tfrac{3}{15} = \tfrac{1}{5}.$$

Hence $c = -\ln(1/5) = \ln 5$. However, it is more convenient to substitute the value of e^{-c} in Eq. (4).

$$N = \frac{800}{1 + 15(\frac{1}{5})^t}.$$

Three years from now, $t = 4$. Thus,

$$N = \frac{800}{1 + 15\left(\frac{1}{5}\right)^4} \approx 781.$$

Let us now turn to a different situation, namely, a simplified model of how a rumor spreads in a population of size M. A similar situation would be the spread of an epidemic or new fad.

Let $N = N(t)$ be the number of persons who know the rumor at time t. We shall assume that those who know the rumor spread it randomly in the population and that those who are told the rumor become spreaders of the rumor. Furthermore, we shall assume that each knower tells the rumor to k individuals per unit of time. (Some of these k individuals may already know the rumor.) We want an expression for the rate of increase of the knowers of the rumor. Over a unit of time, each of approximately N persons will tell the rumor to k persons. Thus the total number of persons who are told the rumor over the unit of time is (approximately) Nk. However, we are interested only in *new* knowers. The proportion of the population who do not know the rumor is $(M - N)/M$. Thus, the total number of new knowers of the rumor is

$$Nk\left(\frac{M - N}{M}\right).$$

Therefore

$$\begin{aligned}\frac{dN}{dt} &= \frac{k}{M}N(M - N)\\ &= KN(M - N), \qquad \left(\frac{k}{M} = K\right),\end{aligned}$$

which has the form of Eq. (1). The solution, from Eq. (3), is a logistic function:

$$N = \frac{M}{1 + be^{-ct}}.$$

EXAMPLE 2

In a large university of 45,000 students, a sociology major is researching the spread of a new campus rumor. When he begins his research, he determines that 300 students know the rumor. After one week he finds that 900 know it.

Estimate the number who know it 4 weeks after the research begins by assuming logistic growth. Give the answer to the nearest thousand.

Let N be the number of students who know the rumor after t weeks. Then

$$N = \frac{M}{1 + be^{-ct}},$$

where M, the size of the population, is 45,000. Hence,

$$N = \frac{45{,}000}{1 + be^{-ct}}.$$

When $t = 0$, then $N = 300$. Thus,

$$300 = \frac{45{,}000}{1 + b},$$

$$b = 149,$$

and

$$N = \frac{45{,}000}{1 + 149e^{-ct}}.$$

When $t = 1$, then $N = 900$. Thus,

$$900 = \frac{45{,}000}{1 + 149e^{-c}},$$

$$1 + 149e^{-c} = \frac{45{,}000}{900} = 50.$$

Therefore, $e^{-c} = \frac{49}{149}$ and so

$$N = \frac{45{,}000}{1 + 149\left(\frac{49}{149}\right)^t}.$$

When $t = 4$,

$$N = \frac{45{,}000}{1 + 149\left(\frac{49}{149}\right)^4} \approx 16{,}000.$$

After 4 weeks, approximately 16,000 students know the rumor.

If a homicide is committed, the temperature of the victim's body will gradually decrease from 37°C (normal body temperature) to the temperature of the surroundings (ambient temperature). In general, the temperature of the cooling body changes at a rate proportional to the difference between the temperature of the body and the ambient temperature. This statement is known as *Newton's law of cooling*. Thus, if

$T(t)$ is the temperature of the body at time t and a is the ambient temperature, then

$$\frac{dT}{dt} = k(T - a),$$

where k is the constant of proportionality. Newton's law of cooling can be applied to determine the time at which a homicide was committed, as the next example illustrates.

EXAMPLE 3

A wealthy industrialist was found murdered in his home. Police arrived on the scene at 11:00 P.M. The temperature of the body at that time was 31 °C, and one hour later it was 30 °C. The temperature of the room in which the body was found was 22°C. Determine the time at which the murder occurred.

Let t be the number of hours after the body was discovered, and $T(t)$ be the temperature (in degrees Celsius) of the body at time t. We want to find the value of t for which $T = 37$ (normal body temperature). This value of t will, of course, be negative. By Newton's law of cooling,

$$\frac{dT}{dt} = k(T - a),$$

where k is a constant and a (the ambient temperature) is 22.

Thus,
$$\frac{dT}{dt} = k(T - 22).$$

Separating variables, we have

$$\frac{dT}{T - 22} = k\,dt,$$

$$\int \frac{dT}{T - 22} = \int k\,dt,$$

$$\ln |T - 22| = kt + C.$$

Since $T - 22 > 0$, we can drop the absolute value symbols:

$$\ln (T - 22) = kt + C.$$

When $t = 0$, then $T = 31$. Thus

$$\ln (31 - 22) = k(0) + C,$$

$$C = \ln 9.$$

Hence,

$$\ln (T - 22) = kt + \ln 9,$$
$$\ln (T - 22) - \ln 9 = kt,$$
$$\ln \frac{T - 22}{9} = kt.$$

When $t = 1$, then $T = 30$ and so

$$\ln \frac{30 - 22}{9} = k(1),$$
$$k = \ln \tfrac{8}{9} \approx -.11778.$$

Thus,

$$\ln \frac{T - 22}{9} \approx -.11778t.$$

Now we find t when $T = 37$:

$$\ln \frac{37 - 22}{9} \approx -.11778t,$$
$$t \approx -\frac{\ln (15/9)}{.11778} \approx -\frac{.51083}{.11778},$$
$$t \approx -4.34.$$

Thus the murder occurred about 4.34 hours *before* the time of discovery (11:00 P.M.). Since 4.34 hours is (approximately) 4 hours 20 minutes, the industrialist was murdered about 6:40 P.M.

Exercise 7-7

1. In a country of 3,000,000 people the prime minister suffers a heart attack, which the government does not officially publicize. Initially 50 governmental personnel know of the attack but are spreading this information as a rumor. At the end of one week 5000 people know the rumor. Assuming logistic growth, find how many people know the rumor after two weeks. Give your answer to the nearest thousand.

2. A new fad is sweeping a college campus of 30,000 students. The college newspaper feels that its readers would be interested in a series on the fad. It assigns a reporter when the number of faddists is 400. One week later there are 1200 faddists. Assuming logistic growth, find a formula for the number N of faddists t weeks later.

3. In a city whose population is 100,000 an outbreak of flu occurs. When the city health department begins its record-keeping, there are 500 infected

persons. One week later there are 1000 infected persons. Assuming logistic growth, estimate the number of infected persons after two weeks.

4. The logistic curve for the United States population from 1790 to 1910 is estimated [24] to be

$$N = \frac{197.30}{1 + 35.60e^{-.031186t}},$$

where N is the population in millions and t is in years counted from 1800. If this logistic function were valid for years after 1910, for what year would the point of inflection occur? Give your answer to one decimal place. Assume $\ln 35.6 = 3.5723$.

5. In the study of growth of a colony of unicellular organisms [31], the following equation was obtained:

$$N = \frac{.2524}{e^{-2.128x} + .005125}, \qquad 0 \leqslant x \leqslant 5,$$

where N is the estimated area of the growth in square centimeters and x is the age of the colony in days after being first observed. (a) Put this equation in the form of a logistic function. (b) Find the area when the age of the colony is 0.

6. In an experiment [14], five Paramecia were placed in a test tube containing a nutritive medium. The number N of Paramecia in the tube at the end of t days is approximately given by

$$N = \frac{375}{1 + e^{5.2 - 2.3t}}.$$

(a) Show that this can be written as

$$N = \frac{375}{1 + 181.27e^{-2.3t}},$$

and hence is a logistic function. (b) Find $\lim_{t\to\infty} N$.

7. A waterfront murder was committed, and the victim's body was discovered at 3:15 A.M. by police. At that time the temperature of the body was 32°C. One hour later the body temperature was 30°C. After checking with the weather bureau, it was determined that the temperature at the waterfront was 10°C from 10:00 P.M. to 5:00 A.M. About what time did the murder occur?

8. An enzyme is a protein that acts as a catalyst for increasing the rate of a chemical reaction that occurs in cells. In a certain reaction an enzyme A is converted to another enzyme B. Enzyme B acts as a catalyst for its own formation. Let p be the amount of enzyme B at time t and I be the total amount of both enzymes when $t = 0$. Suppose the rate of formation of B is proportional to $p(I - p)$. Without directly using calculus, find the value of p for which the rate of formation will be a maximum.

9. A small town decides to conduct a fund-raising drive for a new fire engine whose cost is \$70,000. The initial amount in the fund is \$10,000. On the basis of past drives, it is determined that t months after the beginning of the drive, the rate dx/dt at which people contribute to such a fund is proportional to the difference between the desired goal of \$70,000 and the total amount x in the fund at that time. After one month a total of \$40,000 is in the fund. How much will be in the fund after three months?

10. In a discussion of unexpected properties of mathematical models of population, Bailey [2] considers the case in which the birth rate per *individual* is proportional to the population size N at time t. Since the growth rate per individual is $\frac{1}{N}\frac{dN}{dt}$, this means that

$$\frac{1}{N}\frac{dN}{dt} = kN$$

or

$$\frac{dN}{dt} = kN^2, \qquad \text{(subject to } N = N_0 \text{ at } t = 0)$$

where $k > 0$. Show that

$$N = \frac{N_0}{1 - kN_0t}.$$

Use this result to show that

$$\lim N = \infty \quad \text{as} \quad t \to \left(\frac{1}{kN_0}\right)^-.$$

This means that over a finite interval of time there is an infinite amount of growth. Such a model might be useful only for rapid growth over a short interval of time.

11. Suppose that the rate of growth of a population is proportional to the difference between some maximum size M and the number N of individuals in the population at time t. Suppose that when $t = 0$ the population size is N_0. Find a formula for N.

7-8 REVIEW

IMPORTANT TERMS AND SYMBOLS IN CHAPTER 7

integration by parts *(p. 336)*
continuous annuity
  present value *(p. 345)*
  accumulated amount *(p. 345)*
first-order differential equation *(p. 363)*
exponential growth *(p. 367)*
half-life *(p. 368)*
$\int_a^{\infty} f(x)\,dx$, $\int_{-\infty}^{b} f(x)\,dx$,
  $\int_{-\infty}^{\infty} f(x)\,dx$ *(p. 358)*
improper integral *(p. 358)*
average value of function *(p. 349)*
separation of variables *(p. 364)*
exponential decay *(p. 367)*
Simpson's rule *(p. 355)*
logistic function *(p. 375)*
trapezoidal rule *(p. 354)*

REVIEW SECTION

†**1.** $\int u\,dv = uv - \int v\,du$ is called the ________ formula.

Ans. integration by parts.

†**2.** If $\lim_{r\to\infty} \int_a^r f(x)\,dx$ exists, then $\int_a^{\infty} f(x)\,dx$ is said to be (convergent) (divergent).

Ans. convergent.

†**3.** $\int_0^{\infty} e^x\,dx$ is (convergent) (divergent).

Ans. divergent.

4. The equation $x^3y' + y^2 = 0$ is called a ________ equation.

Ans. first-order differential.

5. Exponential growth of a quantity means that the quantity at any time changes at a rate proportional to ________.

Ans. the amount of the quantity.

†Refers to Sec. 7-1 or Sec. 7-5.

6. The average value of $f(x) = x^3$ over the interval $[a, b]$ is given by the integral ________.

Ans. $\frac{1}{b-a}\int_a^b x^3\,dx$.

7. Two methods of estimating definite integrals are by Simpson's rule and the __(a)__ rule. Of these, the rule that requires an even number of subintervals is __(b)__.

Ans. (a) trapezoidal; (b) Simpson's rule.

8. If the rate of growth of a quantity follows the law $N = N_0e^{kt}$, where $k < 0$, we say that the quantity has exponential (growth) (decay).

Ans. decay.

9. The time it takes for 1/2 of a radioactive substance to decay is called the ________ of the substance.

Ans. half-life.

10. The logistic function $N = \frac{M}{1 + be^{-ct}}$ has the horizontal asymptote $N =$ ________.

Ans. M.

REVIEW PROBLEMS

In Problems **1–18**, *determine the integrals.*

1. $\int x \ln x\,dx$.

2. $\int \frac{1}{\sqrt{4x^2 + 1}}\,dx$.

3. $\int_0^2 \sqrt{4x^2 + 9}\,dx$.

4. $\int \frac{2x}{3 - 4x}\,dx$.

5. $\int \frac{x\,dx}{(2 + 3x)(3 + x)}$.

6. $\int_e^{e^2} \frac{1}{x \ln x}\,dx$.

7. $\int \frac{dx}{x(x + 2)^2}$.

8. $\int \frac{dx}{x^2 - 1}$.

9. $\int \frac{dx}{x^2\sqrt{9 - 16x^2}}$.

10. $\int x^2 \ln 4x\,dx$.

11. $\int \frac{9\,dx}{x^2 - 9}$.

12. $\int \frac{3x}{\sqrt{1 + 3x}}\,dx$.

13. $\int xe^{7x}\,dx$.

14. $\int \frac{dx}{2+3e^{4x}}$.

15. $\int \frac{dx}{2x \ln 2x}$.

16. $\int \frac{dx}{x(2+x)}$.

17. $\int \frac{2x}{3+2x}\,dx$.

18. $\int \frac{dx}{\sqrt{4x^2-9}}$.

19. Find the average value of $f(x) = 3x^2 + 2x$ over the interval $[2, 4]$.

20. Find the average value of $f(t) = te^{t^2}$ over the interval $[2, 5]$.

In Problems **21** *and* **22**, *use (a) the trapezoidal rule and (b) Simpson's rule to estimate the integral. Use the given value of n. Give your answer to three decimal places.*

21. $\int_0^3 \frac{1}{x+1}\,dx,\ n = 6$.

22. $\int_0^1 \frac{1}{2-x^2}\,dx,\ n = 4$.

In Problems **23–26**, *determine the improper integrals, if they exist.*† *Indicate those which are divergent.*

23. $\int_3^{\infty} \frac{1}{x^3}\,dx$.

24. $\int_{-\infty}^{0} e^{3x}\,dx$.

25. $\int_1^{\infty} \frac{1}{2x}\,dx$.

26. $\int_{-\infty}^{\infty} xe^{1-x^2}\,dx$.

In Problems **27** *and* **28**, *solve the differential equations.*

27. $y' = 3x^2y + 2xy,\ y > 0$.

28. $y' - 2xe^{x^2-y+3} = 0,\ y(0) = 3$.

29. The population of a city in 1960 was 100,000 and in 1975 it was 120,000. Assuming exponential growth, project the population in 1990.

30. The population of a city doubles every 10 years due to exponential growth. At a certain time the population is 10,000. Find an expression for the number of people N at time t years later. Assume $\ln 2 = .69$.

31. If 95 percent of a radioactive substance remains after 100 years, find the decay constant and, to the nearest percent, give the percentage of the original amount present after 200 years.

32. For a group of hospitalized individuals, suppose

$$\int_0^t f(x)\,dx, \quad \text{where} \quad f(x) = .008e^{-.01x} + .00004e^{-.0002x},$$

†Refers to Sec. 7-5.

gives the proportion that has been discharged at the end of t days. Evaluate $\int_0^\infty f(x)\,dx$.

33. Two organisms are initially placed in a medium and begin to multiply. The number N of organisms that are present after t days is recorded on a graph with the horizontal axis labeled t and the vertical axis labeled N. It is observed that the points lie on a logistic curve. The number of organisms present after 6 days is 300, and beyond 10 days the number approaches a limit of 450. Find the logistic equation.

34. A coroner is called in on a murder case. He arrives at 6:00 P.M. and finds that the victim's temperature is 35°C. One hour later the body temperature is 34°C. The temperature of the room is 25°C. About what time was the murder committed? (Assume that normal body temperature is 37°C.)

35. Find the present value, to the nearest dollar, of a continuous annuity at an annual rate of 5 percent for 10 years if the payment at time t is at the annual rate of $f(t) = 40t$ dollars.

CHAPTER 8

Multivariable Calculus

8-1 FUNCTIONS OF SEVERAL VARIABLES

Suppose a manufacturer produces two products, X and Y. His total cost is dependent on the levels of production of *both* X and Y. Table 8-1 is a schedule which indicates his total cost at various levels of production.

TABLE 8-1

NO. OF UNITS OF X PRODUCED (x)	NO. OF UNITS OF Y PRODUCED (y)	TOTAL COST OF PRODUCTION (c)
5	6	17
5	7	19
6	6	18
6	7	20

For example, when 5 units of X and 6 units of Y are produced, the total cost is 17. Corresponding to this situation, it seems natural to associate the number 17 with the *ordered pair* (5, 6):

$$(5, 6) \to 17.$$

The first element of the ordered pair, 5, represents the number of units of X produced, while the second element, 6, represents the number of units of Y produced. Corresponding to the other production situations, we have

$$(5, 7) \to 19,$$
$$(6, 6) \to 18,$$
$$\text{and} \qquad (6, 7) \to 20.$$

This correspondence can be considered an input-output relation where the inputs are ordered pairs. Note that with each input we associate exactly one output. Thus the correspondence defines a function f such that

the domain of f consists of (5, 6), (5, 7), (6, 6), (6, 7),

and

the range of f consists of 17, 19, 18, 20.

In function notation,

$$f(5, 6) = 17, \qquad f(6, 6) = 18,$$
$$f(5, 7) = 19, \qquad f(6, 7) = 20.$$

We say that the total cost schedule of this manufacturer can be described by $c = f(x, y)$, a function of the two independent variables x and y. The letter c is the dependent variable.

Turning to another function of two variables, we see that the equation

$$z = \frac{2}{x^2 + y^2}$$

defines z as a function of x and y:

$$z = f(x, y) = \frac{2}{x^2 + y^2}.$$

The domain of f is all ordered pairs of real numbers (x, y) for which the equation has meaning when the first and second elements of (x, y) are substituted for x and y, respectively, in the equation. Thus the domain of f is all ordered pairs except (0, 0). To find $f(2, 3)$, for example, we substitute $x = 2$ and $y = 3$ into $2/(x^2 + y^2)$. Hence we have $f(2, 3) = 2/(2^2 + 3^2) = \frac{2}{13}$.

EXAMPLE 1

a. $f(x, y) = \dfrac{x + 3}{y - 2}$ is a function of two variables. Since $(x + 3)/(y - 2)$ is not defined when $y = 2$, the domain of f is all (x, y) such that $y \neq 2$. Some functional values are

$$f(0, 3) = \frac{0 + 3}{3 - 2} = 3,$$

$$f(3, 0) = \frac{3 + 3}{0 - 2} = -3.$$

Note that $f(0, 3) \neq f(3, 0)$.

b. $g(r, s) = 2r - 3s + 5$ is a function of two variables, r and s. The domain of g is all ordered pairs (r, s). Some functional values are

$$g(4, 7) = 2(4) - 3(7) + 5 = -8,$$

$$g(r + h, s) = 2(r + h) - 3s + 5.$$

c. $h(x, y) = 4x$ defines h as a function of x and y. The domain is all ordered pairs of real numbers. Some functional values are

$$h(2, 5) = 4(2) = 8,$$

$$h(2, 6) = 4(2) = 8.$$

d. If $z^2 = x^2 + y^2$ and $x = 3$ and $y = 4$, then $z^2 = 3^2 + 4^2 = 25$. Consequently $z = \pm 5$. Thus, with the ordered pair (3, 4) we *cannot* associate exactly one output number. Hence z is *not* a function of x and y.

EXAMPLE 2

Under certain conditions, if two brown-eyed parents have exactly k children, the probability $P = P(r, k)$ that there will be exactly r blue-eyed children is given by

$$P(r, k) = \frac{k!\left(\frac{1}{4}\right)^r\left(\frac{3}{4}\right)^{k-r}}{r!\,(k - r)!}, \qquad r = 0, 1, 2, \ldots, k.$$

Find the probability that out of a total of four children exactly three will be blue-eyed.

We want to find $P(3, 4)$:

$$P(3,4) = \frac{4!\left(\frac{1}{4}\right)^3\left(\frac{3}{4}\right)^1}{3!\,(1)!} = \frac{3}{64}.$$

If $y = f(x)$ is a function of one variable, the domain of f can be geometrically represented by points on the real number line. The function itself can be represented by its graph in a coordinate plane, sometimes called a two-dimensional coordinate system. However, for a function of two variables, $z = f(x, y)$, its domain (consisting of ordered pairs of real numbers) can be geometrically represented by a *region* in the plane. The function itself can be geometrically represented in a *three*-dimensional coordinate system. Such a system is formed when three mutually perpendicular real number lines in space intersect at the origin of each line (Fig. 8-1). The three number lines are commonly called the x-, y-, and z-axes, and their point of intersection is called the origin of the system.

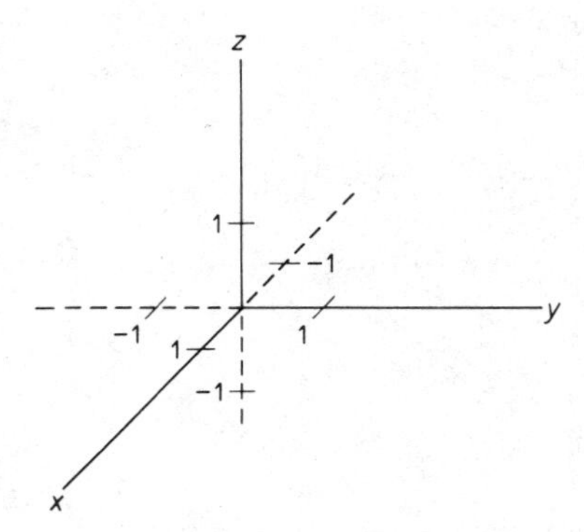

FIG. 8-1

With each point P in space we can associate a unique ordered triple of numbers. To do this [see Fig. 8-2(a)], a perpendicular line is constructed from P to the x, y plane, that is, the plane determined by the x- and y-axes. Letting Q be the point where the line intersects this plane, we construct perpendiculars to the x- and y-axes from Q. These lines intersect the x-axis and the y-axis at points corresponding to the numbers x_0 and y_0, respectively. From P a perpendicular to the z-axis is constructed which intersects the z-axis at a point corresponding to the number z_0. With the point P we associate the ordered triple (x_0, y_0, z_0).

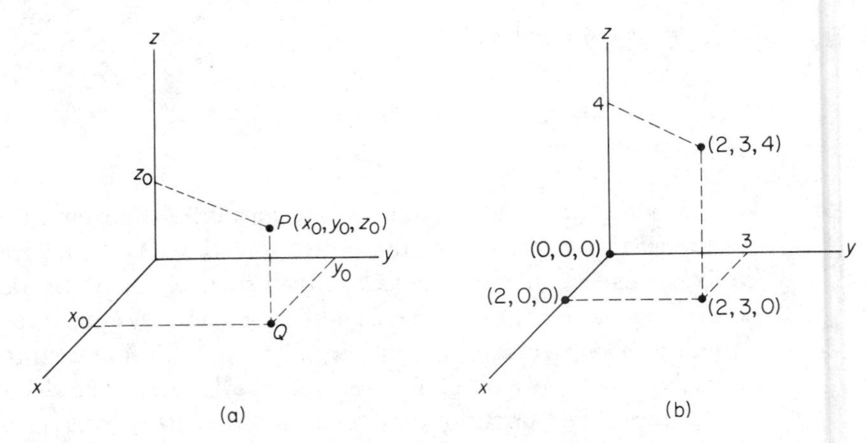

FIG. 8-2

It should also be evident that with each ordered triple of numbers we can associate a unique point in space. Due to this one-to-one correspondence between points in space and ordered triples, an ordered triple may be called a point. In Fig. 8-2(b) the points (2, 0, 0), (2, 3, 0), and (2, 3, 4) are shown. Note that the origin corresponds to (0, 0, 0).

Suppose we wish to represent geometrically a function of two variables, $z = f(x, y)$. Then to each ordered pair (x, y) in the domain of f we assign the point $(x, y, f(x, y))$. The set of all such points is called the *graph* of f. Such a graph appears in Fig. 8-3. You can consider $z = f(x, y)$ as representing a surface in space.†

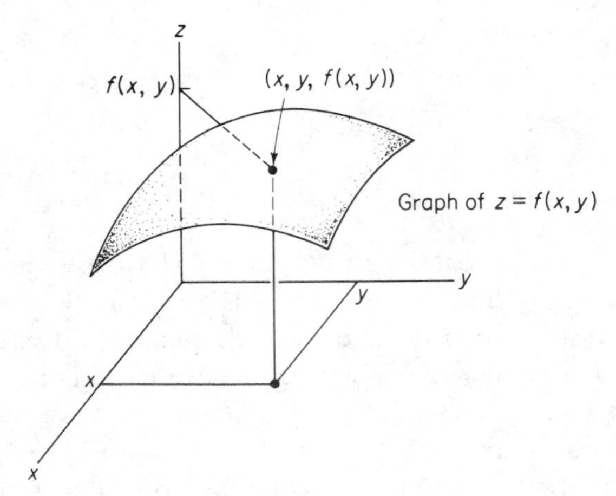

FIG. 8-3

†We shall freely use the term "surface" in the intuitive sense.

In Chapter 2, continuity of a function of one variable was discussed. If $y = f(x)$ is continuous at $x = x_0$, then points near x_0 will have their functional values near $f(x_0)$. Extending this concept to a function of two variables, we say that the function $z = f(x, y)$ is continuous at (x_0, y_0) when points near (x_0, y_0) have their functional values near $f(x_0, y_0)$. Loosely interpreting this and without delving into the concept in great depth, we can say that a function of two variables will be continuous on its domain (that is, continuous at each point in its domain) if its graph is a "connected surface." In the following sections of this chapter we shall see that when a function is continuous, we can make important mathematical generalizations.

Until now, we have considered only functions of either one or two variables. In general, a function of n variables is one whose domain consists of ordered n-tuples $(x_1, x_2, \ldots, x_n)$. For example, $f(x, y, z) = 2x + 3y + 4z$ defines a function of three variables with a domain consisting of all ordered triples. The function $g(x_1, x_2, x_3, x_4) = x_1x_2x_3x_4$ is a function of four variables with a domain consisting of all ordered 4-tuples. Although functions of several variables are extremely important and useful, we cannot geometrically represent functions of more than two variables.

We now give a brief discussion of sketching surfaces in space. We begin with planes that are parallel to a coordinate plane. By a "coordinate plane" we mean a plane containing two coordinate axes. For example, the plane determined by the x- and y-axes is the x, y-plane. Similarly we speak of the x, z-plane and the y, z-plane. The coordinate planes divide space into eight parts, called *octants*. In particular, the part containing all points (x, y, z) where $x, y, z > 0$ is called the **first octant**.

Suppose S is a plane which is parallel to the x, y-plane and which also passes through the point $(0, 0, 5)$ [see Fig. 8-4(a)]. Then the point (x, y, z) will lie on S if and only if $z = 5$; that is, x and y can be any real

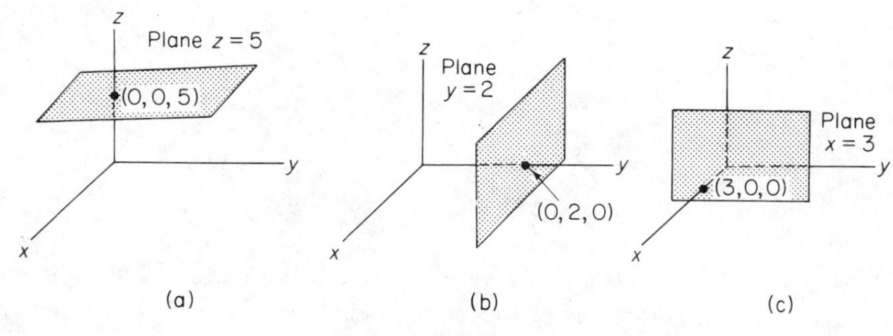

FIG. 8-4

numbers, but z must equal 5. For this reason we say $z = 5$ is an equation of S. Similarly, an equation of the plane parallel to the x, z-plane and passing through the point (0, 2, 0) is $y = 2$ [Fig. 8-4(b)]. The equation $x = 3$ is an equation of the plane passing through (3, 0, 0) and parallel to the y, z-plane [Fig. 8-4(c)]. Now let us look at planes in general.

In space the graph of an equation of the form

$$Ax + By + Cz + D = 0,$$

where A, B, and C are constants, not all zero, is a plane. Since three distinct points determine a plane, a convenient way to sketch a plane is to first determine the points, if any, where the plane intersects the x-, y-, or z-axes. These points are called *intercepts*.

EXAMPLE 3

Sketch the plane $2x + 3y + z = 6$.

The plane intersects the x-axis when $y = 0$ and $z = 0$. Thus $2x = 6$ which gives $x = 3$. Similarly, if $x = z = 0$, then $y = 2$; if $x = y = 0$, then $z = 6$. Thus the intercepts are (3, 0, 0), (0, 2, 0), and (0, 0, 6). The portion of the plane in the first octant is shown in Fig. 8-5(a).

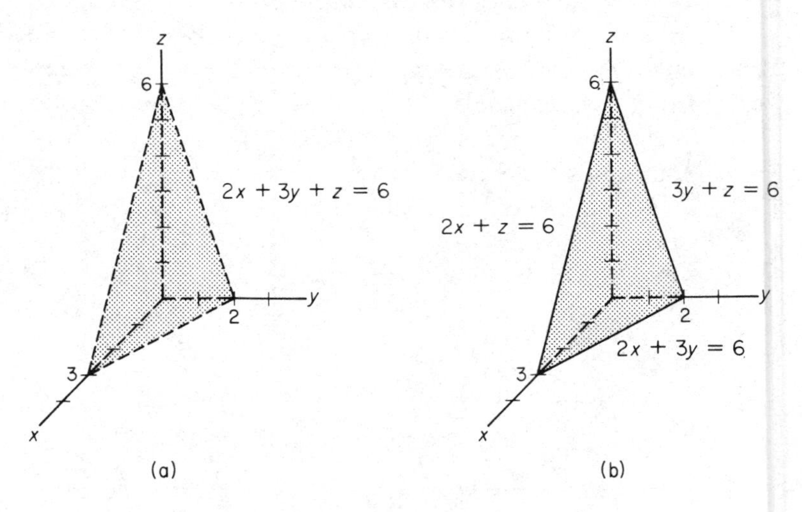

FIG. 8-5

A surface can also be sketched with the aid of its **traces**. These are the intersections of the surface with the coordinate planes. For the plane

$2x + 3y + z = 6$ in Example 3, the trace in the x, y-plane is obtained by setting $z = 0$. This gives the *line* $2x + 3y = 6$. Similarly, setting $x = 0$ gives the trace in the y, z-plane: the line $3y + z = 6$. The x, z-trace is the line $2x + z = 6$. See Fig. 8-5(b).

EXAMPLE 4

Sketch the surface $2x + z = 4$.

This equation has the form of a plane. The x- and z-intercepts are (2, 0, 0) and (0, 0, 4), and there is no y-intercept since x and z cannot both be zero. Setting $y = 0$ gives the x, z-trace $2x + z = 4$, which is a line. In fact, the intersection of the surface with *any* plane $y = k$ is also $2x + z = 4$. Hence the plane appears as in Fig. 8-6.

Our final examples deal with surfaces which are not planes but whose graphs can be easily obtained by previous techniques.

EXAMPLE 5

Sketch the surface $z = x^2$.

The x, z-trace is the parabola $z = x^2$. Moreover, for *any* fixed value of y we get $z = x^2$. Thus the graph appears as in Fig. 8-7.

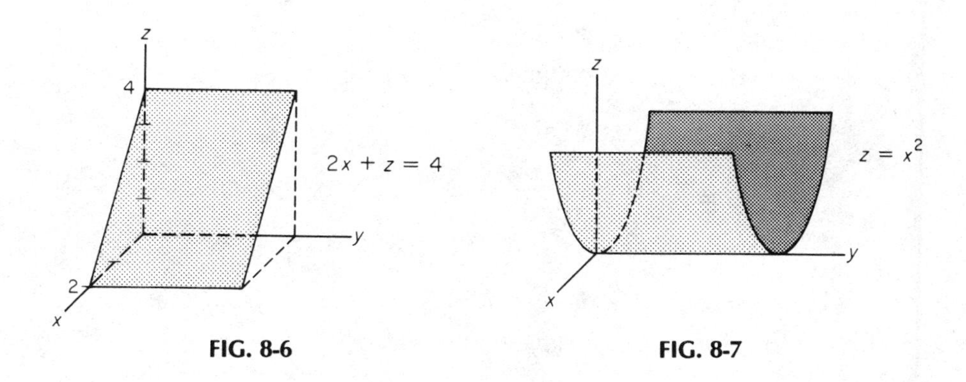

FIG. 8-6 **FIG. 8-7**

EXAMPLE 6

Sketch the surface $x^2 + y^2 + z^2 = 25$.

Setting $z = 0$ gives the x, y-trace $x^2 + y^2 = 25$, which is a circle of radius 5. Similarly, the y, z- and x, z-traces are the circles $y^2 + z^2 = 25$ and $x^2 + z^2 =$

25, respectively. Note also that since $x^2 + y^2 = 25 - z^2$, the intersection of the surface with the plane $z = k$, where $-5 \leqslant k \leqslant 5$, is a circle. For example, if $z = 3$ the intersection is the circle $x^2 + y^2 = 16$. If $z = 4$, the intersection is $x^2 + y^2 = 9$. That is, cross sections of the surface which are parallel to the x, y-plane are circles. A portion of the surface appears in Fig. 8-8. The entire surface is a sphere.

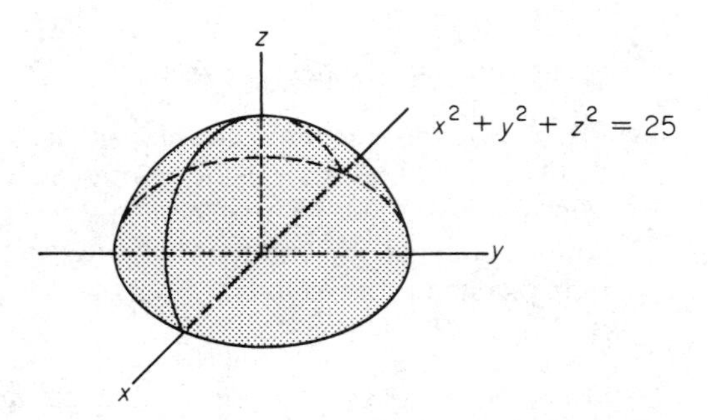

FIG. 8-8

Exercise 8-1

In Problems **1–12**, *determine the given functional values for the indicated functions.*

1. $f(x, y) = 3x + y - 1;\quad f(0, 4)$.

2. $f(x, y) = xy^2 + 2;\quad f(1, -4)$.

3. $g(x, y, z) = ze^{x+y};\quad g(-2, 2, 6)$.

4. $g(x, y, z) = xy + xz + yz;\quad g(1, 2, -3)$.

5. $h(r, s, t, u) = \dfrac{r + s^2}{t - u};\quad h(-3, 3, 5, 4)$.

6. $h(r, s, t, u) = \ln (ru);\quad h(1, 5, 3, 1)$.

7. $g(p_A, p_B) = 2p_A(p_A^2 - 5);\quad g(4, 8)$.

8. $g(p_A, p_B) = p_A\sqrt{p_B} + 10;\quad g(8, 4)$.

9. $F(x, y, z) = 3;\quad F(2, 0, -1)$.

10. $F(x, y, z) = \dfrac{x}{yz};\quad F(0, 0, 3)$.

11. $f(x, y) = 2x - 5y + 4;\quad f(x_0 + h, y_0)$.

12. $f(x, y) = x^2y - 3y^3;\quad f(r + t, r)$.

In Problems **13–16**, *find equations of the planes that satisfy the given conditions.*

13. Parallel to the x, z-plane and passes through the point $(0, -4, 0)$.

14. Parallel to the y, z-plane and passes through the point $(8, 0, 0)$.

15. Parallel to the x, y-plane and passes through the point $(2, 7, 6)$.

16. Parallel to the y, z-plane and passes through the point $(-4, -2, 7)$.

In Problems **17–26**, *sketch the given surfaces.*

17. $x + y + z = 1$.

18. $2x + y + 2z = 6$.

19. $3x + 6y + 2z = 12$.

20. $x + 2y + 3z = 4$.

21. $x + 2y = 2$.

22. $y + z = 1$.

23. $z = 4 - x^2$.

24. $y = x^2$.

25. $x^2 + y^2 + z^2 = 1$.

26. $x^2 + y^2 = 1$.

27. A method of ecological sampling to determine animal populations in a given area involves first marking all the animals obtained in a sample of R animals from the area and then releasing them so that they can mix with unmarked animals. At a later date a second sample is taken of M animals and the number of these which are marked, S, are noted. Based on R, M, and S an estimate of the total population of animals, N, in the sample area is given by

$$N = f(R, M, S) = \frac{RM}{S}.$$

Find $f(400, 400, 80)$. This method is called the *mark and recapture procedure* [35].

28. On hot and humid days, many people tend to feel uncomfortable. The degree of discomfort is numerically given by the temperature-humidity index, THI, which is a function of two variables:

$$\text{THI} = f(t_d, t_w) = 15 + .4(t_d + t_w),$$

where t_d is the dry-bulb temperature (in degrees Fahrenheit) and t_w is the wet-bulb temperature (in degrees Fahrenheit) of the air. When the THI is greater than 75, most people are uncomfortable. In fact, the THI was once called the "discomfort index." Many electric utilities closely follow this index so that they can anticipate the demand of air conditioning on their systems. Evaluate the THI when $t_d = 90$ and $t_w = 80$.

29. In Example 2, find a function which gives the probability that out of a total of five children, exactly r will be blue-eyed.

8-2 PARTIAL DERIVATIVES

Figure 8-9 shows the graph of a function $z = f(x, y)$ and a plane that is parallel to the x, z-plane and that also passes through the point $(x_0, y_0, f(x_0, y_0))$ on the graph. An equation of this plane is $y = y_0$. Hence any point on the curve cut from the surface by the plane must have the form $(x, y_0, f(x, y_0))$. Thus the curve can be described by

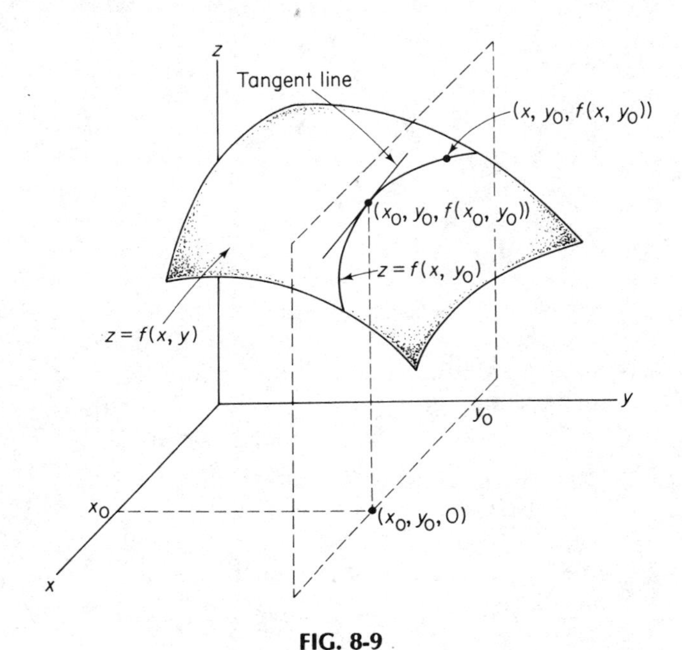

FIG. 8-9

$z = f(x, y_0)$. Since y_0 is constant, $z = f(x, y_0)$ can be considered a function of one variable, x. When the derivative of this function of x is evaluated at x_0, it gives the slope of the tangent line to this curve at $(x_0, y_0, f(x_0, y_0))$. See Fig. 8-9. This slope is called the *partial derivative of f with respect to x at (x_0, y_0)* and is denoted $f_x(x_0, y_0)$. In terms of limits,

$$f_x(x_0, y_0) = \lim_{h \to 0} \frac{f(x_0 + h, y_0) - f(x_0, y_0)}{h}. \qquad (1)$$

On the other hand, in Fig. 8-10 the plane $x = x_0$ is parallel to the y, z-plane and cuts the surface $z = f(x, y)$ in a curve that is given by $z = f(x_0, y)$, a function of y. When the derivative of this function of y is evaluated at y_0, it gives the slope of the tangent line to this curve at the

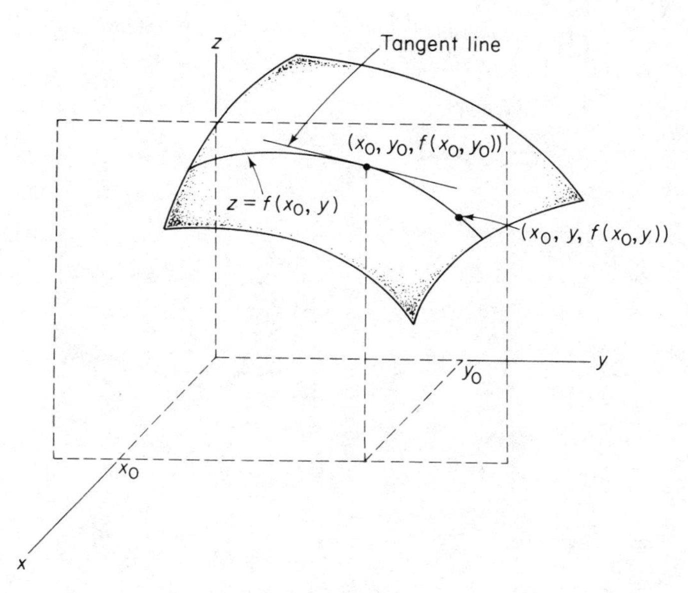

FIG. 8-10

point $(x_0, y_0, f(x_0, y_0))$. This slope is called the *partial derivative of f with respect to y at* (x_0, y_0) and is denoted $f_y(x_0, y_0)$. In terms of limits,

$$f_y(x_0, y_0) = \lim_{h \to 0} \frac{f(x_0, y_0 + h) - f(x_0, y_0)}{h}. \tag{2}$$

Sometimes $f_x(x_0, y_0)$ is said to be the slope at $(x_0, y_0, f(x_0, y_0))$ of the tangent line to the graph of f *in the x-direction*; similarly, $f_y(x_0, y_0)$ is the slope of the tangent line *in the y-direction*.

For generality, by replacing x_0 and y_0 in Eqs. (1) and (2) by x and y, respectively, we get the following definition.

DEFINITION. *If* $z = f(x, y)$, *the* ***partial derivative of f with respect to x****, denoted* f_x, *is*

$$f_x(x, y) = \lim_{h \to 0} \frac{f(x + h, y) - f(x, y)}{h},$$

provided this limit exists.

The ***partial derivative of f with respect to y****, denoted* f_y, *is*

$$f_y(x, y) = \lim_{h \to 0} \frac{f(x, y + h) - f(x, y)}{h},$$

provided this limit exists.

EXAMPLE 1

If $f(x, y) = xy^2 + x^2y$, find $f_x(x, y)$ and $f_y(x, y)$. Also find $f_x(3, 4)$ and $f_y(3, 4)$.

By the definition of $f_x(x, y)$,

$$\begin{aligned} f_x(x, y) &= \lim_{h\to 0} \frac{f(x+h, y) - f(x, y)}{h} \\ &= \lim_{h\to 0} \frac{[(x+h)y^2 + (x+h)^2y] - [xy^2 + x^2y]}{h} \\ &= \lim_{h\to 0} \frac{xy^2 + hy^2 + x^2y + 2xhy + h^2y - xy^2 - x^2y}{h} \\ &= \lim_{h\to 0} (y^2 + 2xy + hy) = y^2 + 2xy. \end{aligned}$$

Therefore,

$$f_x(x, y) = y^2 + 2xy.$$

To find $f_x(3, 4)$ we evaluate $f_x(x, y)$ when $x = 3$ and $y = 4$.

$$f_x(3, 4) = 4^2 + 2(3)(4) = 40.$$

By the definition of $f_y(x, y)$,

$$\begin{aligned} f_y(x, y) &= \lim_{h\to 0} \frac{f(x, y+h) - f(x, y)}{h} \\ &= \lim_{h\to 0} \frac{[x(y+h)^2 + x^2(y+h)] - [xy^2 + x^2y]}{h} \\ &= \lim_{h\to 0} \frac{xy^2 + 2xyh + xh^2 + x^2y + x^2h - xy^2 - x^2y}{h} \\ &= \lim_{h\to 0} (2xy + xh + x^2) = 2xy + x^2. \end{aligned}$$

Therefore,

$$f_y(x, y) = 2xy + x^2.$$

Evaluating when $x = 3$ and $y = 4$, we have

$$f_y(3, 4) = 2(3)(4) + 3^2 = 33.$$

Note that $f_x(x, y)$ and $f_y(x, y)$ are each functions of the two variables x and y.

From its definition we see that to find $f_x(x, y)$ we treat y as a constant and differentiate f with respect to x in the usual way. For example, if $f(x, y) = xy^2 + x^2y$, then by treating y as a constant and differentiating with respect to x we have

$$f_x(x, y) = (1)y^2 + (2x)y = y^2 + 2xy,$$

as was shown in Example 1.

Similarly, to find $f_y(x, y)$ we treat x as a constant and differentiate f with respect to y in the usual way. Thus for $f(x, y) = xy^2 + x^2y$,

$$f_y(x, y) = x(2y) + x^2(1) = 2xy + x^2,$$

as was shown in Example 1.

Notations for partial derivatives of $z = f(x, y)$ are in Table 8-2.

TABLE 8-2

PARTIAL DERIVATIVE OF f (OR z) WITH RESPECT TO x	PARTIAL DERIVATIVE OF f (OR z) WITH RESPECT TO y
$f_x(x, y)$	$f_y(x, y)$
$\frac{\partial}{\partial x}[f(x, y)]$	$\frac{\partial}{\partial y}[f(x, y)]$
$\frac{\partial z}{\partial x}$	$\frac{\partial z}{\partial y}$

Table 8-3 gives notations for partial derivatives evaluated at (x_0, y_0).

TABLE 8-3

PARTIAL DERIVATIVE OF f (OR z) WITH RESPECT TO x EVALUATED AT (x_0, y_0)	PARTIAL DERIVATIVE OF f (OR z) WITH RESPECT TO y EVALUATED AT (x_0, y_0)
$f_x(x_0, y_0)$	$f_y(x_0, y_0)$
$\left.\frac{\partial z}{\partial x}\right\vert_{(x_0, y_0)}$	$\left.\frac{\partial z}{\partial y}\right\vert_{(x_0, y_0)}$

EXAMPLE 2

a. *If* $z = 3x^3y^3 - 9x^2y + xy^2 + 4y$, *find* $\frac{\partial z}{\partial x}, \frac{\partial z}{\partial y}, \left.\frac{\partial z}{\partial x}\right|_{(1,0)}$ *and* $\left.\frac{\partial z}{\partial y}\right|_{(1,0)}$.

To find $\partial z/\partial x$ we differentiate z with respect to x by treating y as a constant:

$$\frac{\partial z}{\partial x} = 3(3x^2)y^3 - 9(2x)y + (1)y^2 + 0$$
$$= 9x^2y^3 - 18xy + y^2.$$

Evaluating at (1, 0), we obtain

$$\left.\frac{\partial z}{\partial x}\right|_{(1,0)} = 9(1)^2(0)^3 - 18(1)(0) + 0^2 = 0.$$

To find $\partial z/\partial y$ we differentiate z with respect to y by treating x as a constant.

$$\frac{\partial z}{\partial y} = 3x^3(3y^2) - 9x^2(1) + x(2y) + 4(1)$$
$$= 9x^3y^2 - 9x^2 + 2xy + 4.$$

Thus,

$$\left.\frac{\partial z}{\partial y}\right|_{(1,0)} = 9(1)^3(0)^2 - 9(1)^2 + 2(1)(0) + 4 = -5.$$

b. *If $g = x^2e^{2x+3y}$, find $\partial g/\partial x$ and $\partial g/\partial y$.*

To find $\partial g/\partial x$, we treat y as a constant and differentiate with respect to x. Since x^2e^{2x+3y} is a product of two functions, each involving x, we use the product rule.

$$\frac{\partial g}{\partial x} = x^2\frac{\partial}{\partial x}(e^{2x+3y}) + e^{2x+3y}\frac{\partial}{\partial x}(x^2)$$
$$= x^2(2e^{2x+3y}) + e^{2x+3y}(2x)$$
$$= 2x(x+1)e^{2x+3y}.$$

To find $\partial g/\partial y$, we treat x as a constant and differentiate with respect to y.

$$\frac{\partial g}{\partial y} = x^2\frac{\partial}{\partial y}(e^{2x+3y}) = 3x^2e^{2x+3y}.$$

We have seen that for a function of two variables, two partial derivatives can be considered. Actually the concept of partial derivatives can be extended to functions of more than two variables. For example, with $w = f(x, y, z)$ we have three partial derivatives:

the partial with respect to x, denoted $f_x(x, y, z)$, $\partial w/\partial x$, etc.;

the partial with respect to y, denoted $f_y(x, y, z)$, $\partial w/\partial y$, etc.;

and the partial with respect to z, denoted $f_z(x, y, z)$, $\partial w/\partial z$, etc.

To determine $\partial w/\partial x$, treat y and z as constants and differentiate with respect to x. For $\partial w/\partial y$, treat x and z as constants and differentiate with respect to y. For $\partial w/\partial z$, treat x and y as constants and differentiate with respect to z. With a function of n variables we have n partial derivatives which are determined in the obvious way.

EXAMPLE 3

a. *If $f(x, y, z) = x^2 + y^2z + z^3$, find $f_x(x, y, z)$, $f_y(x, y, z)$, and $f_z(x, y, z)$.*

Treating y and z as constants and differentiating with respect to x, we have

$$f_x(x, y, z) = 2x.$$

Treating x and z as constants and differentiating with respect to y, we have

$$f_y(x, y, z) = 2yz.$$

Treating x and y as constants and differentiating with respect to z, we have

$$f_z(x, y, z) = y^2 + 3z^2.$$

b. *If $p = g(r, s, t, u) = \dfrac{rsu}{rt^2 + s^2t}$, determine $\dfrac{\partial p}{\partial s}$, $\dfrac{\partial p}{\partial t}$, and $\left.\dfrac{\partial p}{\partial t}\right|_{(0,1,1,1)}$.*

To find $\partial p/\partial s$, first note that p is a quotient of two functions, each involving the variable s. Thus we use the quotient rule and treat r, t, and u as constants.

$$\begin{aligned}\frac{\partial p}{\partial s} &= \frac{(rt^2 + s^2t)\frac{\partial}{\partial s}(rsu) - rsu\frac{\partial}{\partial s}(rt^2 + s^2t)}{(rt^2 + s^2t)^2} \\ &= \frac{(rt^2 + s^2t)(ru) - (rsu)(2st)}{(rt^2 + s^2t)^2}.\end{aligned}$$

Simplifying gives

$$\frac{\partial p}{\partial s} = \frac{ru(rt - s^2)}{t(rt + s^2)^2}.$$

To find $\partial p/\partial t$ we can first write g as

$$g(r, s, t, u) = rsu(rt^2 + s^2t)^{-1}.$$

Next we use the power rule and treat r, s, and u as constants.

$$\frac{\partial p}{\partial t} = rsu(-1)(rt^2 + s^2t)^{-2}\frac{\partial}{\partial t}(rt^2 + s^2t)$$

$$= -rsu(rt^2 + s^2t)^{-2}(2rt + s^2).$$

$$\frac{\partial p}{\partial t} = -\frac{rsu(2rt + s^2)}{(rt^2 + s^2t)^2}.$$

Letting $r = 0$, $s = 1$, $t = 1$ and $u = 1$ gives

$$\left.\frac{\partial p}{\partial t}\right|_{(0,1,1,1)} = -\frac{0(1)(1)[2(0)(1) + (1)^2]}{[0(1)^2 + (1)^2(1)]^2} = 0.$$

Exercise 8-2

In each of Problems **1–26**, *find all partial derivatives.*

1. $f(x, y) = x - 5y + 3.$

2. $f(x, y) = 4 - 5x^2 + 6y^3.$

3. $f(x, y) = 3x - 4.$

4. $f(x, y) = \sqrt{7}.$

5. $g(x, y) = x^5y^4 - 3x^4y^3 + 7x^3 + 2y^2 - 3xy + 4.$

6. $g(x, y) = x^8 - 2x^6y^5 + 3x^5y^3 + x^3y^3 + 3x - 4.$

7. $g(p, q) = \sqrt{pq}.$

8. $g(w, z) = \sqrt[3]{w^2 + z^2}.$

9. $h(s, t) = \dfrac{s^2 + 4}{t - 3}.$

10. $h(u, v) = \dfrac{4uv^2}{u^2 + v^2}.$

11. $u(q_1, q_2) = \frac{3}{4}\ln q_1 + \frac{1}{4}\ln q_2.$

12. $Q(l, k) = 3l^{.41}k^{.59}.$

13. $h(x, y) = \dfrac{x^2 + 3xy + y^2}{\sqrt{x^2 + y^2}}.$

14. $h(x, y) = \dfrac{\sqrt{x + 4}}{x^2y + y^2x}.$

15. $z = e^{5xy}.$

16. $z = (x^2 + y)e^{3x+4y}.$

17. $z = 5x \ln (x^2 + y).$

18. $z = \ln (3x^2 + 4y^4).$

19. $f(r, s) = \sqrt{r + 2s}\,(r^3 - 2rs + s^2).$

20. $f(r, s) = \sqrt{rs}\, e^{2+r}.$

21. $f(r, s) = e^{3-r} \ln (7 - s).$

22. $f(r, s) = (5r^2 + 3s^3)(2r - 5s).$

23. $g(x, y, z) = 3x^2y + 2xy^2z + 3z^3.$

24. $g(x, y, z) = x^2y^3z^5 - 3x^2y^4z^3 + 5xz.$

25. $g(r, s, t) = e^{s+t}(r^2 + 7s^3).$

26. $g(r, s, t, u) = rs \ln (2t + 5u).$

In Problems **27–32**, *evaluate the given partial derivatives.*

27. $f(x, y) = x^3y + 7x^2y^2$; $f_x(1, -2)$.

28. $z = \sqrt{5x^2 + 3xy + 2y}$; $\left.\dfrac{\partial z}{\partial x}\right|_{\substack{x=0 \\ y=2}}$.

29. $g(x, y, z) = e^x\sqrt{y + 2z}$; $g_z(0, 1, 4)$.

30. $g(x, y, z) = \dfrac{3x^2 + 2y}{xy + xz}$; $g_y(1, 1, 1)$.

31. $h(r, s, t, u) = (s^2 + tu) \ln (2r + 7st)$; $h_s(1, 0, 0, 1)$.

32. $h(r, s, t, u) = \dfrac{7r + 3s^2u^2}{s}$; $h_t(4, 3, 2, 1)$.

8-3 APPLICATIONS OF PARTIAL DERIVATIVES

Suppose a manufacturer produces x units of product X and y units of product Y. Then the total cost c of these units is a function of x and y and is called a *joint-cost function*. If such a function is $c = f(x, y)$, then $\partial c/\partial x$ is called the *(partial) marginal cost with respect to* x. It is the rate of change of c with respect to x when y is held fixed. On the other hand, $\partial c/\partial y$ is the *(partial) marginal cost with respect to* y. It is the rate of change of c with respect to y when x is held fixed.

For example, if c is expressed in dollars and $\partial c/\partial y = 2$, then the cost of producing an extra unit of Y when the level of production of X is fixed is approximately two dollars.

If a manufacturer produces n products, his joint-cost function is a function of n variables and there are n (partial) marginal cost functions.

EXAMPLE 1

A company manufactures two types of sleds, the Lightning and the Alaskan models. Suppose the joint-cost function for producing x sleds of the Lightning model and y sleds of the Alaskan model is $c = f(x, y) = .06x^2 + 7x + 15y + 1000$, *where c is expressed in dollars. Determine the marginal costs* $\partial c/\partial x$ *and* $\partial c/\partial y$ *when* $x = 100$ *and* $y = 50$ *and interpret the results.*

Treating y as a constant and differentiating c with respect to x, we obtain

$$\frac{\partial c}{\partial x} = .12x + 7.$$

Treating x as a constant and differentiating c with respect to y gives

$$\frac{\partial c}{\partial y} = 15.$$

Thus,

$$\left.\frac{\partial c}{\partial x}\right|_{(100,\,50)} = .12(100) + 7 = 19,$$

and

$$\left.\frac{\partial c}{\partial y}\right|_{(100,\,50)} = 15.$$

This means that increasing the output of the Lightning model from 100 to 101, while maintaining production of the Alaskan model at 50, increases costs by approximately \$19. On the other hand, increasing output of the Alaskan model from 50 to 51 and holding production of the Lightning model at 100 will increase costs by approximately \$15. In fact, since $\partial c/\partial y$ is a constant function, the marginal cost with respect to y is \$15 at all levels of production.

EXAMPLE 2

On a cold day a person may feel colder when the wind is blowing than when the wind is calm because the rate of heat loss is a function of both temperature and wind speed. The equation

$$H = (10.45 + \sqrt{100w} - w)(33 - t)$$

indicates the rate of heat loss H (in kilocalories per square meter per hour) when the air temperature is t (in degrees Celsius) and the wind speed is w (in meters per second). For $H = 2000$, exposed flesh will freeze in one minute **[13]**. *(a) Evaluate H when $t = 0$ and $w = 4$. (b) Evaluate $\frac{\partial H}{\partial w}$ and $\frac{\partial H}{\partial t}$ when $t = 0$ and $w = 4$, and interpret the results. (c) When $t = 0$ and $w = 4$, which has a greater effect on H: a change in wind speed of 1 m/sec or a change in temperature of 1 °C?*

a. Note that $\sqrt{100w} = 10\sqrt{w}$ and thus H can be written

$$H = (10.45 + 10\sqrt{w} - w)(33 - t).$$

When $t = 0$ and $w = 4$, then

$$H = (10.45 + 10\sqrt{4} - 4)(33 - 0) = 872.85.$$

b. $$\frac{\partial H}{\partial w} = \left(\frac{5}{\sqrt{w}} - 1\right)(33 - t), \qquad \left.\frac{\partial H}{\partial w}\right|_{\substack{t=0\\w=4}} = 49.5;$$

$$\frac{\partial H}{\partial t} = (10.45 + 10\sqrt{w} - w)(-1), \qquad \left.\frac{\partial H}{\partial t}\right|_{\substack{t=0\\w=4}} = -26.45.$$

This means that when $t = 0$ and $w = 4$, then increasing w by a small amount while keeping t fixed will make H increase approximately 49.5 times as much as w increases. Increasing t by a small amount while keeping w fixed will make H *decrease* approximately 26.45 times as much as t increases.

c. Since the partial derivative of H with respect to w is greater in magnitude than the partial with respect to t when $t = 0$ and $w = 4$, a change in wind speed of 1 m/sec has a greater effect on H.

A producer's output of a product depends on many factors of production. Among these may be labor, capital, land, machinery, etc. If the function $P = f(l, k)$ gives the output P when a producer uses l units of labor and k units of capital, then this function is called a *production function*. We define the *marginal productivity with respect to* l to be $\partial P/\partial l$. This is the rate of change of P with respect to l when k is held fixed. Likewise, the *marginal productivity with respect to* k is $\partial P/\partial k$. It is the rate of change of P with respect to k when l is held fixed.

EXAMPLE 3

A manufacturer of a popular toy has determined that its production function is $P = \sqrt{lk}$, where l is the number of man-hours per week and k is the capital (expressed in hundreds of dollars per week) required for a weekly production of P gross of the toy. Determine the marginal productivity functions and evaluate them when $l = 400$ and $k = 16$. Interpret the results.

Since $P = (lk)^{1/2}$,

$$\frac{\partial P}{\partial l} = \frac{1}{2}(lk)^{-1/2}k = \frac{k}{2\sqrt{lk}},$$

and

$$\frac{\partial P}{\partial k} = \frac{1}{2}(lk)^{-1/2}l = \frac{l}{2\sqrt{lk}}.$$

Evaluating when $l = 400$ and $k = 16$, we obtain

$$\left.\frac{\partial P}{\partial l}\right|_{\substack{l=400\\k=16}} = \frac{16}{2\sqrt{400(16)}} = \frac{1}{10},$$

$$\left.\frac{\partial P}{\partial k}\right|_{\substack{l=400\\k=16}} = \frac{400}{2\sqrt{400(16)}} = \frac{5}{2}.$$

Thus if $l = 400$ and $k = 16$, increasing l to 401 and holding k at 16 will increase output by approximately $\frac{1}{10}$ gross. But if k is increased to 17 while l is held at 400, the output increases by approximately $\frac{5}{2}$ gross.

Sometimes two products may be related so that changes in the price of one of them can affect the demand for the other. A typical example is that of butter and margarine. If such a relationship exists between products A and B, then the demand for each product is dependent on the prices of both. Suppose q_A and q_B are the quantities demanded for A and B, respectively, and p_A and p_B are their respective prices. Then both q_A and q_B are functions of p_A and p_B:

$$q_A = f(p_A, p_B), \text{ demand function for } A;$$

$$q_B = g(p_A, p_B), \text{ demand function for } B.$$

We can find four partial derivatives:

$\frac{\partial q_A}{\partial p_A}$, *the marginal demand for A with respect to* p_A,

$\frac{\partial q_A}{\partial p_B}$, *the marginal demand for A with respect to* p_B,

$\frac{\partial q_B}{\partial p_A}$, *the marginal demand for B with respect to* p_A,

$\frac{\partial q_B}{\partial p_B}$, *the marginal demand for B with respect to* p_B.

Under typical conditions, if the price of B is fixed and the price of A increases, then the quantity of A demanded will decrease. Thus $\partial q_A/\partial p_A < 0$. Similarly, $\partial q_B/\partial p_B < 0$. However, $\partial q_A/\partial p_B$ and $\partial q_B/\partial p_A$ may be either positive or negative. If

$$\frac{\partial q_A}{\partial p_B} > 0 \quad \text{and} \quad \frac{\partial q_B}{\partial p_A} > 0,$$

then A and B are said to be **competitive products**. In this situation an increase in the price of B causes an increase in the demand for A, if it is assumed that the price of A does not change. Likewise, an increase in the price of A causes an increase in the demand for B when the price of B is held fixed. Butter and margarine are examples of such competitive commodities.

Proceeding to a different situation, we say that if

$$\frac{\partial q_A}{\partial p_B} < 0 \quad \text{and} \quad \frac{\partial q_B}{\partial p_A} < 0,$$

then A and B are **complementary products.** In this case an increase in the price of B causes a decrease in the demand for A if the price of A does not change. Similarly, an increase in the price of A causes a decrease in the demand for B when the price of B is held fixed. For example, cameras and film are complementary products. An increase in the price of film will make picture-taking more expensive. Hence the demand for cameras will decrease.

EXAMPLE 4

The demand for widgets and the demand for wadgets are each functions of the prices of both widgets and wadgets. If the demand functions for widgets and wadgets are

$$q_A = \frac{50\sqrt[3]{p_B}}{\sqrt{p_A}} \quad \text{and} \quad q_B = \frac{75p_A}{\sqrt[3]{p_B^2}},$$

respectively, find the four marginal demand functions and also determine whether widgets and wadgets are competitive products, complementary products, or neither.

Writing $q_A = 50p_A^{-1/2}p_B^{1/3}$ and $q_B = 75p_Ap_B^{-2/3}$, we have

$$\frac{\partial q_A}{\partial p_A} = 50\left(-\frac{1}{2}\right)p_A^{-3/2}p_B^{1/3} = -25p_A^{-3/2}p_B^{1/3},$$

$$\frac{\partial q_A}{\partial p_B} = 50p_A^{-1/2}\left(\frac{1}{3}\right)p_B^{-2/3} = \frac{50}{3}p_A^{-1/2}p_B^{-2/3},$$

$$\frac{\partial q_B}{\partial p_A} = 75(1)p_B^{-2/3} = 75p_B^{-2/3},$$

$$\frac{\partial q_B}{\partial p_B} = 75p_A\left(-\frac{2}{3}\right)p_B^{-5/3} = -50p_Ap_B^{-5/3}.$$

Since p_A and p_B represent prices, they are both positive. Hence $\partial q_A/\partial p_B > 0$ and $\partial q_B/\partial p_A > 0$. We conclude that widgets and wadgets are competitive products.

Exercise 8-3

For the joint-cost functions in Problems **1–3**, *find the indicated marginal cost at the given production level.*

1. $c = 4x + .3y^2 + 2y + 500;\quad \dfrac{\partial c}{\partial y},\ x = 20, y = 30.$

2. $c = x\sqrt{x+y} + 1000; \quad \dfrac{\partial c}{\partial x}, x = 40, y = 60.$

3. $c = .03(x+y)^3 - .6(x+y)^2 + 4.5(x+y) + 7700; \quad \dfrac{\partial c}{\partial x}, x = 50, y = 50.$

For the production functions in Problems **4** *and* **5**, *find the marginal production functions* $\partial P/\partial k$ *and* $\partial P/\partial l$.

4. $P = 20lk - 2l^2 - 4k^2 + 800.$ **5.** $P = 1.582l^{.192}k^{.764}.$

6. A Cobb-Douglas production function is a production function of the form $P = Al^{\alpha}k^{\beta}$, where A, α, and β are constants and $\alpha + \beta = 1$. For such a function, show that

(a) $\partial P/\partial l = \alpha P/l$. (b) $\partial P/\partial k = \beta P/k$.

(c) $l\dfrac{\partial P}{\partial l} + k\dfrac{\partial P}{\partial k} = P$. This means that summing the marginal productivities of each factor times the amount of each factor results in the total product P.

In Problems **7–9**, q_A *and* q_B *are demand functions for products A and B, respectively. In each case find* $\partial q_A/\partial p_A$, $\partial q_A/\partial p_B$, $\partial q_B/\partial p_A$, $\partial q_B/\partial p_B$ *and determine whether A and B are competitive, complementary, or neither.*

7. $q_A = 1000 - 50p_A + 2p_B; \quad q_B = 500 + 4p_A - 20p_B.$

8. $q_A = 20 - p_A - 2p_B; \quad q_B = 50 - 2p_A - 3p_B.$

9. $q_A = \dfrac{100}{p_A\sqrt{p_B}}; \quad q_B = \dfrac{500}{p_B\sqrt[3]{p_A}}.$

10. The production function for the Canadian manufacturing industries for 1927 is estimated in **[6]** by the equation $P = 33.0l^{.46}k^{.52}$, where P is product, l is labor, and k is capital. Find the marginal productivities for labor and capital and evaluate when $l = 1$ and $k = 1$.

11. An estimate of the production function for dairy farming in Iowa (1939) is given in **[53]** by

$$P = A^{.27}B^{.01}C^{.01}D^{.23}E^{.09}F^{.27},$$

where P is product, A is land, B is labor, C is improvements, D is liquid assets, E is working assets, and F is cash operating expenses. Find the marginal productivities for labor and improvements.

12. A person's general status, S_g, is believed to be a function of status attributable to education, S_e, and status attributable to income, S_i, where S_g, S_e, and S_i are represented numerically. If $S_g = 7\sqrt[3]{S_e}\sqrt{S_i}$, determine $\dfrac{\partial S_g}{\partial S_e}$ and

$\dfrac{\partial S_g}{\partial S_i}$ when $S_e = 125$ and $S_i = 100$, and interpret your results (adapted from [28]).

13. In a study of success among master of business administration (MBA) graduates [55], it was estimated that for staff managers (which includes accountants, analysts, etc.) current annual compensation z (in dollars) was given by

$$z = 10{,}990 + 1120x + 873y,$$

where x and y are the number of years of work experience before and after receiving the MBA degree, respectively. Find $\partial z/\partial x$ and interpret your result.

14. The study of frequency of vibrations of a taut wire is useful in considering such things as an individual's voice. Suppose [52]

$$\omega = \frac{1}{bL}\sqrt{\frac{\tau}{\pi\rho}},$$

where ω (a Greek letter read "omega") is frequency, b is diameter, L is length, ρ (a Greek letter read "rho") is density, and τ (a Greek letter read "tau") is tension. Find $\partial\omega/\partial b$, $\partial\omega/\partial L$, $\partial\omega/\partial\rho$, and $\partial\omega/\partial\tau$.

15. Sometimes we want to evaluate the degree of readability of a piece of writing. Rudolf Flesch [12] developed a function of two variables that will do this:

$$R = f(w, s) = 206.835 - (1.015w + .846s),$$

where R is called the *reading ease score*, w is the average number of words per sentence in 100-word samples, and s is the average number of syllables in such samples. Flesch says that an article for which $R = 0$ is "practically unreadable," but one with $R = 100$ is "easy for any literate person." (a) Find $\partial R/\partial w$ and $\partial R/\partial s$. (b) Which is "easier" to read: an article for which $w = w_0$ and $s = s_0$, or one for which $w = w_0 + 1$ and $s = s_0$?

16. Consider the following traffic-flow situation. On a highway where two lanes of traffic flow in the same direction, there is a maintenance vehicle blocking the left lane. See Fig. 8-11. Two vehicles (*lead* and *following*) are in

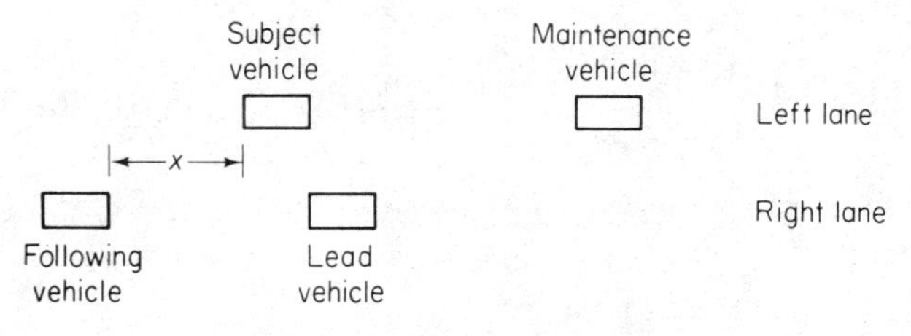

FIG. 8-11

the right lane with a gap between them. The *subject* vehicle can choose either to fill or not to fill the gap. That decision may be based not only on the distance x shown in the diagram, but on other factors (such as the velocity of the *following* vehicle). A *gap index*, g, has been used in analyzing such a decision [21, 37]. The greater the g-value, the greater is the propensity for the *subject* vehicle to fill the gap. Suppose

$$g = \frac{x}{V_F} - \left(.75 + \frac{V_F - V_S}{19.2}\right),$$

where x (in feet) is as before, V_F is the velocity of the *following* vehicle (in feet per second), and V_S is the velocity of the *subject* vehicle (in feet per second). From the diagram it seems reasonable that if both V_F and V_S are fixed and x increases, then g should increase. Also, if x and V_s are fixed and V_F increases, then g should decrease. Assume that x, V_F, and V_S are positive and show that these observations are indeed true by applying calculus to the function g given above.

17. For the congressional elections of 1974, the Republican percentage, R, of the Republican-Democratic vote in a district is given (approximately) in [44] by

$$\begin{aligned} R &= f(E_r, E_d, I_r, I_d, N) \\ &= 15.4725 + 2.5945E_r - .0804E_r^2 - 2.3648E_d + \\ &\quad .0687E_d^2 + 2.1914I_r - .0912I_r^2 - .8096I_d + \\ &\quad .0081I_d^2 - .0277E_rI_r + .0493E_dI_d + \\ &\quad .8579N - .0061N^2. \end{aligned}$$

Here E_r and E_d are the campaign expenditures (in units of \$10,000) by Republicans and Democrats, respectively; I_r and I_d are the number of terms served in Congress, *plus* one, for the Republican and Democratic candidates, respectively; and N is the percentage of the two-party presidential vote that Richard Nixon received in the district for 1968. The variable N gives a measure of Republican strength in the district.

(a) In the Federal Election Campaign Act of 1974, Congress set a limit of \$188,000 on campaign expenditures. By analyzing $\partial R/\partial E_r$, would you have advised a Republican candidate who served nine terms in Congress to spend \$188,000 on his campaign?

(b) Determine the percentage above which the Nixon vote had a negative effect on R; that is, determine when $\partial R/\partial N < 0$. Give your answer to the nearest percent.

8-4 IMPLICIT PARTIAL DIFFERENTIATION†

In the equation

$$z^2 - x^2 - y^2 = 0, \tag{1}$$

if $x = 1$ and $y = 1$, then $z^2 - 1 - 1 = 0$ and so $z = \pm\sqrt{2}$. Thus Eq. (1) does not define z as a function of x and y. However, solving Eq. (1) for z gives

$$z = \sqrt{x^2 + y^2} \quad \text{or} \quad z = -\sqrt{x^2 + y^2},$$

each of which defines z as a function of x and y. Although Eq. (1) does not *explicitly* express z as a function of x and y, it can be thought of as expressing z *implicitly* as two different functions of x and y. Note that $z^2 - x^2 - y^2 = 0$ has the form $F(x, y, z) = 0$. Any equation of the form $F(x, y, z) = 0$ can be thought of as expressing z implicitly as one or more functions of x and y.

To find $\partial z/\partial x$ where

$$z^2 - x^2 - y^2 = 0, \tag{2}$$

we first differentiate both sides of Eq. (2) with respect to x while treating z as a function of x and y, and treating y as a constant.

$$\frac{\partial}{\partial x}(z^2 - x^2 - y^2) = \frac{\partial}{\partial x}(0),$$

$$\frac{\partial}{\partial x}(z^2) - \frac{\partial}{\partial x}(x^2) - \frac{\partial}{\partial x}(y^2) = 0,$$

$$2z\frac{\partial z}{\partial x} - 2x - 0 = 0.$$

Solving for $\partial z/\partial x$, we obtain

$$2z\frac{\partial z}{\partial x} = 2x,$$

$$\frac{\partial z}{\partial x} = \frac{x}{z}.$$

To find $\partial z/\partial y$ we differentiate both sides of Eq. (2) with respect to y

†May be omitted without loss of continuity.

while treating z as a function of x and y, and treating x as a constant.

$$\frac{\partial}{\partial y}(z^2 - x^2 - y^2) = \frac{\partial}{\partial y}(0),$$

$$2z\frac{\partial z}{\partial y} - 0 - 2y = 0,$$

$$2z\frac{\partial z}{\partial y} = 2y.$$

Hence,

$$\frac{\partial z}{\partial y} = \frac{y}{z}.$$

The method we used to find $\partial z/\partial x$ and $\partial z/\partial y$ is called *implicit (partial) differentiation*.

EXAMPLE 1

a. *If* $\dfrac{xz^2}{x+y} + y^2 = 0$, *evaluate* $\partial z/\partial x$ *when* $x = -1$, $y = 2$, *and* $z = 2$.

We treat z as a function of x and y and differentiate both sides of the equation with respect to x.

$$\frac{\partial}{\partial x}\left(\frac{xz^2}{x+y}\right) + \frac{\partial}{\partial x}(y^2) = \frac{\partial}{\partial x}(0).$$

Using the quotient rule for the first term on the left side, we have

$$\frac{(x+y)\frac{\partial}{\partial x}(xz^2) - xz^2\frac{\partial}{\partial x}(x+y)}{(x+y)^2} + 0 = 0.$$

Using the product rule for $\frac{\partial}{\partial x}(xz^2)$ gives

$$\frac{(x+y)\left[x\left(2z\frac{\partial z}{\partial x}\right) + z^2(1)\right] - xz^2(1)}{(x+y)^2} = 0.$$

Solving for $\partial z/\partial x$, we obtain

$$2xz(x+y)\frac{\partial z}{\partial x} + z^2(x+y) - xz^2 = 0,$$

$$\frac{\partial z}{\partial x} = \frac{xz^2 - z^2(x+y)}{2xz(x+y)} = -\frac{yz}{2x(x+y)}, \qquad z \neq 0.$$

When $x = -1$, $y = 2$, and $z = 2$, then $\partial z/\partial x = 2$.

b. *If* $se^{r^2+u^2} = u \ln(t^2+1)$, *determine* $\partial t/\partial u$.

We consider t as a function of r, s, and u. By differentiating both sides with respect to u while treating r and s as constants, we get

$$\frac{\partial}{\partial u}(se^{r^2+u^2}) = \frac{\partial}{\partial u}[u \ln(t^2+1)],$$

$$2sue^{r^2+u^2} = u\frac{\partial}{\partial u}[\ln(t^2+1)] + \ln(t^2+1)\frac{\partial}{\partial u}(u),$$

$$2sue^{r^2+u^2} = u\frac{2t}{t^2+1}\frac{\partial t}{\partial u} + \ln(t^2+1).$$

Thus,

$$\frac{\partial t}{\partial u} = \frac{(t^2+1)\left[2sue^{r^2+u^2} - \ln(t^2+1)\right]}{2ut}.$$

Exercise 8-4

In Problems **1–11**, *by the method of implicit partial differentiation find the indicated partial derivatives.*

1. $x^2 + y^2 + z^2 = 9$; $\partial z/\partial x$.

2. $z^2 - 3x^2 + y^2 = 0$; $\partial z/\partial x$.

3. $2z^3 - x^2 - 4y^2 = 0$; $\partial z/\partial y$.

4. $3x^2 + y^2 + 2z^3 = 9$; $\partial z/\partial y$.

5. $x^2 - 2y - z^2 + x^2yz^2 = 20$; $\partial z/\partial x$.

6. $z^3 - xz - y = 0$; $\partial z/\partial x$.

7. $e^x + e^y + e^z = 10$; $\partial z/\partial y$.

8. $xyz + 2y^2x - z^3 = 0$; $\partial z/\partial x$.

9. $\ln(z) + z - xy = 1$; $\partial z/\partial x$.

10. $\ln x + \ln y - \ln z = e^y$; $\partial z/\partial x$.

11. $(z^2 + 6xy)\sqrt{x^3+5} = 2$; $\partial z/\partial y$.

In Problems **12–18**, *evaluate the indicated partial derivatives for the given values of the variables.*

12. $xz + xyz - 5 = 0$; $\partial z/\partial x$, $x = 1, y = 4, z = 1$.

13. $xz^2 + yz - 12 = 0$; $\partial z/\partial x$, $x = 2, y = -2, z = 3$.

14. $e^{zx} = xyz$; $\partial z/\partial y$, $x = 1, y = -e^{-1}, z = -1$.

15. $\ln z = x + y$; $\partial z/\partial x$, $x = 5, y = -5, z = 1$.

16. $\sqrt{xz + y^2} - xy = 0$; $\partial z/\partial y$, $x = 2, y = 2, z = 6$.

17. $\dfrac{s^2 + t^2}{rs} = 10$; $\partial t/\partial r$, $r = 1, s = 2, t = 4$.

18. $\dfrac{rs}{s^2 + t^2} = t$; $\partial r/\partial t$, $r = 0, s = 1, t = 0$.

8-5 HIGHER-ORDER PARTIAL DERIVATIVES

If $z = f(x, y)$, then both f_x and f_y are also functions of x and y. Hence we may differentiate them and obtain so-called second-order partial derivatives of f. Symbolically,

$$f_{xx} \text{ means } (f_x)_x, \qquad f_{xy} \text{ means } (f_x)_y,$$

$$f_{yx} \text{ means } (f_y)_x, \qquad f_{yy} \text{ means } (f_y)_y.$$

In terms of ∂-notation,

$$\frac{\partial^2 z}{\partial x^2} \text{ means } \frac{\partial}{\partial x}\left[\frac{\partial z}{\partial x}\right], \qquad \frac{\partial^2 z}{\partial y\, \partial x} \text{ means } \frac{\partial}{\partial y}\left[\frac{\partial z}{\partial x}\right],$$

$$\frac{\partial^2 z}{\partial x\, \partial y} \text{ means } \frac{\partial}{\partial x}\left[\frac{\partial z}{\partial y}\right], \qquad \frac{\partial^2 z}{\partial y^2} \text{ means } \frac{\partial}{\partial y}\left[\frac{\partial z}{\partial y}\right].$$

Note that to find f_{xy}, first differentiate f with respect to x. For $\partial^2 z/\partial x\, \partial y$, first differentiate with respect to y.

We can extend our notation beyond second-order partial derivatives. For example, f_{xyx} (or $\partial^3 z/\partial x\, \partial y\, \partial x$) is a third-order partial derivative of f. It is the partial derivative of f_{xy} (or $\partial^2 z/\partial y\, \partial x$) with respect to x. A generalization regarding higher-order partial derivatives to functions of more than two variables should be obvious.

EXAMPLE 1

Find the four second-order partial derivatives of $f(x, y) = x^2y + x^2y^2$.

Since

$$f_x(x, y) = 2xy + 2xy^2,$$

then

$$f_{xx}(x, y) = \frac{\partial}{\partial x}(2xy + 2xy^2) = 2y + 2y^2,$$

and

$$f_{xy}(x, y) = \frac{\partial}{\partial y}(2xy + 2xy^2) = 2x + 4xy.$$

Since

$$f_y(x, y) = x^2 + 2x^2y,$$

then

$$f_{yy}(x, y) = \frac{\partial}{\partial y}(x^2 + 2x^2y) = 2x^2,$$

and

$$f_{yx}(x, y) = \frac{\partial}{\partial x}(x^2 + 2x^2y) = 2x + 4xy.$$

Observe in Example 1 that $f_{xy}(x, y) = f_{yx}(x, y)$. This equality did not occur by chance. It can be shown that for any function f, if f_{xy} and f_{yx} are both continuous, then $f_{xy} = f_{yx}$; that is, the order of differentiation is of no concern.

EXAMPLE 2

Determine the value of $\left.\frac{\partial^3 w}{\partial z\, \partial y\, \partial x}\right|_{(1, 2, 3)}$ *if* $w = (2x + 3y + 4z)^3$.

$$\frac{\partial w}{\partial x} = 3(2x + 3y + 4z)^2 \frac{\partial}{\partial x}(2x + 3y + 4z)$$

$$= 6(2x + 3y + 4z)^2.$$

$$\frac{\partial^2 w}{\partial y\, \partial x} = 6 \cdot 2(2x + 3y + 4z) \frac{\partial}{\partial y}(2x + 3y + 4z)$$

$$= 36(2x + 3y + 4z).$$

$$\frac{\partial^3 w}{\partial z\, \partial y\, \partial x} = 36 \cdot 4 = 144.$$

Thus,

$$\left.\frac{\partial^3 w}{\partial z\,\partial y\,\partial x}\right|_{(1,\,2,\,3)} = 144.$$

Exercise 8-5

In Problems **1–10**, *find the indicated partial derivatives.*

1. $f(x, y) = 3x^2y^2$; $f_x(x, y), f_{xy}(x, y)$.

2. $f(x, y) = 3x^2y + 2xy^2 - 7y$; $f_x(x, y), f_{xx}(x, y)$.

3. $f(x, y) = e^{3xy} + 4x^2y$; $f_y(x, y), f_{yx}(x, y), f_{yxy}(x, y)$.

4. $f(x, y) = 7x^2 + 3y$; $f_y(x, y), f_{yy}(x, y), f_{yyx}(x, y)$.

5. $f(x, y) = (x^2 + xy + y^2)(x^2 + xy + 1)$; $f_x(x, y), f_{xy}(x, y)$.

6. $f(x, y) = \ln (x^2 + y^2) + 2$; $f_x(x, y), f_{xx}(x, y), f_{xy}(x, y)$.

7. $f(x, y) = (x + y)^2(xy)$; $f_x(x, y), f_y(x, y), f_{xx}(x, y), f_{yy}(x, y)$.

8. $f(x, y, z) = xy^2z^3$; $f_x(x, y, z), f_{xz}(x, y, z), f_{xy}(x, y, z)$.

9. $z = \sqrt{x^2 + y^2}$; $\dfrac{\partial z}{\partial x}, \dfrac{\partial^2 z}{\partial x^2}$.

10. $z = \dfrac{\ln (x^2 + 5)}{y}$; $\dfrac{\partial z}{\partial x}, \dfrac{\partial^2 z}{\partial y\,\partial x}$.

11. If $f(x, y, z) = 7$, find $f_{yxx}(4, 3, -2)$.

12. If $f(x, y, z) = z^2(3x^2 - 4xy^3)$, find $f_{xyz}(1, 2, 3)$.

13. If $f(l, k) = 5l^3k^6 - lk^7$, find $f_{kkl}(2, 1)$.

14. If $f(x, y) = 2x^2y + xy^2 - x^2y^2$, find $f_{xxy}(0, 1)$.

15. If $f(x, y) = y^2e^x + \ln (xy)$, find $f_{xyy}(1, 1)$.

16. If $f(x, y) = x^3 - 3xy^2 + x^2 - y^3$, find $f_{xy}(1, -1)$.

17. For $f(x, y) = 8x^3 + 2x^2y^2 + 5y^4$, show $f_{xy}(x, y) = f_{yx}(x, y)$.

18. For $f(x, y) = x^4y^4 + 3x^3y^2 - 7x + 4$, show $f_{xyx}(x, y) = f_{xxy}(x, y)$.

19. For $z = \ln (x^2 + y^2)$, show that

$$\frac{\partial^2 z}{\partial x^2} + \frac{\partial^2 z}{\partial y^2} = 0.$$

8-6 CHAIN RULE†

Suppose a manufacturer of two related products A and B has a joint-cost function given by

$$c = f(q_A, q_B),$$

where c is the total cost of producing quantities q_A and q_B of A and B, respectively. Furthermore, suppose the demand functions for his products are

$$q_A = g(p_A, p_B) \quad \text{and} \quad q_B = h(p_A, p_B),$$

where p_A and p_B are the prices per unit of A and B, respectively. Since c is a function of q_A and q_B and both q_A and q_B are themselves functions of p_A and p_B, then c can be viewed as a function of p_A and p_B. (Appropriately, the variables q_A and q_B are called *intermediate variables* of c.) Consequently, we should be able to determine $\partial c/\partial p_A$, the rate of change of total cost with respect to the price of A. One way to do this is to substitute the expressions $g(p_A, p_B)$ and $h(p_A, p_B)$ for q_A and q_B, respectively, into $c = f(q_A, q_B)$. Then c is a function of p_A and p_B and we can differentiate c with respect to p_A directly. This approach has some drawbacks—especially when f, g, or h is given by a complicated expression. Another way to approach the problem would be to use the chain rule (actually *a* chain rule) which we now state without proof.

CHAIN RULE. *Let $z = f(x, y)$, where both x and y are functions of r and s given by $x = x(r, s)$ and $y = y(r, s)$. If f, x, and y have continuous partial derivatives, then z is a function of r and s, and*

$$\frac{\partial z}{\partial r} = \frac{\partial z}{\partial x}\frac{\partial x}{\partial r} + \frac{\partial z}{\partial y}\frac{\partial y}{\partial r},$$

and

$$\frac{\partial z}{\partial s} = \frac{\partial z}{\partial x}\frac{\partial x}{\partial s} + \frac{\partial z}{\partial y}\frac{\partial y}{\partial s}.$$

Note that in the chain rule the number of intermediate variables of z (two) is the same as the number of terms that compose each of $\partial z/\partial r$ and $\partial z/\partial s$.

†May be omitted without loss of continuity.

Returning to the original situation concerning the manufacturer, we see that if f, q_A, and q_B have continuous partial derivatives, then by the chain rule

$$\frac{\partial c}{\partial p_A} = \frac{\partial c}{\partial q_A}\frac{\partial q_A}{\partial p_A} + \frac{\partial c}{\partial q_B}\frac{\partial q_B}{\partial p_A}.$$

EXAMPLE 1

For a manufacturer of cameras and film, the total cost of producing q_C cameras and q_F units of film is given by

$$c = 30q_C + .015q_Cq_F + q_F + 900.$$

The demand functions for the cameras and film are given by

$$q_C = \frac{9000}{p_C\sqrt{p_F}} \quad \text{and} \quad q_F = 2000 - p_C - 400p_F,$$

where p_C is the price per camera and p_F is the price per unit of film. Find the rate of change of total cost with respect to the price of the camera when $p_C = 50$ and $p_F = 2$.

We must first determine $\partial c/\partial p_C$. By the chain rule,

$$\begin{aligned}\frac{\partial c}{\partial p_C} &= \frac{\partial c}{\partial q_C}\frac{\partial q_C}{\partial p_C} + \frac{\partial c}{\partial q_F}\frac{\partial q_F}{\partial p_C}\\ &= (30 + .015q_F)\left[\frac{-9000}{p_C^2\sqrt{p_F}}\right] + (.015q_C + 1)(-1).\end{aligned}$$

When $p_C = 50$ and $p_F = 2$, then $q_C = 90\sqrt{2}$ and $q_F = 1150$. Substituting these values into $\partial c/\partial p_C$ and simplifying, we have

$$\left.\frac{\partial c}{\partial p_C}\right|_{\substack{p_C=50\\ p_F=2}} = -123.2 \text{ (approximately)}.$$

The chain rule can be extended. For example, suppose $z = f(v, w, x, y)$ and v, w, x, and y are all functions of r, s, and t. Then, if certain conditions of continuity are assumed, z is a function of r, s, and t and

$$\frac{\partial z}{\partial r} = \frac{\partial z}{\partial v}\frac{\partial v}{\partial r} + \frac{\partial z}{\partial w}\frac{\partial w}{\partial r} + \frac{\partial z}{\partial x}\frac{\partial x}{\partial r} + \frac{\partial z}{\partial y}\frac{\partial y}{\partial r},$$

$$\frac{\partial z}{\partial s} = \frac{\partial z}{\partial v}\frac{\partial v}{\partial s} + \frac{\partial z}{\partial w}\frac{\partial w}{\partial s} + \frac{\partial z}{\partial x}\frac{\partial x}{\partial s} + \frac{\partial z}{\partial y}\frac{\partial y}{\partial s},$$

and

$$\frac{\partial z}{\partial t} = \frac{\partial z}{\partial v}\frac{\partial v}{\partial t} + \frac{\partial z}{\partial w}\frac{\partial w}{\partial t} + \frac{\partial z}{\partial x}\frac{\partial x}{\partial t} + \frac{\partial z}{\partial y}\frac{\partial y}{\partial t}.$$

Observe that the number of intermediate variables of z (four) is the same as the number of terms that form each of $\partial z/\partial r$, $\partial z/\partial s$, and $\partial z/\partial t$.

Now consider the situation where $z = f(x, y)$ and $x = x(t)$ and $y = y(t)$. Then

$$\frac{dz}{dt} = \frac{\partial z}{\partial x}\frac{dx}{dt} + \frac{\partial z}{\partial y}\frac{dy}{dt}.$$

Here we use the symbol dz/dt rather than $\partial z/\partial t$, since z can be considered a function of the *one* variable t. Likewise, the symbols dx/dt and dy/dt are used rather than $\partial x/\partial t$ and $\partial y/\partial t$. As is typical, the number of terms that compose dz/dt equals the number of intermediate variables of z. Other situations would be treated in a similar way.

EXAMPLE 2

a. *If* $w = f(x, y, z) = 3x^2y + xyz - 4y^2z^3$ *where* $x = 2r - 3s$, $y = 6r + s$, *and* $z = r - s$, *determine* $\partial w/\partial r$ *and* $\partial w/\partial s$.

Since x, y, and z are functions of r and s, then by the chain rule,

$$\begin{aligned}\frac{\partial w}{\partial r} &= \frac{\partial w}{\partial x}\frac{\partial x}{\partial r} + \frac{\partial w}{\partial y}\frac{\partial y}{\partial r} + \frac{\partial w}{\partial z}\frac{\partial z}{\partial r}\\ &= (6xy + yz)(2) + (3x^2 + xz - 8yz^3)(6) + (xy - 12y^2z^2)(1)\\ &= x(18x + 13y + 6z) + 2yz(1 - 24z^2 - 6yz),\end{aligned}$$

and

$$\begin{aligned}\frac{\partial w}{\partial s} &= \frac{\partial w}{\partial x}\frac{\partial x}{\partial s} + \frac{\partial w}{\partial y}\frac{\partial y}{\partial s} + \frac{\partial w}{\partial z}\frac{\partial z}{\partial s}\\ &= (6xy + yz)(-3) + (3x^2 + xz - 8yz^3)(1) + (xy - 12y^2z^2)(-1)\\ &= x(3x - 19y + z) - yz(3 + 8z^2 - 12yz).\end{aligned}$$

b. *If* $z = \dfrac{x + e^y}{y}$ *where* $x = rs + se^{rt}$ *and* $y = 9 + rt$, *evaluate* $\partial z/\partial s$ *when* $r = -2$, $s = 5$, *and* $t = 4$.

Since x and y are functions of r, s, and t (note that we can write

$y = 9 + rt + 0 \cdot s$), by the chain rule,

$$\frac{\partial z}{\partial s} = \frac{\partial z}{\partial x}\frac{\partial x}{\partial s} + \frac{\partial z}{\partial y}\frac{\partial y}{\partial s}$$

$$= \left(\frac{1}{y}\right)(r + e^{rt}) + \frac{\partial z}{\partial y} \cdot (0) = \frac{r + e^{rt}}{y}.$$

If $r = -2$, $s = 5$, and $t = 4$, then $y = 1$. Thus,

$$\left.\frac{\partial z}{\partial s}\right|_{\substack{r=-2\\ s=5\\ t=4}} = \frac{-2 + e^{-8}}{1} = -2 + e^{-8}.$$

c. *Determine $\partial y/\partial r$ if $y = x^2 \ln (x^4 + 6)$ and $x = (r + 3s)^6$.*

By the chain rule,

$$\frac{\partial y}{\partial r} = \frac{dy}{dx}\frac{\partial x}{\partial r}$$

$$= \left[x^2 \cdot \frac{4x^3}{x^4 + 6} + 2x \cdot \ln (x^4 + 6)\right]\left[6(r + 3s)^5\right]$$

$$= 12x(r + 3s)^5\left[\frac{2x^4}{x^4 + 6} + \ln (x^4 + 6)\right].$$

EXAMPLE 3

Given that $z = e^{xy}$, $x = r - 4s$, and $y = r - s$, find $\partial z/\partial r$ in terms of r and s.

$$\frac{\partial z}{\partial r} = \frac{\partial z}{\partial x}\frac{\partial x}{\partial r} + \frac{\partial z}{\partial y}\frac{\partial y}{\partial r}$$

$$= (ye^{xy})(1) + (xe^{xy})(1)$$

$$= (x + y)e^{xy}.$$

Since $x = r - 4s$ and $y = r - s$,

$$\frac{\partial z}{\partial r} = [(r - 4s) + (r - s)]e^{(r-4s)(r-s)}$$

$$= (2r - 5s)e^{r^2 - 5rs + 4s^2}.$$

Exercise 8-6

In Problems **1–12**, *find the indicated derivatives by using the chain rule.*

1. $z = 5x + 3y$, $x = 2r + 3s$, $y = r - 2s$; $\partial z/\partial r$, $\partial z/\partial s$.

2. $z = x^2 + 3xy + 7y^3$, $x = r^2 - 2s$, $y = 5s^2$; $\partial z/\partial r$, $\partial z/\partial s$.

3. $z = e^{x+y}$, $x = t^2 + 3$, $y = \sqrt{t^3}$; dz/dt.

4. $z = \sqrt{8x + y}$, $x = t^2 + 3t + 4$, $y = t^3 + 4$; dz/dt.

5. $w = x^2z^2 + xyz + yz^2$, $x = 5t$, $y = 2t + 3$, $z = 6 - t$; dw/dt.

6. $w = \ln(x^2 + y^2 + z^2)$, $x = 2 - 3t$, $y = t^2 + 3$, $z = 4 - t$; dw/dt.

7. $z = (x^2 + xy^2)^3$, $x = r + s + t$, $y = 2r - 3s + t$; $\partial z/\partial t$.

8. $z = \sqrt{x^2 + y^2}$, $x = r^2 + s - t$, $y = r - s + t$; $\partial z/\partial r$.

9. $w = x^2 + xyz + y^3z^2$, $x = r - s^2$, $y = rs$, $z = 2r - 5s$; $\partial w/\partial s$.

10. $w = e^{xyz}$, $x = r^2s^3$, $y = r - s$, $z = rs^2$; $\partial w/\partial r$.

11. $y = x^2 - 7x + 5$, $x = 15rs + 2s^2t^2$; $\partial y/\partial r$.

12. $y = 4 - x^2$, $x = 2r + 3s - 4t$; $\partial y/\partial t$.

13. If $z = (4x + 3y)^3$ where $x = r^2s$ and $y = r - 2s$, evaluate $\partial z/\partial r$ when $r = 0$ and $s = 1$.

14. If $z = \sqrt{5x + 2y}$ where $x = 4t + 7$ and $y = t^2 - 3t + 4$, evaluate dz/dt when $t = 1$.

15. If $w = e^{3x-y}(x^2 + 4z^3)$ where $x = rs$, $y = 2s - r$, and $z = r + s$, evaluate $\partial w/\partial s$ when $r = 1$ and $s = -1$.

16. If $y = x/(x - 5)$ where $x = 2t^2 - 3rs - r^2t$, evaluate $\partial y/\partial t$ when $r = 0$, $s = 2$, and $t = -1$.

8-7 MAXIMA AND MINIMA FOR FUNCTIONS OF TWO VARIABLES

We now extend to functions of two variables the notion of relative maxima and minima (or relative extrema) which was introduced in Chapter 4.

DEFINITION. *A function $z = f(x, y)$ is said to have a* ***relative maximum*** *at the point (x_0, y_0), that is, when $x = x_0$ and $y = y_0$, if for all points (x, y) in the plane which are sufficiently "close" to (x_0, y_0) we have*

$$f(x_0, y_0) \geqslant f(x, y). \tag{1}$$

For a ***relative minimum*** *we replace $\geqslant$ by $\leqslant$ in (1).*

To say that $z = f(x, y)$ has a relative maximum at (x_0, y_0) means geometrically that the point (x_0, y_0, z_0) on the graph of f is higher than (or is as high as) all other points on the surface which are "near" (x_0, y_0, z_0). In Fig. 8-12(a), f has a relative maximum at (x_1, y_1). Similarly, the function f in Fig. 8-12(b) has a relative minimum when $x = y = 0$ which corresponds to a *low* point on the surface.

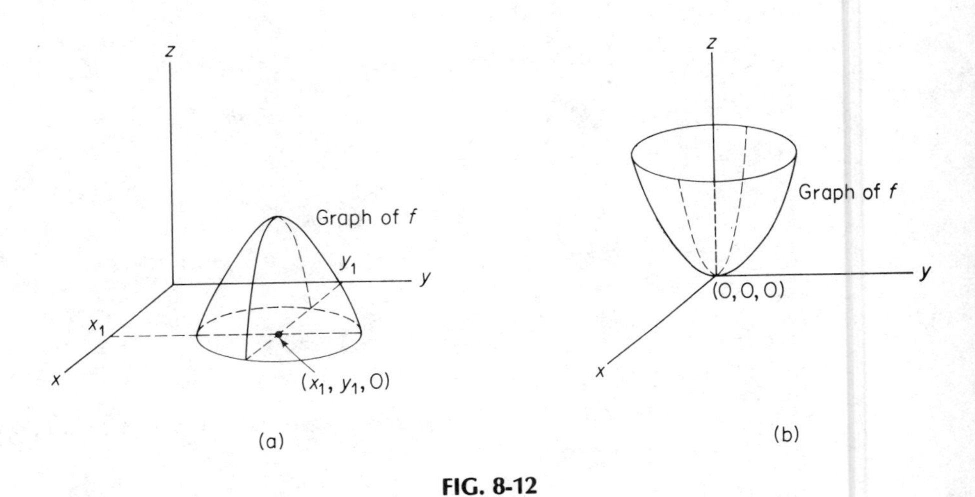

FIG. 8-12

Recall that in locating extrema for a function $y = f(x)$ of one variable, we examined those values of x for which $f'(x) = 0$ or $f'(x)$ does not exist. For functions of two (or more) variables, a similar procedure can be followed. However, for the functions that we shall work with, extrema will not occur where a derivative does not exist, and such situations will be excluded from considerations.

Suppose $z = f(x, y)$ has a relative maximum at (x_0, y_0), as indicated in Fig. 8-13(a). Then the curve where the plane $y = y_0$ intersects the surface must have a relative maximum when $x = x_0$. Hence the slope of the tangent line to the surface in the x-direction must be 0 at (x_0, y_0). Equivalently, $f_x(x, y) = 0$ at (x_0, y_0). Similarly, on the curve where the plane $x = x_0$ intersects the surface [Fig. 8-13(b)], there must be a relative maximum when $y = y_0$. Thus in the y-direction, the slope of the tangent to the surface must be 0 at (x_0, y_0). Equivalently, $f_y(x, y) = 0$ at (x_0, y_0). Since a similar discussion can be given for a relative minimum, we can combine these results as follows.

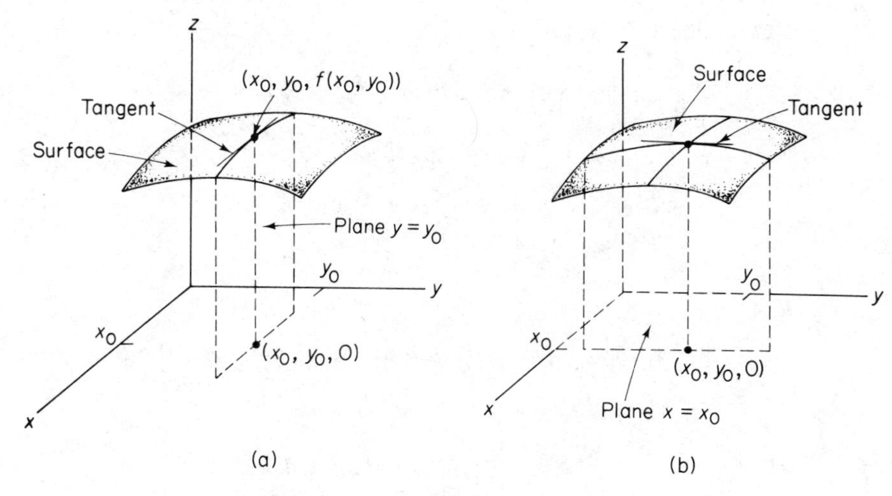

FIG. 8-13

RULE 1. *If $z = f(x, y)$ has a relative maximum or minimum at (x_0, y_0) and if both f_x and f_y are defined for all points close to (x_0, y_0), it is necessary that (x_0, y_0) be a solution of the system*

$$\begin{cases} f_x(x, y) = 0, \\ f_y(x, y) = 0. \end{cases}$$

A point (x_0, y_0) for which $f_x(x, y) = f_y(x, y) = 0$ is called a **critical point** of f. Thus from Rule 1 we infer that to locate relative extrema for a function we should examine its critical points.

PITFALL. *Rule 1 does not imply that there must be an extremum at a critical point. Just as in the case of functions of one variable, a critical point can give rise to a relative maximum, a relative minimum, or neither.*

Two additional comments: First, Rule 1, as well as the notion of a critical point, can be extended to functions of more than two variables. Thus, to locate possible extrema for $w = f(x, y, z)$ we would examine those points for which $w_x = w_y = w_z = 0$. Second, for a function whose domain is restricted, a thorough examination for *absolute* maxima and minima would include consideration of boundary points.

EXAMPLE 1

Examine each of the following for critical points.

a. $f(x, y) = 2x^2 + y^2 - 2xy + 5x - 3y + 1$.

Since $f_x(x, y) = 4x - 2y + 5$ and $f_y(x, y) = 2y - 2x - 3$, we solve the system

$$\begin{cases} 4x - 2y + 5 = 0, \\ -2x + 2y - 3 = 0. \end{cases}$$

This gives $x = -1$ and $y = \frac{1}{2}$. Thus, $(-1, \frac{1}{2})$ is the only critical point.

b. $f(l, k) = l^3 + k^3 - lk$.

$$\begin{cases} f_l(l, k) = 3l^2 - k = 0, & (2) \\ f_k(l, k) = 3k^2 - l = 0. & (3) \end{cases}$$

From Eq. (2), $k = 3l^2$. Substituting for k in Eq. (3) gives $0 = 27l^4 - l = l(27l^3 - 1)$. Hence, $l = 0$ or $l = \frac{1}{3}$. If $l = 0$, then $k = 0$; if $l = \frac{1}{3}$, then $k = \frac{1}{3}$. The critical points are thus $(0, 0)$ and $(\frac{1}{3}, \frac{1}{3})$.

c. $f(x, y, z) = 2x^2 + xy + y^2 + 100 - z(x + y - 100)$.

Solving the system

$$\begin{cases} f_x(x, y, z) = 4x + y - z = 0, \\ f_y(x, y, z) = x + 2y - z = 0, \\ f_z(x, y, z) = -x - y + 100 = 0 \end{cases}$$

gives the critical point (25, 75, 175), as you may verify.

EXAMPLE 2

Find the critical points of $f(x, y) = x^2 - 4x + 2y^2 + 4y + 7$.

We have $f_x(x, y) = 2x - 4$ and $f_y(x, y) = 4y + 4$. The system

$$\begin{cases} 2x - 4 = 0, \\ 4y + 4 = 0 \end{cases}$$

gives the critical point $(2, -1)$. Observe that the given function can be written

$$\begin{aligned} f(x, y) &= x^2 - 4x + 4 + 2(y^2 + 2y + 1) + 1 \\ &= (x - 2)^2 + 2(y + 1)^2 + 1, \end{aligned}$$

and $f(2, -1) = 1$. Clearly, if $(x, y) \neq (2, -1)$, then $f(x, y) > 1$. Hence a relative minimum occurs at $(2, -1)$. Moreover, there is an *absolute minimum* at $(2, -1)$ since $f(x, y) > f(2, -1)$ for *all* $(x, y) \neq (2, -1)$.

Although in Example 2 we were able to show that the critical point gave rise to a relative minimum, in many cases this is not so easy to do. There is, however, a second-derivative test which gives conditions under which a critical point will be a relative maximum or minimum. We state it now, omitting the proof.

RULE 2. ***Second-Derivative Test for Functions of Two Variables.*** *Suppose $z = f(x, y)$ has continuous partial derivatives f_{xx}, f_{yy}, and f_{xy} at all points (x, y) near the critical point (x_0, y_0). Let Δ be defined by*

$$\Delta = f_{xx}(x_0, y_0) f_{yy}(x_0, y_0) - [f_{xy}(x_0, y_0)]^2.$$

Then

a. *if $\Delta > 0$ and $f_{xx}(x_0, y_0) < 0$, f has a relative maximum at (x_0, y_0);*
b. *if $\Delta > 0$ and $f_{xx}(x_0, y_0) > 0$, f has a relative minimum at (x_0, y_0);*
c. *if $\Delta < 0$, f has neither a relative maximum nor a relative minimum at (x_0, y_0);*
d. *if $\Delta = 0$, the test fails, that is, no conclusion can be drawn and further analysis is required.*

EXAMPLE 3

Examine $f(x, y) = x^3 + y^3 - xy$ for relative maxima or minima by using the second-derivative test.

$$f_x(x, y) = 3x^2 - y, \qquad f_y(x, y) = 3y^2 - x.$$

In the same manner as in Example 1(b), solving $f_x(x, y) = f_y(x, y) = 0$ gives the critical points $(0, 0)$ and $(\frac{1}{3}, \frac{1}{3})$.

At $(0, 0)$,

$$f_{xx}(x, y) = 6x = 0, \qquad f_{xy}(x, y) = -1, \qquad f_{yy}(x, y) = 6y = 0,$$

and

$$\Delta = 0(0) - (-1)^2 = -1.$$

Since $\Delta < 0$, there is no relative extremum at (0, 0).

At $(\frac{1}{3}, \frac{1}{3})$,

$$f_{xx}(x, y) = 6\left(\tfrac{1}{3}\right) = 2, \qquad f_{xy}(x, y) = -1, \qquad f_{yy}(x, y) = 6\left(\tfrac{1}{3}\right) = 2,$$

and

$$\Delta = 2(2) - (-1)^2 = 3.$$

Since $\Delta > 0$ and $f_{xx}(\frac{1}{3}, \frac{1}{3}) > 0$, there is a relative minimum at $(\frac{1}{3}, \frac{1}{3})$. At this point the value of the given function is

$$f\left(\tfrac{1}{3}, \tfrac{1}{3}\right) = \left(\tfrac{1}{3}\right)^3 + \left(\tfrac{1}{3}\right)^3 - \left(\tfrac{1}{3}\right)\left(\tfrac{1}{3}\right) = -\tfrac{1}{27}.$$

EXAMPLE 4

Examine $f(x, y) = y^2 - x^2$ for relative extrema.

Solving

$$f_x(x, y) = -2x = 0 \quad \text{and} \quad f_y(x, y) = 2y = 0,$$

we get the critical point (0, 0). Moreover, at (0, 0), and indeed at any point,

$$f_{xx}(x, y) = -2, \qquad f_{yy}(x, y) = 2, \qquad f_{xy}(x, y) = 0.$$

Hence $\Delta = (-2)(2) - (0)^2 = -4 < 0$ and no relative extrema exist. A sketch of $z = f(x, y) = y^2 - x^2$ appears in Fig. 8-14. Note that for the surface curve cut by the plane $y = 0$, there is a *maximum* at (0, 0); but for the surface curve cut by the plane $x = 0$, there is a *minimum* at (0, 0). Thus, on the *surface* no relative extremum can exist at the origin, although (0, 0) is a critical point. Around the origin the curve is saddle-shaped and (0, 0) is called a *saddle-point* of *f*.

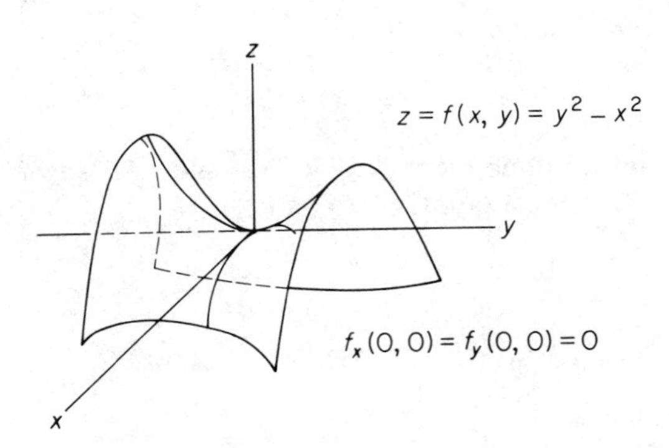

FIG. 8-14

EXAMPLE 5

Examine $f(x, y) = x^4 + (x - y)^4$ for relative extrema.

If we set

$$f_x(x, y) = 4x^3 + 4(x - y)^3 = 0, \tag{4}$$

$$f_y(x, y) = -4(x - y)^3 = 0, \tag{5}$$

then from Eq. (5) we have $x - y = 0$ or $x = y$. Substitution into Eq. (4) gives $4x^3 = 0$ or $x = 0$. Thus, $x = y = 0$ and $(0, 0)$ is the only critical point. But at $(0, 0)$, $f_{xx}(x, y) = 12x^2 + 12(x - y)^2 = 0$, $f_{yy}(x, y) = 12(x - y)^2 = 0$, and $f_{xy}(x, y) = -12(x - y)^2 = 0$. Hence $\Delta = 0$ and the second-derivative test fails. However, for all $(x, y) \neq (0, 0)$ we have $f(x, y) > 0$ while $f(0, 0) = 0$. Hence at $(0, 0)$ the graph of f has a low point and we conclude that f has a relative (and absolute) minimum at $(0, 0)$.

In many situations involving functions of two variables, and especially in their applications, the nature of the given problem is an indicator of whether a critical point is in fact a relative (or absolute) maximum or a relative (or absolute) minimum. In such cases the second-derivative test is not needed. Often, in mathematical studies of applied areas the appropriate second-order conditions are assumed to hold.

EXAMPLE 6

Let P be a production function given by

$$P = f(l, k) = .54l^2 - .02l^3 + 1.89k^2 - .09k^3,$$

where l and k are the amounts of labor and capital, respectively, and P is the quantity of output produced. Find the values of l and k that maximize P.

$$P_l = 1.08l - .06l^2 \qquad P_k = 3.78k - .27k^2$$

$$= .06l(18 - l) = 0. \qquad = .27k(14 - k) = 0.$$

$$l = 0, l = 18. \qquad k = 0, k = 14.$$

The critical points are $(0, 0)$, $(0, 14)$, $(18, 0)$ and $(18, 14)$.

At $(0, 0)$,

$$P_{ll} = 1.08 - .12l = 1.08, \quad P_{lk} = 0, \quad P_{kk} = 3.78 - .54k = 3.78,$$

$$\Delta = 1.08(3.78) - 0^2 > 0.$$

Since $\Delta > 0$ and $P_{ll} > 0$, there is a relative minimum at (0, 0).

At (0, 14),

$$P_{ll} = 1.08, \qquad P_{lk} = 0, \qquad P_{kk} = -3.78,$$
$$\Delta = 1.08(-3.78) - 0^2 < 0.$$

Since $\Delta < 0$, there is no relative extremum at (0, 14).

At (18, 0),

$$P_{ll} = -1.08, \qquad P_{lk} = 0, \qquad P_{kk} = 3.78,$$
$$\Delta = (-1.08)(3.78) - 0^2 < 0.$$

Since $\Delta < 0$, there is no relative extremum at (18, 0).

At (18, 14),

$$P_{ll} = -1.08, \qquad P_{lk} = 0, \qquad P_{kk} = -3.78,$$
$$\Delta = (-1.08)(-3.78) - 0^2 > 0.$$

Since $\Delta > 0$ and $P_{ll} < 0$, there is a relative maximum at (18, 14). The maximum output is obtained when $l = 18$ and $k = 14$.

EXAMPLE 7

A food manufacturer produces two types of candy, A and B, for which the average costs of production are constant at 70 and 80 cents per pound, respectively. The quantities q_A, q_B (in pounds) of A and B which can be sold each week are given by the joint-demand functions

$$q_A = 240(p_B - p_A)$$

and

$$q_B = 240(150 + p_A - 2p_B),$$

where p_A and p_B are the selling prices (in cents per pound) of A and B, respectively. Determine the selling prices that will maximize the manufacturer's profit P.

For A and B the profits per pound are $(p_A - 70)$ and $(p_B - 80)$, respectively. Hence, total profit P is

$$\begin{aligned} P &= (p_A - 70)q_A + (p_B - 80)q_B \\ &= (p_A - 70)[240(p_B - p_A)] + (p_B - 80)[240(150 + p_A - 2p_B)]. \end{aligned}$$

To maximize P, we set its partial derivatives equal to 0:

$$\frac{\partial P}{\partial p_A} = (p_A - 70)[240(-1)] + 240(p_B - p_A)(1) + (p_B - 80)(240)(1) = 0,$$
$$\frac{\partial P}{\partial p_B} = (p_A - 70)[240(1)] + (p_B - 80)[240(-2)] +$$
$$[240(150 + p_A - 2p_B)](1) = 0.$$

Simplifying the two preceding equations gives

$$\begin{cases} p_B - p_A - 5 = 0, \\ -2p_B + p_A + 120 = 0, \end{cases}$$

whose solution is $p_A = 110$ and $p_B = 115$ (both in cents). Moreover,

$$\frac{\partial^2 P}{\partial p_A^2} = -480 < 0, \qquad \frac{\partial^2 P}{\partial p_B^2} = -960, \qquad \frac{\partial^2 P}{\partial p_B \partial p_A} = 480.$$

Thus $\Delta = (-480)(-960) - (480)^2 > 0$ and we indeed have a maximum. The manufacturer should sell candy A at \$1.10 per pound and B at \$1.15 per pound.

Exercise 8-7

In Problems **1–6**, *find the critical points of the functions.*

1. $f(x, y) = x^2 + y^2 - 5x + 4y + xy.$

2. $f(x, y) = x^2 + 4y^2 - 6x + 16y.$

3. $f(x, y) = 2x^3 + y^3 - 3x^2 + 1.5y^2 - 12x - 90y.$

4. $f(x, y) = xy - \frac{1}{x} - \frac{1}{y}.$

5. $f(x, y, z) = 2x^2 + xy + y^2 + 100 - z(x + y - 200).$

6. $f(x, y, z, w) = x^2 + y^2 + z^2 - w(x - y + 2z - 6).$

In Problems **7–18**, *find the critical points of the functions. Determine, by the second-derivative test, whether these points correspond to a relative maximum, a relative minimum, neither, or whether the test fails.*

7. $f(x, y) = x^2 + 3y^2 + 4x - 9y + 3.$

8. $f(x, y) = -2x^2 + 8x - 3y^2 + 24y + 7.$

9. $f(x, y) = y - y^2 - 3x - 6x^2.$

10. $f(x, y) = x^2 + y^2 + xy - 9x + 1.$

11. $f(x, y) = x^3 - 3xy + y^2 + y - 5.$

12. $f(x, y) = \frac{x^3}{3} + y^2 - 2x + 2y - 2xy.$

13. $f(x, y) = \frac{1}{3}(x^3 + 8y^3) - 2(x^2 + y^2) + 1.$

14. $f(x, y) = x^2 + y^2 - xy + x^3.$

15. $f(l, k) = 2lk - l^2 + 264k - 10l - 2k^2$.

16. $f(l, k) = l^3 + k^3 - 3lk$.

17. $f(p, q) = pq - \frac{1}{p} - \frac{1}{q}$.

18. $f(x, y) = (x - 3)(y - 3)(x + y - 3)$.

In Problems **19–26**, *unless otherwise indicated the variables p_A and p_B denote selling prices of products A and B, respectively. Similarly, q_A and q_B denote quantities of A and B which are produced and sold during some time period. In all cases, the variables employed will be assumed to be units of output, input, money, etc.*

19. Suppose $P = f(l, k) = 1.08l^2 - .03l^3 + 1.68k^2 - .08k^3$ is a production function for a firm. Find the quantities of input, l and k, so as to maximize output P.

20. In a certain automated manufacturing process, machines M and N are utilized for m and n hours, respectively. If daily output Q is a function of m and n, namely $Q = 4.5m + 5n - .5m^2 - n^2 - .25mn$, find the values of m and n which will maximize Q.

21. A candy company produces two varieties of candy, A and B, for which the constant average costs of production are 60 and 70 (cents per pound), respectively. The demand functions for A and B are respectively given by $q_A = 5(p_B - p_A)$ and $q_B = 500 + 5(p_A - 2p_B)$. Find the selling prices, p_A and p_B, which would maximize the company's profit.

22. Repeat Problem 21 if the constant costs of production of A and B are a and b (cents per pound), respectively.

23. Suppose a monopolist is practicing price discrimination in the sale of his product by charging different prices in two separate markets. In market A the demand function is $p_A = 100 - q_A$ and in B it is $p_B = 84 - q_B$, where q_A and q_B are the quantities sold per week in A and B, and p_A and p_B are the respective prices per unit. If the monopolist's cost function is $c = 600 + 4(q_A + q_B)$, how much should be sold in each market to maximize profit? What selling prices would give this maximum profit? Find the maximum profit.

24. A monopolist sells two competitive products, A and B, for which the demand functions are $q_A = 1 - 2p_A + 4p_B$ and $q_B = 11 + 2p_A - 6p_B$. If the constant average cost of producing a unit of A is 4 and for B it is 1, how many units of A and B should be sold to maximize the monopolist's profit?

25. For products A and B, the joint-cost function for a manufacturer is $c = 1.5q_A^2 + 4.5q_B^2$ and the demand functions are $p_A = 36 - q_A^2$ and $p_B = 30 - q_B^2$. Find the level of production which will maximize profit.

26. For a monopolist's products, A and B, the joint-cost function is $c = (q_A + q_B)^2$ and the demand functions are $q_A = 26 - p_A$ and $q_B = 10 - .25p_B$. Find the values of p_A and p_B which will maximize profit. What are the quantities of A and B which correspond to these prices? What is the total profit?

27. An open-top rectangular box is to have a volume of 6 cubic feet. The cost per square foot of materials is \$3 for the bottom, \$1 for the front and back, and \$0.50 for the other two sides. Find the dimensions of the box so that the cost of materials is minimized. See Fig. 8-15.

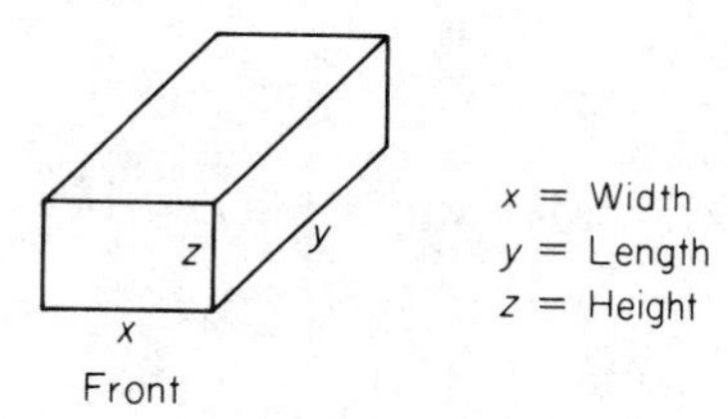

FIG. 8-15

28. Suppose A and B are the only two firms in the market selling the same product (we say they are *duopolists*). The industry demand function for the product is $p = 92 - q_A - q_B$, where q_A and q_B denote the output produced and sold by A and B, respectively. For A the cost function is $c_A = 10q_A$ and for B it is $c_B = .5q_B^2$. Suppose the firms decide to enter into an agreement on output and price control by jointly acting as a monopoly. In this case we say they enter into *collusion*. Show that the profit function for the monopoly is given by

$$P = pq_A - c_A + pq_B - c_B.$$

Express P as a function of q_A and q_B and determine how output should be allocated so as to maximize the profit of the monopoly.

29. Suppose $f(x, y) = -2x^2 + 5y^2 + 7$, where x and y must satisfy the equation $3x - 2y = 7$. Find the relative extrema of f subject to the given conditions on x and y by first solving the second equation for y. Substitute the result for y in the given equation. Thus f is expressed as a function of one variable for which extrema may be found in the usual way.

30. Repeat Problem 29 if $f(x, y) = x^2 + 4y^2 + 6$ subject to the condition that $2x - 8y = 20$.

8-8 LAGRANGE MULTIPLIERS

We shall now find relative maxima and minima for a function on which certain *constraints* are imposed. Such a situation could arise if a manufacturer wished to minimize the total cost of factors of input and yet obtain a particular level of output.

Suppose we find the relative extrema of

$$w = x^2 + y^2 + z^2, \tag{1}$$

subject to the constraint that x, y, and z must satisfy

$$x - y + 2z = 6. \tag{2}$$

Solving Eq. (2) for x, we get

$$x = y - 2z + 6, \tag{3}$$

which when substituted for x in Eq. (1) gives

$$w = (y - 2z + 6)^2 + y^2 + z^2. \tag{4}$$

Since w in Eq. (4) is expressed as a function of two variables, to find relative extrema we follow the usual procedure of setting its partial derivatives equal to 0:

$$\frac{\partial w}{\partial y} = 2(y - 2z + 6) + 2y = 4y - 4z + 12 = 0, \tag{5}$$

$$\frac{\partial w}{\partial z} = -4(y - 2z + 6) + 2z = -4y + 10z - 24 = 0. \tag{6}$$

Solving Eqs. (5) and (6) simultaneously gives $y = -1$ and $z = 2$. Substituting in Eq. (3), we get $x = 1$. Hence, the only critical point of (1) subject to constraint (2) is $(1, -1, 2)$. Evaluating the second-order derivatives of (4) when $y = -1$ and $z = 2$ gives

$$\frac{\partial^2 w}{\partial y^2} = 4, \qquad \frac{\partial^2 w}{\partial z^2} = 10, \qquad \frac{\partial^2 w}{\partial z \partial y} = -4,$$

$$\Delta = 4(10) - (-4)^2 = 24 > 0.$$

Thus w, subject to the constraint, has a relative minimum at $(1, -1, 2)$.

This solution was found by using the constraint to express one of the variables in the original function in terms of the other variables. Often this is not practical, but there is another technique, called the method of **Lagrange multipliers**†, which avoids this step and yet allows us to obtain critical points.

The method is as follows. Suppose we have a function $f(x, y, z)$ subject to the constraint $g(x, y, z) = 0$. We construct a new function F of four variables defined by the following (where λ is a Greek letter read "lambda"):

$$F(x, y, z, \lambda) = f(x, y, z) - \lambda g(x, y, z).$$

It can be shown that if (x_0, y_0, z_0) is a critical point of f subject to the constraint $g(x, y, z) = 0$, there exists a value of λ, say λ_0, such that $(x_0, y_0, z_0, \lambda_0)$ is a critical point of F. Also, if $(x_0, y_0, z_0, \lambda_0)$ is a critical point of F, then (x_0, y_0, z_0) is a critical point of f subject to the constraint. Thus to find critical points of f subject to $g(x, y, z) = 0$, we instead find critical points of F. These are obtained by solving the simultaneous equations

$$\begin{cases} F_x(x, y, z, \lambda) = 0, \\ F_y(x, y, z, \lambda) = 0, \\ F_z(x, y, z, \lambda) = 0, \\ F_\lambda(x, y, z, \lambda) = 0. \end{cases}$$

At times ingenuity must be used to do this. Once we obtain a critical point $(x_0, y_0, z_0, \lambda_0)$ of F, we can conclude that (x_0, y_0, z_0) is a critical point of f subject to the constraint.

Let us illustrate for the original situation:

$$f(x, y, z) = x^2 + y^2 + z^2 \quad \text{subject to} \quad x - y + 2z = 6.$$

First, we write the constraint as $g(x, y, z) = x - y + 2z - 6 = 0$. Second, we form the function

$$\begin{aligned} F(x, y, z, \lambda) &= f(x, y, z) - \lambda g(x, y, z) \\ &= x^2 + y^2 + z^2 - \lambda(x - y + 2z - 6). \end{aligned}$$

Next we set each partial derivative of F equal to 0. For convenience, we

†After the French mathematician, Joseph-Louis Lagrange (1736–1813).

shall write $F_x(x, y, z, \lambda)$ as F_x, etc.

$$\begin{cases} F_x = 2x - \lambda = 0, & (7) \\ F_y = 2y + \lambda = 0, & (8) \\ F_z = 2z - 2\lambda = 0, & (9) \\ F_\lambda = -x + y - 2z + 6 = 0. & (10) \end{cases}$$

From Eqs. (7)–(9) we see immediately that

$$x = \frac{\lambda}{2}, \qquad y = -\frac{\lambda}{2}, \quad \text{and} \quad z = \lambda. \tag{11}$$

Substituting these values in Eq. (10), we obtain

$$-\frac{\lambda}{2} - \frac{\lambda}{2} - 2\lambda + 6 = 0,$$

$$\lambda = 2.$$

Thus from Eq. (11), $x = 1, y = -1$, and $z = 2$. Hence the only critical point of f subject to the constraint is $(1, -1, 2)$ at which there may exist a relative maximum, a relative minimum, or neither of these. The method of Lagrange multipliers does not directly indicate which of these possibilities occur, although from our previous work we saw that it is indeed a relative minimum. In applied problems, the nature of the problem itself may give a clue as to how a critical point is to be regarded. Often the existence of either a relative minimum or a relative maximum is assumed and a critical point is treated accordingly. Actually, sufficient second-order conditions for relative extrema are available, but we shall not consider them.

EXAMPLE 1

Find the critical points for $z = f(x, y) = 3x - y + 6$ subject to the constraint $x^2 + y^2 = 4$.

We write the constraint as $g(x, y) = x^2 + y^2 - 4 = 0$ and construct the function

$$F(x, y, \lambda) = f(x, y) - \lambda g(x, y) = 3x - y + 6 - \lambda(x^2 + y^2 - 4).$$

Setting $F_x = F_y = F_\lambda = 0$, we have

$$\begin{cases} 3 - 2x\lambda = 0, & (12) \\ -1 - 2y\lambda = 0, & (13) \\ -x^2 - y^2 + 4 = 0. & (14) \end{cases}$$

From Eqs. (12) and (13),

$$x = \frac{3}{2\lambda} \quad \text{and} \quad y = -\frac{1}{2\lambda}.$$

Substituting in Eq. (14), we obtain

$$-\frac{9}{4\lambda^2} - \frac{1}{4\lambda^2} + 4 = 0,$$

$$\lambda = \pm\frac{\sqrt{10}}{4}.$$

If $\lambda = \sqrt{10}/4$,

$$x = \frac{3}{2\left(\frac{\sqrt{10}}{4}\right)} = \frac{3\sqrt{10}}{5}, \qquad y = -\frac{1}{2\left(\frac{\sqrt{10}}{4}\right)} = -\frac{\sqrt{10}}{5}.$$

Similarly, if $\lambda = -\sqrt{10}/4$,

$$x = -\frac{3\sqrt{10}}{5}, \qquad y = \frac{\sqrt{10}}{5}.$$

Thus the critical points of $f(x, y)$ subject to the constraint are $(3\sqrt{10}/5, -\sqrt{10}/5)$ and $(-3\sqrt{10}/5, \sqrt{10}/5)$. Note that the values of λ do not appear in the answer. They are simply a means to obtain it.

EXAMPLE 2

Find critical points for $f(x, y, z) = xyz$, where $xyz \neq 0$, subject to the constraint $x + 2y + 3z = 36$.

Set

$$F(x, y, z, \lambda) = xyz - \lambda(x + 2y + 3z - 36).$$

Then

$$\begin{cases} F_x = yz - \lambda = 0, \\ F_y = xz - 2\lambda = 0, \\ F_z = xy - 3\lambda = 0, \\ F_\lambda = -x - 2y - 3z + 36 = 0. \end{cases}$$

We can write the system as

$$\begin{cases} yz = \lambda, & (15) \\ xz = 2\lambda, & (16) \\ xy = 3\lambda, & (17) \\ x + 2y + 3z - 36 = 0. & (18) \end{cases}$$

Dividing each side of Eq. (15) by the corresponding side of Eq. (16), we get

$$\frac{yz}{xz} = \frac{\lambda}{2\lambda} \quad \text{or} \quad y = \frac{x}{2}.$$

This division is valid since $xyz \neq 0$. Similarly, from Eqs. (15) and (17) we get

$$z = \frac{x}{3}.$$

Substituting into Eq. (18) gives

$$x + 2\left(\frac{x}{2}\right) + 3\left(\frac{x}{3}\right) - 36 = 0,$$

and so

$$x = 12.$$

Thus $y = 6$ and $z = 4$. Hence (12, 6, 4) is the only critical point satisfying the given conditions.

EXAMPLE 3

Suppose a firm has an order for 200 units of its product and wishes to distribute their manufacture between two of its plants, Plant 1 and Plant 2. Let q_1 and q_2 denote the outputs of Plants 1 and 2, respectively, and suppose the total cost function is given by $c = f(q_1, q_2) = 2q_1^2 + q_1q_2 + q_2^2 + 200$. How should the output be distributed in order to minimize costs?

We must minimize $c = f(q_1, q_2)$ subject to the constraint $q_1 + q_2 = 200$.

$$F(q_1, q_2, \lambda) = 2q_1^2 + q_1q_2 + q_2^2 + 200 - \lambda(q_1 + q_2 - 200).$$

$$\frac{\partial F}{\partial q_1} = 4q_1 + q_2 - \lambda = 0, \quad (19)$$

$$\frac{\partial F}{\partial q_2} = q_1 + 2q_2 - \lambda = 0, \quad (20)$$

$$\frac{\partial F}{\partial \lambda} = -q_1 - q_2 + 200 = 0. \quad (21)$$

Solving Eqs. (19) and (20) simultaneously gives

$$q_1 = \frac{\lambda}{7}, \qquad q_2 = \frac{3\lambda}{7}.$$

Substituting these values in Eq. (21) gives $\lambda = 350$. Thus $q_1 = 50$ and $q_2 = 150$. Plant 1 should produce 50 units and Plant 2, 150 units, in order to minimize costs.

An interesting observation can be made concerning Example 3. From Eq. (19), $\lambda = 4q_1 + q_2 = \partial c/\partial q_1$, the marginal cost of Plant 1. From Eq. (20), $\lambda = q_1 + 2q_2 = \partial c/\partial q_2$, the marginal cost of Plant 2. Hence, $\partial c/\partial q_1 = \partial c/\partial q_2$ and we conclude that to minimize cost it is necessary that the marginal costs of each plant be equal to each other.

EXAMPLE 4

Suppose a firm must produce a given quantity P_0 of output in the cheapest possible manner. If there are two input factors l and k, and their prices per unit are fixed at p_l and p_k respectively, discuss the economic significance of combining input to achieve least cost. That is, describe the least-cost input combination.

Let $P = f(l, k)$ be the production function. Then we must minimize the cost function

$$c = lp_l + kp_k,$$

subject to

$$P_0 = f(l, k).$$

We construct

$$F(l, k, \lambda) = lp_l + kp_k - \lambda[f(l, k) - P_0].$$

We have

$$\frac{\partial F}{\partial l} = p_l - \lambda\frac{\partial}{\partial l}[f(l, k)] = 0, \tag{22}$$

$$\frac{\partial F}{\partial k} = p_k - \lambda\frac{\partial}{\partial k}[f(l, k)] = 0, \tag{23}$$

$$\frac{\partial F}{\partial \lambda} = -f(l, k) + P_0 = 0.$$

From Eqs. (22) and (23),

$$\lambda = \frac{p_l}{\frac{\partial}{\partial l}[f(l, k)]} = \frac{p_k}{\frac{\partial}{\partial k}[f(l, k)]}. \tag{24}$$

Hence,

$$\frac{p_l}{p_k} = \frac{\frac{\partial}{\partial l}[f(l, k)]}{\frac{\partial}{\partial k}[f(l, k)]}.$$

We conclude that when the least-cost combination of factors is used, the ratio of the marginal products of the input factors must be equal to the ratio of their corresponding prices.

The method of Lagrange multipliers is by no means restricted to problems involving a single constraint. For example, suppose $f(x, y, z, w)$ were subject to constraints $g_1(x, y, z, w) = 0$ and $g_2(x, y, z, w) = 0$. Then there would be two Lagrange multipliers, λ_1 and λ_2 (one for each constraint), and we would construct the function $F = f - \lambda_1 g_1 - \lambda_2 g_2$. We would then solve the system $F_x = F_y = F_z = F_w = F_{\lambda_1} = F_{\lambda_2} = 0$.

EXAMPLE 5

Find critical points for $f(x, y, z) = xy + yz$ subject to the constraints $x^2 + y^2 = 8$ and $yz = 8$.

Set

$$F(x, y, z, \lambda_1, \lambda_2) = xy + yz - \lambda_1(x^2 + y^2 - 8) - \lambda_2(yz - 8).$$

Then

$$\begin{cases} F_x = y - 2x\lambda_1 = 0, \\ F_y = x + z - 2y\lambda_1 - z\lambda_2 = 0, \\ F_z = y - y\lambda_2 = 0, \\ F_{\lambda_1} = -x^2 - y^2 + 8 = 0, \\ F_{\lambda_2} = -yz + 8 = 0. \end{cases}$$

We can write the system as

$$\begin{cases} \dfrac{y}{2x} = \lambda_1, & (25) \\ x + z - 2y\lambda_1 - z\lambda_2 = 0, & (26) \\ \lambda_2 = 1, & (27) \\ x^2 + y^2 = 8, & (28) \\ z = \dfrac{8}{y}. & (29) \end{cases}$$

Substituting $\lambda_2 = 1$ from Eq. (27) into Eq. (26) and simplifying, we have $x - 2y\lambda_1 = 0$, and so

$$\lambda_1 = \frac{x}{2y}.$$

Substituting into Eq. (25) gives

$$\frac{y}{2x} = \frac{x}{2y},$$

$$y^2 = x^2. \tag{30}$$

Substituting into Eq. (28) gives $x^2 + x^2 = 8$ from which $x = \pm 2$. If $x = 2$, then from Eq. (30) we have $y = \pm 2$. Similarly, if $x = -2$, then $y = \pm 2$. Thus if $x = 2$ and $y = 2$, then from Eq. (29) we obtain $z = 4$. Continuing in this manner we obtain four critical points:

$$(2, 2, 4), (2, -2, -4), (-2, 2, 4), (-2, -2, -4).$$

Exercise 8-8

In Problems **1–12** *find, by the method of Lagrange multipliers, the critical points of the functions subject to the given constraints.*

1. $f(x, y) = x^2 + 4y^2 + 6; \quad 2x - 8y = 20.$

2. $f(x, y) = -2x^2 + 5y^2 + 7; \quad 3x - 2y = 7.$

3. $f(x, y, z) = x^2 + y^2 + z^2; \quad 2x + y - z = 9.$

4. $f(x, y, z) = x + y + z; \quad xyz = 27.$

5. $f(x, y, z) = x^2 + xy + 2y^2 + z^2; \quad x - 3y - 4z = 16.$

6. $f(x, y, z) = xyz^2; \quad x - y + z = 20, \quad (xyz^2 \neq 0).$

7. $f(x, y, z) = xyz; \quad x + 2y + 3z = 18, \quad (xyz \neq 0).$

8. $f(x, y, z) = x^2 + y^2 + z^2; \quad x + y + z = 1.$

9. $f(x, y, z) = x^2 + 2y - z^2; \quad 2x - y = 0, y + z = 0.$

10. $f(x, y, z) = x^2 + y^2 + z^2; \quad x + y + z = 1, x - y + z = 1.$

11. $f(x, y, z) = xyz; \quad x + y + z = 12, x + y - z = 0, \quad (xyz \neq 0).$

12. $f(x, y, z, w) = 2x^2 + 2y^2 + 3z^2 - 4w^2; \quad 4x - 8y + 6z + 16w = 6.$

13. To fill an order for 100 units of its product, a firm wishes to distribute the production between its two plants, Plant 1 and Plant 2. The total cost function is given by $c = f(q_1, q_2) = .1q_1^2 + 7q_1 + 15q_2 + 1000$, where q_1 and q_2 are the number of units produced at Plants 1 and 2, respectively. How should the output be distributed in order to minimize costs?

14. Repeat Problem 13 if the cost function is $c = 3q_1^2 + q_1q_2 + 2q_2^2$ and a total of 200 units are to be produced.

15. The production function for a firm is $f(l, k) = 12l + 20k - l^2 - 2k^2$. The cost to the firm of l and k is 4 and 8 per unit, respectively. If the firm wants the total cost of input to be 88, find the greatest output possible subject to this budget constraint.

16. Repeat Problem 15 given that $f(l, k) = 60l + 30k - 2l^2 - 3k^2$ and the budget constraint is $2l + 3k = 30$.

Problems **17–20** *refer to the following definition. A* ***utility function*** *is a function which attaches a measure to the satisfaction or utility a consumer gets from the consumption of products per unit of time. Suppose* $U = f(x, y)$ *is such a function, where* x *and* y *are the amounts of two products,* X *and* Y. *The* ***marginal utility*** *of* X *is* $\partial U/\partial x$ *and approximately represents the change in total utility resulting from a one-unit change in consumption of product* X *per unit of time. We define the marginal utility of* Y *in similar fashion. If the prices of* X *and* Y *are* p_x *and* p_y, *respectively, and the consumer has an income or budget of* I *to spend on these products, then his budget constraint is* $xp_x + yp_y = I$. *In the following problems you are asked to find the quantities of each product which the consumer should buy, subject to his budget, which will allow him to maximize his satisfaction. That is, you are to maximize* $U = f(x, y)$ *subject to* $xp_x + yp_y = I$. *Assume that such a maximum exists.*

17. $U = x^3y^3$; $p_x = 2, p_y = 3, I = 48$, $(x^3y^3 \neq 0)$.

18. $U = 46x - (5x^2/2) + 34y - 2y^2$; $p_x = 5, p_y = 2, I = 30$.

19. $U = f(x, y, z) = xyz$; $p_x = 2, p_y = 1, p_z = 4, I = 60$, $(xyz \neq 0)$.

20. Let $U = f(x, y)$ be a utility function subject to the budget constraint $xp_x + yp_y = I$ where p_x, p_y, and I are constant. Show that to maximize satisfaction it is necessary that

$$\lambda = \frac{f_x(x, y)}{p_x} = \frac{f_y(x, y)}{p_y},$$

where $f_x(x, y)$ and $f_y(x, y)$ are the marginal utilities of X and Y, respectively. Deduce that $f_x(x, y)/p_x$ is the marginal utility of one dollar's worth of X. Hence, maximum satisfaction is obtained when the consumer allocates his budget so that the marginal utility of a dollar's worth of X is equal to the marginal utility per dollar's worth of Y. Performing the same procedure as above, verify that this is true for $U = f(x, y, z, w)$ subject to the corresponding budget equation. In each case, λ is called the *marginal utility of income*.

8-9 LINES OF REGRESSION

To study the influence of advertising on sales, a firm compiled the data in Table 8-4. The variable x denotes advertising expenditures in

TABLE 8-4

expenditures (x)	2	3	4.5	5.5	7
revenue (y)	3	6	8	10	11

hundreds of dollars, and the variable y denotes the resulting sales revenue in thousands of dollars. If each pair (x, y) of data is plotted, the result is called a *scatter diagram* [Fig. 8-16(a)].

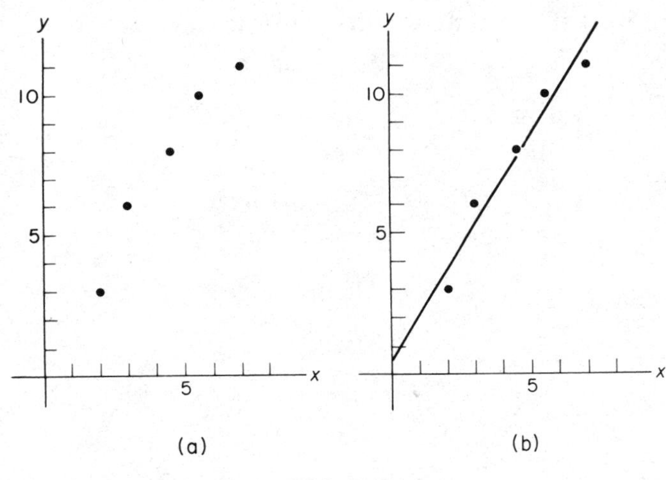

FIG. 8-16

From an observation of the distribution of the points, it is reasonable to assume that a relationship exists between x and y and that it is approximately linear. On this basis we may fit a straight line "by eye" to the data [Fig. 8-16(b)], and from this line predict a value of y for a given value of x. This line seems consistent with the trend of the data, although other lines could be drawn as well. Unfortunately, determining a line "by eye" is not very objective. We want to apply criteria in specifying what we shall call a line of "best fit." A frequently used technique is called the **method of least squares.**

To apply the method of least squares to the data in Table 8-4, we first assume that x and y are approximately linearly related and that we can fit a straight line

$$\hat{y} = \hat{a} + \hat{b}x \tag{1}$$

to the given points by a suitable objective choice of the constants $\hat{a}$ and $\hat{b}$ (read "a hat" and "b hat," respectively). For a given value of x in Eq.

(1), $\hat{y}$ is the corresponding predicted value of y, and $(x, \hat{y})$ will be on the line. Our aim is that $\hat{y}$ be near y.

When $x = 2$, the observed value of y is 3. Our predicted value of y is obtained by substituting $x = 2$ in Eq. (1), which gives $\hat{y} = \hat{a} + 2\hat{b}$. The error of estimation, or vertical deviation of the point (2, 3) from the line, is $\hat{y} - y$, or

$$\hat{a} + 2\hat{b} - 3.$$

This vertical deviation is indicated (although exaggerated for clarity) in Fig. 8-17. Similarly, the vertical deviation of (3, 6) from the line is

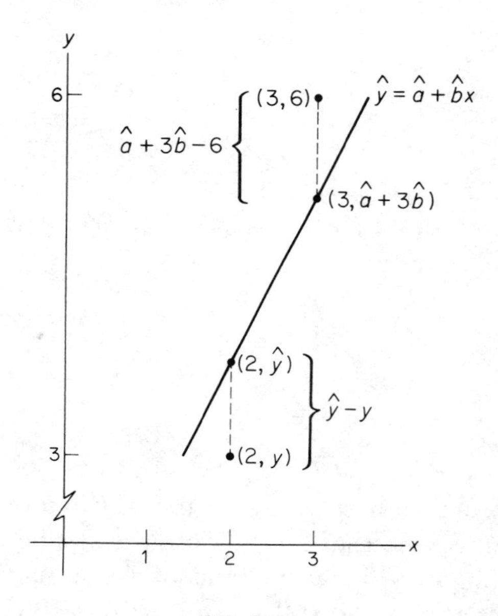

FIG. 8-17

$\hat{a} + 3\hat{b} - 6$, as is also illustrated. To avoid possible difficulties associated with positive and negative deviations, we shall consider the squares of the deviations and shall form the sum S of all such squares for the given data.

$$S = (\hat{a} + 2\hat{b} - 3)^2 + (\hat{a} + 3\hat{b} - 6)^2 + (\hat{a} + 4.5\hat{b} - 8)^2 + (\hat{a} + 5.5\hat{b} - 10)^2 + (\hat{a} + 7\hat{b} - 11)^2.$$

The method of least squares requires that we choose as the line of "best fit" the one obtained by selecting $\hat{a}$ and $\hat{b}$ so as to minimize S. We can minimize S with respect to $\hat{a}$ and $\hat{b}$ by solving the system

$$\begin{cases} \dfrac{\partial S}{\partial \hat{a}} = 0, \\ \dfrac{\partial S}{\partial \hat{b}} = 0. \end{cases}$$

We have

$$\frac{\partial S}{\partial \hat{a}} = 2(\hat{a} + 2\hat{b} - 3) + 2(\hat{a} + 3\hat{b} - 6) + 2(\hat{a} + 4.5\hat{b} - 8) + 2(\hat{a} + 5.5\hat{b} - 10) + 2(\hat{a} + 7\hat{b} - 11) = 0,$$

$$\frac{\partial S}{\partial \hat{b}} = 4(\hat{a} + 2\hat{b} - 3) + 6(\hat{a} + 3\hat{b} - 6) + 9(\hat{a} + 4.5\hat{b} - 8) + 11(\hat{a} + 5.5\hat{b} - 10) + 14(\hat{a} + 7\hat{b} - 11) = 0,$$

which when simplified gives

$$\begin{cases} 5\hat{a} + 22\hat{b} = 38, \\ 44\hat{a} + 225\hat{b} = 384. \end{cases}$$

Solving for $\hat{a}$ and $\hat{b}$, we obtain

$$\hat{a} = \tfrac{102}{157} \approx .65, \qquad \hat{b} = \tfrac{248}{157} \approx 1.58.$$

It can be shown that the values of $\hat{a}$ and $\hat{b}$ obtained this way always lead to a minimum value of S. Hence, in the sense of least squares, the line of best fit is

$$\hat{y} = .65 + 1.58x. \tag{2}$$

This is, in fact, the line indicated in Fig. 8-16(b). It is called the ***least squares line of y on x*** or the ***linear regression line of y on x***. The constants $\hat{a}$ and $\hat{b}$ are called ***linear regression coefficients***. With Eq. (2) we would predict that when $x = 5$, the corresponding value of y is $\hat{y} = .65 + 1.58(5) = 8.55$.

More generally, suppose we are given the following n pairs of observations:

$$(x_1, y_1), (x_2, y_2), \ldots, (x_n, y_n).$$

If we assume that x and y are approximately linearly related and that we can fit a straight line $\hat{y} = \hat{a} + \hat{b}x$ to the data, the sum of the squares of the errors $\hat{y} - y$ is

$$S = (\hat{a} + \hat{b}x_1 - y_1)^2 + (\hat{a} + \hat{b}x_2 - y_2)^2 + \ldots + (\hat{a} + \hat{b}x_n - y_n)^2.$$

Since S must be minimized with respect to $\hat{a}$ and $\hat{b}$,

$$\begin{cases} \dfrac{\partial S}{\partial \hat{a}} = 2(\hat{a} + \hat{b}x_1 - y_1) + 2(\hat{a} + \hat{b}x_2 - y_2) + \ldots + 2(\hat{a} + \hat{b}x_n - y_n) = 0, \\ \dfrac{\partial S}{\partial \hat{b}} = 2x_1(\hat{a} + \hat{b}x_1 - y_1) + 2x_2(\hat{a} + \hat{b}x_2 - y_2) + \ldots \\ \qquad\qquad + 2x_n(\hat{a} + \hat{b}x_n - y_n) = 0. \end{cases}$$

Dividing both equations by 2 and using sigma notation, we have

$$\begin{cases} \hat{a}n + \hat{b}\sum_{i=1}^{n} x_i - \sum_{i=1}^{n} y_i = 0, \\ \hat{a}\sum_{i=1}^{n} x_i + \hat{b}\sum_{i=1}^{n} x_i^2 - \sum_{i=1}^{n} x_i y_i = 0. \end{cases}$$

Equivalently, we have the system of so-called *normal equations*:

$$\begin{cases} \sum_{i=1}^{n} y_i = \hat{a}n + \hat{b}\sum_{i=1}^{n} x_i, & (3) \\ \sum_{i=1}^{n} x_i y_i = \hat{a}\sum_{i=1}^{n} x_i + \hat{b}\sum_{i=1}^{n} x_i^2. & (4) \end{cases}$$

To solve for $\hat{b}$ we first multiply Eq. (3) by $\sum_{i=1}^{n} x_i$ and Eq. (4) by n:

$$\begin{cases} \left(\sum_{i=1}^{n} x_i\right)\left(\sum_{i=1}^{n} y_i\right) = \hat{a}n\sum_{i=1}^{n} x_i + \hat{b}\left(\sum_{i=1}^{n} x_i\right)^2, & (5) \\ n\sum_{i=1}^{n} x_i y_i = \hat{a}n\sum_{i=1}^{n} x_i + \hat{b}n\sum_{i=1}^{n} x_i^2. & (6) \end{cases}$$

Subtracting Eq. (5) from Eq. (6), we obtain

$$n \sum_{i=1}^{n} x_i y_i - \left(\sum_{i=1}^{n} x_i\right)\left(\sum_{i=1}^{n} y_i\right) = \hat{b}n \sum_{i=1}^{n} x_i^2 - \hat{b}\left(\sum_{i=1}^{n} x_i\right)^2$$

$$= \hat{b}\left[n \sum_{i=1}^{n} x_i^2 - \left(\sum_{i=1}^{n} x_i\right)^2\right].$$

Thus,

$$\boxed{\hat{b} = \frac{n \sum_{i=1}^{n} x_i y_i - \left(\sum_{i=1}^{n} x_i\right)\left(\sum_{i=1}^{n} y_i\right)}{n \sum_{i=1}^{n} x_i^2 - \left(\sum_{i=1}^{n} x_i\right)^2}} \tag{7}$$

It can also be shown that

$$\boxed{\hat{a} = \frac{\left(\sum_{i=1}^{n} x_i^2\right)\left(\sum_{i=1}^{n} y_i\right) - \left(\sum_{i=1}^{n} x_i\right)\left(\sum_{i=1}^{n} x_i y_i\right)}{n \sum_{i=1}^{n} x_i^2 - \left(\sum_{i=1}^{n} x_i\right)^2}} \tag{8}$$

Computing the linear regression coefficients $\hat{a}$ and $\hat{b}$ by the formulas of Eqs. (7) and (8) gives the linear regression line of y on x, namely, $\hat{y} = \hat{a} + \hat{b}x$, which can be used to estimate y for a given value of x.

In the next example, as well as in the exercises, you will encounter *index numbers*. They are used to relate a variable in one period of time to the same variable in another period, this latter period called the *base period*. An index number is a *relative* number which describes data that are changing over time. Such data are referred to as *time series*.

For example, consider the time series data of total production of widgets in the United States for 1975–1979 indicated in Table 8-5. If we choose 1976 as the base year and assign to it the index number 100, then the other index numbers are obtained by dividing each year's production by the 1976 production of 900 and multiplying the result by 100. We can, for example, interpret the index 106 for 1979 as meaning that production for that year was 106 percent of the production in 1976.

TABLE 8-5

YEAR	PRODUCTION IN THOUSANDS	INDEX [1976 = 100]
1975	828	92
1976	900	100
1977	936	104
1978	891	99
1979	954	106

In time series analysis, index numbers are obviously of great advantage if the data involve numbers of great magnitude. But regardless of the magnitude of the data, index numbers simplify the task of comparing changes in data over periods of time.

EXAMPLE 1

By means of the least squares linear regression line, represent the trend for the Index of Industrial Production from 1963 *to* 1968 (*Table* 8-6 [1967 = 100]).

TABLE 8-6

Year	1963	1964	1965	1966	1967	1968
Index	79	84	91	99	100	105

Source: *Economic Report of the President*, 1971, U.S. Government Printing Office, Washington, D.C., 1971.

We shall let x denote time, y denote the index, and treat y as a linear function of x. Also, we shall designate 1963 by $x = 1$, 1964 by $x = 2$, etc. There are $n = 6$ pairs of measurements. To use the formulas of Eqs. (7) and (8) we first perform the arithmetic shown in Table 8-7. Hence by Eq. (8),

$$\hat{a} = \frac{91(558) - 21(2046)}{6(91) - (21)^2} = \frac{7812}{105} = 74.4,$$

TABLE 8-7

Year	x_i	y_i	$x_i y_i$	x_i^2
1963	1	79	79	1
1964	2	84	168	4
1965	3	91	273	9
1966	4	99	396	16
1967	5	100	500	25
1968	6	105	630	36
Total	21	558	2046	91
	$=\sum_{i=1}^{6} x_i$	$=\sum_{i=1}^{6} y_i$	$=\sum_{i=1}^{6} x_i y_i$	$=\sum_{i=1}^{6} x_i^2$

and by Eq. (7),

$$\hat{b} = \frac{6(2046) - 21(558)}{6(91) - (21)^2} = \frac{558}{105} = 5.31 \text{ (approximately)}.$$

Thus the regression line of y on x is

$$\hat{y} = 74.4 + 5.31x,$$

whose graph, as well as a scatter diagram, appears in Fig. 8-18.

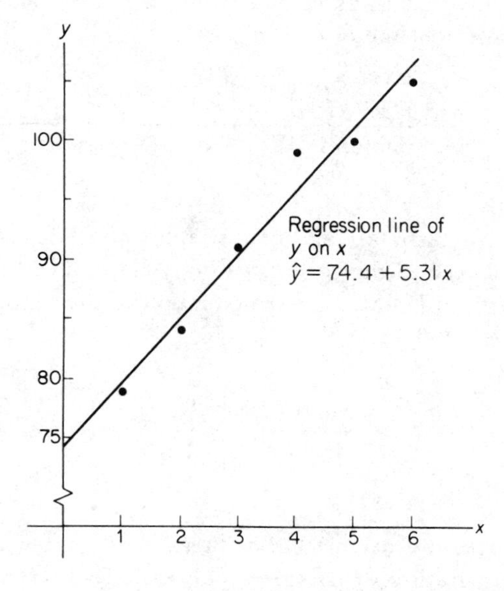

FIG. 8-18

Exercise 8-9

In Problems **1–4**, *find an equation of the least squares linear regression line of y on x for the given data and sketch both the line and the data. Predict the value of y corresponding to* $x = 3.5$.

1.

x	1	2	3	4	5	6
y	1.5	2.3	2.6	3.7	4.0	4.5

2.

x	1	2	3	4	5	6	7
y	1	1.8	2	4	4.5	7	9

3.

x	2	3	4.5	5.5	7
y	3	5	8	10	11

4.

x	2	3	4	5	6	7
y	2.4	2.9	3.3	3.8	4.3	4.9

5. A firm finds that when the price of its product is p dollars per unit, the number of units sold is q, as indicated below. Find an equation of the regression line of q on p.

price (p)	10	30	40	50	60	70
demand (q)	70	68	63	50	46	32

6. On a farm an agronomist finds that the amount of water applied (in inches) and the corresponding yield of a certain crop (in tons per acre) are as given below. Find an equation of the regression line of y on x. Predict y when $x = 12$.

water (x)	8	16	24	32
yield (y)	4.1	4.5	5.1	6.1

7. A rabbit was injected with a virus, and x hours after the injection the temperature y (in degrees Fahrenheit) of the rabbit was measured **[47]**. The data are given below. Find an equation of the regression line of y on x, and estimate the rabbit's temperature 40 hours after the injection.

elapsed time (x)	24	32	48	56
temperature (y)	102.8	104.5	106.5	107.0

.

8. In a psychological experiment, four persons were subjected to a stimulus. Both before and after the stimulus, the systolic blood pressure (in millimeters of mercury) of each subject was measured. The data are given below. Find an equation of the regression line of y on x where x and y are defined below.

	blood pressure			
before stimulus (x)	130	132	136	140
after stimulus (y)	139	140	144	146

.

For the time series in Problems **9** *and* **10** *fit a linear regression line by least squares; that is, find an equation of the linear regression line of y on x. In each case let the first year in the table correspond to $x = 1$.*

9.

Production of Product A, 1975–1979 (in thousands of units)

YEAR	PRODUCTION
1975	10
1976	15
1977	16
1978	18
1979	21

10. In the following, let 1961 correspond to $x = 1$, 1963 correspond to $x = 3$, etc.

Wholesale Price Index—
Lumber and Wood Products
[1967 = 100]

YEAR	INDEX
1961	91
1963	94
1965	96
1967	100

Source: *Economic Report of the President*, 1971, U. S. Government Printing Office, Washington, D.C., 1971.

11. (a) By the method of least squares find an equation of the linear regression line of y on x for the following data. Refer to 1971 as year $x = 1$, etc.

Overseas Shipments of Buttons by Acme Button Co., Inc.
(in millions)

YEAR	QUANTITY
1971	35
1972	31
1973	26
1974	24
1975	26

(b) For the data in part (a), refer to 1971 as year $x = -2$, 1972 as year $x = -1$, 1973 as year $x = 0$, etc. Then $\sum_{i=1}^{5} x_i = 0$. Fit a least squares line and observe how the calculation is simplified.

12. For the following time series, find an equation of the linear regression line that best fits the data. Refer to 1965 as year $x = -2$, 1966 as year $x = -1$, etc.

Consumer Price Index—Medical Care
1965–1969, [1967 = 100]

YEAR	INDEX
1965	90
1966	93
1967	100
1968	106
1969	113

Source: *Economic Report of the President*, 1971, U.S. Government Printing Office, Washington, D.C., 1971.

8-10 MULTIPLE INTEGRALS

Just as we were able to differentiate functions of two variables, we are also able to integrate them. The symbol

$$\int_0^1 \int_{x^3}^{x^2} (x^3 - xy)\, dy\, dx \quad \text{or} \quad \int_0^1 \left[\int_{x^3}^{x^2} (x^3 - xy)\, dy \right] dx$$

is called a (definite) *double integral*. To evaluate it we use successive integrations starting with the innermost integral.

First, we evaluate

$$\int_{x^3}^{x^2} (x^3 - xy)\, dy$$

by treating x as a constant and integrating with respect to y between the limits x^3 and x^2:

$$\int_{x^3}^{x^2} (x^3 - xy)\, dy = \left(x^3y - \frac{xy^2}{2}\right)\Bigg|_{x^3}^{x^2}.$$

Substituting the limits for the variable y, we have

$$\left[x^3(x^2) - \frac{x(x^2)^2}{2}\right] - \left[x^3(x^3) - \frac{x(x^3)^2}{2}\right]$$

$$= x^5 - \frac{x^5}{2} - x^6 + \frac{x^7}{2} = \frac{x^5}{2} - x^6 + \frac{x^7}{2}.$$

Now we integrate this result with respect to x between the limits 0 and 1.

$$\int_0^1 \left(\frac{x^5}{2} - x^6 + \frac{x^7}{2}\right) dx = \left(\frac{x^6}{12} - \frac{x^7}{7} + \frac{x^8}{16}\right)\Bigg|_0^1$$

$$= \left(\frac{1}{12} - \frac{1}{7} + \frac{1}{16}\right) - 0 = \frac{1}{336}.$$

Thus,

$$\int_0^1 \int_{x^3}^{x^2} (x^3 - xy)\, dy\, dx = \frac{1}{336}.$$

Similarly, the integral

$$\int_0^2 \int_3^4 xy\, dx\, dy \quad \text{or} \quad \int_0^2 \left[\int_3^4 xy\, dx\right] dy$$

is evaluated by first treating y as a constant and then integrating xy with respect to x between 3 and 4. Then we integrate the result with respect to y between 0 and 2.

$$\int_0^2 \int_3^4 xy\, dx\, dy = \int_0^2 \left[\int_3^4 xy\, dx\right] dy = \int_0^2 \left(\frac{x^2y}{2}\right)\Bigg|_3^4 dy$$

$$= \int_0^2 \left(8y - \frac{9}{2}y\right) dy = \int_0^2 \left(\frac{7}{2}y\right) dy$$

$$= \frac{7y^2}{4}\Bigg|_0^2 = 7 - 0 = 7.$$

EXAMPLE 1

Evaluate $\int_{-1}^{1}\int_{0}^{1-x}(2x+1)\,dy\,dx.$

$$\int_{-1}^{1}\int_{0}^{1-x}(2x+1)\,dy\,dx = \int_{-1}^{1}\left[\int_{0}^{1-x}(2x+1)\,dy\right]dx$$

$$= \int_{-1}^{1}(2xy+y)\Big|_{0}^{1-x}dx = \int_{-1}^{1}\{[2x(1-x)+(1-x)]-0\}\,dx$$

$$= \int_{-1}^{1}(-2x^2+x+1)\,dx = \left(-\frac{2x^3}{3}+\frac{x^2}{2}+x\right)\Big|_{-1}^{1}$$

$$= \left(-\frac{2}{3}+\frac{1}{2}+1\right)-\left(\frac{2}{3}+\frac{1}{2}-1\right)=\frac{2}{3}.$$

EXAMPLE 2

Evaluate $\int_{1}^{\ln 2}\int_{e^y}^{2}dx\,dy.$

$$\int_{1}^{\ln 2}\int_{e^y}^{2}dx\,dy = \int_{1}^{\ln 2}\left[\int_{e^y}^{2}dx\right]dy = \int_{1}^{\ln 2}x\Big|_{e^y}^{2}dy$$

$$= \int_{1}^{\ln 2}(2-e^y)\,dy = (2y-e^y)\Big|_{1}^{\ln 2}$$

$$= (2\ln 2-2)-(2-e) = 2\ln 2-4+e$$

$$= \ln 4 - 4 + e.$$

A double integral can be interpreted in terms of the volume of the region between the x, y-plane and a surface $z = f(x, y)$ if $z \geqslant 0$. In Fig. 8-19 is a region whose volume we shall consider. The element of volume for this region is a vertical column. It has height $z = f(x, y)$ and a base area of $\Delta y\,\Delta x$. Thus its volume is $f(x, y)\,\Delta y\,\Delta x$. The double integral sums up all such volumes for $a \leqslant x \leqslant b$ and $c \leqslant y \leqslant d$. Thus

$$\text{Volume} = \int_{a}^{b}\int_{c}^{d}f(x, y)\,dy\,dx.$$

Triple integrals are handled by successively evaluating three integrals, as the next example shows.

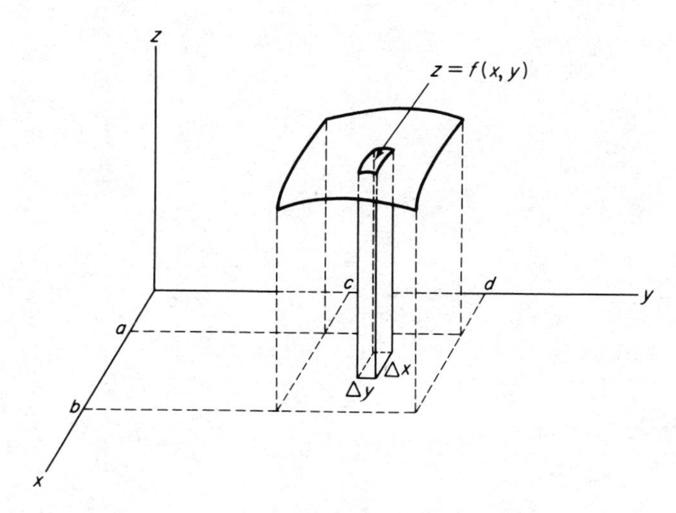

FIG. 8-19

EXAMPLE 3

Evaluate $\int_0^1 \int_0^x \int_0^{x-y} x\, dz\, dy\, dx.$

$$\int_0^1 \int_0^x \int_0^{x-y} x\, dz\, dy\, dx = \int_0^1 \int_0^x \left[\int_0^{x-y} x\, dz\right] dy\, dx$$

$$= \int_0^1 \int_0^x (xz)\Big|_0^{x-y} dy\, dx = \int_0^1 \int_0^x [x(x-y) - 0]\, dy\, dx$$

$$= \int_0^1 \int_0^x (x^2 - xy)\, dy\, dx = \int_0^1 \left[\int_0^x (x^2 - xy)\, dy\right] dx$$

$$= \int_0^1 \left(x^2 y - \frac{xy^2}{2}\right)\Big|_0^x dx = \int_0^1 \left[\left(x^3 - \frac{x^3}{2}\right) - 0\right] dx$$

$$= \int_0^1 \frac{x^3}{2}\, dx = \frac{x^4}{8}\Big|_0^1 = \frac{1}{8}.$$

Exercise 8-10

In Problems **1–22**, *evaluate the given multiple integrals.*

1. $\int_0^3 \int_0^4 x\, dy\, dx.$

2. $\int_0^2 \int_1^2 y\, dy\, dx.$

3. $\int_0^1 \int_0^1 xy\, dx\, dy.$

4. $\int_0^2 \int_0^3 x^2\, dy\, dx.$

5. $\int_1^3 \int_1^2 (x^2 - y)\, dx\, dy.$

6. $\int_{-1}^2 \int_1^4 (x^2 - 2xy)\, dy\, dx.$

7. $\int_0^1 \int_0^2 (x + y)\, dy\, dx.$

8. $\int_0^3 \int_0^x (x^2 + y^2)\, dy\, dx.$

9. $\int_0^6 \int_0^{3x} y\, dy\, dx.$

10. $\int_1^2 \int_0^{x-1} y\, dy\, dx.$

11. $\int_0^1 \int_{3x}^{x^2} 2x^2y\, dy\, dx.$

12. $\int_0^2 \int_0^{x^2} xy\, dy\, dx.$

13. $\int_0^2 \int_0^{\sqrt{4-y^2}} x\, dx\, dy.$

14. $\int_0^1 \int_y^{\sqrt{y}} y\, dx\, dy.$

15. $\int_{-1}^1 \int_x^{1-x} (x + y)\, dy\, dx.$

16. $\int_0^3 \int_{y^2}^{3y} x\, dx\, dy.$

17. $\int_0^1 \int_0^y e^{x+y}\, dx\, dy.$

18. $\int_2^3 \int_0^2 e^{x-y}\, dx\, dy.$

19. $\int_{-1}^0 \int_{-1}^2 \int_1^2 6xy^2z^3\, dx\, dy\, dz.$

20. $\int_0^1 \int_0^x \int_0^{x-y} x\, dz\, dy\, dx.$

21. $\int_0^1 \int_{x^2}^x \int_0^{xy} dz\, dy\, dx.$

22. $\int_0^2 \int_{y^2}^{3y} \int_0^x dz\, dx\, dy.$

23. In statistics a joint density function $z = f(x, y)$ defined on a region in the x, y–plane is represented by a surface in space. The probability that $a \leqslant x \leqslant b$ and $c \leqslant y \leqslant d$ is given by

$$P(a \leqslant x \leqslant b, c \leqslant y \leqslant d) = \int_c^d \int_a^b f(x, y)\, dx\, dy,$$

and is represented by the volume between the graph of f and the rectangular region determined by $a \leqslant x \leqslant b$ and $c \leqslant y \leqslant d$. If $f(x, y) = e^{-(x+y)}$ is a density function, where $x \geqslant 0$ and $y \geqslant 0$, find $P(0 \leqslant x \leqslant 2, 1 \leqslant y \leqslant 2)$ and give your answer in terms of e.

24. In Problem 23, let $f(x, y) = 12e^{-4x-3y}$ for $x, y \geqslant 0$. Find $P(3 \leqslant x \leqslant 4, 2 \leqslant y \leqslant 6)$ and give your answer in terms of e.

25. In Problem 23, let $f(x, y) = \frac{x}{8}$ where $0 \leqslant x \leqslant 2$ and $0 \leqslant y \leqslant 4$. Find $P(x \geqslant 1, y \geqslant 2)$.

26. In Problem 23, let f be the uniform distribution $f(x, y) = 1$ defined over the unit square $0 \leqslant x \leqslant 1, 0 \leqslant y \leqslant 1$. Find the probability that $0 \leqslant x \leqslant \frac{1}{3}$ and $\frac{1}{4} \leqslant y \leqslant \frac{3}{4}$.

8-11 REVIEW

IMPORTANT TERMS AND SYMBOLS IN CHAPTER 8

function of n variables (*p. 393*)

partial derivative (*p. 399*)

$\partial z/\partial x$, $f_x(x, y)$ (*p. 401*)

$\left.\dfrac{\partial z}{\partial x}\right|_{(x_0, y_0)}$, $f_x(x_0, y_0)$ (*p. 401*)

3-dimensional coordinate system (*p. 391*)

$\dfrac{\partial^2 z}{\partial x \partial y}, \dfrac{\partial^2 z}{\partial x^2}, \dfrac{\partial^2 z}{\partial y^2}$ (*p. 416*)

f_{xy}, f_{xx}, f_{yy} (*p. 416*)

intermediate variables (*p. 419*)

chain rule (*p. 419*)

joint-cost function (*p. 405*)

implicit partial differentiation (*p. 414*)

double integral (*p. 452*)

triple integral (*p. 454*)

x, y-plane; x, z-plane; y, z-plane (*p. 393*)

production function (*p. 407*)

marginal products (*p. 407*)

competitive products (*p. 408*)

complementary products (*p. 409*)

relative maxima and minima (*p. 423*)

critical point (*p. 425*)

second-derivative test (*p. 427*)

method of Lagrange multipliers (*p. 435*)

method of least squares (*p. 443*)

index numbers (*p. 447*)

linear regression line of y on x (*p. 445*)

REVIEW SECTION

1. If $f(x, y) = 3x^2y^3$, then to find $f_x(x, y)$, which do we think of as a constant, x or y? ________

Ans. y.

2. If $f(x, y) = x^2y + 3y$, then $f_x(x, y) =$ __(a)__ and $f_y(x, y) =$ __(b)__ .

Ans. (a) $2xy$; (b) $x^2 + 3$.

3. If $f(x, y) = 2$, then $f_{xy}(x, y) =$ ________ .

Ans. 0.

†4. If $w = f(x, y, z)$, $x = g(r, s)$, $y = h(r, s)$, and $z = k(r, s)$, then the number of terms in $\partial w/\partial s$ is ________.

Ans. 3.

5. If (1, 2) is a critical point of $z = f(x, y)$ and $f_{xx}(1, 2) = f_{yy}(1, 2) = 2$ and $f_{xy}(1, 2) = 1$, then f has a relative (maximum) (minimum) at (1, 2).

Ans. minimum.

6. If $f(x, y, z) = x^2 + y^2 + z^2 + 6z$, then f has a critical point at ________.

Ans. $(0, 0, -3)$.

†7. If $yz - x + y = 0$, then $\partial z/\partial x =$ ________.

Ans. $1/y$.

†8. Let $z = x^2 + y^2$, $x = r^2 + rs$, $y = s^2 + s$. Then $\partial z/\partial r =$ ________.

Ans. $2x(2r + s)$.

9. If $z = xyw^2$, then $\dfrac{\partial^2 z}{\partial x \partial w} =$ ________.

Ans. $2yw$.

10. True or false: If $f(x, y) = 2x^2 + xy + y^2$, then $f_{xy}(1, 2) = f_{yx}(1, 2)$. ________

Ans. true.

11. In finding critical points of $f(x, y, z) = x^2 - xy + 2z^2$ subject to $x - y + z = 6$ by the method of Lagrange multipliers, we examine the function $F(x, y, z, \lambda) =$ ________.

Ans. $x^2 - xy + 2z^2 - \lambda(x - y + z - 6)$ or
$x^2 - xy + 2z^2 - \lambda(6 - x + y - z)$.

†12. If $z = f(x, y)$, $x = g(s, t)$, and $y = h(s, t)$, then by the chain rule, $\partial z/\partial s =$ ________.

Ans. $(\partial z/\partial x)(\partial x/\partial s) + (\partial z/\partial y)(\partial y/\partial s)$.

13. A critical point of $w = f(x, y, z)$ is a point where $w_x = w_y = w_z =$ ________.

Ans. 0.

14. In a three-dimensional coordinate system, the graph of $y = 2$ is a (a) (line) (plane) parallel to the (b) (x, y) (x, z)-plane.

Ans. (a) plane; (b) x, z.

†Refers to Sec. 8-4 or Sec. 8-6.

15. There is a natural one-to-one correspondence between all points in space and all ordered (pairs) (triples) of real numbers.

Ans. triples.

16. If $P = f(l, k)$ is a production function, then in terms of partial derivatives the marginal product of l is ________.

Ans. $\partial P/\partial l$.

†17. True or false: If $z = f(x)$ and $x = g(y, w)$, then $\partial z/\partial y = (dz/dx)(\partial x/\partial y)$.

____

Ans. true.

18. In evaluating $\int_0^3 \int_0^2 xy\, dx\, dy$, we integrate first with respect to ________.

Ans. x.

REVIEW PROBLEMS

In Problems **1–12**, *find the indicated partial derivatives.*

1. $f(x, y) = 2x^2 + 3xy + y^2 - 1$; $f_x(x, y), f_y(x, y)$.
2. $P = l^3 + k^3 - lk$; $\partial P/\partial l, \partial P/\partial k$.
3. $z = x/(x + y)$; $\partial z/\partial x, \partial z/\partial y$.
4. $w = \dfrac{\sqrt{x^2 + y^2}}{y}$; $\partial w/\partial x$.
5. $w = e^{x^3yz}$; $w_{xy}(x, y)$.
6. $f(x, y) = xy \ln (xy)$; $f_{xy}(x, y)$.
7. $f(x, y) = \ln\sqrt{x^2 + y^2}$; $\dfrac{\partial}{\partial y}[f(x, y)]$.
8. $f(p_A, p_B) = (p_A - 20)q_A + (p_B - 30)q_B$; $f_{p_A}(p_A, p_B)$.
9. $f(x, y, z) = (x + y)(y + z^2)$; $\dfrac{\partial^2}{\partial z^2}[f(x, y, z)]$.
10. $z = (x^2 - y)(y^2 - 2xy)$; $\partial^2 z/\partial y^2$.
11. $w = xe^{yz} \ln z$; $\partial w/\partial y, \partial^2 w/\partial x\, \partial z$.
12. $P = 2.4l^{.11}k^{.89}$; $\partial P/\partial k$.

†Refers to Sec. 8-4 or Sec. 8-6.

†13. If $w = x^2 + 2xy + 3y^2$, $x = e^r$, and $y = \ln(r + s)$, find $\partial w/\partial r$ and $\partial w/\partial s$.

†14. If $z = \ln(x/y) + e^y - xy$, $x = r^2s^2$, and $y = r + s$, find $\partial z/\partial s$.

†15. If $x^2 + 2xy - 2z^2 + xz + 2 = 0$, find $\partial z/\partial x$.

†16. If $z^2 - e^{yz} + \ln z + e^{xz} = 0$, find $\partial z/\partial y$.

17. Examine $f(x, y) = x^2 + 2y^2 - 2xy - 4y + 3$ for relative extrema.

18. Examine $f(w, z) = 2w^3 + 2z^3 - 6wz + 7$ for relative extrema.

19. Find all critical points of $f(x, y, z) = x^2 + y^2 + z^2$ subject to the constraint $3x + 2y + z = 14$.

20. Find all critical points of $f(x, y, z) = xyz$ subject to the condition that $3x + 2y + 4z - 120 = 0$ $(xyz \neq 0)$.

21. In an experiment‡ a group of fish were injected with living bacteria. Of those fish maintained at 28° C, the percentage p which survived the infection t hours after the injection is given below. Find the linear regression line of p on t.

t	8	10	18	20	48
p	82	79	78	78	64

22. A manufacturer's cost for producing x units of product X and y units of product Y is given by $c = 5x + .03xy + 7y + 200$. Determine the (partial) marginal cost with respect to x when $x = 100$ and $y = 200$.

23. If a manufacturer's production function is defined by $P = 20l^{.7}k^{.3}$, determine his marginal productivity functions.

24. If $q_A = 200 - 3p_A + p_B$ and $q_B = 50 - 5p_B + p_A$, where q_A and q_B are the number of units demanded of products A and B, respectively, and p_A and p_B are their respective prices per unit, determine whether A and B are competitive or complementary products.

25. For industry there is a model that describes the rate α (a Greek letter read "alpha") at which a new innovation substitutes for an established process [22]. It is given by

$$\alpha = Z + .530P - .027S,$$

†Refers to Sec. 8–4 or Sec. 8–6.

‡J. B. Covert, W. W. Reynolds. "Survival value of fever in fish," *Nature*, vol. 267, no. 5606 (1977), 43–45.

where Z is a constant that depends on the particular industry, P is an index of profitability of the new innovation, and S is an index of the extent of the investment necessary to make use of the innovation. Find $\partial\alpha/\partial P$ and $\partial\alpha/\partial S$.

26. Find the least squares linear regression line of y on x for the data given below. Refer to year 1973 as year $x = 1$, etc.

Equipment Expenditures of Allied Computer Company, 1973–1978 (in millions of dollars)

YEAR	EXPENDITURES
1973	15
1974	22
1975	21
1976	27
1977	26
1978	34

In Problems **27–30**, *sketch the given surfaces.*

27. $2x + 3y + z = 9$.

28. $z = x$.

29. $z = y^2$.

30. $x^2 + z^2 = 1$.

In Problems **31–34**, *evaluate the double integrals.*

31. $\int_1^2 \int_0^y x^2y^2 \, dx \, dy$.

32. $\int_0^4 \int_{y/2}^2 xy \, dx \, dy$.

33. $\int_0^3 \int_{y^2}^{3y} x \, dx \, dy$.

34. $\int_0^1 \int_{\sqrt{x}}^{x^2} (x^2 + 2xy - 3y^2) \, dy \, dx$.

35. An open–top rectangular cardboard box is to have a volume of 32 cubic feet. Find the dimensions of the box so that the amount of cardboard used is minimized.

References

[1] Babkoff, H. "Magnitude estimation of short electrocutaneous pulses," *Psychological Research*, vol. 39, no. 1 (1976), 39–49.

[2] Bailey, N. T. J. *The Mathematical Approach to Biology and Medicine*. New York: John Wiley, 1967.

[3] Barbosa, L. C., and M. Friedman. "Deterministic inventory lot size models—a general root law," *Management Science*, vol. 24, no. 8 (1978), 819–826.

[4] Bressani, R. "The use of yeast in human foods," Mateles and Tannenbaum (eds.), *Single-Cell Protein*, Cambridge, Mass.: M. I. T. Press, 1968.

[5] Bullen, K. E. *An Introduction to the Theory of Seismology*. Cambridge at the University Press, 1963.

[6] Daly, P., and P. Douglas. "The production function for Canadian manufactures," *Journal of the American Statistical Association*, vol. 38 (1943), 178–186.

[7] Dean, J. "Statistical cost functions of a hosiery mill," *Studies in Business Administration*, vol. XI, no. 4. Chicago: University of Chicago Press, 1941.

[8] Doelle, L. L. *Environmental Acoustics*. New York: McGraw-Hill, 1972.

[9] Eaton, W. W., and G. A. Whitmore. "Length of stay as a stocastic process: a general approach and application to hospitalization for schizophrenia," *Journal of Mathematical Sociology*, vol. 5 (1977), 273–292.

[10] Embree, D. G. "The population dynamics of the winter moth in Nova Scotia, 1954–1962," *Memoirs of the Entomological Society of Canada*, no. 46 (1965).

[11] Ewens, W. J. *Population Genetics*. London: Methuen and Co. Ltd., 1969.

[12] Flesch, R. *The Art of Readable Writing*. New York: Harper and Row, 1949.

[13] Folk, G. E., Jr. *Textbook of Environmental Physiology*, 2nd ed. Philadelphia: Lea and Febiger, 1974.

[14] Gause, G. F. *The Struggle for Existence*. New York: Hafner Publishing Co., 1964.

[15] Graesser, A., and G. Mandler. "Limited processing capacity constrains the storage of unrelated sets of words and retrieval from natural categories," *Human Learning and Memory*, vol. 4, no. 1 (1978), 86–100.

[16] Haavelmo, T. "Methods of measuring the marginal propensity to consume," *Journal of the American Statistical Association*, vol. XLII (1947), 105–122.

[17] Hintzman, D. L. "Repetition and learning," G. H. Bower (ed.), *The Psychology of Learning and Motivation*, vol. 10 (1976), Academic Press, p. 77.

[18] Holling, C. S. "The functional response of invertebrate predators to prey density," *Memoirs of the Etomological Society of Canada*, no. 48 (1966).

[19] Holling, C. S. "Some characteristics of simple types of predation and parasitism," *The Canadian Entomologist*, vol. XCI, no. 7 (1959), 385–398.

[20] Hovland, C. I. "The generalization of conditioned responses: I. The sensory generalization of conditioned responses with varying frequencies of tone," *Journal of General Psychology*, vol. 17 (1937), 125–148.

[21] Hurst, P. M.; K. Perchonok; and E. L. Seguin. "Vehicle kinematics and gap acceptance," *Journal of Applied Psychology*, vol. 52, no. 4 (1968), 321–324.

[22] Hurter, A. P., Jr., A. H. Rubenstein, et al. "Market penetration by new innovations: the technological literature," *Technological Forecasting and Social Change*, vol. 11 (1978), 197–221.

[23] Imrie, F. K. E., and A. J. Vlitos. "Production of fungal protein from carob," S. R. Tannenbaum and D. I. C. Wang (eds.), *Single-Cell Protein II*, Cambridge, Mass.: M. I. T. Press, 1975.

[24] Keyfitz, N. *Introduction to the Mathematics of Population*. Reading, Mass.: Addison-Wesley, 1968.

[25] Keyfitz, N., and W. Flieger. *Population Facts and Methods of Demography*. San Francisco: W. H. Freeman, 1971.

[26] Laming, D. *Mathematical Psychology*. London-New York: Academic Press, 1973.

[27] Lancaster, P. *Mathematics: Models of the Real World*. Englewood Cliffs, N.J.: Prentice-Hall, 1976.

[28] Leik, R. K., and B. F. Meeker. *Mathematical Sociology*. Englewood Cliffs, N.J.: Prentice-Hall, 1975.

[29] Lewis, A. E. *Biostatistics*. New York: Reinhold, 1966.

[30] Loftus, G. R., and E. F. Loftus. *Human Memory: the Processing of Information*. New York: L. Erlbaum Associates, distributed by the Halsted Press Division of John Wiley, 1976.

[31] Lotka, A. J. *Elements of Mathematical Biology*. New York: Dover, 1956.

[32] Mantell, L. H., and F. P. Sing. *Economics for Business Decisions*. New York: McGraw-Hill, 1972.

[33] Mather, W. B. *Principles of Quantitative Genetics*. Minneapolis: Burgess Publishing Co., 1964.

[34] Nordin, J. A. "Note on a light plant's cost curves," *Econometrica*, vol. 15 (1947), 231–235.

[35] Odum, E. P. *Ecology*. New York: Holt, Rinehart and Winston, 1966.

[36] Odum, H. T. "Biological circuits and the marine systems of Texas," T. A. Olsen and F. J. Burgess (eds.), *Pollution and Marine Biology*, New York: Interscience Publishers, 1967.

[37] Perchonok, K., and P. M. Hurst. "Effect of lane-closure signals upon driver decision making and traffic flow," *Journal of Applied Psychology*, vol. 52, no. 5 (1968), 410–413.

[38] Peterson, L. R., and M. J. Peterson. "Short-term retention of individual verbal items," *Journal of Experimental Psychology*, vol. 58 (1959), 193–198.

[39] Poole, R. W. *An Introduction to Quantitative Ecology*. New York: McGraw-Hill, 1974.

[40] Rha, C. "Utilization of single-cell protein for human food," S. R. Tannenbaum and D. I. C. Wang (eds.), *Single-Cell Protein II*, Cambridge, Mass.: M. I. T. Press, 1975.

[41] Richter, C. F. *Elementary Seismology*. San Francisco: W. H. Freeman, 1958.

[42] Roberts, A. M. "The origins of fluctuations in the human secondary sex ratio," *Journal of Biosocial Science*, vol. 10, no. 2 (1978), 169–182.

[43] Shonle, J. I. *Environmental Applications of General Physics*. Reading, Mass.: Addison-Wesley, 1975.

[44] Silberman, J., and G. Yochum. "The role of money in determining election outcomes," *Social Science Quarterly*, vol. 58, no. 4 (1978), 671–682.

[45] Simon, W. *Mathematical Techniques for Physiology and Medicine*. New York: Academic Press, 1972.

[46] Smith, J. M. *Mathematical Ideas in Biology*. London: Cambridge University Press, 1968.

[47] Sokal, R. R., and F. J. Rohlf. *Introduction to Biostatistics*. San Francisco: W. H. Freeman, 1973.

[48] Stacy, R. W.; D. T. Williams; R. E. Worden; and R. O. McMorris. *Essentials of Biological and Medical Physics*. New York: McGraw-Hill, 1955.

[49] Stigler, G. *The Theory of Price*, 3rd ed. New York: Macmillan, 1966.

[50] Taagepera, R. "Why the trade/GNP ratio decreases with country size," *Social Science Research*, vol. 5 (1976), 385–404.

[51] Taagepera, R., and J. P. Hayes. "How trade/GNP ratio decreases with country size," *Social Science Research*, vol. 6 (1977), 108–132.

[52] Thrall, R. M.; J. A. Mortimer; K. R. Rebman; and R. F. Baum, eds. *Some Mathematical Models in Biology*, rev. ed. Report No. 40241-R-7. Prepared at University of Michigan, 1967.

[53] Tintner, G., and O. H. Brownlee. "Production functions derived from farm records," *American Journal of Agricultural Economics*, vol. 26 (1944), 566–571.

[54] Tintner, G. *Methodology of Economics and Econometrics*. Chicago: University of Chicago Press, 1967.

[55] Weinstein, A. G., and V. Srinivasen. "Predicting managerial success of master of business administration (MBA) graduates," *Journal of Applied Psychology*, vol. 59, no. 2 (1974), 207–212.

[56] Wilson, E. O., and W. H. Bossert. *A Primer of Population Biology*. Stamford, Conn.: Sinauer Associates, 1971.

APPENDIX A

Algebra Review

A-1 REAL NUMBERS

The numbers 1, 2, 3, etc. form the set of **positive integers**:

$$\text{set of positive integers} = \{1, 2, 3, \ldots\}.$$

The positive integers together with 0 and the **negative integers** $-1, -2, -3, \ldots$ form the set of **integers**:

$$\text{set of integers} = \{\ldots, -3, -2, -1, 0, 1, 2, 3, \ldots\}.$$

The set of **rational numbers** consists of numbers, such as $\frac{1}{2}$ and $\frac{5}{3}$, which can be written as a ratio (quotient) of two integers. That is, a rational number is one that can be written as p/q, where p and q are integers and $q \neq 0$. (The symbol "$\neq$" is read "is not equal to.") **We never divide by zero.** Other rational numbers are $\frac{19}{20}$, $\frac{-2}{7}$, and $\frac{-6}{-2}$. The integer 2 is rational since $2 = \frac{2}{1}$. In fact, every integer is rational.

All rational numbers can be represented by decimal numbers that *terminate*, such as $\frac{3}{4} = .75$ and $\frac{3}{2} = 1.5$, or by *nonterminating repeating* decimals (a group of digits repeats without end), such as $\frac{2}{3} = .666\ldots$ and $\frac{-4}{11} = -.3636\ldots$. Numbers represented by *nonterminating nonrepeating* decimals are called **irrational numbers**. An irrational number cannot be written as an integer divided by an integer. The numbers π (pi) and $\sqrt{2}$ are irrational.

Together, the rational numbers and irrational numbers form the set of **real numbers**. Real numbers can be represented by points on a line. In Fig. 1 some points and their associated real numbers are

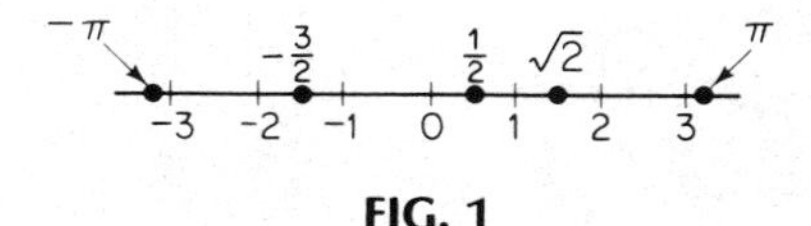

FIG. 1

identified. To each point on the line there corresponds a unique real number, and to each real number there corresponds a unique point on the line. We call this line the **real number line**.

Three important properties of the real numbers are the *commutative, associative, and distributive properties.*

THE COMMUTATIVE PROPERTIES. *If a and b are real numbers, then*

$$a + b = b + a \quad \textit{and} \quad ab = ba.$$

This means that we can add or multiply two real numbers in any order. For example, $3 + 4 = 4 + 3$ and $7(-4) = (-4)(7)$.

THE ASSOCIATIVE PROPERTIES. *If a, b, and c are real numbers, then*

$$a + (b + c) = (a + b) + c \quad \textit{and} \quad a(bc) = (ab)c.$$

This means that in addition or multiplication, numbers can be grouped in any order. For example, $2 + (3 + 4) = (2 + 3) + 4$, $6(\frac{1}{3} \cdot 5) = (6 \cdot \frac{1}{3}) \cdot 5$, and $2x + (x + y) = (2x + x) + y$.

THE DISTRIBUTIVE PROPERTIES. *If a, b, and c are real numbers, then*

$$a(b + c) = ab + ac \quad \text{and} \quad (b + c)a = ba + ca.$$

For example,

$$x(z + 4) = x(z) + x(4) = xz + 4x.$$

The distributive property can be extended to the form $a(b + c + d) = ab + ac + ad$. In fact, it can be extended to sums involving any other number of terms.

Listed below are other properties of real numbers along with numerical examples. All denominators are different from zero.

Property	*Example*
1. $a - b = a + (-b)$.	$2 - 7 = 2 + (-7) = -5$.
2. $a - (-b) = a + b$.	$2 - (-7) = 2 + 7 = 9$.
3. $-a = (-1)(a)$.	$-7 = (-1)(7)$.
4. $a(b + c) = ab + ac$.	$6(7 + 2) = 6 \cdot 7 + 6 \cdot 2 = 54$.
5. $a(b - c) = ab - ac$.	$6(7 - 2) = 6 \cdot 7 - 6 \cdot 2 = 30$.
6. $-(a + b) = -a - b$.	$-(7 + 2) = -7 - 2 = -9$.
7. $-(a - b) = -a + b$.	$-(2 - 7) = -2 + 7 = 5$.
8. $-(-a) = a$.	$-(-2) = 2$.
9. $a(0) = (-a)(0) = 0$.	$2(0) = (-2)(0) = 0$.
10. $(-a)(b) = -(ab) = a(-b)$.	$(-2)(7) = -(2 \cdot 7) = 2(-7)$.
11. $(-a)(-b) = ab$.	$(-2)(-7) = 2 \cdot 7 = 14$.
12. $\frac{a}{1} = a$.	$\frac{7}{1} = 7, \ \frac{-2}{1} = -2$.
13. $\frac{a}{b} = a\left(\frac{1}{b}\right)$.	$\frac{2}{7} = 2\left(\frac{1}{7}\right)$.
14. $\frac{1}{-a} = -\frac{1}{a} = \frac{-1}{a}$.	$\frac{1}{-4} = -\frac{1}{4} = \frac{-1}{4}$.
15. $\frac{a}{-b} = -\frac{a}{b} = \frac{-a}{b}$.	$\frac{2}{-7} = -\frac{2}{7} = \frac{-2}{7}$.

16. $\dfrac{-a}{-b} = \dfrac{a}{b}$. $\qquad \dfrac{-2}{-7} = \dfrac{2}{7}$.

17. $\dfrac{0}{a} = 0$ when $a \neq 0$. $\qquad \dfrac{0}{7} = 0$.

18. $\dfrac{a}{a} = 1$ when $a \neq 0$. $\qquad \dfrac{2}{2} = 1,\ \dfrac{-5}{-5} = 1$.

19. $a\left(\dfrac{b}{a}\right) = b$. $\qquad 2\left(\dfrac{7}{2}\right) = 7$.

20. $a \cdot \dfrac{1}{a} = 1$ when $a \neq 0$. $\qquad 2 \cdot \dfrac{1}{2} = 1$.

21. $\dfrac{1}{a} \cdot \dfrac{1}{b} = \dfrac{1}{ab}$. $\qquad \dfrac{1}{2} \cdot \dfrac{1}{7} = \dfrac{1}{2 \cdot 7} = \dfrac{1}{14}$.

22. $\dfrac{ab}{c} = \left(\dfrac{a}{c}\right)b = a\left(\dfrac{b}{c}\right)$. $\qquad \dfrac{2 \cdot 7}{3} = \dfrac{2}{3} \cdot 7 = 2 \cdot \dfrac{7}{3}$.

23. $\dfrac{a}{bc} = \left(\dfrac{a}{b}\right)\left(\dfrac{1}{c}\right) = \left(\dfrac{1}{b}\right)\left(\dfrac{a}{c}\right)$. $\qquad \dfrac{2}{3 \cdot 7} = \dfrac{2}{3} \cdot \dfrac{1}{7} = \dfrac{1}{3} \cdot \dfrac{2}{7}$.

24. $\dfrac{a}{b} = \left(\dfrac{a}{b}\right)\left(\dfrac{c}{c}\right) = \dfrac{ac}{bc}$ when $c \neq 0$. $\qquad \dfrac{2}{7} = \left(\dfrac{2}{7}\right)\left(\dfrac{5}{5}\right) = \dfrac{2 \cdot 5}{7 \cdot 5}$.

25. $\dfrac{a}{b(-c)} = \dfrac{a}{(-b)(c)} = \dfrac{-a}{bc}$ $\qquad \dfrac{2}{3(-5)} = \dfrac{2}{(-3)(5)} = \dfrac{-2}{3(5)}$

$= \dfrac{-a}{(-b)(-c)} = -\dfrac{a}{bc}$. $\qquad = \dfrac{-2}{(-3)(-5)} = -\dfrac{2}{3(5)} = -\dfrac{2}{15}$.

26. $\dfrac{a}{c} + \dfrac{b}{c} = \dfrac{a+b}{c}$. $\qquad \dfrac{2}{9} + \dfrac{3}{9} = \dfrac{2+3}{9} = \dfrac{5}{9}$.

27. $\dfrac{a}{c} - \dfrac{b}{c} = \dfrac{a-b}{c}$. $\qquad \dfrac{2}{9} - \dfrac{3}{9} = \dfrac{2-3}{9} = \dfrac{-1}{9}$.

28. $\dfrac{a}{b} + \dfrac{c}{d} = \dfrac{ad+bc}{bd}$. $\qquad \dfrac{4}{5} + \dfrac{2}{3} = \dfrac{4 \cdot 3 + 5 \cdot 2}{5 \cdot 3} = \dfrac{22}{15}$.

29. $\dfrac{a}{b} - \dfrac{c}{d} = \dfrac{ad-bc}{bd}$. $\qquad \dfrac{4}{5} - \dfrac{2}{3} = \dfrac{4 \cdot 3 - 5 \cdot 2}{5 \cdot 3} = \dfrac{2}{15}$.

30. $\dfrac{a}{b} \cdot \dfrac{c}{d} = \dfrac{ac}{bd}$. $\qquad \dfrac{2}{3} \cdot \dfrac{4}{5} = \dfrac{2 \cdot 4}{3 \cdot 5} = \dfrac{8}{15}$.

31. $\dfrac{a}{\dfrac{b}{c}} = a \div \dfrac{b}{c} = \dfrac{ac}{b}$. $\qquad \dfrac{2}{\dfrac{3}{5}} = 2 \div \dfrac{3}{5} = \dfrac{2 \cdot 5}{3} = \dfrac{10}{3}$.

32. $\dfrac{\dfrac{a}{b}}{c} = \dfrac{a}{b} \div c = \dfrac{a}{bc}$. $\qquad \dfrac{\dfrac{2}{3}}{5} = \dfrac{2}{3} \div 5 = \dfrac{2}{3 \cdot 5} = \dfrac{2}{15}$.

33. $\dfrac{\frac{a}{b}}{\frac{c}{d}} = \dfrac{a}{b} \div \dfrac{c}{d} = \dfrac{a}{b} \cdot \dfrac{d}{c} = \dfrac{ad}{bc}$. $\qquad \dfrac{\frac{2}{3}}{\frac{7}{5}} = \dfrac{2}{3} \div \dfrac{7}{5} = \dfrac{2}{3} \cdot \dfrac{5}{7} = \dfrac{10}{21}$.

Property 24 is essentially the **fundamental principle of fractions,** which states that *multiplying or dividing both the numerator and denominator of a fraction by the same number, except* 0, *results in a fraction which is equivalent to the original fraction.* Thus,

$$\frac{7}{\frac{1}{8}} = \frac{7 \cdot 8}{\frac{1}{8} \cdot 8} = \frac{56}{1} = 56.$$

By properties 28 and 24 we have

$$\frac{2}{5} + \frac{4}{15} = \frac{2 \cdot 15 + 5 \cdot 4}{5 \cdot 15} = \frac{50}{75} = \frac{2 \cdot 25}{3 \cdot 25} = \frac{2}{3}.$$

We can do this problem another way. For the fractions $\frac{2}{5}$ and $\frac{4}{15}$, a common denominator is $5 \cdot 15$. However, 15 is the *least common denominator* (L.C.D.), and we can write

$$\frac{2}{5} + \frac{4}{15} = \frac{2 \cdot 3}{5 \cdot 3} + \frac{4}{15} = \frac{6}{15} + \frac{4}{15} = \frac{10}{15} = \frac{2}{3}.$$

Similarly,

$$\frac{3}{8} - \frac{5}{12} = \frac{3 \cdot 3}{8 \cdot 3} - \frac{5 \cdot 2}{12 \cdot 2}$$

$$= \frac{9}{24} - \frac{10}{24} = \frac{9 - 10}{24} = -\frac{1}{24}.$$

Exercise A-1

Find each of the following if possible.

1. $-2 + (-4)$. **2.** $-6 + 2$. **3.** $6 + (-4)$.

4. $7 - 2$. **5.** $7 - (-4)$. **6.** $-7 - (-4)$.

7. $-8 - (-6)$. **8.** $(-2)(9)$. **9.** $7(-9)$.

10. $(-2)(-12)$. **11.** $(-1)6$. **12.** $-(-9)$.

13. $-(-6 + x)$.
14. $-7(x)$.
15. $-12(x - y)$.
16. $-[-6 + (-y)]$.
17. $-2 \div 6$.
18. $-2 \div (-4)$.
19. $4 \div (-2)$.
20. $2(-6 + 2)$.
21. $3[-2(3) + 6(2)]$.
22. $(-2)(-4)(-1)$.
23. $(-5)(-5)$.
24. $x(0)$.
25. $3(x - 4)$.
26. $4(5 + x)$.
27. $-(x - 2)$.
28. $0(-x)$.
29. $8\left(\frac{1}{11}\right)$.
30. $\frac{7}{1}$.
31. $\frac{-5x}{7y}$.
32. $\frac{3}{-2x}$.
33. $\frac{2}{3} \cdot \frac{1}{x}$.
34. $\frac{x}{y}(2z)$.
35. $(2x)\left(\frac{3}{2x}\right)$.
36. $\frac{-15x}{-3y}$.
37. $\frac{7}{y} \cdot \frac{1}{x}$.
38. $\frac{2}{x} \cdot \frac{5}{y}$.
39. $\frac{1}{2} + \frac{1}{3}$.
40. $\frac{5}{12} + \frac{3}{4}$.
41. $\frac{3}{10} - \frac{7}{15}$.
42. $\frac{2}{3} + \frac{7}{3}$.
43. $\frac{x}{9} - \frac{y}{9}$.
44. $\frac{3}{2} - \frac{1}{4} + \frac{1}{6}$.
45. $\frac{2}{3} - \frac{5}{8}$.
46. $\frac{6}{\frac{x}{y}}$.
47. $\frac{\frac{x}{6}}{y}$.
48. $\frac{\frac{-7}{2}}{\frac{5}{8}}$.
49. $\frac{7}{0}$.
50. $\frac{0}{7}$.
51. $\frac{0}{0}$.
52. $0 \cdot 0$.

A-2 EXPONENTS AND RADICALS

The product

$$x \cdot x \cdot x$$

is abbreviated x^3. In general, for n a positive integer, x^n is the product of n x's. The letter n in x^n is called the *exponent* and x is called the *base*.

More specifically, if n is a positive integer we have

1. $x^n = \underbrace{x \cdot x \cdot x \cdot \ldots \cdot x}_{n \text{ factors}}$.

2. $x^{-n} = \dfrac{1}{x^n} = \dfrac{1}{\underbrace{x \cdot x \cdot x \cdot \ldots \cdot x}_{n \text{ factors}}}$.

3. $\dfrac{1}{x^{-n}} = x^n$.

4. $x^0 = 1$ if $x \neq 0$. 0^0 is not defined.

EXAMPLE 1

a. $3^{-5} = \dfrac{1}{3^5} = \dfrac{1}{3 \cdot 3 \cdot 3 \cdot 3 \cdot 3} = \dfrac{1}{243}$.

b. $\dfrac{1}{3^{-5}} = 3^5 = 243$.

c. $2^0 = 1$, $\pi^0 = 1$, $(-5)^0 = 1$.

If $r^n = x$, where n is a positive integer, then r is an nth *root* of x. For example, $3^2 = 9$ and so 3 is a second root (usually called a *square root*) of 9. Since $(-3)^2 = 9$, -3 is also a square root of 9. Similarly, -2 is a *cube root* of -8 since $(-2)^3 = -8$.

The **principal *n*th root** of x is that nth root of x which is positive if x is positive, and is negative if x is negative and n is odd. We denote it by $\sqrt[n]{x}$. Thus,

$$\sqrt[n]{x} \text{ is } \begin{cases} \text{positive if } x \text{ is positive,} \\ \text{negative if } x \text{ is negative and } n \text{ is odd.} \end{cases}$$

For example, $\sqrt[2]{9} = 3$, $\sqrt[3]{-8} = -2$, and $\sqrt[3]{\frac{1}{27}} = \frac{1}{3}$. We define $\sqrt[n]{0}$ to be 0.

The symbol $\sqrt[n]{x}$ is called a **radical**. Here n is the *index*, x is the *radicand*, and $\sqrt{}$ is the *radical sign*. With principal square roots we usually drop the index and write $\sqrt{x}$ instead of $\sqrt[2]{x}$. Thus $\sqrt{9} = 3$.

PITFALL. *Although 2 and -2 are square roots of 4, the* ***principal*** *square root of 4 is 2, not -2. Hence $\sqrt{4} = 2$.*

If x is positive, the expression $x^{p/q}$ where p and q are integers and q is positive is defined to be $\sqrt[q]{x^p}$. Thus,

$$x^{3/4} = \sqrt[4]{x^3}\,; \qquad 8^{2/3} = \sqrt[3]{8^2} = \sqrt[3]{64} = 4;$$
$$4^{-1/2} = \sqrt[2]{4^{-1}} = \sqrt{\tfrac{1}{4}} = \tfrac{1}{2}.$$

Below are the basic laws of exponents and radicals.†

Law	*Example*
1. $x^m \cdot x^n = x^{m+n}$.	$2^3 \cdot 2^5 = 2^8 = 256$; $x^2 \cdot x^3 = x^5$.
2. $x^0 = 1$ if $x \neq 0$.	$2^0 = 1$.
3. $x^{-n} = \dfrac{1}{x^n}$.	$2^{-3} = \dfrac{1}{2^3} = \dfrac{1}{8}$.
4. $\dfrac{1}{x^{-n}} = x^n$.	$\dfrac{1}{2^{-3}} = 2^3 = 8$; $\dfrac{1}{x^{-5}} = x^5$.
5. $\dfrac{x^m}{x^n} = x^{m-n} = \dfrac{1}{x^{n-m}}$.	$\dfrac{2^{12}}{2^8} = 2^4 = 16$; $\dfrac{x^8}{x^{12}} = \dfrac{1}{x^4}$.
6. $\dfrac{x^m}{x^m} = 1$ if $x \neq 0$.	$\dfrac{2^4}{2^4} = 1$.
7. $(x^m)^n = x^{mn}$.	$(2^3)^5 = 2^{15}$; $(x^2)^3 = x^6$.
8. $(xy)^n = x^n y^n$.	$(2 \cdot 4)^3 = 2^3 \cdot 4^3 = 8 \cdot 64$.
9. $\left(\dfrac{x}{y}\right)^n = \dfrac{x^n}{y^n}$.	$\left(\dfrac{2}{3}\right)^3 = \dfrac{2^3}{3^3}$; $\left(\dfrac{1}{3}\right)^5 = \dfrac{1^5}{3^5} = \dfrac{1}{3^5} = 3^{-5}$.
10. $\left(\dfrac{x}{y}\right)^{-n} = \left(\dfrac{y}{x}\right)^n$.	$\left(\dfrac{3}{4}\right)^{-2} = \left(\dfrac{4}{3}\right)^2 = \dfrac{4^2}{3^2} = \dfrac{16}{9}$.
11. $x^{1/n} = \sqrt[n]{x}$.	$3^{1/5} = \sqrt[5]{3}$.

†Although some laws involve restrictions, they are not vital to our discussion.

12. $x^{-1/n} = \frac{1}{x^{1/n}} = \frac{1}{\sqrt[n]{x}}$. $\qquad 4^{-1/2} = \frac{1}{4^{1/2}} = \frac{1}{\sqrt{4}} = \frac{1}{2}$.

13. $\sqrt[n]{x}\sqrt[n]{y} = \sqrt[n]{xy}$. $\qquad \sqrt[3]{9}\sqrt[3]{2} = \sqrt[3]{18}$.

14. $\frac{\sqrt[n]{x}}{\sqrt[n]{y}} = \sqrt[n]{\frac{x}{y}}$. $\qquad \frac{\sqrt[3]{90}}{\sqrt[3]{10}} = \sqrt[3]{\frac{90}{10}} = \sqrt[3]{9}$.

15. $\sqrt[m]{\sqrt[n]{x}} = \sqrt[mn]{x}$. $\qquad \sqrt[3]{\sqrt[4]{2}} = \sqrt[12]{2}$.

16. $x^{m/n} = \sqrt[n]{x^m} = (\sqrt[n]{x})^m$. $\qquad 8^{2/3} = \sqrt[3]{8^2} = (\sqrt[3]{8})^2 = 2^2 = 4$.

17. $(\sqrt[m]{x})^m = x$. $\qquad (\sqrt[8]{7})^8 = 7$.

EXAMPLE 2

a. By Law 1,

$$\begin{aligned} x^6x^8 &= x^{6+8} = x^{14}, \\ a^3b^2a^5b &= a^3a^5b^2b = a^8b^3, \\ x^{11}x^{-5} &= x^{11-5} = x^6, \\ z^{2/5}z^{3/5} &= z^1 = z, \\ xx^{1/2} &= x^1x^{1/2} = x^{3/2}. \end{aligned}$$

b. By Law 16,

$$\left(\frac{1}{4}\right)^{3/2} = \left(\sqrt{\frac{1}{4}}\right)^3 = \left(\frac{1}{2}\right)^3 = \frac{1}{8}.$$

c. $$\left(-\frac{8}{27}\right)^{4/3} = \left(\sqrt[3]{\frac{-8}{27}}\right)^4 = \left(\frac{\sqrt[3]{-8}}{\sqrt[3]{27}}\right)^4 \qquad \text{(Laws 16 and 14)}$$
$$= \left(\frac{-2}{3}\right)^4$$
$$= \frac{(-2)^4}{3^4} = \frac{16}{81} \qquad \text{(Law 9).}$$

d. $$(64a^3)^{2/3} = 64^{2/3}(a^3)^{2/3} \qquad \text{(Law 8)}$$
$$= (\sqrt[3]{64})^2a^2 \qquad \text{(Laws 16 and 7)}$$
$$= (4)^2a^2 = 16a^2.$$

Rationalizing the denominator of a fraction is a procedure in which a fraction having a radical in its denominator is expressed as an equivalent fraction without a radical in its denominator. We use the fundamental principle of fractions.

EXAMPLE 3

Rationalize the denominators.

a. $$\frac{2}{\sqrt{5}} = \frac{2}{5^{1/2}} = \frac{2 \cdot 5^{1/2}}{5^{1/2} \cdot 5^{1/2}} = \frac{2 \cdot 5^{1/2}}{5^1} = \frac{2\sqrt{5}}{5}.$$

b. $$\frac{2}{\sqrt[6]{3x^5}} = \frac{2}{\sqrt[6]{3} \cdot \sqrt[6]{x^5}} = \frac{2}{3^{1/6}x^{5/6}} = \frac{2 \cdot 3^{5/6}x^{1/6}}{3^{1/6}x^{5/6} \cdot 3^{5/6}x^{1/6}}$$

$$= \frac{2(3^5x)^{1/6}}{3x} = \frac{2\sqrt[6]{3^5x}}{3x}.$$

The following examples illustrate various applications of the laws of exponents and radicals.

EXAMPLE 4

a. *Eliminate negative exponents in* $\frac{x^{-2}y^3}{z^{-2}}$.

$$\frac{x^{-2}y^3}{z^{-2}} = \frac{y^3z^2}{x^2}.$$

Thus we can bring a factor of the numerator down to the denominator by changing the sign of the exponent, and vice versa.

b. *Simplify* $\frac{x^2y^7}{x^3y^5}$.

$$\frac{x^2y^7}{x^3y^5} = \frac{y^{7-5}}{x^{3-2}} = \frac{y^2}{x}.$$

c. *Eliminate negative exponents in* $x^{-1} + y^{-1}$ *and simplify.*

$$x^{-1} + y^{-1} = \frac{1}{x} + \frac{1}{y} = \frac{y + x}{xy}. \qquad \left(\text{Note: } x^{-1} + y^{-1} \neq \frac{1}{x + y}.\right)$$

d. *Simplify* $x^{3/2} - x^{1/2}$ *by using the distributive property.*

$$x^{3/2} - x^{1/2} = x^{1/2}(x - 1).$$

e. *Simplify* $(x^5y^8)^5$.

$$(x^5y^8)^5 = x^{25}y^{40}.$$

f. *Simplify* $(x^{5/9}y^{4/3})^{18}$.

$$(x^{5/9}y^{4/3})^{18} = (x^{5/9})^{18}(y^{4/3})^{18} = x^{10}y^{24}.$$

g. *Simplify* $\left(\frac{x^{1/5}y^{6/5}}{z^{2/5}}\right)^5$.

$$\left(\frac{x^{1/5}y^{6/5}}{z^{2/5}}\right)^5 = \frac{(x^{1/5}y^{6/5})^5}{(z^{2/5})^5} = \frac{xy^6}{z^2}.$$

h. *Eliminate negative exponents in* $7x^{-2} + (7x)^{-2}$.

$$7x^{-2} + (7x)^{-2} = \frac{7}{x^2} + \frac{1}{(7x)^2} = \frac{7}{x^2} + \frac{1}{49x^2}.$$

i. *Eliminate negative exponents in* $(x^{-1} - y^{-1})^{-2}$.

$$(x^{-1} - y^{-1})^{-2} = \left(\frac{1}{x} - \frac{1}{y}\right)^{-2} = \left(\frac{y - x}{xy}\right)^{-2}$$

$$= \left(\frac{xy}{y - x}\right)^2 = \frac{x^2y^2}{(y - x)^2}.$$

j. *Apply the distributive law to* $x^{2/5}(y^{1/2} + 2z^{6/5})$.

$$x^{2/5}(y^{1/2} + 2z^{6/5}) = x^{2/5}y^{1/2} + 2x^{2/5}z^{6/5}.$$

k. *Simplify* $\frac{x^3}{y^2} \div \frac{x^6}{y^5}$.

$$\frac{x^3}{y^2} \div \frac{x^6}{y^5} = \frac{x^3}{y^2} \cdot \frac{y^5}{x^6} = \frac{y^3}{x^3}.$$

EXAMPLE 5

a. *Simplify* $\sqrt[4]{48}$.

$$\sqrt[4]{48} = \sqrt[4]{16 \cdot 3} = \sqrt[4]{16}\,\sqrt[4]{3} = 2\sqrt[4]{3}.$$

b. *Use exponents to rewrite* $\sqrt{2 + 5x}$.

$$\sqrt{2 + 5x} = (2 + 5x)^{1/2}.$$

c. *Rationalize the denominator of* $\frac{\sqrt[5]{2}}{\sqrt[3]{6}}$ *and simplify.*

$$\frac{\sqrt[5]{2}}{\sqrt[3]{6}} = \frac{2^{1/5} \cdot 6^{2/3}}{6^{1/3} \cdot 6^{2/3}} = \frac{2^{3/15}6^{10/15}}{6} = \frac{(2^3 6^{10})^{1/15}}{6} = \frac{\sqrt[15]{2^3 6^{10}}}{6}.$$

d. *Simplify* $\dfrac{\sqrt{20}}{\sqrt{5}}$.

$$\frac{\sqrt{20}}{\sqrt{5}} = \sqrt{\frac{20}{5}} = \sqrt{4} = 2.$$

e. *Simplify* $\sqrt[3]{x^6y^4}$.

$$\sqrt[3]{x^6y^4} = \sqrt[3]{(x^2)^3y^3y} = \sqrt[3]{(x^2)^3} \cdot \sqrt[3]{y^3} \cdot \sqrt[3]{y}$$
$$= x^2y\sqrt[3]{y}.$$

f. *Simplify* $\sqrt{\dfrac{2}{7}}$.

$$\sqrt{\frac{2}{7}} = \sqrt{\frac{2}{7}\cdot\frac{7}{7}} = \sqrt{\frac{14}{7^2}} = \frac{\sqrt{14}}{\sqrt{7^2}} = \frac{\sqrt{14}}{7}.$$

g. *Simplify* $\sqrt{250} - \sqrt{50} + 15\sqrt{2}$.

$$\sqrt{250} - \sqrt{50} + 15\sqrt{2} = \sqrt{25 \cdot 10} - \sqrt{25 \cdot 2} + 15\sqrt{2}$$
$$= 5\sqrt{10} - 5\sqrt{2} + 15\sqrt{2}$$
$$= 5\sqrt{10} + 10\sqrt{2}.$$

h.
$$\sqrt{x^2} = \begin{cases} x, & \text{if } x \text{ is positive,} \\ -x, & \text{if } x \text{ is negative,} \\ 0, & \text{if } x = 0. \end{cases}$$

Thus, $\sqrt{2^2} = 2$ and $\sqrt{(-3)^2} = -(-3) = 3$.

Exercise A-2

Evaluate the following expressions.

1. $\sqrt{25}$.
2. $\sqrt[3]{64}$.
3. $\sqrt[5]{-32}$.
4. $\sqrt{.04}$.
5. $\sqrt[4]{\frac{1}{16}}$.
6. $\sqrt[3]{-\frac{8}{27}}$.
7. $(100)^{1/2}$.
8. $(64)^{1/3}$.
9. $4^{3/2}$.
10. $(25)^{-3/2}$.
11. $(32)^{-2/5}$.
12. $(.09)^{-1/2}$.
13. $\left(\dfrac{1}{16}\right)^{5/4}$.
14. $\left(-\dfrac{27}{64}\right)^{2/3}$.

Simplify the following expressions.

15. $\sqrt{32}$.
16. $\sqrt[3]{24}$.
17. $\sqrt[3]{2x^3}$.
18. $\sqrt{4x}$.
19. $\sqrt{16x^4}$.
20. $\sqrt[4]{x/16}$.
21. $(9z^4)^{1/2}$.
22. $(16y^8)^{3/4}$.
23. $\left(\frac{27t^3}{8}\right)^{2/3}$.
24. $\left(\frac{1000}{a^9}\right)^{-2/3}$.

Write the following expressions in terms of positive exponents only. Avoid all radicals in the final form. For example, $y^{-1}\sqrt{x} = \frac{x^{1/2}}{y}$.

25. $\frac{x^3y^{-2}}{z^2}$.
26. $\sqrt[5]{x^2y^3z^{-10}}$.
27. $2x^{-1}x^{-3}$.
28. $x + y^{-1}$.
29. $(3t)^{-2}$.
30. $(3 - z)^{-4}$.
31. $\sqrt[3]{7s^2}$.
32. $(x^{-2}y^2)^{-2}$.
33. $\sqrt{x} - \sqrt{y}$.
34. $\frac{x^{-2}y^{-6}z^2}{xy^{-1}}$.
35. $x^2\sqrt[4]{xy^{-2}z^3}$.
36. $\left(\sqrt[5]{xy^{-3}}\right)x^{-1}y^{-2}$.

Write the following exponential forms in equivalent forms involving radicals.

37. $(8x - y)^{4/5}$.
39. $(ab^2c^3)^{3/4}$.
39. $x^{-4/5}$.
40. $2x^{1/2} - (2y)^{1/2}$.
41. $2x^{-2/5} - (2x)^{-2/5}$.
42. $[(x^{-4})^{1/5}]^{1/6}$.

Simplify the following. Express all answers in terms of positive exponents. Rationalize the denominator where necessary to avoid fractional exponents in the denominator.

43. $2x^2y^{-3}x^4$.
44. $\frac{2}{x^{3/2}y^{1/3}}$.
45. $\sqrt{\sqrt[3]{t^4}}$.
46. $\{[(2x^2)^3]^{-4}\}^{-1}$.
47. $\frac{2^0}{(2^{-2}x^{1/2}y^{-2})^3}$.
48. $\frac{\sqrt{s^5}}{\sqrt[3]{s^2}}$.
49. $\sqrt[3]{x^2yz^3}\,\sqrt[3]{xy^2}$.
50. $(\sqrt[5]{2})^{10}$.
51. $3^2(27)^{-4/3}$.
52. $\left(\sqrt[5]{x^2y}\right)^{2/5}$.

53. $(2x^{-1}y^2)^2$.

54. $\dfrac{3}{\sqrt[3]{y}\,\sqrt[4]{x}}$.

55. $\sqrt{x}\,\sqrt{x^2y^3}\,\sqrt{xy^2}$.

56. $\sqrt{75k^4}$.

57. $\dfrac{(x^2y^{-1}z)^{-2}}{(xy^2)^{-4}}$.

58. $\sqrt{6(6)}$.

59. $\dfrac{(x^2)^3}{x^4} \div \left[\dfrac{x^3}{(x^3)^2}\right]^{-2}$.

60. $\sqrt{(-6)(-6)}$.

61. $-\dfrac{8s^{-2}}{2s^3}$.

62. $(x^{-1}y^{-2}\sqrt{z}\,)^4$.

63. $(2x^2y \div 3y^3z^{-2})^2$.

64. $\dfrac{1}{\left(\dfrac{\sqrt{2}\,x^{-2}}{\sqrt{16}\,x^3}\right)^2}$.

A-3 OPERATIONS WITH ALGEBRAIC EXPRESSIONS

The algebraic expression $5ax^3 - 2bx + 3$ consists of three *terms*: $+5ax^3$, $-2bx$, and $+3$. Some of the *factors* of the first term $5ax^3$ are 5, a, x, x^2, x^3, $5ax$, and ax^2. Also, $5a$ is the *coefficient* of x^3 and 5 is the *numerical coefficient* of ax^3. If a and b represent fixed numbers throughout a discussion, then a and b are called *constants*.

Algebraic expressions with exactly one term are called *monomials*. Those having exactly two terms are *binomials*, and those with exactly three terms are *trinomials*. Algebraic expressions with more than one term are called *multinomials*. Thus the multinomial $2x - 5$ is a binomial, and the multinomial $3\sqrt{y} + 2y - 4y^2$ is a trinomial.

A *polynomial in x* is an algebraic expression of the form

$$c_0x^n + c_1x^{n-1} + \cdots + c_{n-1}x + c_n^{\dagger}$$

where n is a positive integer and $c_0, c_1, \ldots, c_n$ are constants with $c_0 \neq 0$. We call n the degree of the polynomial. Hence $4x^3 - 5x^2 + x - 2$ is a

†The three dots indicate the terms that are understood to be included in the sum.

polynomial in x of degree 3, and $y^5 - 2$ is a polynomial in y of degree 5. A nonzero constant is treated as a polynomial of degree zero; thus 5 is a polynomial of degree zero. The constant 0 is also a polynomial; however, no degree is assigned to it.

Below is a list of special products which may be obtained from the distributive property and are useful in multiplying algebraic expressions.

SPECIAL PRODUCTS

(I) $x(y + z) = xy + xz$ (Distributive property).

(II) $(x + a)(x + b) = x^2 + (a + b)x + ab$.

(III) $(ax + c)(bx + d) = abx^2 + (ad + cb)x + cd$.

(IV) $(x + a)^2 = x^2 + 2ax + a^2$ (Square of a binomial).

(V) $(x - a)^2 = x^2 - 2ax + a^2$ (Square of a binomial).

(VI) $(x + a)(x - a) = x^2 - a^2$ (Product of sum and difference).

EXAMPLE 1

a. By II, $(x + 2)(x - 5) = [x + 2][x + (-5)]$
$$= x^2 + (2 - 5)x + 2(-5)$$
$$= x^2 - 3x - 10.$$

b. By III, $(3z + 5)(7z + 4) = 3 \cdot 7z^2 + (3 \cdot 4 + 5 \cdot 7)z + 5 \cdot 4$
$$= 21z^2 + 47z + 20.$$

c. By V, $(x - 4)^2 = x^2 - 2(4)x + 4^2$
$$= x^2 - 8x + 16.$$

d. By VI,

$$(\sqrt{y^2 + 1} + 3)(\sqrt{y^2 + 1} - 3) = \left[(y^2 + 1)^{1/2} + 3\right]\left[(y^2 + 1)^{1/2} - 3\right]$$
$$= \left[(y^2 + 1)^{1/2}\right]^2 - 3^2$$
$$= (y^2 + 1) - 9$$
$$= y^2 - 8.$$

EXAMPLE 2

Multiply $(2t - 3)(5t^2 + 3t - 1)$.

We treat $2t - 3$ as a single number and apply the distributive property.

$$\begin{aligned}(2t-3)(5t^2+3t-1) &= (2t-3)5t^2 + (2t-3)3t - (2t-3)1\\ &= 10t^3 - 15t^2 + 6t^2 - 9t - 2t + 3\\ &= 10t^3 - 9t^2 - 11t + 3.\end{aligned}$$

To divide a multinomial by a monomial, divide each term in the multinomial by the monomial.

EXAMPLE 3

a. $\dfrac{x^3+3x}{x} = \dfrac{x^3}{x} + \dfrac{3x}{x} = x^2 + 3.$

b. $$\begin{aligned}\frac{4z^3 - 8z^2 + 3z - 6}{2z} &= \frac{4z^3}{2z} - \frac{8z^2}{2z} + \frac{3z}{2z} - \frac{6}{2z}\\ &= 2z^2 - 4z + \frac{3}{2} - \frac{3}{z}.\end{aligned}$$

To divide a polynomial by a polynomial we use so called "long division."

EXAMPLE 4

Divide $2x^3 - 14x - 5$ *by* $x - 3$.

Here $2x^3 - 14x - 5$ is the *dividend* and $x - 3$ is the *divisor.* To avoid errors it is best to write the dividend as $2x^3 + 0x^2 - 14x - 5$. Note that the powers of x are in decreasing order.

$$\begin{array}{r|l} & 2x^2 + 6x + 4 \leftarrow \text{Quotient}\\ \hline x-3 & 2x^3 + 0x^2 - 14x - 5\\ & 2x^3 - 6x^2\\ \hline & \qquad 6x^2 - 14x\\ & \qquad 6x^2 - 18x\\ \hline & \qquad\qquad 4x - 5\\ & \qquad\qquad 4x - 12\\ \hline & \qquad\qquad\quad 7 \leftarrow \text{Remainder.}\end{array}$$

Here we divided x into $2x^3$ and got $2x^2$. Then we multiplied $2x^2$ by $x - 3$, getting $2x^3 - 6x^2$. After subtracting $2x^3 - 6x^2$ from $2x^3 + 0x^2$, we obtained $6x^2$ and then "brought down" the term $-14x$. This process is continued until we arrive at 7, the *remainder.* We always stop when the remainder is a polynomial whose degree is less than the degree of the divisor. Our answer may be written as

$$2x^2 + 6x + 4 + \frac{7}{x - 3}.$$

A way of checking a division is to verify that

$$\text{(Quotient)(Divisor)} + \text{Remainder} = \text{Dividend}.$$

By using this equation you should verify the result of the example.

Exercise A-3

Simplify the following expressions.

1. $(x + 4)(x + 5)$.

2. $(x + 3)(x + 2)$.

3. $(x + 3)(x - 2)$.

4. $(z - 7)(z - 3)$.

5. $(2x + 3)(5x + 2)$.

6. $(y - 4)(2y + 3)$.

7. $(x + 3)^2$.

8. $(2x - 1)^2$.

9. $(x - 5)^2$.

10. $(\sqrt{x} - 1)(2\sqrt{x} + 5)$.

11. $(\sqrt{2y} + 3)^2$.

12. $(y - 3)(y + 3)$.

13. $(2s - 1)(2s + 1)$.

14. $(z^2 - 3w)(z^2 + 3w)$.

15. $(x^2 - 3)(x + 4)$.

16. $(x + 1)(x^2 + x + 3)$.

17. $(x^2 - 1)(2x^2 + 2x - 3)$.

18. $(2x - 1)(3x^3 + 7x^2 - 5)$.

19. $x\{3(x - 1)(x - 2) + 2[x(x + 7)]\}$.

20. $[(2z + 1)(2z - 1)](4z^2 + 1)$.

21. $(x + y + 2)(3x + 2y - 4)$.

22. $(x^2 + x + 1)^2$.

23. $\dfrac{z^2 - 4z}{z}$.

24. $\dfrac{2x^3 - 7x + 4}{x}$.

25. $\dfrac{6x^5 + 4x^3 - 1}{2x^2}$.

26. $\dfrac{(3x - 4) - (x + 8)}{4x}$.

27. $(x^2 + 3x - 1) \div (x + 3)$.

28. $(x^2 - 5x + 4) \div (x - 4)$.

29. $(3x^3 - 2x^2 + x - 3) \div (x + 2)$.

30. $(x^4 + 2x^2 + 1) \div (x - 1)$.

31. $t^2 \div (t - 8)$.

32. $(4x^2 + 6x + 1) \div (2x - 1)$.

33. $(3x^2 - 4x + 3) \div (3x + 2)$.

34. $(z^3 + z^2 + z) \div (z^2 - z + 1)$.

A-4 FACTORING

If two or more expressions are multiplied together, the expressions are called *factors* of the product. Thus if $c = ab$, then a and b are both factors of the product c. The process by which an expression is written as a product of its factors is called *factoring*.

Listed below are the special products discussed in Section A-3. The right side of each identity is the factored form of the left side.

(I) $xy + xz = x(y + z)$ (Common factor).

(II) $x^2 + (a + b)x + ab = (x + a)(x + b)$.

(III) $abx^2 + (ad + cb)x + cd = (ax + c)(bx + d)$.

(IV) $x^2 + 2ax + a^2 = (x + a)^2$ (Perfect-square trinomial).

(V) $x^2 - 2ax + a^2 = (x - a)^2$ (Perfect-square trinomial).

(VI) $x^2 - a^2 = (x + a)(x - a)$ (Difference of two squares).

When factoring a polynomial we usually choose factors which themselves are polynomials. For example, $x^2 - 4 = (x + 2)(x - 2)$. We shall not write $x - 4$ as $(\sqrt{x} + 2)(\sqrt{x} - 2)$.

Always factor completely. For example,

$$2x^2 - 8 = 2(x^2 - 4) = 2(x + 2)(x - 2).$$

EXAMPLE 1

a. *Completely factor* $3k^2x^2 + 9k^3x$.

Since $3k^2x^2 = (3k^2x)(x)$ and $9k^3x = (3k^2x)(3k)$, each term of the original expression contains the common factor $3k^2x$. Thus by Rule I, $3k^2x^2 + 9k^3x = 3k^2x(x + 3k)$. Note that although $3k^2x^2 + 9k^3x = 3(k^2x^2 + 3k^3x)$, we do not say that the expression is completely factored since $k^2x^2 + 3k^3x$ can yet be factored.

b. *Completely factor* $8a^5x^2y^3 - 6a^2b^3yz - 2a^4b^4xy^2z^2$.

$$8a^5x^2y^3 - 6a^2b^3yz - 2a^4b^4xy^2z^2 = 2a^2y(4a^3x^2y^2 - 3b^3z - a^2b^4xyz^2).$$

c. *Completely factor* $3x^2 + 6x + 3$.

$$\begin{aligned} 3x^2 + 6x + 3 &= 3(x^2 + 2x + 1) \\ &= 3(x + 1)^2 \qquad \text{(Rule IV).} \end{aligned}$$

EXAMPLE 2

a. *Completely factor* $x^2 - x - 6$.

If this trinomial factors into the form $x^2 - x - 6 = (x + a)(x + b)$, which is a product of two binomials, then we must determine a and b. Since $(x + a)(x + b) = x^2 + (a + b)x + ab$, then

$$x^2 + (-1)x + (-6) = x^2 + (a + b)x + ab.$$

By equating corresponding coefficients, we want

$$a + b = -1 \quad \text{and} \quad ab = -6.$$

If $a = -3$ and $b = 2$, then both conditions are met and hence

$$x^2 - x - 6 = (x - 3)(x + 2).$$

b. *Completely factor* $x^2 - 7x + 12$.

$$x^2 - 7x + 12 = (x - 3)(x - 4).$$

EXAMPLE 3

Listed below are expressions that are completely factored. The numbers in parentheses refer to the rules used.

a. $x^2 + 8x + 16 = (x + 4)^2$ (IV).

b. $9x^2 + 9x + 2 = (3x + 1)(3x + 2)$ (III).

c. $6y^3 + 3y^2 - 18y = 3y(2y^2 + y - 6)$ (I).

$= 3y(2y - 3)(y + 2)$ (III).

d. $x^2 - 6x + 9 = (x - 3)^2$ (V).

e. $z^{1/4} + z^{5/4} = z^{1/4}(1 + z)$ (I).

f. $x^4 - 1 = (x^2 + 1)(x^2 - 1)$ (VI).

$= (x^2 + 1)(x + 1)(x - 1)$ (VI).

g. $x^{2/3} - 5x^{1/3} + 4 = (x^{1/3} - 1)(x^{1/3} - 4)$ (II).

h. $ax^2 - ay^2 + bx^2 - by^2 = (ax^2 - ay^2) + (bx^2 - by^2)$

$= a(x^2 - y^2) + b(x^2 - y^2)$ (I).

$= (a + b)(x^2 - y^2)$ (I).

$= (a + b)(x + y)(x - y)$ (VI).

Note in Example 3(f) that $x^2 - 1$ is factorable but $x^2 + 1$ is not. In 3(h) we factored by making use of grouping.

Exercise A-4

Completely factor the expressions.

1. $6x + 4$.

2. $6y^2 - 4y$.

3. $10xy + 5xz$.

4. $3x^2y - 9x^3y^3$.

5. $8a^3bc - 12ab^3cd + 4b^4c^2d^2$.

6. $6z^2t^3 + 3zst^4 - 12z^2t^3$.

7. $x^2 - 25$.

8. $x^2 + 3x - 4$.

9. $p^2 + 4p + 3$.

10. $s^2 - 6s + 8$.

11. $16x^2 - 9$.

12. $x^2 + 5x - 24$.

13. $z^2 + 6z + 8$.

14. $4t^2 - 9s^2$.

15. $x^2 + 6x + 9$.

16. $y^2 - 15y + 50$.

17. $2x^2 + 12x + 16$.

18. $2x^2 + 7x - 15$.

19. $3x^2 - 3$.

20. $4y^2 - 8y + 3$.

21. $6y^2 + 13y + 2$.

22. $4x^2 - x - 3$.

23. $12s^3 + 10s^2 - 8s$.

24. $9z^2 + 24z + 16$.

25. $x^{2/3}y - 4x^{8/3}y^3$.

26. $9x^{4/7} - 1$.

27. $2x^3 + 2x^2 - 12x$.

28. $x^2y^2 - 4xy + 4$.

29. $(4x + 2)^2$.

30. $3s^2(3s - 9s^2)^2$.

31. $(x + 3)^3(x - 1) + (x + 3)^2(x - 1)^2$.

32. $(x + 5)^2(x + 1)^3 + (x + 5)^3(x + 1)^2$.

33. $(x + 4)(2x + 1) + (x + 4)$.

34. $(x - 3)(2x + 3) - (2x + 3)(x + 5)$.

35. $x^4 - 16$.

36. $81x^4 - y^4$.

37. $y^8 - 1$.

38. $t^4 - 4$.

39. $x^4 + x^2 - 2$.

40. $x^4 - 5x^2 + 4$.

41. $x^5 - 2x^3 + x$.

42. $4x^3 - 6x^2 - 4x$.

43. $x^3y^2 - 10x^2y + 25x$.

44. $(3x^2 + x) + (6x + 2)$.

45. $(x^3 - 4x) + (8 - 2x^2)$.

46. $(x^2 - 1) + (x^2 - x - 2)$.

47. $(y^{10} + 8y^6 + 16y^2) - (y^8 + 8y^4 + 16)$.

48. $x^3y - xy + z^2x^2 - z^2$.

A-5 FRACTIONS

By using the fundamental principle of fractions (Sec. A-1), we may be able to simplify fractions. That principle allows us to multiply or divide both numerator and denominator of a fraction by the same nonzero quantity. The resulting fraction will be equivalent to the original one. The fractions that we shall consider are assumed to have nonzero denominators.

EXAMPLE 1

a. *Simplify* $\frac{x^2 - x - 6}{x^2 - 7x + 12}$.

First, we completely factor the numerator and denominator:

$$\frac{x^2 - x - 6}{x^2 - 7x + 12} = \frac{(x - 3)(x + 2)}{(x - 3)(x - 4)}.$$

Dividing both numerator and denominator by the common factor $x - 3$, we have

$$\frac{(x - 3)(x + 2)}{(x - 3)(x - 4)} = \frac{1(x + 2)}{1(x - 4)} = \frac{x + 2}{x - 4}.$$

Usually we just write

$$\frac{x^2 - x - 6}{x^2 - 7x + 12} = \frac{(x - 3)(x + 2)}{(x - 3)(x - 4)} = \frac{x + 2}{x - 4}.$$

b. *Simplify* $\dfrac{2x^2 + 6x - 8}{8 - 4x - 4x^2}$.

$$\frac{2x^2 + 6x - 8}{8 - 4x - 4x^2} = \frac{2(x - 1)(x + 4)}{4(1 - x)(2 + x)}$$

$$= \frac{2(x - 1)(x + 4)}{2(2)[(-1)(x - 1)](2 + x)}$$

$$= \frac{x + 4}{-2(x + 2)}$$

$$= -\frac{x + 4}{2(x + 2)}.$$

If we wish to multiply $\frac{a}{b}$ by $\frac{c}{d}$, then

$$\frac{a}{b} \cdot \frac{c}{d} = \frac{ac}{bd}.$$

To divide $\frac{a}{b}$ by $\frac{c}{d}$, where $c \neq 0$, we have

$$\frac{a}{b} \div \frac{c}{d} = \frac{\frac{a}{b}}{\frac{c}{d}} = \frac{a}{b} \cdot \frac{d}{c}.$$

EXAMPLE 2

a. $\dfrac{x}{x + 2} \cdot \dfrac{x + 3}{x - 5} = \dfrac{x(x + 3)}{(x + 2)(x - 5)}$.

b. $$\frac{x^2 - 4x + 4}{x^2 + 2x - 3} \cdot \frac{6x^2 - 6}{x^2 + 2x - 8} = \frac{[(x - 2)^2][6(x + 1)(x - 1)]}{[(x + 3)(x - 1)][(x + 4)(x - 2)]}$$

$$= \frac{6(x - 2)(x + 1)}{(x + 3)(x + 4)}.$$

c. $$\frac{\frac{x - 5}{x - 3}}{2x} = \frac{\frac{x - 5}{x - 3}}{\frac{2x}{1}} = \frac{x - 5}{x - 3} \cdot \frac{1}{2x} = \frac{x - 5}{2x(x - 3)}.$$

d. $$\frac{\frac{4x}{x^2-1}}{\frac{2x^2+8x}{x-1}} = \frac{4x}{x^2-1} \cdot \frac{x-1}{2x^2+8x} = \frac{4x(x-1)}{[(x+1)(x-1)][2x(x+4)]}$$

$$= \frac{2}{(x+1)(x+4)}.$$

e. *Rationalize the denominator of* $\frac{x}{\sqrt{2}-6}$.

$$\frac{x}{\sqrt{2}-6} = \frac{x}{\sqrt{2}-6} \cdot \frac{\sqrt{2}+6}{\sqrt{2}+6}$$

$$= \frac{x(\sqrt{2}+6)}{(\sqrt{2})^2-6^2} = \frac{x(\sqrt{2}+6)}{2-36}$$

$$= -\frac{x(\sqrt{2}+6)}{34}.$$

To add two fractions having a common denominator, use the rule $\frac{a}{c} + \frac{b}{c} = \frac{a+b}{c}$. Similarly, $\frac{a}{c} - \frac{b}{c} = \frac{a-b}{c}$.

EXAMPLE 3

$$\frac{p^2-5}{p-2} + \frac{3p+2}{p-2} = \frac{(p^2-5)+(3p+2)}{p-2} = \frac{p^2+3p-3}{p-2}.$$

To add (or subtract) two fractions with different denominators, transform the fractions by the fundamental principle of fractions into equivalent fractions that have the same denominator. Then proceed with the addition (or subtraction) by the method described above.

For example, to find

$$\frac{2}{x^3(x-3)} + \frac{3}{x(x-3)^2},$$

we can convert the first fraction into the equivalent fraction

$$\frac{2(x-3)}{x^3(x-3)^2},$$

and we can convert the second fraction into

$$\frac{3x^2}{x^3(x-3)^2}.$$

These fractions have the same denominator. Hence,

$$\frac{2}{x^3(x-3)} + \frac{3}{x(x-3)^2} = \frac{2(x-3)}{x^3(x-3)^2} + \frac{3x^2}{x^3(x-3)^2}$$

$$= \frac{3x^2 + 2x - 6}{x^3(x-3)^2}.$$

We could have converted the original fractions into equivalent fractions with any common denominator. However, we chose to convert them into fractions with the denominator $x^3(x-3)^2$. This is the **least common denominator** (L.C.D.) of the fractions $2/[x^3(x-3)]$ and $3/[x(x-3)^2]$.

In general, to find the L.C.D. of two or more fractions, first factor each denominator completely. *The L.C.D. is the product of each of the distinct factors appearing in the denominators, each raised to the highest power to which it occurs in any one denominator.*

EXAMPLE 4

a. *Subtract:* $\frac{t}{3t+2} - \frac{4}{t-1}$.

Here the denominators are already factored. The L.C.D. is $(3t+2)(t-1)$.

$$\frac{t}{3t+2} - \frac{4}{t-1} = \frac{t(t-1)}{(3t+2)(t-1)} - \frac{4(3t+2)}{(3t+2)(t-1)}$$

$$= \frac{t(t-1) - 4(3t+2)}{(3t+2)(t-1)}$$

$$= \frac{t^2 - t - 12t - 8}{(3t+2)(t-1)}$$

$$= \frac{t^2 - 13t - 8}{(3t+2)(t-1)}.$$

b. $\frac{4}{q-1} + 3 = \frac{4}{q-1} + \frac{3(q-1)}{q-1}$

$$= \frac{4 + 3(q-1)}{q-1} = \frac{3q+1}{q-1}.$$

EXAMPLE 5

$$\frac{x-2}{x^2+6x+9} - \frac{x+2}{2(x^2-9)}$$

$$= \frac{x-2}{(x+3)^2} - \frac{x+2}{2(x+3)(x-3)} \qquad [\text{L.C.D.} = 2(x+3)^2(x-3)]$$

$$= \frac{(x-2)(2)(x-3)}{(x+3)^2(2)(x-3)} - \frac{(x+2)(x+3)}{2(x+3)(x-3)(x+3)}$$

$$= \frac{(x-2)(2)(x-3) - (x+2)(x+3)}{2(x+3)^2(x-3)}$$

$$= \frac{2(x^2-5x+6) - [x^2+5x+6]}{2(x+3)^2(x-3)}$$

$$= \frac{2x^2-10x+12-x^2-5x-6}{2(x+3)^2(x-3)}$$

$$= \frac{x^2-15x+6}{2(x+3)^2(x-3)}.$$

EXAMPLE 6

$$\frac{\frac{1}{x+h} - \frac{1}{x}}{h} = \frac{\frac{x}{x(x+h)} - \frac{x+h}{x(x+h)}}{h} = \frac{\frac{x-(x+h)}{x(x+h)}}{h}$$

$$= \frac{\frac{-h}{x(x+h)}}{\frac{h}{1}} = \frac{-h}{x(x+h)h} = -\frac{1}{x(x+h)}.$$

The original fraction can also be simplified by multiplying the numerator and denominator by $x(x+h)$:

$$\frac{\frac{1}{x+h} - \frac{1}{x}}{h} = \frac{x(x+h)\left[\frac{1}{x+h} - \frac{1}{x}\right]}{x(x+h)h}$$

$$= \frac{x-(x+h)}{x(x+h)h} = \frac{-h}{x(x+h)h}$$

$$= -\frac{1}{x(x+h)}.$$

Exercise A-5

*In Problems **1–28**, perform the operations and simplify as much as possible.*

1. $\frac{x^2}{x+3} + \frac{5x+6}{x+3}.$

2. $\frac{2}{x+2} + \frac{x}{x+2}.$

3. $\frac{1}{t} + \frac{2}{3t}$.

4. $\frac{4}{x^2} - \frac{1}{x}$.

5. $1 - \frac{p^2}{p^2 - 1}$.

6. $\frac{4}{s + 4} + s$.

7. $\frac{4}{2x - 1} + \frac{x}{x + 3}$.

8. $\frac{x + 1}{x - 1} - \frac{x - 1}{x + 1}$.

9. $\frac{1}{x^2 - x - 2} + \frac{1}{x^2 - 1}$.

10. $\frac{y}{3y^2 - 5y - 2} - \frac{2}{3y^2 - 7y + 2}$.

11. $\frac{4}{x - 1} - 3 + \frac{-3x^2}{5 - 4x - x^2}$.

12. $\frac{2x - 3}{2x^2 + 11x - 6} - \frac{3x + 1}{3x^2 + 16x - 12} + \frac{1}{3x - 2}$.

13. $\frac{y^2}{y - 3} \cdot \frac{-1}{y + 2}$.

14. $\frac{z^2 - 4}{z^2 + 2z} \cdot \frac{z^2}{z - 2}$.

15. $\frac{2x - 3}{x - 2} \cdot \frac{2 - x}{2x + 3}$.

16. $\frac{x^2 - y^2}{x + y} \cdot \frac{x^2 + 2xy + y^2}{y - x}$.

17. $\dfrac{\frac{x^2}{6}}{\frac{x}{3}}$.

18. $\dfrac{\frac{4x^3}{9x}}{\frac{x}{18}}$.

19. $\dfrac{\frac{4x}{3}}{2x}$.

20. $\dfrac{-9x^3}{\frac{x}{3}}$.

21. $\frac{2x - 2}{x^2 - 2x - 8} \div \frac{x^2 - 1}{x^2 + 5x + 4}$.

22. $\frac{x^2 + 2x}{3x^2 - 18x + 24} \div \frac{x^2 - x - 6}{x^2 - 4x + 4}$.

23. $\dfrac{\frac{4x^2 - 9}{x^2 + 3x - 4}}{\frac{2x - 3}{1 - x^2}}$.

24. $\dfrac{\frac{6x^2y + 7xy - 3y}{xy - x + 5y - 5}}{\frac{x^3y + 4x^2y}{xy - x + 4y - 4}}$.

25. $\dfrac{1 + \frac{1}{x}}{3}$.

26. $\dfrac{\frac{x + 3}{x}}{x - \frac{9}{x}}$.

27. $\dfrac{3 - \dfrac{1}{2x}}{x + \dfrac{x}{x + 2}}$.

28. $\dfrac{\dfrac{x - 1}{x^2 + 5x + 6} - \dfrac{1}{x + 2}}{3 + \dfrac{x - 7}{3}}$.

In Problems **29** *annd* **30**, *simplify and express your answer in a form that is free of radicals in the denominator.*

29. $\dfrac{1}{x + \sqrt{5}}$.

30. $\dfrac{4}{\sqrt{x} + 2} \cdot \dfrac{x^2}{3}$.

A-6 EQUATIONS

The basic operations that are used to solve equations are:

(I) Adding (or subtracting) the same number to (or from) both sides of an equation;

(II) Multiplying (or dividing) both sides of an equation by the same *constant*, except zero.

EXAMPLE 1

Solve the following equations for p or x.

a. $2(p + 4) = 7p + 2$.

$$2(p + 4) = 7p + 2,$$

$$2p + 8 = 7p + 2 \quad \text{(distributive property)},$$

$$2p = 7p - 6 \quad \text{(subtracting 8 from both sides)},$$

$$-5p = -6 \quad \text{(subtracting } 7p \text{ from both sides)},$$

$$p = \frac{-6}{-5} \quad \text{(dividing both sides by } -5\text{)},$$

$$p = \frac{6}{5}.$$

b. $\dfrac{7x + 3}{2} - \dfrac{9x - 8}{4} = 6$.

We first clear the equation of fractions by multiplying *both* sides by the least common denominator (L.C.D.), which is 4.

$$4\left(\frac{7x + 3}{2} - \frac{9x - 8}{4}\right) = 4(6),$$
$$2(7x + 3) - (9x - 8) = 24,$$
$$14x + 6 - 9x + 8 = 24,$$
$$5x + 14 = 24,$$
$$5x = 10,$$
$$x = 2.$$

c. $(a + c)x + x^2 = (x + a)^2$.

$$(a + c)x + x^2 = (x + a)^2,$$
$$ax + cx + x^2 = x^2 + 2ax + a^2,$$
$$cx - ax = a^2,$$
$$x(c - a) = a^2,$$
$$x = \frac{a^2}{c - a}.$$

Actually, we assumed $c - a \neq 0$ to avoid division by 0.

EXAMPLE 2

Solve $\frac{5}{x - 4} = \frac{6}{x - 3}$.

To solve this equation we first clear it of fractions. Multiplying both sides by the L.C.D., $(x - 4)(x - 3)$, we have

$$(x - 4)(x - 3)\left(\frac{5}{x - 4}\right) = (x - 4)(x - 3)\left(\frac{6}{x - 3}\right),$$
$$5(x - 3) = 6(x - 4),$$
$$5x - 15 = 6x - 24,$$
$$9 = x.$$

In the first step we multiplied each side by an expression involving a *variable*, x. This means that we must check whether or not 9 satisfies the *original* equation. If 9 is substituted for x in that equation, the left side is

$$\frac{5}{9 - 4} = \frac{5}{5} = 1,$$

and the right side is

$$\frac{6}{9 - 3} = \frac{6}{6} = 1.$$

Since both sides are equal, 9 is a root (solution).

Quadratic equations have the form

$$ax^2 + bx + c = 0, \tag{1}$$

where a, b, and c are constants and $a \neq 0$. Equation (1) may be solved by first factoring the left side of (1) and then setting each factor equal to 0. This method also applies to many equations that are not quadratic. The roots of (1) are also given by the **quadratic formula**:

$$\boxed{x = \frac{-b \pm \sqrt{b^2 - 4ac}}{2a}.}$$

EXAMPLE 3

a. *Solve* $3x^2 - x - 4 = -2$.

$$3x^2 - x - 4 = -2,$$
$$3x^2 - x - 2 = 0,$$
$$(3x + 2)(x - 1) = 0.$$

Setting each factor equal to 0, we have

$$3x + 2 = 0, \quad 3x = -2, \quad x = -\frac{2}{3}. \qquad \Bigg| \qquad x - 1 = 0, \quad x = 1.$$

Thus $x = -\frac{2}{3}, 1$.

b. *Solve* $4x - 4x^3 = 0$.

Although the equation is not quadratic, the method of factoring applies.

$$4x - 4x^3 = 0,$$
$$4x(1 - x^2) = 0,$$
$$4x(1 - x)(1 + x) = 0.$$

Setting each factor equal to 0, we get

$$x = 0, 1, -1,$$

which we can write as $x = 0, \pm 1$.

c. *Solve* $x^2 = 3$.

$$x^2 - 3 = 0,$$
$$(x - \sqrt{3})(x + \sqrt{3}) = 0.$$

Thus $x - \sqrt{3} = 0$ or $x + \sqrt{3} = 0$. The roots are $\pm\sqrt{3}$. More generally,

$$\text{if } u^2 = k, \quad \text{then } u = \pm\sqrt{k}.$$

For example, the solution of $y^2 = 4$ is $y = \pm\sqrt{4} = \pm 2$.

EXAMPLE 4

a. *Solve* $4x^2 - 17x + 15 = 0$ *by the quadratic formula.*

Here $a = 4$, $b = -17$, and $c = 15$.

$$x = \frac{-b \pm \sqrt{b^2 - 4ac}}{2a} = \frac{-(-17) \pm \sqrt{(-17)^2 - 4(4)(15)}}{2(4)}$$

$$= \frac{17 \pm \sqrt{49}}{8} = \frac{17 \pm 7}{8}.$$

The roots are $\frac{17 + 7}{8} = \frac{24}{8} = 3$ and $\frac{17 - 7}{8} = \frac{10}{8} = \frac{5}{4}$.

b. *Solve* $2 + 6\sqrt{2}\,y + 9y^2 = 0$ *by the quadratic formula.*

Look at the arrangement of the terms. Here $a = 9$, $b = 6\sqrt{2}$, and $c = 2$.

$$y = \frac{-b \pm \sqrt{b^2 - 4ac}}{2a} = \frac{-6\sqrt{2} \pm \sqrt{0}}{2(9)}.$$

Thus, $y = \frac{-6\sqrt{2} + 0}{18} = -\frac{\sqrt{2}}{3}$ or $y = \frac{-6\sqrt{2} - 0}{18} = -\frac{\sqrt{2}}{3}$. The only root is $-\frac{\sqrt{2}}{3}$.

c. *Solve* $z^2 + z + 1 = 0$ *by the quadratic formula.*

Here $a = 1$, $b = 1$, and $c = 1$. The roots are

$$\frac{-b \pm \sqrt{b^2 - 4ac}}{2a} = \frac{-1 \pm \sqrt{-3}}{2}.$$

Since $\sqrt{-3}$ is not a real number, there are no real roots.†

† $\frac{-1 \pm \sqrt{-3}}{2}$ can be expressed as $\frac{-1 \pm i\sqrt{3}}{2}$, where $i (= \sqrt{-1})$ is called the *imaginary unit.*

Exercise A-6

In Problems **1–22**, *solve the equations.*

1. $6x = 45.$

2. $3y = 0.$

3. $5x - 3 = 9.$

4. $7x + 7 = 2(x + 1).$

5. $2(p - 1) - 3(p - 4) = 4p.$

6. $\frac{x}{5} = 2x - 6.$

7. $5 + \frac{4x}{9} = \frac{x}{2}.$

8. $q = \frac{3}{2}q - 4.$

9. $3x + \frac{x}{5} - 5 = \frac{1}{5} + 5x.$

10. $w + \frac{w}{2} - \frac{w}{3} + \frac{w}{4} = 5.$

11. $\frac{2y - 3}{4} = \frac{6y + 7}{3}.$

12. $\frac{x + 2}{3} - \frac{2 - x}{6} = x - 2.$

13. $\frac{9}{5}(3 - x) = \frac{3}{4}(x - 3).$

14. $\frac{3}{2}(4x - 3) = 2[x - (4x - 3)].$

15. $\frac{4}{x} = 16.$

16. $\frac{4}{8 - x} = \frac{3}{4}.$

17. $\frac{x}{3x - 4} = 3.$

18. $\frac{1}{p - 1} = \frac{2}{p - 2}.$

19. $\frac{1}{x} + \frac{1}{5} = \frac{4}{5}.$

20. $\frac{3x - 2}{2x + 3} = \frac{3x - 1}{2x + 1}.$

21. $\frac{y - 6}{y} - \frac{6}{y} = \frac{y + 6}{y - 6}.$

22. $\frac{9}{x - 3} = \frac{3x}{x - 3}.$

In Problems **23–28**, *express the indicated symbol in terms of the remaining symbols.*

23. $I = Prt; \quad P.$

24. $ax + b = 0; \quad x.$

25. $p = 6x - 1; \quad x.$

26. $p = -3x + 6; \quad x.$

27. $S = P(1 + rt); \quad r.$

28. $S = \frac{R[(1 + i)^n - 1]}{i}; \quad R.$

In Problems **29–48**, *solve by factoring.*

29. $x^2 + 3x + 2 = 0.$

30. $x^2 - 2x - 3 = 0.$

31. $y^2 - 7y + 12 = 0.$

32. $x^2 - 12x = -36.$

33. $x^2 - 4 = 0.$

34. $y(2y + 3) = 5.$

35. $z^2 - 8z = 0.$

36. $-x^2 + 3x + 10 = 0.$

37. $4x^2 + 1 = 4x.$

38. $2p^2 = 3p.$

39. $x^3 - 64x = 0.$

40. $6x^3 + 5x^2 - 4x = 0.$

41. $x(x - 1)(x + 2) = 0.$

42. $x^2(x - 4) = 0.$

43. $(x - 2)^2(x + 1)^2 = 0.$

44. $x(x - 1)(x + 1) = 0.$

45. $7x^2(x - 2)^2(x + 3)(x - 4) = 0.$

46. $x(x^2 - 1)(x^2 - 4) = 0.$

47. $x(x^2 - 1)(x^2 - 1) = 0.$

48. $x^4 - x^2 = 0.$

In Problems **49–62**, *find all real roots by using the quadratic formula.*

49. $x^2 + 2x - 15 = 0.$

50. $x^2 - 2x - 24 = 0.$

51. $4x^2 - 12x + 9 = 0.$

52. $p^2 + 2p = 0.$

53. $p^2 - 5p + 3 = 0.$

54. $2 - 2x + x^2 = 0.$

55. $4 - 2n + n^2 = 0.$

56. $2x^2 + x = 5.$

57. $6x^2 + 7x - 5 = 0.$

58. $w^2 - 2\sqrt{2}\, w + 2 = 0.$

59. $2x^2 - 3x = 20.$

60. $.01x^2 + .2x - .6 = 0.$

61. $2x^2 + 4x = 5.$

62. $-2x^2 - 6x + 5 = 0.$

A-7 INEQUALITIES

The basic rules that are used to solve inequalities are

(I) If $a < b$, then $a + c < b + c$ and $a - c < b - c$.

(II) If $a < b$ and c is a positive number, then $ac < bc$ and $\frac{a}{c} < \frac{b}{c}$.

(III) If $a < b$ and $-c$ is a negative number, then $a(-c) > b(-c)$ and $\frac{a}{-c} > \frac{b}{-c}$.

These rules are also true if you replace "$<$" by "$\leqslant$," "$>$," or "$\geqslant$." In (III), however, if you started with the symbol "$\leqslant$," it would change to "$\geqslant$," etc.

In the following examples, the rule used is indicated to the right.

EXAMPLE 1

a. *Solve* $2(x-3)<4$.

$$\begin{aligned} 2(x-3) &< 4, \\ 2x-6 &< 4 \quad \text{(distributive property)}, \\ 2x-6+6 &< 4+6 \qquad \text{(I)}, \\ 2x &< 10, \\ \frac{2x}{2} &< \frac{10}{2} \qquad \text{(II)}, \\ x &< 5. \end{aligned}$$

Thus the original inequality is true for *all* real numbers x such that $x<5$.

b. *Solve* $3-2x \leqslant 6$.

$$\begin{aligned} 3-2x &\leqslant 6, \\ -2x &\leqslant 3 \qquad \text{(I)}, \\ x &\geqslant -\frac{3}{2} \qquad \text{(III)}. \end{aligned}$$

The solution is $x \geqslant -3/2$.

c. *Solve* $\frac{3}{2}(s-2)+1 > -2(s-4)$.

$$\begin{aligned} \frac{3}{2}(s-2)+1 &> -2(s-4), \\ 3(s-2)+2 &> -4(s-4) \qquad \text{(II)}, \\ 3s-4 &> -4s+16, \\ 7s &> 20 \qquad \text{(I)}, \\ s &> \frac{20}{7} \qquad \text{(II)}. \end{aligned}$$

EXAMPLE 2

a. *Solve* $2(x-4)-3>2x-1$.

$$\begin{aligned} 2(x-4)-3 &> 2x-1, \\ 2x-8-3 &> 2x-1, \\ -11 &> -1 \qquad \text{(I)}. \end{aligned}$$

Since $-11 > -1$ is never true, there is no solution.

b. *Solve* $2(x-4)-3<2x-1$.

Proceeding in the same manner as in (a), we obtain $-11 < -1$. This inequality is true for all real numbers x. We write our solution as $-\infty < x < \infty$. The symbols $-\infty$ and ∞ are not numbers, but are merely a convenience for indicating that the solution is all real numbers.

Suppose that $a < b$ and x is between a and b. Then not only is $a < x$, but $x < b$. We indicate this by writing $a < x < b$. For example, $0 < 7 < 9$. Similarly, $-1 \leqslant x < 3$ represents all real numbers x such that $x \geqslant -1$ and $x < 3$ simultaneously.

Exercise A-7

Solve the following inequalities.

1. $4x > 8$.
2. $8x < -2$.
3. $3x - 4 \leqslant 2$.
4. $5x \geqslant 0$.
5. $3 < 2y + 3$.
6. $2y + 1 > 0$.
7. $-4x \geqslant 2$.
8. $6 \leqslant 5 - 3y$.
9. $3 - 5s > 5$.
10. $4s - 1 < -5$.
11. $2x - 3 \leqslant 4 + 7x$.
12. $-3 \geqslant 8(2 - x)$.
13. $3(2 - 3x) > 4(1 - 4x)$.
14. $8(x + 1) + 1 < 3(2x) + 1$.
15. $2(3x - 2) > 3(2x - 1)$.
16. $3 - 2(x - 1) \leqslant 2(4 + x)$.
17. $x + 2 < \sqrt{3} - x$.
18. $\sqrt{2}\,(x + 2) > \sqrt{8}\,(3 - x)$.
19. $\dfrac{9y + 1}{4} \leqslant 2y - 1$.
20. $\dfrac{4y - 3}{2} \geqslant \dfrac{1}{3}$.
21. $4x - 1 \geqslant 4(x - 2) + 7$.
22. $0x \leqslant 0$.
23. $\dfrac{1 - t}{2} < \dfrac{3t - 7}{3}$.
24. $\dfrac{3(2t - 2)}{2} > \dfrac{6t - 3}{5} + \dfrac{t}{10}$.
25. $2x + 3 \geqslant \frac{1}{2}x - 4$.
26. $4x - \frac{1}{2} \leqslant \frac{3}{2}x$.
27. $\frac{2}{3}r < \frac{5}{6}r$.
28. $\frac{7}{4}t > -\frac{2}{3}t$.
29. $.1(.03x + 4) \geqslant .02x + .434$.
30. $9 - .1x \leqslant (2 - .01x)/(.2)$.
31. $\dfrac{y}{2} + \dfrac{y}{3} > y + \dfrac{y}{5}$.
32. $\dfrac{5y - 1}{-3} < \dfrac{7(y + 1)}{-2}$.

APPENDIX B

Tables of Powers – Roots – Reciprocals

n	n^2	$\sqrt{n}$	$\sqrt{10n}$	n^3	$\sqrt[3]{n}$	$\sqrt[3]{10n}$	$\sqrt[3]{100n}$	$1/n$
1.0	1.0000	1.0000	3.1623	1.0000	1.0000	2.1544	4.6416	1.0000
1.1	1.2100	1.0488	3.3166	1.3310	1.0323	2.2240	4.7914	0.9091
1.2	1.4400	1.0954	3.4641	1.7280	1.0627	2.2894	4.9324	0.8333
1.3	1.6900	1.1402	3.6056	2.1970	1.0914	2.3513	5.0658	0.7692
1.4	1.9600	1.1832	3.7417	2.7440	1.1187	2.4101	5.1925	0.7143
1.5	2.2500	1.2247	3.8730	3.3750	1.1447	2.4662	5.3133	0.6667
1.6	2.5600	1.2649	4.0000	4.0960	1.1696	2.5198	5.4288	0.6250
1.7	2.8900	1.3038	4.1231	4.9130	1.1935	2.5713	5.5397	0.5882
1.8	3.2400	1.3416	4.2426	5.8320	1.2164	2.6207	5.6462	0.5556
1.9	3.6100	1.3784	4.3589	6.8590	1.2386	2.6684	5.7489	0.5263
2.0	4.0000	1.4142	4.4721	8.0000	1.2599	2.7144	5.8480	0.5000
2.1	4.4100	1.4491	4.5826	9.2610	1.2806	2.7589	5.9439	0.4762
2.2	4.8400	1.4832	4.6904	10.6480	1.3006	2.8020	6.0368	0.4545
2.3	5.2900	1.5166	4.7958	12.1670	1.3200	2.8439	6.1269	0.4348

n	n^2	$\sqrt{n}$	$\sqrt{10n}$	n^3	$\sqrt[3]{n}$	$\sqrt[3]{10n}$	$\sqrt[3]{100n}$	$1/n$
2.4	5.7600	1.5492	4.8990	13.8240	1.3389	2.8845	6.2145	0.4167
2.5	6.2500	1.5811	5.0000	15.6250	1.3572	2.9240	6.2996	0.4000
2.6	6.7600	1.6125	5.0990	17.5760	1.3751	2.9625	6.3825	0.3846
2.7	7.2900	1.6432	5.1962	19.6830	1.3925	3.0000	6.4633	0.3704
2.8	7.8400	1.6733	5.2915	21.9520	1.4095	3.0366	6.5421	0.3571
2.9	8.4100	1.7029	5.3852	24.3890	1.4260	3.0723	6.6191	0.3448
3.0	9.0000	1.7321	5.4772	27.0000	1.4422	3.1072	6.6943	0.3333
3.1	9.6100	1.7607	5.5678	29.7910	1.4581	3.1414	6.7679	0.3226
3.2	10.2400	1.7889	5.6569	32.7680	1.4736	3.1748	6.8399	0.3125
3.3	10.8900	1.8166	5.7446	35.9370	1.4888	3.2075	6.9104	0.3030
3.4	11.5600	1.8439	5.8310	39.3040	1.5037	3.2396	6.9795	0.2941
3.5	12.2500	1.8708	5.9161	42.8750	1.5183	3.2711	7.0473	0.2857
3.6	12.9600	1.8974	6.0000	46.6560	1.5326	3.3019	7.1138	0.2778
3.7	13.6900	1.9235	6.0828	50.6530	1.5467	3.3322	7.1791	0.2703
3.8	14.4400	1.9494	6.1644	54.8720	1.5605	3.3620	7.2432	0.2632
3.9	15.2100	1.9748	6.2450	59.3190	1.5741	3.3912	7.3061	0.2564
4.0	16.0000	2.0000	6.3246	64.0000	1.5874	3.4200	7.3681	0.2500
4.1	16.8100	2.0248	6.4031	68.9210	1.6005	3.4482	7.4290	0.2439
4.2	17.6400	2.0494	6.4807	74.0880	1.6134	3.4760	7.4889	0.2381
4.3	18.4900	2.0736	6.5574	79.5070	1.6261	3.5034	7.5478	0.2326
4.4	19.3600	2.0976	6.6333	85.1840	1.6386	3.5303	7.6059	0.2273
4.5	20.2500	2.1213	6.7082	91.1250	1.6510	3.5569	7.6631	0.2222
4.6	21.1600	2.1448	6.7823	97.3360	1.6631	3.5830	7.7194	0.2174
4.7	22.0900	2.1679	6.8557	103.823	1.6751	3.6088	7.7750	0.2128
4.8	23.0400	2.1909	6.9282	110.592	1.6869	3.6342	7.8297	0.2083
4.9	24.0100	2.2136	7.0000	117.649	1.6985	3.6593	7.8837	0.2041
5.0	25.0000	2.2361	7.0711	125.000	1.7100	3.6840	7.9370	0.2000
5.1	26.0100	2.2583	7.1414	132.651	1.7213	3.7084	7.9896	0.1961
5.2	27.0400	2.2804	7.2111	140.608	1.7325	3.7325	8.0415	0.1923
5.3	28.0900	2.3022	7.2801	148.877	1.7435	3.7563	8.0927	0.1887
5.4	29.1600	2.3238	7.3485	157.464	1.7544	3.7798	8.1433	0.1852
5.5	30.2500	2.3452	7.4162	166.375	1.7652	3.8030	8.1932	0.1818
5.6	31.3600	2.3664	7.4833	175.616	1.7758	3.8259	8.2426	0.1786
5.7	32.4900	2.3875	7.5498	185.193	1.7863	3.8485	8.2913	0.1754
5.8	33.6400	2.4083	7.6158	195.112	1.7967	3.8709	8.3396	0.1724
5.9	34.8100	2.4290	7.6811	205.379	1.8070	3.8930	8.3872	0.1695
6.0	36.0000	2.4495	7.7460	216.000	1.8171	3.9149	8.4343	0.1667
6.1	37.2100	2.4698	7.8102	226.981	1.8272	3.9365	8.4809	0.1639
6.2	38.4400	2.4900	7.8740	238.328	1.8371	3.9579	8.5270	0.1613
6.3	39.6900	2.5100	7.9372	250.047	1.8469	3.9791	8.5726	0.1587
6.4	40.9600	2.5298	8.0000	262.144	1.8566	4.0000	8.6177	0.1563
6.5	42.2500	2.5495	8.0623	274.625	1.8663	4.0207	8.6624	0.1538
6.6	43.5600	2.5690	8.1240	287.496	1.8758	4.0412	8.7066	0.1515

n	n^2	$\sqrt{n}$	$\sqrt{10n}$	n^3	$\sqrt[3]{n}$	$\sqrt[3]{10n}$	$\sqrt[3]{100n}$	$1/n$
6.7	44.8900	2.5884	8.1854	300.763	1.8852	4.0615	8.7503	0.1493
6.8	46.2400	2.6077	8.2462	314.432	1.8945	4.0817	8.7937	0.1471
6.9	47.6100	2.6268	8.3066	328.509	1.9038	4.1016	8.8366	0.1449
7.0	49.0000	2.6458	8.3666	343.000	1.9129	4.1213	8.8790	0.1429
7.1	50.4100	2.6646	8.4261	357.911	1.9220	4.1408	8.9211	0.1408
7.2	51.8400	2.6833	8.4853	373.248	1.9310	4.1602	8.9628	0.1389
7.3	53.2900	2.7019	8.5440	389.017	1.9399	4.1793	9.0041	0.1370
7.4	54.7600	2.7203	8.6023	405.224	1.9487	4.1983	9.0450	0.1351
7.5	56.2500	2.7386	8.6603	421.875	1.9574	4.2172	9.0856	0.1333
7.6	57.7600	2.7568	8.7178	438.976	1.9661	4.2358	9.1258	0.1316
7.7	59.2900	2.7749	8.7750	456.533	1.9747	4.2543	9.1657	0.1299
7.8	60.8400	2.7928	8.8318	474.552	1.9832	4.2727	9.2052	0.1282
7.9	62.4100	2.8107	8.8882	493.039	1.9916	4.2908	9.2443	0.1266
8.0	64.0000	2.8284	8.9443	512.000	2.0000	4.3089	9.2832	0.1250
8.1	65.6100	2.8460	9.0000	531.441	2.0083	4.3267	9.3217	0.1235
8.2	67.2400	2.8636	9.0554	551.368	2.0165	4.3445	9.3599	0.1220
8.3	68.8900	2.8810	9.1104	571.787	2.0247	4.3621	9.3978	0.1205
8.4	70.5600	2.8983	9.1652	592.704	2.0328	4.3795	9.4354	0.1190
8.5	72.2500	2.9155	9.2195	614.125	2.0408	4.3968	9.4727	0.1176
8.6	73.9600	2.9326	9.2736	636.056	2.0488	4.4140	9.5097	0.1163
8.7	75.6900	2.9496	9.3274	658.503	2.0567	4.4310	9.5464	0.1149
8.8	77.4400	2.9665	9.3808	681.472	2.0646	4.4480	9.5828	0.1136
8.9	79.2100	2.9833	9.4340	704.969	2.0723	4.4647	9.6190	0.1124
9.0	81.000	3.0000	9.4868	729.000	2.0801	4.4814	9.6549	0.1111
9.1	82.8100	3.0166	9.5394	753.571	2.0878	4.4979	9.6905	0.1099
9.2	84.6400	3.0332	9.5917	778.688	2.0954	4.5144	9.7259	0.1087
9.3	86.4900	3.0496	9.6436	804.357	2.1029	4.5307	9.7610	0.1075
9.4	88.3600	3.0659	9.6954	830.584	2.1105	4.5468	9.7959	0.1064
9.5	90.2500	3.0822	9.7468	857.375	2.1179	4.5629	9.8305	0.1053
9.6	92.1600	3.0984	9.7980	884.736	2.1253	4.5789	9.8648	0.1042
9.7	94.0900	3.1145	9.8489	912.673	2.1327	4.5947	9.8990	0.1031
9.8	96.0400	3.1305	9.8995	941.192	2.1400	4.6104	9.9329	0.1020
9.9	98.0100	3.1464	9.9499	970.299	2.1472	4.6261	9.9666	0.1010
10.0	100.000	3.1623	10.000	1000.00	2.1544	4.6416	10.0000	0.1000

APPENDIX C

Table of e^x and e^{-x}

x	e^x	e^{-x}	x	e^x	e^{-x}
0.00	1.0000	1.00000	0.15	1.1618	.86071
0.01	1.0101	0.99005	0.16	1.1735	.85214
0.02	1.0202	.98020	0.17	1.1853	.84366
0.03	1.0305	.97045	0.18	1.1972	.83527
0.04	1.0408	.96079	0.19	1.2092	.82696
0.05	1.0513	.95123	0.20	1.2214	.81873
0.06	1.0618	.94176	0.21	1.2337	.81058
0.07	1.0725	.93239	0.22	1.2461	.80252
0.08	1.0833	.92312	0.23	1.2586	.79453
0.09	1.0942	.91393	0.24	1.2712	.78663
0.10	1.1052	.90484	0.25	1.2840	.77880
0.11	1.1163	.89583	0.26	1.2969	.77105
0.12	1.1275	.88692	0.27	1.3100	.76338
0.13	1.1388	.87809	0.28	1.3231	.75578
0.14	1.1503	.86936	0.29	1.3364	.74826

x	e^x	e^{-x}
0.30	1.3499	.74082
0.31	1.3634	.73345
0.32	1.3771	.72615
0.33	1.3910	.71892
0.34	1.4049	.71177
0.35	1.4191	.70469
0.36	1.4333	.69768
0.37	1.4477	.69073
0.38	1.4623	.68386
0.39	1.4770	.67706
0.40	1.4918	.67032
0.41	1.5068	.66365
0.42	1.5220	.65705
0.43	1.5373	.65051
0.44	1.5527	.64404
0.45	1.5683	.63763
0.46	1.5841	.63128
0.47	1.6000	.62500
0.48	1.6161	.61878
0.49	1.6323	.61263
0.50	1.6487	.60653
0.51	1.6653	.60050
0.52	1.6820	.59452
0.53	1.6989	.58860
0.54	1.7160	.58275
0.55	1.7333	.57695
0.56	1.7507	.57121
0.57	1.7683	.56553
0.58	1.7860	.55990
0.59	1.8040	.55433
0.60	1.8221	.54881
0.61	1.8404	.54335
0.62	1.8589	.53794
0.63	1.8776	.53259
0.64	1.8965	.52729
0.65	1.9155	.52205
0.66	1.9348	.51685
0.67	1.9542	.51171
0.68	1.9739	.50662
0.69	1.9937	.50158

x	e^x	e^{-x}
0.70	2.0138	.49659
0.71	2.0340	.49164
0.72	2.0544	.48675
0.73	2.0751	.48191
0.74	2.0959	.47711
0.75	2.1170	.47237
0.76	2.1383	.46767
0.77	2.1598	.46301
0.78	2.1815	.45841
0.79	2.2034	.45384
0.80	2.2255	.44933
0.81	2.2479	.44486
0.82	2.2705	.44043
0.83	2.2933	.43605
0.84	2.3164	.43171
0.85	2.3396	.42741
0.86	2.3632	.42316
0.87	2.3869	.41895
0.88	2.4109	.41478
0.89	2.4351	.41066
0.90	2.4596	.40657
0.91	2.4843	.40252
0.92	2.5093	.39852
0.93	2.5345	.39455
0.94	2.5600	.39063
0.95	2.5857	.38674
0.96	2.6117	.38289
0.97	2.6379	.37908
0.98	2.6645	.37531
0.99	2.6912	.37158
1.00	2.7183	.36788
1.10	3.0042	.33287
1.20	3.3201	.30119
1.30	3.6693	.27253
1.40	4.0552	.24660
1.50	4.4817	.22313
1.60	4.9530	.20190
1.70	5.4739	.18268
1.80	6.0496	.16530
1.90	6.6859	.14957

x	e^x	e^{-x}	x	e^x	e^{-x}
2.00	7.3891	.13534	4.50	90.017	.01111
2.10	8.1662	.12246	4.60	99.484	.01005
2.20	9.0250	.11080	4.70	109.95	.00910
2.30	9.9742	.10026	4.80	121.51	.00823
2.40	11.023	.09072	4.90	134.29	.00745
2.50	12.182	.08208	5.00	148.41	.00674
2.60	13.464	.07427	5.10	164.02	.00610
2.70	14.880	.06721	5.20	181.27	.00552
2.80	16.445	.06081	5.30	200.34	.00499
2.90	18.174	.05502	5.40	221.41	.00452
3.00	20.086	.04979	5.50	244.69	.00409
3.10	22.198	.04505	5.60	270.43	.00370
3.20	24.533	.04076	5.70	298.87	.00335
3.30	27.113	.03688	5.80	330.30	.00303
3.40	29.964	.03337	5.90	365.04	.00274
3.50	33.115	.03020	6.00	403.43	.00248
3.60	36.598	.02732	6.25	518.01	.00193
3.70	40.447	.02472	6.50	665.14	.00150
3.80	44.701	.02237	6.75	854.06	.00117
3.90	49.402	.02024	7.00	1096.6	.00091
4.00	54.598	.01832	7.50	1808.0	.00055
4.10	60.340	.01657	8.00	2981.0	.00034
4.20	66.686	.01500	8.50	4914.8	.00020
4.30	73.700	.01357	9.00	8103.1	.00012
4.40	81.451	.01227	9.50	13360.	.00007
			10.00	22026.	.00005

APPENDIX D

Table of Natural Logarithms

In the body of the table the first two digits (and decimal point) of most entries are carried over from a preceding entry in the first column. For example, $\ln 3.32 \approx 1.19996$. However, an asterisk (*) indicates that the first two digits are those of a following entry in the first column. For example, $\ln 3.33 \approx 1.20297$.

To extend this table for a number less than 1.0 or greater than 10.09, write the number in the form $x = y \cdot 10^n$, where $1.0 \leqslant y < 10$, and use the fact that $\ln x = \ln y + n \ln 10$. Some values of $n \ln 10$ are:

$1 \ln 10 \approx 2.30259$,	$6 \ln 10 \approx 13.81551$,
$2 \ln 10 \approx 4.60517$,	$7 \ln 10 \approx 16.11810$,
$3 \ln 10 \approx 6.90776$,	$8 \ln 10 \approx 18.42068$,
$4 \ln 10 \approx 9.21034$,	$9 \ln 10 \approx 20.72327$,
$5 \ln 10 \approx 11.51293$,	$10 \ln 10 \approx 23.02585$.

For example,

$$\ln 332 = \ln[(3.32)(10^2)] = \ln 3.32 + 2 \ln 10$$
$$\approx 1.19996 + 4.60517 = 5.80513,$$

and

$$\ln .0332 = \ln[(3.32)(10^{-2})] = \ln 3.32 - 2 \ln 10$$
$$\approx 1.19996 - 4.60517 = -3.40521.$$

Properties of logarithms may be used to find the logarithm of a number such as $\frac{3}{8}$:

$$\ln \tfrac{3}{8} = \ln 3 - \ln 8 \approx 1.09861 - 2.07944$$
$$= -.98083.$$

N	0	1	2	3	4	5	6	7	8	9
1.0	0.0 0000	0995	1980	2956	3922	4879	5827	6766	7696	8618
1.1	9531	*0436	*1333	*2222	*3103	*3976	*4842	*5700	*6551	*7395
1.2	0.1 8232	9062	9885	*0701	*1511	*2314	*3111	*3902	*4686	*5464
1.3	0.2 6236	7003	7763	8518	9267	*0010	*0748	*1481	*2208	*2930
1.4	0.3 3647	4359	5066	5767	6464	7156	7844	8526	9204	9878
1.5	0.4 0547	1211	1871	2527	3178	3825	4469	5108	5742	6373
1.6	7000	7623	8243	8858	9470	*0078	*0672	*1282	*1879	*2473
1.7	0.5 3063	3649	4232	4812	5389	5962	6531	7098	7661	8222
1.8	8779	9333	9884	*0432	*0977	*1519	*2058	*2594	*3127	*3658
1.9	0.6 4185	4710	5233	5752	6269	6783	7294	7803	8310	8813
2.0	9315	9813	*0310	*0804	*1295	*1784	*2271	*2755	*3237	*3716
2.1	0.7 4194	4669	5142	5612	6081	6547	7011	7473	7932	8390
2.2	8846	9299	9751	*0200	*0648	*1093	*1536	*1978	*2418	*2855
2.3	0.8 3291	3725	4157	4587	5015	5442	5866	6289	6710	7129
2.4	7547	7963	8377	8789	9200	9609	*0016	*0422	*0826	*1228
2.5	0.9 1629	2028	2426	2822	3216	3609	4001	4391	4779	5166
2.6	5551	5935	6317	6698	7078	7456	7833	8208	8582	8954
2.7	9325	9695	*0063	*0430	*0796	*1160	*1523	*1885	*2245	*2604
2.8	1.0 2962	3318	3674	4028	4380	4732	5082	5431	5779	6126
2.9	6471	6815	7158	7500	7841	8181	8519	8856	9192	9527
3.0	9861	*0194	*0526	*0856	*1186	*1514	*1841	*2168	*2493	*2817
3.1	1.1 3140	3462	3783	4103	4422	4740	5057	5373	5688	6002
3.2	6315	6627	6938	7248	7557	7865	8173	8479	8784	9089
3.3	9392	9695	9996	*0297	*0597	*0896	*1194	*1491	*1788	*2083
3.4	1.2 2378	2671	2964	3256	3547	3837	4127	4415	4703	4990
3.5	5276	5562	5846	6130	6413	6695	6976	7257	7536	7815
3.6	8093	8371	8647	8923	9198	9473	9746	*0019	*0291	*0563
3.7	1.3 0833	1103	1372	1641	1909	2176	2442	2708	2972	3237
3.8	3500	3763	4025	4286	4547	4807	5067	5325	5584	5841
3.9	6098	6354	6609	6864	7118	7372	7624	7877	8128	8379
4.0	8629	8879	9128	9377	9624	9872	*0118	*0364	*0610	*0854
4.1	1.4 1099	1342	1585	1828	2070	2311	2552	2792	3031	3270
4.2	3508	3746	3984	4220	4456	4692	4927	5161	5395	5629
4.3	5862	6094	6326	6557	6787	7018	7247	7476	7705	7933
4.4	8160	8387	8614	8840	9065	9290	9515	9739	9962	*0185
4.5	1.5 0408	0630	0851	1072	1293	1513	1732	1951	2170	2388
4.6	2606	2823	3039	3256	3471	3687	3902	4116	4330	4543
4.7	4756	4969	5181	5393	5604	5814	6025	6235	6444	6653
4.8	6862	7070	7277	7485	7691	7898	8104	8309	8515	8719
4.9	8924	9127	9331	9534	9737	9939	*0141	*0342	*0543	*0744
5.0	1.6 0944	1144	1343	1542	1741	1939	2137	2334	2531	2728
5.1	2924	3120	3315	3511	3705	3900	4094	4287	4481	4673
5.2	4866	5058	5250	5441	5632	5823	6013	6203	6393	6582
5.3	6771	6959	7147	7335	7523	7710	7896	8083	8269	8455
5.4	8640	8825	9010	9194	9378	9562	9745	9928	*0111	*0293
N	0	1	2	3	4	5	6	7	8	9

N	0	1	2	3	4	5	6	7	8	9
5.5	1.7 0475	0656	0838	1019	1199	1380	1560	1740	1919	2098
5.6	2277	2455	2633	2811	2988	3166	3342	3519	3695	3871
5.7	4047	4222	4397	4572	4746	4920	5094	5267	5440	5613
5.8	5786	5958	6130	6302	6473	6644	6815	6985	7156	7326
5.9	7495	7665	7834	8002	8171	8339	8507	8675	8842	9009
6.0	1.7 9176	9342	9509	9675	9840	*0006	*0171	*0336	*0500	*0665
6.1	1.8 0829	0993	1156	1319	1482	1645	1808	1970	2132	2294
6.2	2455	2616	2777	2938	3098	3258	3418	3578	3737	3896
6.3	4055	4214	4372	4530	4688	4845	5003	5160	5317	5473
6.4	5630	5786	5942	6097	6253	6408	6563	6718	6872	7026
6.5	7180	7334	7487	7641	7794	7947	8099	8251	8403	8555
6.6	8707	8858	9010	9160	9311	9462	9612	9762	9912	*0061
6.7	1.9 0211	0360	0509	0658	0806	0954	1102	1250	1398	1545
6.8	1692	1839	1986	2132	2279	2425	2571	2716	2862	3007
6.9	3152	3297	3442	3586	3730	3874	4018	4162	4305	4448
7.0	4591	4734	4876	5019	5161	5303	5445	5586	5727	5869
7.1	6009	6150	6291	6431	6571	6711	6851	6991	7130	7269
7.2	7408	7547	7685	7824	7962	8100	8238	8376	8513	8650
7.3	8787	8924	9061	9198	9334	9470	9606	9742	9877	*0013
7.4	2.0 0148	0283	0418	0553	0687	0821	0956	1089	1223	1357
7.5	1490	1624	1757	1890	2022	2155	2287	2419	2551	2683
7.6	2815	2946	3078	3209	3340	3471	3601	3732	3862	3992
7.7	4122	4252	4381	4511	4640	4769	4898	5027	5156	5284
7.8	5412	5540	5668	5796	5924	6051	6179	6306	6433	6560
7.9	6686	6813	6939	7065	7191	7317	7443	7568	7694	7819
8.0	7944	8069	8194	8318	8443	8567	8691	8815	8939	9063
8.1	9186	9310	9433	9556	9679	9802	9924	*0047	*0169	*0291
8.2	2.1 0413	0535	0657	0779	0900	1021	1142	1263	1384	1505
8.3	1626	1746	1866	1986	2106	2226	2346	2465	2585	2704
8.4	2823	2942	3061	3180	3298	3417	3535	3653	3771	3889
8.5	4007	4124	4242	4359	4476	4593	4710	4827	4943	5060
8.6	5176	5292	5409	5524	5640	5756	5871	5987	6102	6217
8.7	6332	6447	6562	6677	6791	6905	7020	7134	7248	7361
8.8	7475	7589	7702	7816	7929	8042	8155	8267	8380	8493
8.9	8605	8717	8830	8942	9054	9165	9277	9389	9500	9611
9.0	9722	9834	9944	*0055	*0166	*0276	*0387	*0497	*0607	*0717
9.1	2.2 0827	0937	1047	1157	1266	1375	1485	1594	1703	1812
9.2	1920	2029	2138	2246	2354	2462	2570	2678	2786	2894
9.3	3001	3109	3216	3324	3431	3538	3645	3751	3858	3965
9.4	4071	4177	4284	4390	4496	4601	4707	4813	4918	5024
9.5	5129	5234	5339	5444	5549	5654	5759	5863	5968	6072
9.6	6176	6280	6384	6488	6592	6696	6799	6903	7006	7109
9.7	7213	7316	7419	7521	7624	7727	7829	7932	8034	8136
9.8	8238	8340	8442	8544	8646	8747	8849	8950	9051	9152
9.9	9253	9354	9455	9556	9657	9757	9858	9958	*0058	*0158
10.0	2.3 0259	0358	0458	0558	0658	0757	0857	0956	1055	1154
N	0	1	2	3	4	5	6	7	8	9

APPENDIX E

Table of Selected Integrals

Rational Forms Containing $(a + bu)$

1. $\int u^n \, du = \frac{u^{n+1}}{n+1} + C, \; n \neq -1.$

2. $\int \frac{du}{a+bu} = \frac{1}{b} \ln |a + bu| + C.$

3. $\int \frac{u \, du}{a+bu} = \frac{u}{b} - \frac{a}{b^2} \ln |a + bu| + C.$

4. $\int \frac{u^2 \, du}{a+bu} = \frac{u^2}{2b} - \frac{au}{b^2} + \frac{a^2}{b^3} \ln |a + bu| + C.$

5. $\int \frac{du}{u(a+bu)} = \frac{1}{a} \ln \left| \frac{u}{a+bu} \right| + C.$

6. $\int \frac{du}{u^2(a+bu)} = -\frac{1}{au} + \frac{b}{a^2} \ln \left| \frac{a+bu}{u} \right| + C.$

7. $\int \frac{u\,du}{(a+bu)^2} = \frac{1}{b^2}\left(\ln|a+bu| + \frac{a}{a+bu}\right) + C.$

8. $\int \frac{u^2\,du}{(a+bu)^2} = \frac{u}{b^2} - \frac{a^2}{b^3(a+bu)} - \frac{2a}{b^3}\ln|a+bu| + C.$

9. $\int \frac{du}{u(a+bu)^2} = \frac{1}{a(a+bu)} + \frac{1}{a^2}\ln\left|\frac{u}{a+bu}\right| + C.$

10. $\int \frac{du}{u^2(a+bu)^2} = -\frac{a+2bu}{a^2u(a+bu)} + \frac{2b}{a^3}\ln\left|\frac{a+bu}{u}\right| + C.$

11. $\int \frac{du}{(a+bu)(c+ku)} = \frac{1}{bc-ak}\ln\left|\frac{a+bu}{c+ku}\right| + C.$

12. $\int \frac{u\,du}{(a+bu)(c+ku)} = \frac{1}{bc-ak}\left[\frac{c}{k}\ln|c+ku| - \frac{a}{b}\ln|a+bu|\right] + C.$

Forms Containing $\sqrt{a+bu}$

13. $\int u\sqrt{a+bu}\,du = \frac{2(3bu-2a)(a+bu)^{3/2}}{15b^2} + C.$

14. $\int u^2\sqrt{a+bu}\,du = \frac{2(8a^2-12abu+15b^2u^2)(a+bu)^{3/2}}{105b^3} + C.$

15. $\int \frac{u\,du}{\sqrt{a+bu}} = \frac{2(bu-2a)\sqrt{a+bu}}{3b^2} + C.$

16. $\int \frac{u^2\,du}{\sqrt{a+bu}} = \frac{2(3b^2u^2-4abu+8a^2)\sqrt{a+bu}}{15b^3} + C.$

17. $\int \frac{du}{u\sqrt{a+bu}} = \frac{1}{\sqrt{a}}\ln\left|\frac{\sqrt{a+bu}-\sqrt{a}}{\sqrt{a+bu}+\sqrt{a}}\right| + C, \quad a > 0.$

18. $\int \frac{\sqrt{a+bu}\,du}{u} = 2\sqrt{a+bu} + a\int \frac{du}{u\sqrt{a+bu}}.$

Forms Containing $\sqrt{a^2-u^2}$

19. $\int \frac{du}{(a^2-u^2)^{3/2}} = \frac{u}{a^2\sqrt{a^2-u^2}} + C.$

20. $\int \frac{du}{u\sqrt{a^2 - u^2}} = -\frac{1}{a}\ln\left|\frac{a + \sqrt{a^2 - u^2}}{u}\right| + C.$

21. $\int \frac{du}{u^2\sqrt{a^2 - u^2}} = -\frac{\sqrt{a^2 - u^2}}{a^2u} + C.$

22. $\int \frac{\sqrt{a^2 - u^2}\ du}{u} = \sqrt{a^2 - u^2} - a\ln\left|\frac{a + \sqrt{a^2 - u^2}}{u}\right| + C,\quad a > 0.$

Forms Containing $\sqrt{u^2 \pm a^2}$

23. $\int \sqrt{u^2 \pm a^2}\ du = \frac{1}{2}(u\sqrt{u^2 \pm a^2} \pm a^2 \ln |u + \sqrt{u^2 \pm a^2}|) + C.$

24. $\int u^2\sqrt{u^2 \pm a^2}\ du = \frac{u}{8}(2u^2 \pm a^2)\sqrt{u^2 \pm a^2} - \frac{a^4}{8}\ln |u + \sqrt{u^2 \pm a^2}| + C.$

25. $\int \frac{\sqrt{u^2 + a^2}\ du}{u} = \sqrt{u^2 + a^2} - a\ln\left|\frac{a + \sqrt{u^2 + a^2}}{u}\right| + C.$

26. $\int \frac{\sqrt{u^2 \pm a^2}\ du}{u^2} = -\frac{\sqrt{u^2 \pm a^2}}{u} + \ln |u + \sqrt{u^2 \pm a^2}| + C.$

27. $\int \frac{du}{\sqrt{u^2 \pm a^2}} = \ln |u + \sqrt{u^2 \pm a^2}| + C.$

28. $\int \frac{du}{u\sqrt{u^2 + a^2}} = \frac{1}{a}\ln\left|\frac{\sqrt{u^2 + a^2} - a}{u}\right| + C.$

29. $\int \frac{u^2\ du}{\sqrt{u^2 \pm a^2}} = \frac{1}{2}(u\sqrt{u^2 \pm a^2} \mp a^2 \ln |u + \sqrt{u^2 \pm a^2}|) + C.$

30. $\int \frac{du}{u^2\sqrt{u^2 \pm a^2}} = -\frac{\pm\sqrt{u^2 \pm a^2}}{a^2u} + C.$

31. $\int (u^2 \pm a^2)^{3/2}\ du = \frac{u}{8}(2u^2 \pm 5a^2)\sqrt{u^2 \pm a^2} + \frac{3a^4}{8}\ln |u + \sqrt{u^2 \pm a^2}| + C.$

32. $\int \frac{du}{(u^2 \pm a^2)^{3/2}} = \frac{\pm u}{a^2\sqrt{u^2 \pm a^2}} + C.$

33. $\int \frac{u^2\ du}{(u^2 \pm a^2)^{3/2}} = \frac{-u}{\sqrt{u^2 \pm a^2}} + \ln |u + \sqrt{u^2 \pm a^2}| + C.$

Rational Forms Containing $a^2 - u^2$ and $u^2 - a^2$

34. $\displaystyle\int \frac{du}{a^2 - u^2} = \frac{1}{2a}\ln\left|\frac{a + u}{a - u}\right| + C.$

35. $\displaystyle\int \frac{du}{u^2 - a^2} = \frac{1}{2a}\ln\left|\frac{u - a}{u + a}\right| + C.$

Exponential and Logarithmic Forms

36. $\displaystyle\int e^u\, du = e^u + C.$

37. $\displaystyle\int a^u\, du = \frac{a^u}{\ln a} + C, \quad a > 0, a \neq 1.$

38. $\displaystyle\int ue^{au}\, du = \frac{e^{au}}{a^2}(au - 1) + C.$

39. $\displaystyle\int u^n e^{au}\, du = \frac{u^n e^{au}}{a} - \frac{n}{a}\int u^{n-1}e^{au}\, du.$

40. $\displaystyle\int \frac{e^{au}\, du}{u^n} = -\frac{e^{au}}{(n - 1)u^{n-1}} + \frac{a}{n - 1}\int \frac{e^{au}\, du}{u^{n-1}}.$

41. $\displaystyle\int \ln u\, du = u \ln u - u + C.$

42. $\displaystyle\int u^n \ln u\, du = \frac{u^{n+1} \ln u}{n + 1} - \frac{u^{n+1}}{(n + 1)^2} + C, \quad n \neq -1.$

43. $\displaystyle\int u^n \ln^m u\, du = \frac{u^{n+1}}{n + 1}\ln^m u - \frac{m}{n + 1}\int u^n \ln^{m-1} u\, du, \quad m, n \neq -1.$

44. $\displaystyle\int \frac{du}{u \ln u} = \ln |\ln u| + C.$

45. $\displaystyle\int \frac{du}{a + be^{cu}} = \frac{1}{ac}(cu - \ln |a + be^{cu}|) + C.$

Miscellaneous Forms

46. $\displaystyle\int \sqrt{\frac{a + u}{b + u}}\, du =$

$$\sqrt{(a + u)(b + u)} + (a - b) \ln (\sqrt{a + u} + \sqrt{b + u}) + C.$$

47. $$\int \frac{du}{\sqrt{(a+u)(b+u)}} = \ln\left|\frac{a+b}{2} + u + \sqrt{(a+u)(b+u)}\right| + C.$$

48. $$\int \sqrt{a+bu+cu^2}\,du = \frac{2cu+b}{4c}\sqrt{a+bu+cu^2} - \frac{b^2-4ac}{8c^{3/2}}\ln\left|2cu+b+2\sqrt{c}\sqrt{a+bu+cu^2}\right| + C, \quad c > 0.$$

Answers to Odd-Numbered Problems

Exercise 1-1

1. x; all real numbers; 0, 12, -1, $4t$.

3. x; all real numbers; 1, $1 + 2u$, -13, $1 + 4x$, $1 - 2(x + h) = 1 - 2x - 2h$.

5. t; all real numbers; 1.02, 1.02, 1.02, 1.02.

7. p; all real numbers; $\frac{15}{2}$, $\frac{3(2p + 1)}{2}$, $\frac{3(4 + p)}{2p}$.

9. p; all real numbers; 1, 9, $x_1^2 + 2x_1 + 1$, $w^2 + 2w + 1$, $p^2 + 2ph + h^2 + 2p + 2h + 1$.

11. t; all real numbers; 16, 36, $h^2 + 12h + 36$, $h + 12$.

13. q; all real numbers; 5, 3, 0, $|2x^2 + 3| = 2x^2 + 3$.

15. x; all $x \geqslant -4$; 2, 0, 1, $\sqrt{5 + x} - \sqrt{4 + x}$.

17. s; all $s \neq \pm 3$; $-\frac{1}{2}, -\frac{1}{2}, \frac{4}{4w^2-9}, \frac{4}{s^2-2s-8}, \frac{13-s^2}{s^2-9}$.

19. x; all real numbers; 4, 3, 4, 3.

21. r; all r such that $r < -2$ or $r > 2$; 8, 28, 14, 52.

23. x; all real numbers; 0, 256, $\frac{1}{16}$, $32\sqrt[3]{2}$.

25. x; all $x > 0$; 1, $\frac{1}{2}$.

27. Yes; no, since if $z = 4$, then $x = \pm 1$.

29. 3. **31.** $2x + h + 2$. **33.** yes; P; q. **35.** $C = 850 + 3q$. **37.** $\frac{9}{64}$.

39. (a) All T such that $30 \leqslant T \leqslant 39$. (b) 4, $\frac{17}{4}$, $\frac{33}{4}$.

41. (a) 4. (b) $8\sqrt[3]{2}$. (c) $2\sqrt[3]{2}\, f(I_0)$; doubling the intensity increases the response by a factor of $2\sqrt[3]{2}$.

Exercise 1-2

1. (a) $2x + 9$. (b) 1. (c) $x^2 + 9x + 20$.
(d) $\frac{x+5}{x+4}$. (e) $x + 9$. (f) $x + 9$.

3. (a) $x^2 + 3x + 3$. (b) $3x + 5 - x^2$. (c) $3x^3 + 4x^2 - 3x - 4$.
(d) $\frac{3x+4}{x^2-1}$. (e) $3x^2 + 1$. (f) $9x^2 + 24x + 15$.

5. (a) $3x^2 + 3x - 11$. (b) $-x^2 + 3x + 3$.
(c) $2x^4 + 6x^3 - 15x^2 - 21x + 28$. (d) $\frac{x^2+3x-4}{2x^2-7}$.
(e) $4x^4 - 22x^2 + 24$. (f) $2x^4 + 12x^3 + 2x^2 - 48x + 25$.

7. (a) 26. (b) $20x - 4$. (c) -8. (d) $-2w - 6$. (e) -6.
(f) $24x^2 + 34x + 5$. (g) -5. (h) $\frac{6t^2-5}{4t^2+1}$.

9. (a) 18. (b) $2x^2 + 4xh + 2h^2 - 3x - 3h + 4$.
(c) 4. (d) $2x^2 + 4xh + 2h^2 + 3x + 3h + 2$.
(e) 3. (f) $-6x^{3/2} + 2x - 9x^{1/2} + 3$.
(g) $-\frac{5}{2}$. (h) $\frac{8p^2+27}{9(1-2p)}$.

11. 53; -32.

13. $\dfrac{4}{(t-1)^2} + \dfrac{6}{t-1} + 1;\quad \dfrac{2}{t^2+3t}.$

15. $\dfrac{1}{v+3};\quad \sqrt{\dfrac{2w^2+3}{w^2+1}}.$

17. (a) 14; all real numbers; 14.

(b) 2; all real numbers; 2.

19. $f(x) = \sqrt{x}$, $g(x) = x - 2$.

21. $f(x) = x^{2/3}$, $g(x) = \dfrac{4x-5}{x^2+1}$.

23. $f(x) = x^3 - x^2 + 7$, $g(x) = 3x^3 - 2x$.

25. $.45(6202 + .29E^{3.68})^{.53}$; function describes status based on years of education.

Exercise 1-3

1. (a) 1, 2, 3, 0. (b) All real numbers. (c) All real numbers.

3. (a) 0, -1, -1. (b) All real numbers. (c) All nonpositive reals.

5. All real numbers; all real numbers.

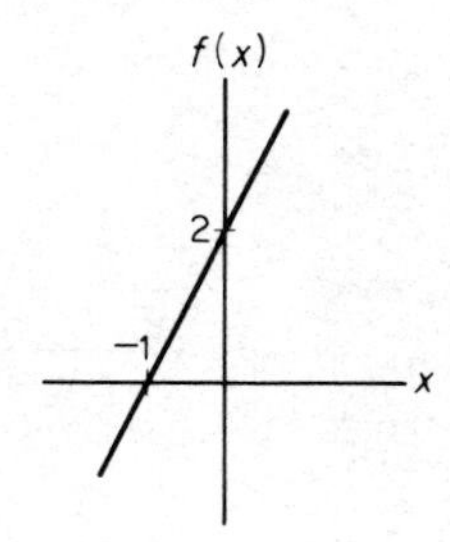

7. All real numbers; all nonnegative reals.

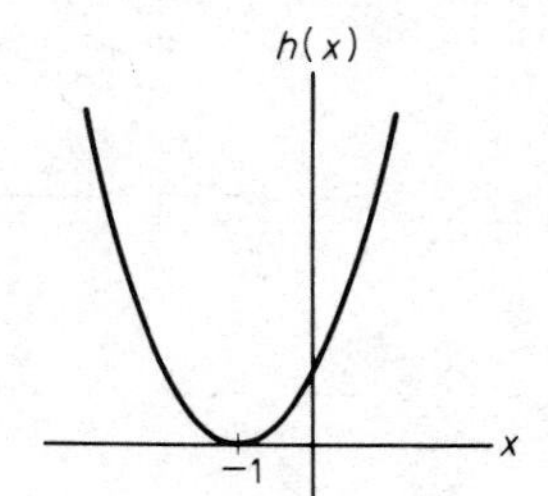

9. All real numbers; 2.

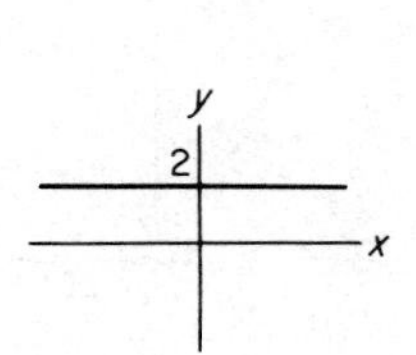

11. All real numbers; all reals $\geqslant -3$.

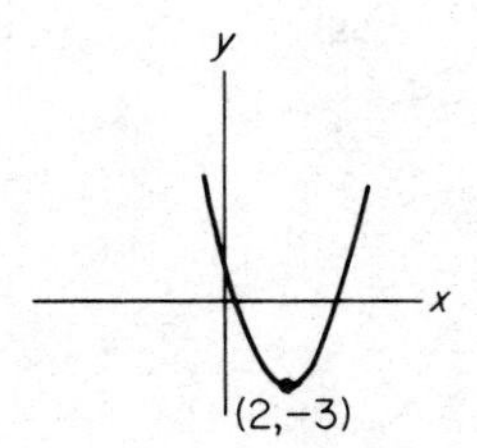

13. All real numbers; all real numbers.

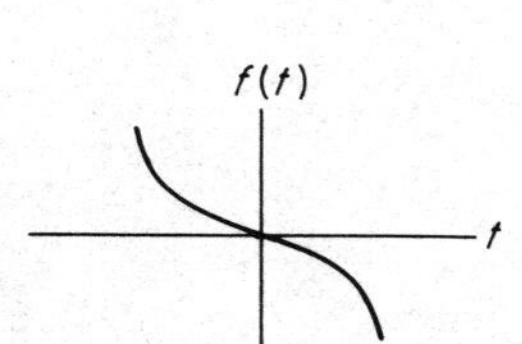

15. All real numbers $\geqslant 5$; all nonnegative reals.

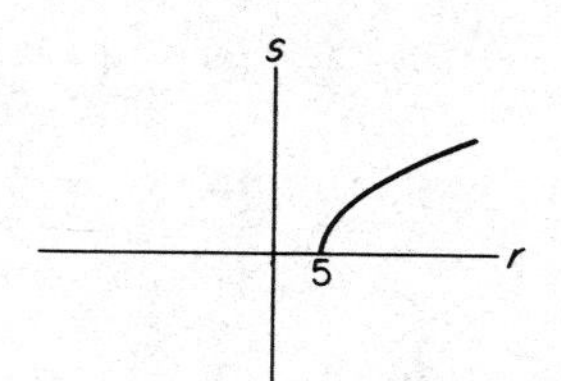

17. All real numbers; all nonnegative reals.

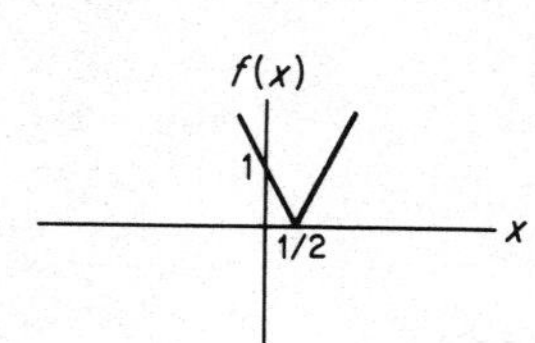

19. All nonzero real numbers; all positive real numbers.

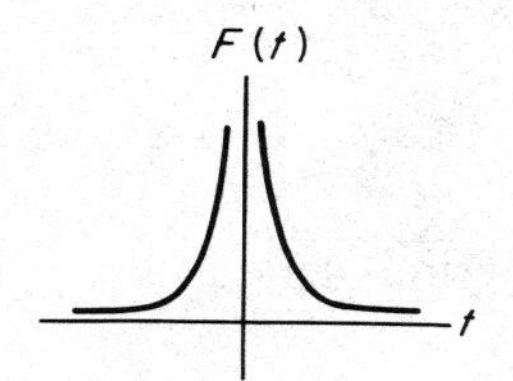

21.

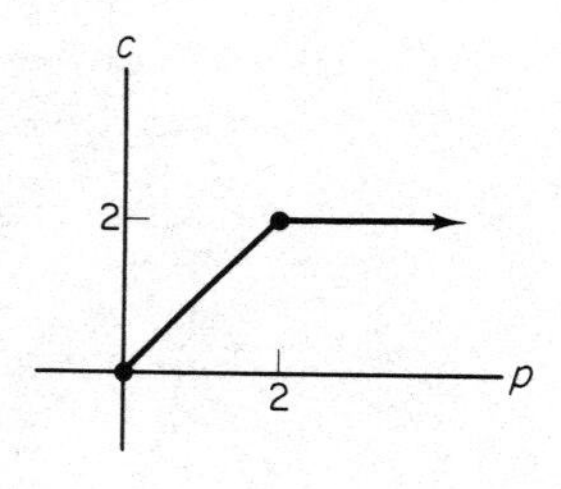

23.

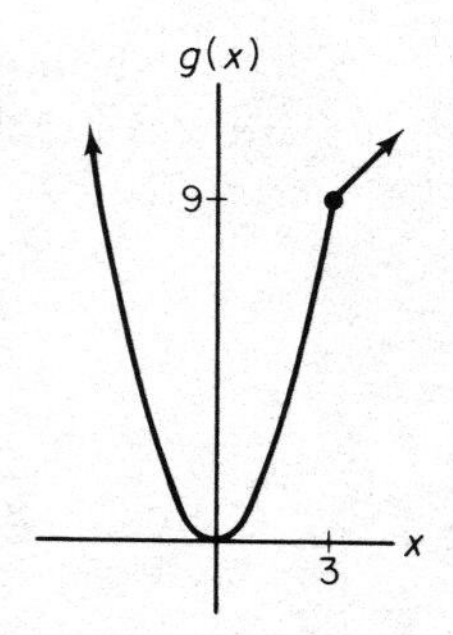

25. Domain: all T such that $20 \leqslant T \leqslant 90$; range: all y such that $70 \leqslant y \leqslant 100$.

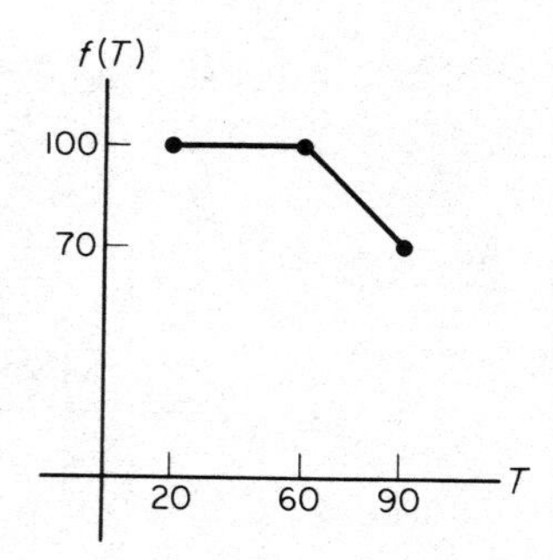

27.

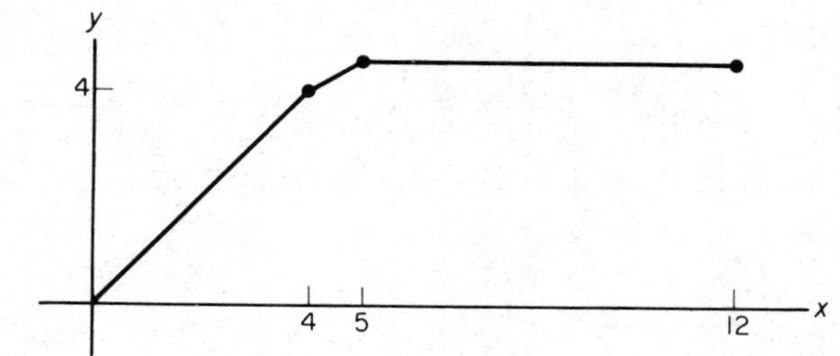

29. As price increases, quantity supplied increases; p is a function of q.

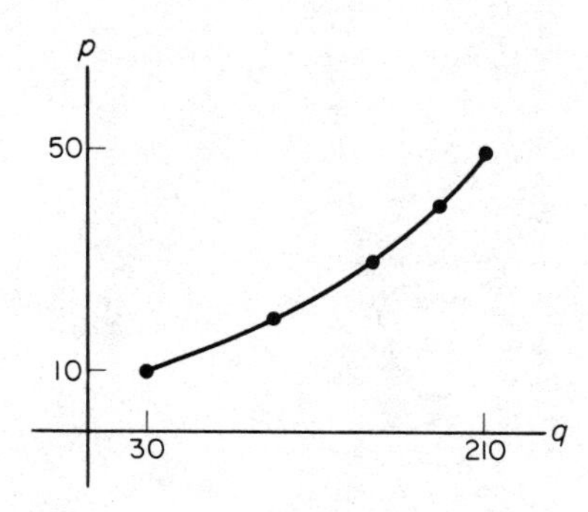

Exercise 1-4

1. -4; $(0, 0)$. **3.** 2; $(0, -4)$. **5.** $-\frac{1}{2}$, $(0, \frac{7}{2})$.

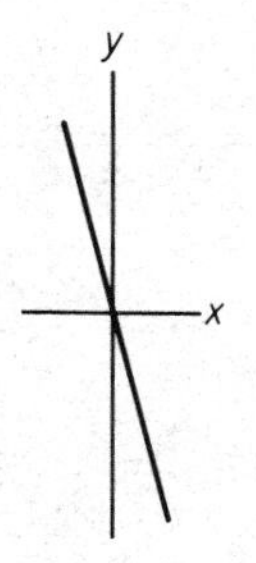

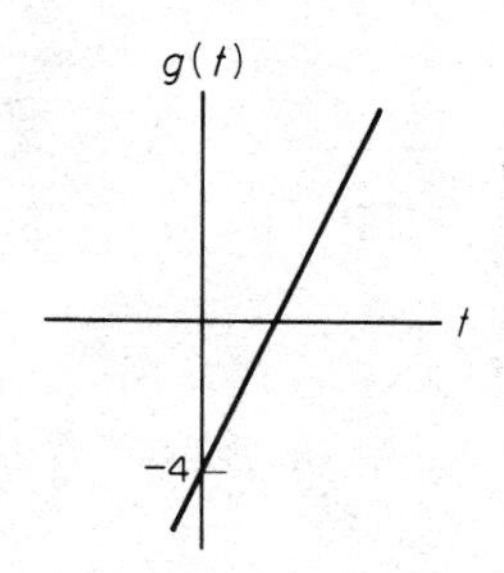

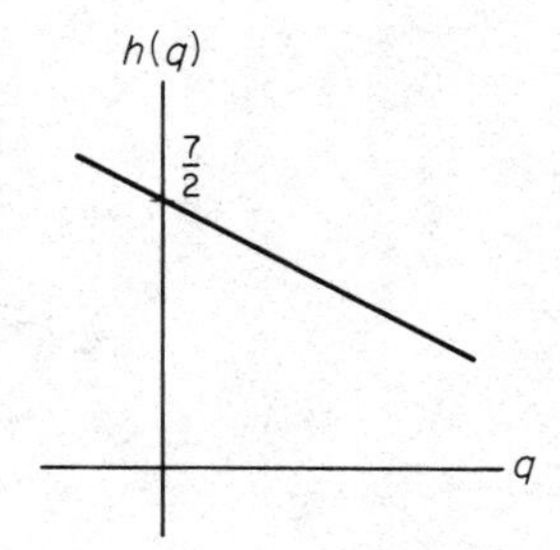

7. $f(x) = 5x - 14$. **9.** $f(x) = -3x + 9$. **11.** $f(x) = -\frac{1}{2}x + \frac{15}{4}$.

13. $f(x) = x - 1$. **15.** (a) $r = -\frac{35}{2}l + 195$. (b) 177.5.

17. (a) $p = .059t + .025$. (b) .556.

Exercise 1-5

1. Linear. **3.** Quadratic. **5.** Neither. **7.** Quadratic.

9. (a) (1, 11). (b) Highest.

11. (a) (0, −8).
(b) (−4, 0), (2, 0).
(c) (−1, −9).

13. Intercepts: (1, 0), (5, 0), (0, 5); vertex: (3, −4); range: all $y \geqslant -4$.

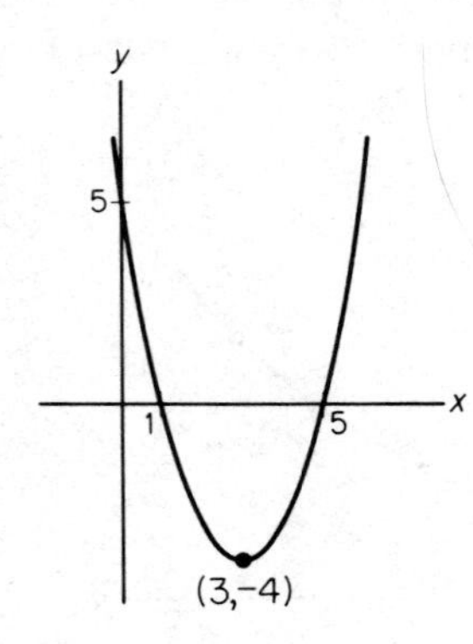

15. $m = 4$; y-intercept (0, −3).

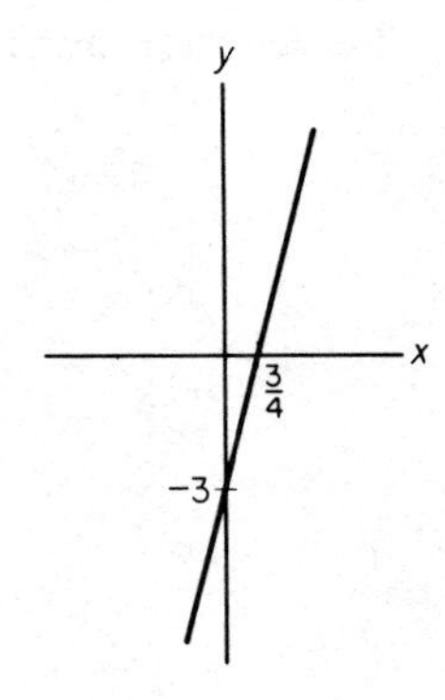

17. Intercepts: (0, 0), (−3, 0); vertex: $(-\frac{3}{2}, \frac{9}{2})$; range: all $y \leqslant \frac{9}{2}$.

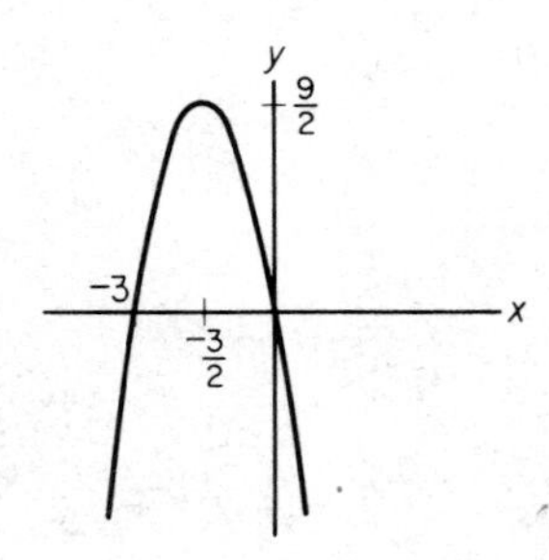

19. Intercepts: (−1, 0), (0, 1); vertex: (−1, 0); range: all $s \geqslant 0$.

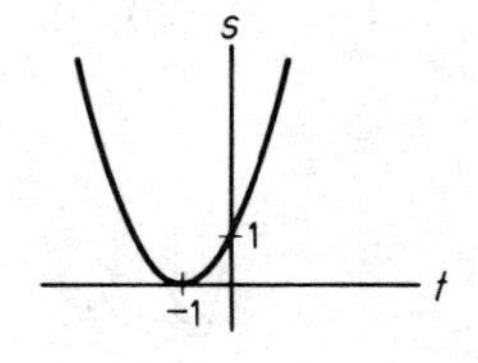

21. $m = -\frac{3}{2}$; y-intercept (0, 15).

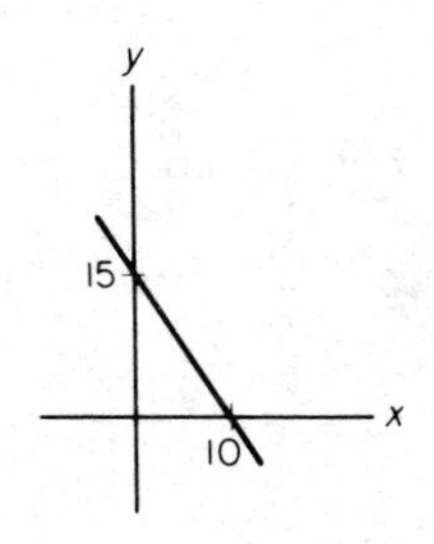

23. Intercept: (0, −9); vertex: (2, −1); range: all $y \leqslant -1$.

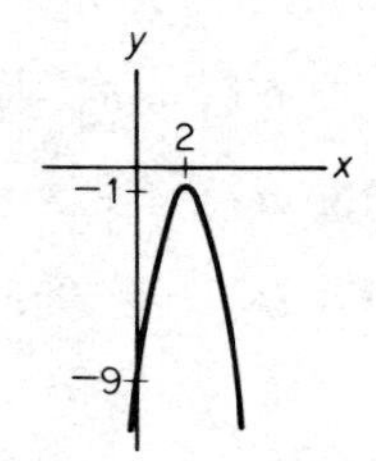

25. Intercepts: $(4 + \sqrt{3}, 0)$, $(4 - \sqrt{3}, 0)$, $(0, 13)$; vertex: $(4. -3)$; range: all $t \geqslant -3$.

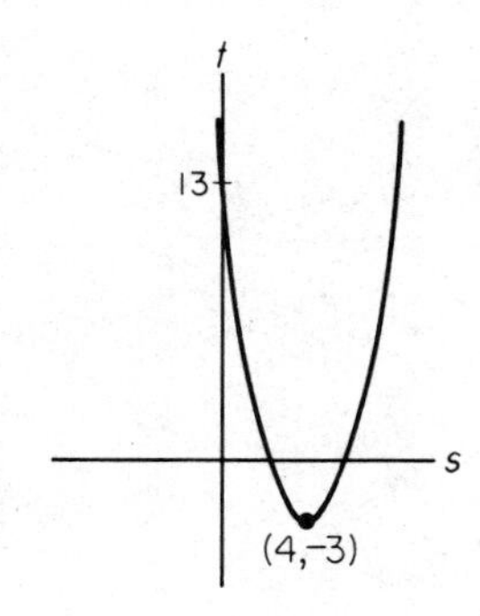

27. 24.

29. $q = 200$; $r = \$120{,}000$.

31. 70 grams.

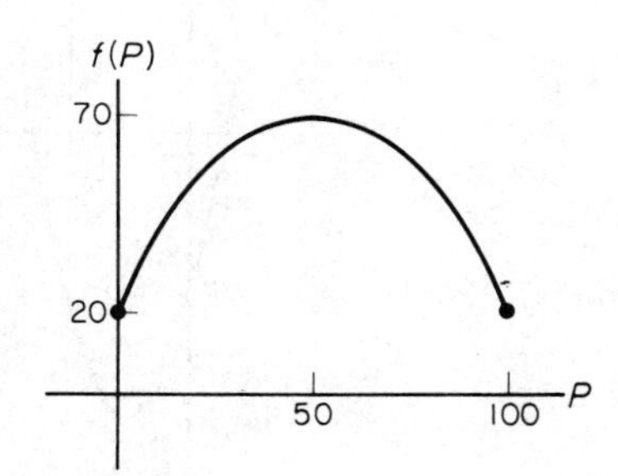

Review Problems—Chapter 1

1. $7, 46, 62, 3x^4 - 4x^2 + 7$; all real numbers.

3. $\frac{3}{2}, 18, 6\sqrt{u}, 12uh + 6h^2$; all real numbers.

5. $2, \frac{1}{2}, 8, \dfrac{1}{2\sqrt[3]{100}}$; all nonzero real numbers.

7. $-8, 4, 4, -92$; all real numbers $\neq 2$.

9. (a) $-x - 4$. (b) $12 - 5x$. (c) $-6x^2 + 32x - 32$.
(d) $\dfrac{4 - 3x}{2x - 8}$. (e) $28 - 6x$. (f) $-6x$.

11. $20 + 5\sqrt{23}$; 2.

13. All real numbers; all real numbers; -2; $(0, 4)$.

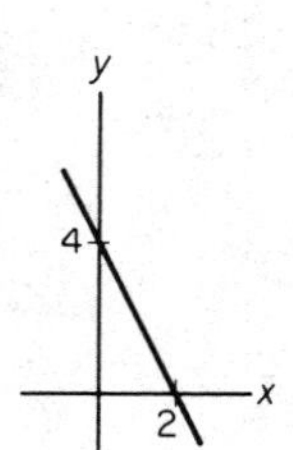

15. All $t \neq 4$; all nonzero real numbers.

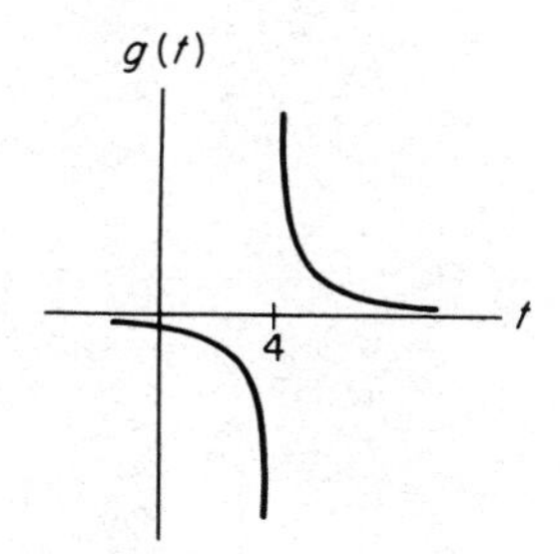

17. All real numbers; all reals $\leqslant 9$; $(3, 0)$, $(-3, 0)$, $(0, 9)$; $(0, 9)$.

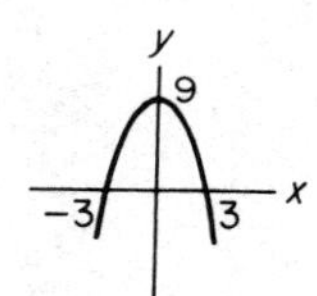

19. All real numbers; all reals $\geqslant -9$; $(5, 0)$, $(-1, 0)$, $(0, -5)$; $(2, -9)$.

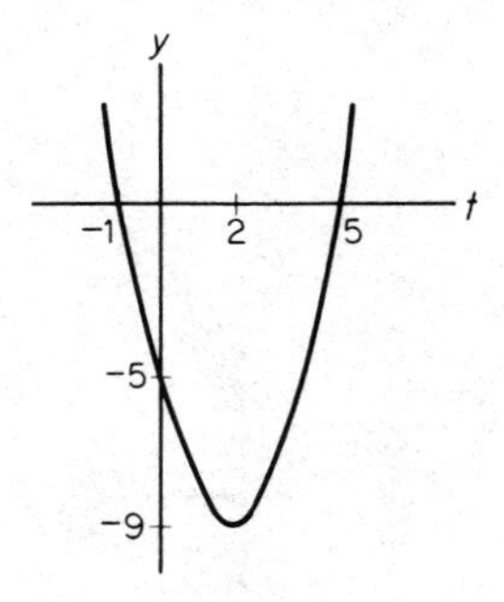

21. All real numbers; all real numbers; 3; $(0, 0)$.

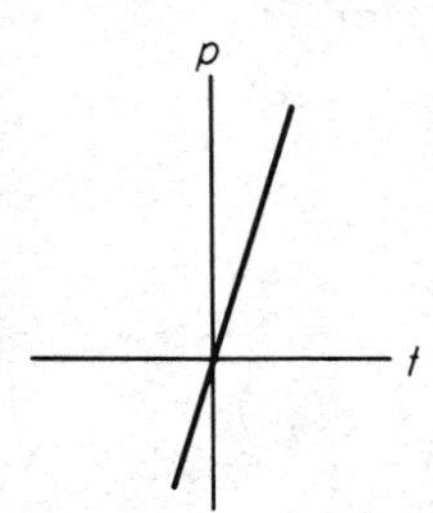

23. All real numbers; all reals $\leqslant -2$; $(0, -3)$; $(-1, -2)$.

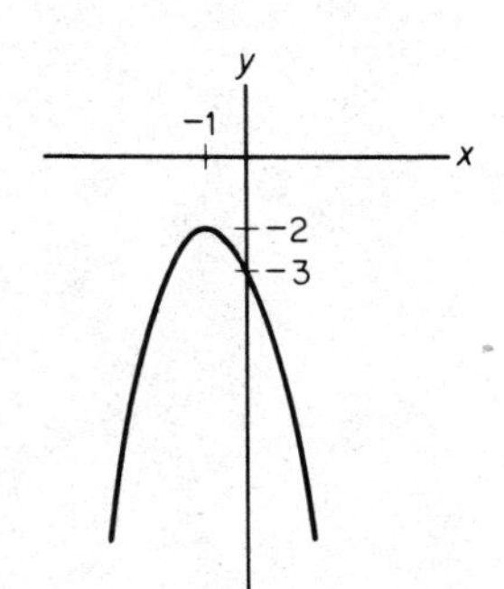

25. All real numbers; all reals $\geqslant 1$.

27. $f(x) = -\frac{4}{3}x + \frac{19}{3}$.

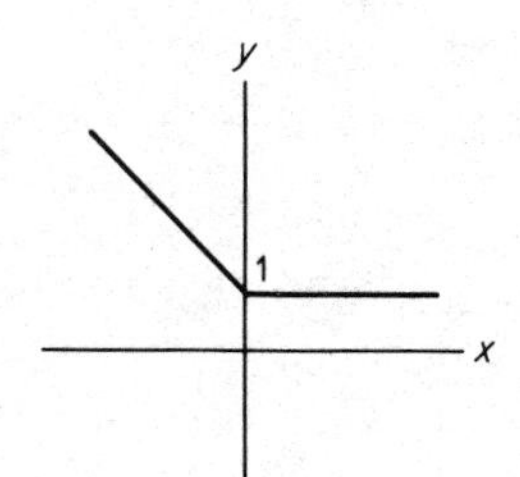

29. (a) $R = 75L + 1310$. (b) 1385 milliseconds. (c) The slope is 75. The time necessary to travel from one level to the next level is 75 milliseconds.

31. 50 units; \$5000.

Exercise 2-1

1. 14. **3.** -11. **5.** 5. **7.** 2.82. **9.** 0. **11.** -1.

13. $-\frac{5}{2}$. **15.** 0. **17.** 5. **19.** 3. **21.** $-\dfrac{5\sqrt{3}}{6}$. **23.** 1.

25. 0. **27.** $\frac{1}{6}$. **29.** $-\frac{1}{5}$. **31.** $\frac{7}{2}$. **33.** $\frac{11}{9}$. **35.** 4.

37. $2x$. **41.** (a) 1. (b) 0.

Exercise 2-2

1. (a) 2. (b) 3. (c) Does not exist. (d) $-\infty$. (e) ∞. (f) ∞. (g) ∞. (h) 0. (i) 1. (j) 1. (k) 1.

3. 1. **5.** $-\infty$. **7.** $-\infty$. **9.** ∞. **11.** 0.

13. Does not exist. **15.** 0. **17.** 0. **19.** 1. **21.** 0.

23. ∞. **25.** $-\frac{2}{5}$. **27.** $-\infty$. **29.** $\frac{2}{5}$. **31.** $\frac{11}{5}$. **33.** $-\frac{1}{2}$.

35. ∞. **37.** ∞. **39.** Does not exist. **41.** $-\infty$. **43.** 0. **45.** 1.

47. (a) 1. (b) 2. (c) Does not exist. (d) 1. (e) 2.

49. (a) 0. (b) 0. (c) 0. (d) $-\infty$. (e) $-\infty$.

51. 20. **53.**

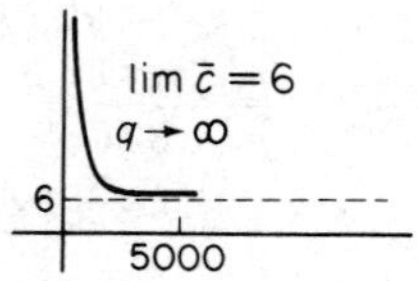

$\lim_{q \to \infty} \bar{c} = 6$

55. 2. **57.** $2x + 1$.

59. 1, .5, .525, .631, .912, .986, .998; conclude limit is 1.

Exercise 2-3

7. Continuous at -2 and 0.

9. Discontinuous at ± 3.

11. Continuous at 2 and 0.

13. f is a polynomial function.

15. f is a quotient of polynomial functions and the denominator is never 0.

17. None.

19. $x = 4$.

21. None.

23. $x = -5, 3$.

25. $x = 0, \pm 1$.

27. None.

29. $x = 2$.

31. None.

33. $x = 0, 2$.

37. Discontinuities at $t = 3, 4, 5$.

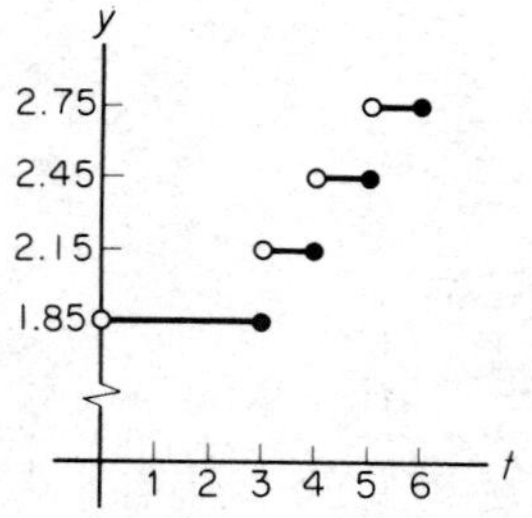

Exercise 2-4

1. $x < -1, x > 4$.

3. $2 \leqslant x \leqslant 3$.

5. $-\frac{7}{2} < x < -2$.

7. $-1 < x < 0, 0 < x < 1$.

9. $x \leqslant -6, -2 \leqslant x \leqslant 3$.

11. $x < -4, 0 < x < 5$.

13. $x \geqslant 0$.

15. $-3 < x < 0, x > 1$.

17. $x < -1, 0 < x < 1$.

19. $x > 1$.

21. $x < -5,\ -2 \leqslant x < 1, x \geqslant 3$.

23. $-4 < x < -2$.

25. $x \leqslant -1 - \sqrt{3}\,, x \geqslant -1 + \sqrt{3}\,$.

27. Between 50 and 150, inclusive.

Review Problems—Chapter 2

1. -5. **3.** 2. **5.** x. **7.** 0. **9.** $\frac{3}{5}$.

11. Does not exist. **13.** -1. **15.** 0. **17.** $\frac{1}{9}$.

19. $-\infty$. **21.** 100. **25.** Continuous everywhere. **27.** $x = -3$.

29. $x = -2$. **31.** $x < -6, x > 2$. **33.** $x \geqslant 2, x = 0$.

35. $x < -5,\ -1 < x < 1$. **37.** $x < -4,\ -3 \leqslant x \leqslant 0, x > 2$.

Exercise 3-1

1. 1. **3.** 2. **5.** -2. **7.** 0. **9.** $2x + 4$. **11.** $4p + 5$.

13. $-\dfrac{1}{x^2}$. **15.** $\dfrac{1}{2\sqrt{x + 2}}$. **17.** -4. **19.** 0.

21. $y = -4x + 2$. **23.** $y = -\frac{1}{3}x + 2$. **25.** $y = -3x - 7$.

Exercise 3-2

1. 0. **3.** $8x^7$. **5.** $45x^4$. **7.** $-7w^{-8}$.

9. $-\frac{56}{5}x^{-19/5}$. **11.** 3. **13.** $\frac{13}{5}$.

15. $6x - 5$. **17.** $-26t + 14 = 2(7 - 13t)$.

19. $42x^2 - 12x + 7$. **21.** $-9q^2 + 9q + 9 = 9(1 + q - q^2)$.

23. $8x^7 - 42x^5 + 6x = 2x(4x^6 - 21x^4 + 3)$.

25. $1002x^{500} - 12{,}500x^{99} + .68x^{2.4}$. **27.** $-8x^3$. **29.** $-\frac{4}{3}x^3$.

31. $-4x^{-5} - 3x^{-2/3} - 2x^{-7/5}$.

33. $-2(27 - 70x^4) = 2(70x^4 - 27)$. **35.** $-4x + \frac{3}{2} + x^3$.

37. $-x^{-2} = -\dfrac{1}{x^2}$. **39.** $-\frac{5}{4}s^{-6}$. **41.** $2t^{-1/2} = \dfrac{2}{\sqrt{t}}$.

43. $-\frac{1}{5}x^{-6/5}$. **45.** $9x^2 - 14x + 7$. **47.** $t + 4t^{-3}$.

49. $45x^4$. **51.** $\frac{1}{3}x^{-2/3} - \frac{10}{3}x^{-5/3} = \frac{1}{3}x^{-5/3}(x - 10)$.

53. $8q + \dfrac{4}{q^2}$. **55.** $2(x + 2)$. **57.** 1.

59. 4, 16, -14. **61.** 0, 0, 0. **63.** $y = 13x - 2$.

65. $y = x + 3$. **67.** $(0, 0), (2, -\frac{4}{3})$.

Exercise 3-3

1. (a) 1. (b) $\dfrac{1}{x + 4}$. (c) 1. (d) $\frac{1}{9} \approx .111$. (e) 11.1%.

3. (a) $6x$. (b) $\dfrac{2x}{x^2 + 2}$. (c) 12. (d) $\frac{2}{3} \approx .667$. (e) 66.7%.

5. (a) $-3x^2$. (b) $-\dfrac{3x^2}{8 - x^3}$. (c) -3. (d) $-\frac{3}{7} \approx -.429$.

(e) -42.9%. **7.** $\dfrac{dc}{dq} = 10$; 10. **9.** $\dfrac{dc}{dq} = .6q + 2$; 3.8.

11. $\dfrac{dc}{dq} = 2q + 50$; 80, 82, 84.

13. $\dfrac{dc}{dq} = 6.750 - .000656q$; 3.47. **15.** 3.2.

17. $\dfrac{dc}{dq} = .02q + 5$; 6, 7. **19.** $\dfrac{dc}{dq} = .00006q^2 - .02q + 6$; 4.6, 11.

21. .27. **23.** (a) $\dfrac{dr}{dq} = 30 - .6q$. (b) $\frac{4}{45} \approx .089$. (c) 9%.

25. -1; -10%. **27.** -7.5; 4.5. **29.** $\dfrac{.432}{t}$.

Exercise 3-5

1. $(3x - 1)(7) + (7x + 2)(3) = 42x - 1$.

3. $(5 - 2x)(2x) + (x^2 + 1)(-2) = -2(3x^2 - 5x + 1)$.

5. $(3r^2 - 4)(2r - 5) + (r^2 - 5r + 1)(6r) = 12r^3 - 45r^2 - 2r + 20$.

7. $(x^2 + 3x - 2)(4x - 1) + (2x^2 - x - 3)(2x + 3) =$
$8x^3 + 15x^2 - 20x - 7$.

9. $(8w^2 + 2w - 3)(15w^2) + (5w^3 + 2)(16w + 2) =$
$200w^4 + 40w^3 - 45w^2 + 32w + 4$.

11. $3[(x^3 - 2x^2 + 5x - 4)(4x^3 - 6x^2 + 7) +$
$(x^4 - 2x^3 + 7x + 1)(3x^2 - 4x + 5)] =$
$3(7x^6 - 24x^5 + 45x^4 - 28x^3 - 15x^2 + 66x - 23)$.

13. $(x^2 - 1)(9x^2 - 6) + (3x^3 - 6x + 5)(2x) -$
$[(x + 4)(8x + 2) + (4x^2 + 2x + 1)(1)] = 15x^4 - 39x^2 - 26x - 3$.

15. $\frac{3}{2}\left[(p^{1/2} - 4)(4) + (4p - 5)\left(\frac{1}{2}p^{-1/2}\right)\right] = \frac{3}{4}(12p^{1/2} - 5p^{-1/2} - 32)$.

17. $(2x^{.45} - 3)(1.3x^{.3} - 7) + (x^{1.3} - 7x)(.9x^{-.55}) =$
$3.5x^{.75} - 20.3x^{.45} - 3.9x^{.3} + 21$.

19. $18x^2 + 94x + 31$.

21. 0.

23. $\dfrac{(x - 1)(1) - (x)(1)}{(x - 1)^2} = -\dfrac{1}{(x - 1)^2}$.

25. $\dfrac{(x + 2)(1) - (x - 1)(1)}{(x + 2)^2} = \dfrac{3}{(x + 2)^2}$.

27. $\dfrac{(z^2 - 4)(-2) - (5 - 2z)(2z)}{(z^2 - 4)^2} = \dfrac{2(z - 4)(z - 1)}{(z^2 - 4)^2}$.

29. $\dfrac{(x^2 - 5x)(16x - 2) - (8x^2 - 2x + 1)(2x - 5)}{(x^2 - 5x)^2} = \dfrac{-38x^2 - 2x + 5}{(x^2 - 5x)^2}$.

31. $\dfrac{(2x^2 - 3x + 2)(2x - 4) - (x^2 - 4x + 3)(4x - 3)}{(2x^2 - 3x + 2)^2} = \dfrac{5x^2 - 8x + 1}{(2x^2 - 3x + 2)^2}$.

33. $\dfrac{-100x^{99}}{(x^{100} + 1)^2}$.

35. $\dfrac{4(v^5 + 2)}{v^2}$.

37. $\dfrac{15x^2 - 2x + 1}{3x^{4/3}}$.

39. $\dfrac{4}{(x - 8)^2} + \dfrac{2}{(3x + 1)^2}$.

41. $\dfrac{(s - 5)(2s - 2) - [(s + 2)(s - 4)](1)}{(s - 5)^2} = \dfrac{s^2 - 10s + 18}{(s - 5)^2}$.

43. $\dfrac{[(x + 2)(x - 4)](1) - (x - 5)(2x - 2)}{[(x + 2)(x - 4)]^2} = \dfrac{-(x^2 - 10x + 18)}{[(x + 2)(x - 4)]^2}$.

45. $\dfrac{[(t^2-1)(t^3+7)](2t+3)-(t^2+3t)(5t^4-3t^2+14t)}{[(t^2-1)(t^3+7)]^2}=$

$\dfrac{-3t^6-12t^5+t^4+6t^3-21t^2-14t-21}{[(t^2-1)(t^3+7)]^2}.$

47. $\dfrac{(x^2-7x+12)(2x-3)-(x^2-3x+2)(2x-7)}{[(x-3)(x-4)]^2}=$

$\dfrac{-2(2x^2-10x+11)}{[(x-3)(x-4)]^2}.$

49. -6. **51.** $y=-\frac{3}{2}x+\frac{15}{2}$. **53.** $y=16x+24$.

55. $\dfrac{dr}{dq}=.7$; $.7$, $.7$, $.7$.

57. $\dfrac{dr}{dq}=250+90q-3q^2$; 625, 850, 625.

59. $\dfrac{dr}{dq}=25-.04q$. **61.** $\dfrac{dr}{dq}=\dfrac{216}{(q+2)^2}-3$.

63. $\dfrac{dC}{dI}=.672$. **65.** $\frac{1}{3}$; $\frac{2}{3}$. **67.** .393; .607.

69. $\dfrac{.7355}{(1+.02744x)^2}$. **71.** $\dfrac{dc}{dq}=\dfrac{5q(q+6)}{(q+3)^2}$.

Exercise 3-6

1. $(2u-2)(2x-1)=4x^3-6x^2-2x+2$.

3. $\left(-\dfrac{2}{w^3}\right)(-1)=\dfrac{2}{(2-x)^3}$.

5. -2. **7.** 0. **9.** $56(7x+4)^7$. **11.** $-56p(3-2p^2)^{13}$.

13. $\dfrac{40(3x^2-2)(4x^3-8x+2)^9}{3}$. **15.** $-30(4r-5)(4r^2-10r+3)^{-16}$.

17. $-49x(3x-2)(x^3-x^2+2)^{-8}$. **19.** $6z(6z-1)(4z^3-z^2+2)^{-4/5}$.

21. $\frac{1}{2}(4x-1)(2x^2-x+3)^{-1/2}$. **23.** $\frac{6}{5}x(x^2+1)^{-2/5}$.

25. $-6.5(5x^4+21x^2-8)(x^5+7x^3-8x+4)^{-7.5}$.

27. $10\left(\frac{x-7}{x+4}\right)^9\left[\frac{(x+4)(1)-(x-7)(1)}{(x+4)^2}\right]=\frac{110(x-7)^9}{(x+4)^{11}}$.

29. $10\left(\frac{q^3-2q+4}{5q^2+1}\right)^4\left[\frac{5q^4+13q^2-40q-2}{(5q^2+1)^2}\right]$

$=\frac{10(5q^4+13q^2-40q-2)(q^3-2q+4)^4}{(5q^2+1)^6}$.

31. $\frac{5}{2(x+3)^2}\left(\frac{x-2}{x+3}\right)^{-1/2}=\frac{5}{2(x+3)^2}\sqrt{\frac{x+3}{x-2}}$.

33. $(x^2+2x-1)^3(5)+(5x+7)[3(2x+2)(x^2+2x-1)^2]$
$=(x^2+2x-1)^2(35x^2+82x+37)$.

35. $8[(4x+3)(6x^2+x+8)]^7[(4x+3)(12x+1)+(6x^2+x+8)(4)]$
$=8(72x^2+44x+35)[(4x+3)(6x^2+x+8)]^7$.

37. $\frac{(w^2+4)[6(2w+3)^2]-(2w+3)^3(2w)}{(w^2+4)^2}=$
$\frac{2(w^2-3w+12)(2w+3)^2}{(w^2+4)^2}$.

39. $6\{(5x^2+2)[2x^3(x^4+5)^{-1/2}]+(x^4+5)^{1/2}(10x)\}$
$=12x(x^4+5)^{-1/2}(10x^4+2x^2+25)$.

41. $(4-3x^2)^2[3(2-3x)^2(-3)]+(2-3x)^3[2(4-3x^2)(-6x)]$
$=3(4-3x^2)(2-3x)^2(21x^2-8x-12)$.

43. $8+\frac{5}{(t+4)^2}-(8t-7)=15-8t+\frac{5}{(t+4)^2}$.

45. $\frac{(3x-1)^3[40(8x-1)^4]-(8x-1)^5[9(3x-1)^2]}{(3x-1)^6}=$
$\frac{(8x-1)^4(48x-31)}{(3x-1)^4}$.

47. 0. **49.** 0. **51.** $y=4x-11$.

53. $y=-\frac{1}{6}x+\frac{5}{3}$. **55.** -325. **57.** (a) 10; (b) 12.

59. (a) $-\frac{1000}{\sqrt{100-x}}$, -125; (b) $-\frac{1}{128}$.

61. $\frac{1}{1600}$; when t increases from 300 to 301, the proportion discharged increases by approximately 1/1600.

63. $\frac{dr}{dq} = 100 - \frac{q^2}{\sqrt{q^2 + 20}} - \sqrt{q^2 + 20}$.

65. .443; .557.

Exercise 3-7

1. $-\frac{x}{4y}$. **3.** $-\frac{y}{x}$. **5.** $\frac{4 - y}{x - 1}$.

7. $\frac{4y - x^2}{y^2 - 4x}$. **9.** $-\frac{y^{1/4}}{x^{1/4}}$. **11.** $\frac{5}{12y^3}$.

13. $-\frac{\sqrt{y}}{\sqrt{x}}$. **15.** $\frac{6y^{2/3}}{3y^{1/6} + 2}$. **17.** $\frac{1 - 6xy^3}{1 + 9x^2y^2}$.

19. $-\frac{3}{5}$. **21.** $y = -\frac{3}{4}x + \frac{5}{4}$.

23. $\frac{dq}{dp} = -\frac{1}{2q}$. **25.** $\frac{dq}{dp} = -\frac{(q + 5)^3}{40}$.

Exercise 3-8

1. 24. **3.** 0. **5.** $\frac{105}{16}x^{-9/2}$.

7. $-\frac{10}{p^6}$. **9.** $-\frac{1}{4(1 - r)^{3/2}}$. **11.** $\frac{50}{(5x - 6)^3}$.

13. $20(x^2 - 4)^8(19x^2 - 4)$. **15.** $\frac{4}{(x - 1)^3}$.

17. $12(1 - 2x)(4x - 1)$. **19.** $-\frac{1}{y^3}$. **21.** $-\frac{4}{y^3}$.

23. $\frac{1}{8x^{3/2}}$. **25.** $\frac{2(y - 1)}{(1 + x)^2}$. **27.** $300(5x - 3)^2$. **29.** .6.

Review Problems—Chapter 3

1. 0. **3.** $28x^3 - 18x^2 + 10x = 2x(14x^2 - 9x + 5)$.

5. $4s^3 + 4s = 4s(s^2 + 1)$. **7.** $\frac{2x}{5}$.

9. $(x^2 + 6x)(3x^2 - 12x) + (x^3 - 6x^2 + 4)(2x + 6)$
$= 5x^4 - 108x^2 + 8x + 24$.

11. $100(2x^2 + 4x)^{99}(4x + 4) = 400(x + 1)[(2x)(x + 2)]^{99}$.

13. $-\frac{2}{(2x + 1)^2}$.

15. $(8 + 2x)(4)(x^2 + 1)^3(2x) + (x^2 + 1)^4(2)$
$= 2(x^2 + 1)^3(9x^2 + 32x + 1)$.

17. $\frac{(z^2 + 1)(2z) - (z^2 - 1)(2z)}{(z^2 + 1)^2} = \frac{4z}{(z^2 + 1)^2}$.

19. $\frac{4}{3}(4x - 1)^{-2/3}$. **21.** $-\frac{y}{x + y}$.

23. $-\frac{1}{2}(1 - x)^{-3/2}(-1) = \frac{1}{2}(1 - x)^{-3/2}$.

25. $(x - 6)^4[3(x + 5)^2] + (x + 5)^3[4(x - 6)^3]$
$= (x - 6)^3(x + 5)^2(7x + 2)$.

27. $\frac{(x + 1)(5) - (5x - 4)(1)}{(x + 1)^2} = \frac{9}{(x + 1)^2}$.

29. $2(-\frac{3}{8})x^{-11/8} + (-\frac{3}{8})(2x)^{-11/8}(2) = -\frac{3}{4}(1 + 2^{-11/8})x^{-11/8}$.

31. $\frac{\sqrt{x^2 + 5}\,(2x) - (x^2 + 6)(1/2)(x^2 + 5)^{-1/2}(2x)}{x^2 + 5} = \frac{x(x^2 + 4)}{(x^2 + 5)^{3/2}}$.

33. $(\frac{3}{5})(x^3 + 6x^2 + 9)^{-2/5}(3x^2 + 12x) = \frac{9}{5}x(x + 4)(x^3 + 6x + 9)^{-2/5}$.

35. $-\frac{y}{x}$. **37.** $7(1 - 2z)$.

39. 12. **41.** $-\frac{1}{128}$. **43.** $\frac{4}{9}$.

45. $-\frac{1}{4}$. **47.** $y = -4x + 3$. **49.** $y = \frac{1}{12}x + \frac{4}{3}$.

51. $7x - 2\sqrt{10}\,y - 9 = 0$. **53.** $\frac{5}{7} \approx .714$; 71.4%.

55. $\frac{dr}{dq} = 20 - .2q$. **57.** $\frac{dc}{dq} = .125 + .00878q$; .7396.

59. 84. **61.** .569, .431. **63.** $\frac{dr}{dq} = 450 - q$.

Exercise 4-1

1. (0, 0); sym. to origin.

3. (±2, 0), (0, 8); sym. to y-axis.

5. (±3, 0); sym. to x-axis, y-axis, origin.

7. (−2, 0); sym. to x-axis.

9. Sym. to x-axis.

11. (−21, 0), (0, −7), (0, 3).

13. (0, 0); sym. to origin.

15. (0, 0); sym. to x-axis, y-axis, origin.

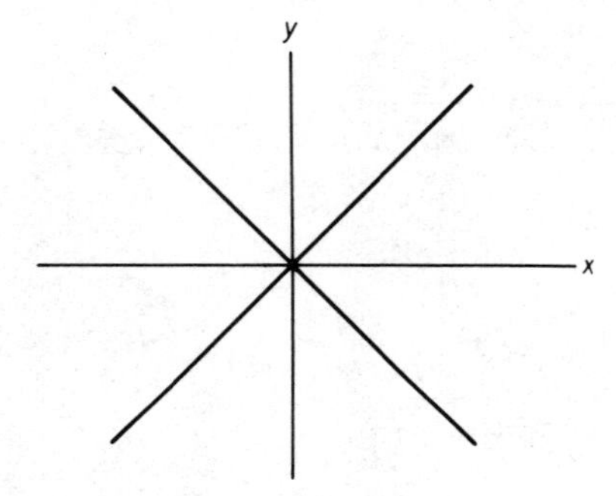

17. (2, 0), (0, ±2); sym. to x-axis.

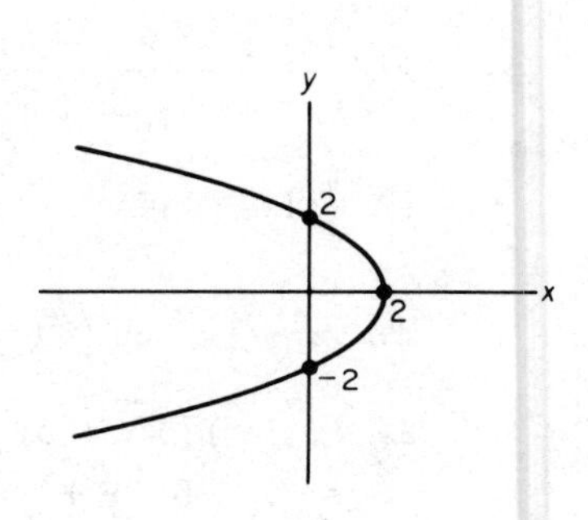

19. (±2, 0), (0, 0); sym. to origin.

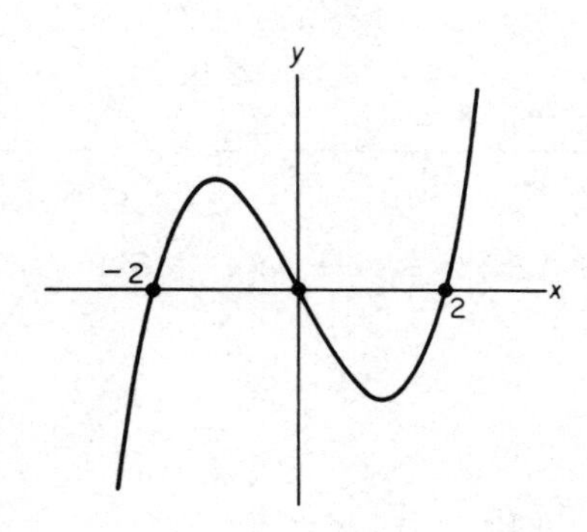

21. (±2, 0), (0, ±4); sym. to x-axis, y-axis, origin.

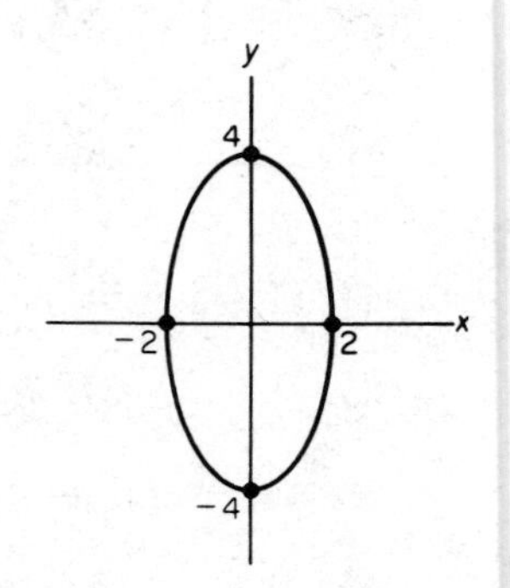

Exercise 4-2

1. $y = 0, x = 0$. **3.** $y = 4, x = 6$. **5.** None.

7. $y = \frac{1}{2}, x = -\frac{3}{2}$. **9.** $y = 2, x = -3, x = 2$. **11.** None.

13. $(0, -3)$; $y = 0$, $x = 1$.

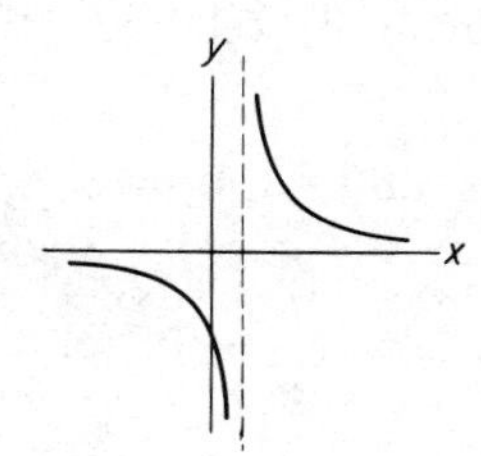

15. Sym. to origin;
$y = 0, x = 0$.

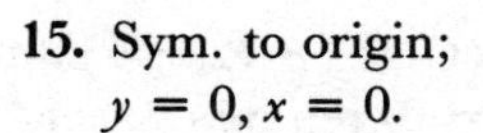

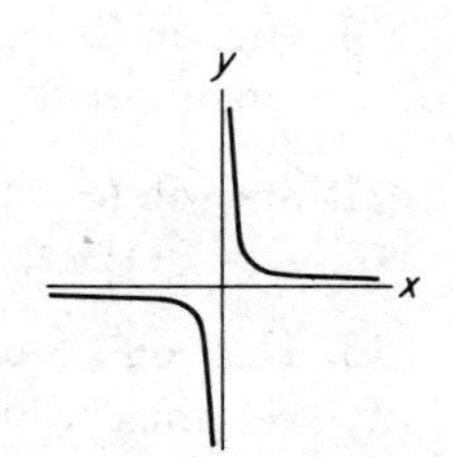

17. $(0, -1)$; sym. to y-axis;
$y = 0$, $x = 1$, $x = -1$.

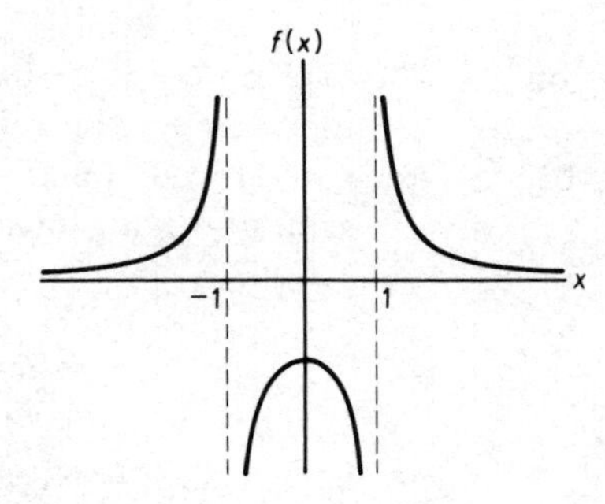

19. $(\pm 1, 0)$, $(0, \frac{1}{4})$;
sym. to y-axis;
$y = 1$, $x = 2$, $x = -2$.

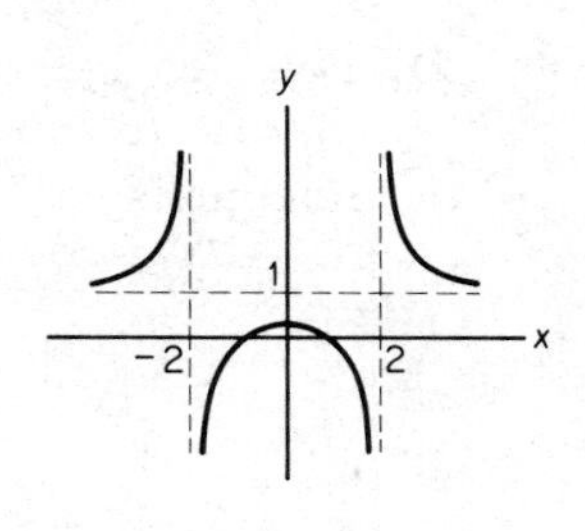

21. $(\pm 3, 0)$; sym. to y-axis.

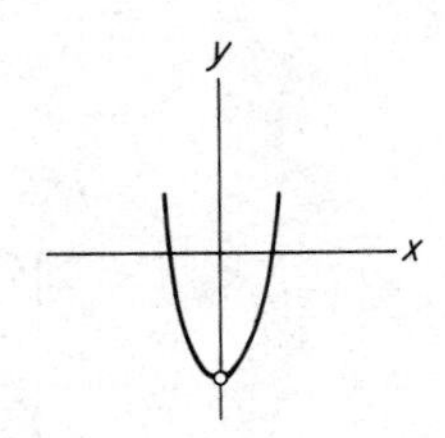

Exercise 4-3

1. Dec. on $(-\infty, 0)$; inc. on $(0, \infty)$; rel. min. when $x = 0$.

3. Inc. on $(-\infty, \frac{1}{2})$; dec. on $(\frac{1}{2}, \infty)$; rel. max. when $x = \frac{1}{2}$.

5. Dec. on $(-\infty, -5)$ and $(1, \infty)$; inc. on $(-5, 1)$; rel. min. when $x = -5$; rel. max. when $x = 1$.

7. Dec. on $(-\infty, -1)$ and $(0, 1)$; inc. on $(-1, 0)$ and $(1, \infty)$; rel max. when $x = 0$; rel. min. when $x = \pm 1$.

9. Inc. on $(-\infty, 1)$ and $(3, \infty)$; dec. on $(1, 3)$; rel. max. when $x = 1$; rel. min. when $x = 3$.

11. Inc. on $(-\infty, -1)$ and $(1, \infty)$; dec. on $(-1, 0)$ and $(0, 1)$; rel. max. when $x = -1$; rel. min. when $x = 1$.

13. Dec. on $(-\infty, -4)$ and $(0, \infty)$; inc. on $(-4, 0)$; rel. min. when $x = -4$; rel. max. when $x = 0$.

15. Dec. on $(-\infty, 1)$ and $(1, \infty)$; no rel. max. or min.

17. Dec. on $(0, \infty)$; no rel. max. or min.

19. Dec. on $(-\infty, 0)$ and $(2, \infty)$; inc. on $(0, 1)$ and $(1, 2)$; rel. min. when $x = 0$; rel. max. when $x = 2$.

21. Dec. on $(-\infty, 3)$; inc. on $(3, \infty)$; rel. min. when $x = 3$; intercepts: $(7, 0)$, $(-1, 0)$, $(0, -7)$.

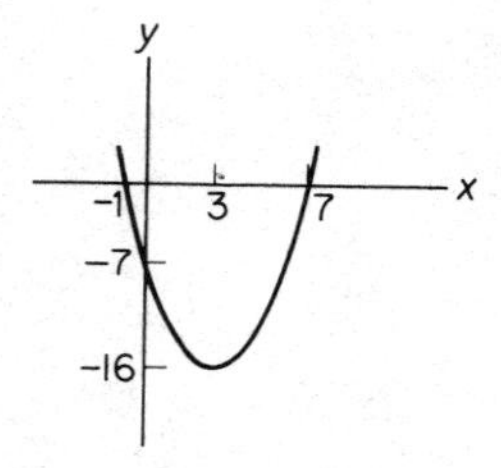

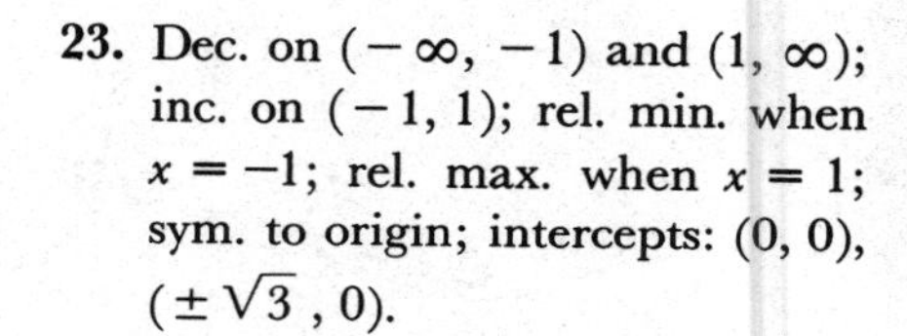

23. Dec. on $(-\infty, -1)$ and $(1, \infty)$; inc. on $(-1, 1)$; rel. min. when $x = -1$; rel. max. when $x = 1$; sym. to origin; intercepts: $(0, 0)$, $(\pm\sqrt{3}, 0)$.

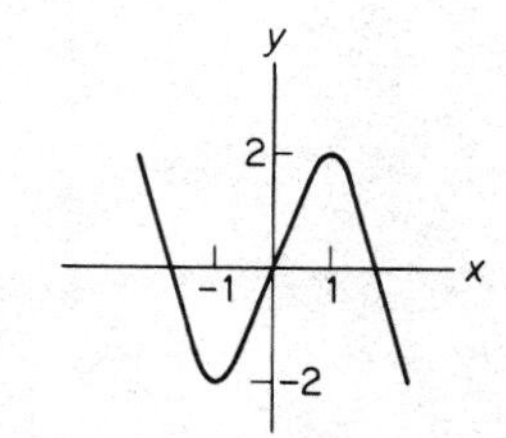

25. Inc. on $(-\infty, 1)$ and $(2, \infty)$; dec. on $(1, 2)$; rel. max. when $x = 1$; rel. min. when $x = 2$; intercept: $(0, 0)$.

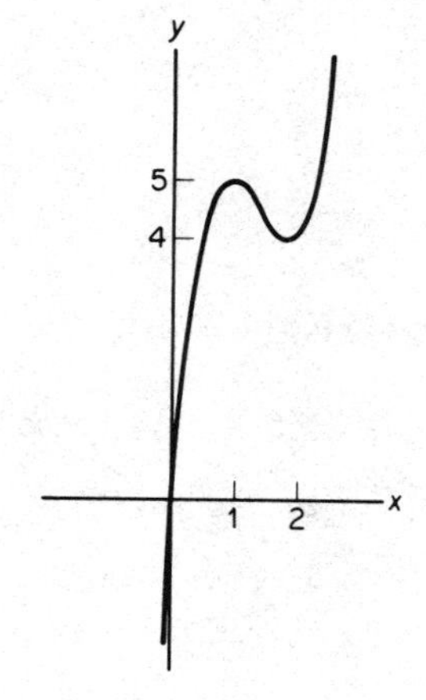

27. Inc. on $(-2, -1)$, $(0, \infty)$; dec. on $(-\infty, -2)$, $(-1, 0)$; rel. max. when $x = -1$; rel. min. when $x = -2, 0$; intercepts: $(0, 0)$, $(-2, 0)$.

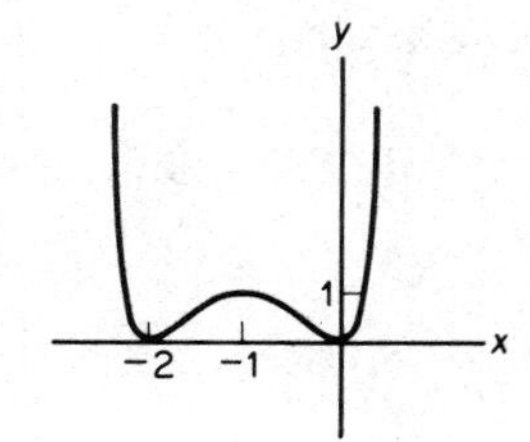

29. Dec. on $(-\infty, 1)$, $(1, \infty)$; asym. $y = 1$, $x = 1$; intercepts: $(0, -1)$, $(-1, 0)$.

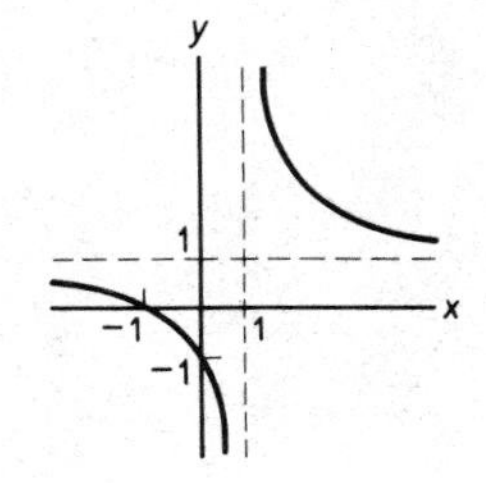

31. Inc. on $(-\infty, -6)$, $(0, \infty)$; dec. on $(-6, -3)$, $(-3, 0)$; rel. max. when $x = -6$; rel. min. when $x = 0$; asym. $x = -3$; intercept $(0, 0)$.

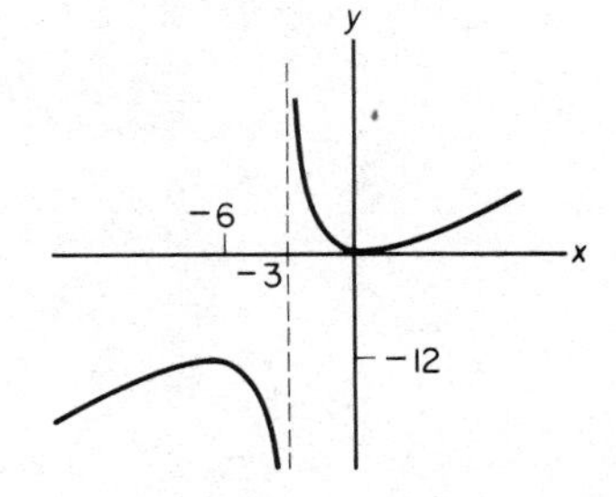

33. Abs. max. when $x = -1$; abs. min. when $x = 1$.

35. Abs. max. when $x = 0$; abs. min. when $x = 2$.

37. Abs. max. when $x = 3$; abs. min. when $x = 1$.

41. Never. **45.**

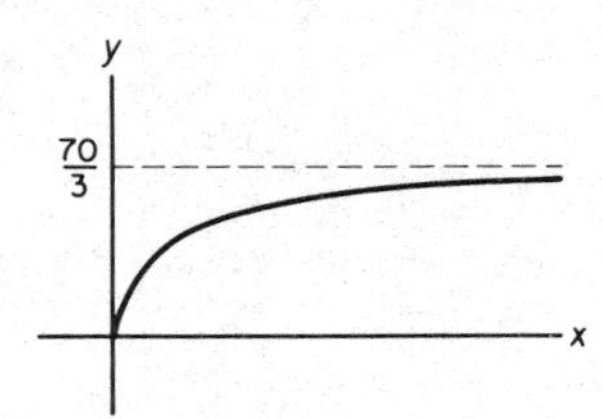

47. (a) 5; $\frac{1}{5}$.

Exercise 4-4

1. Conc. down $(-\infty, \infty)$.

3. Conc. down $(-\infty, -1)$; conc. up $(-1, \infty)$; inf. pt. when $x = -1$.

5. Conc. up $(-\infty, -1)$, $(1, \infty)$; conc. down $(-1, 1)$; inf. pt. when $x = \pm 1$.

7. Conc. down $(-\infty, 1)$; conc. up $(1, \infty)$.

9. Conc. down $\left(-\infty, -\frac{1}{\sqrt{3}}\right)$, $\left(\frac{1}{\sqrt{3}}, \infty\right)$; conc. up $\left(-\frac{1}{\sqrt{3}}, \frac{1}{\sqrt{3}}\right)$; inf. pt. when $x = \pm \frac{1}{\sqrt{3}}$.

11. Int. $(-3, 0)$, $(-1, 0)$, $(0, 3)$; dec. $(-\infty, -2)$; inc. $(-2, \infty)$; rel. min. when $x = -2$; conc. up $(-\infty, \infty)$.

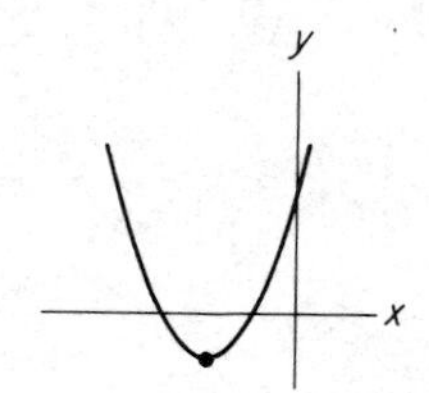

13. Int. $(0, 0)$, $(4, 0)$; inc. $(-\infty, 2)$; dec. $(2, \infty)$; rel. max. when $x = 2$; conc. down $(-\infty, \infty)$.

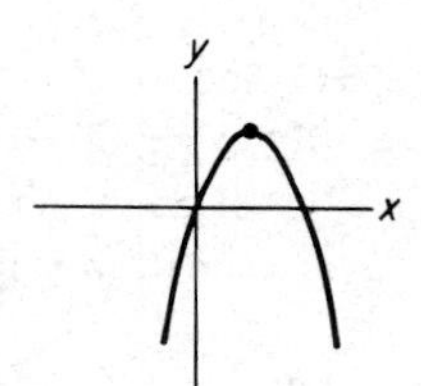

15. Int. $(-3/2, 0)$, $(4, 0)$, $(0, -12)$; dec. $(-\infty, 5/4)$; inc. $(5/4, \infty)$; rel. min. when $x = 5/4$; conc. up $(-\infty, \infty)$.

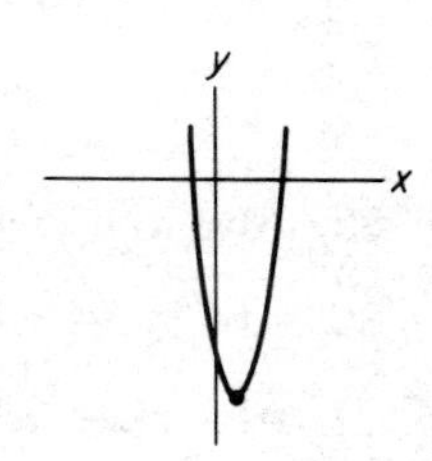

17. Int. $(0, -19)$; inc. $(-\infty, 2)$, $(4, \infty)$; dec. $(2, 4)$; rel. max. when $x = 2$; rel. min. when $x = 4$; conc. down $(-\infty, 3)$; conc. up $(3, \infty)$; inf. pt. when $x = 3$.

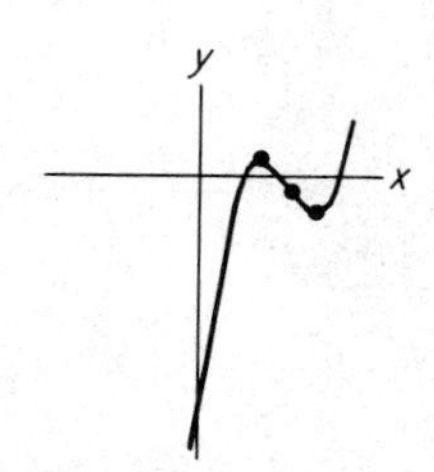

19. Int. (0, 0), (± 3, 0); inc. $(-\infty, -\sqrt{3})$, $(\sqrt{3}, \infty)$; dec. $(-\sqrt{3}, \sqrt{3})$; rel. max. when $x = -\sqrt{3}$; rel. min. when $x = \sqrt{3}$; conc. down $(-\infty, 0)$; conc. up $(0, \infty)$; inf. pt. when $x = 0$; sym. to origin.

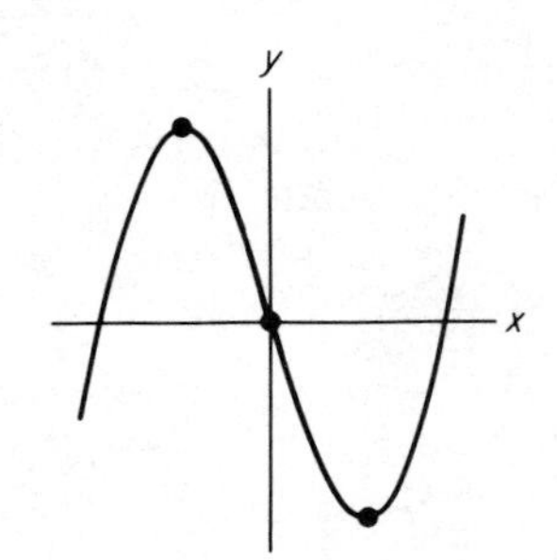

21. Int. $(0, -3)$; inc. $(-\infty, 1)$, $(1, \infty)$; no rel. max. or min.; conc. down $(-\infty, 1)$; conc. up $(1, \infty)$; inf. pt. when $x = 1$.

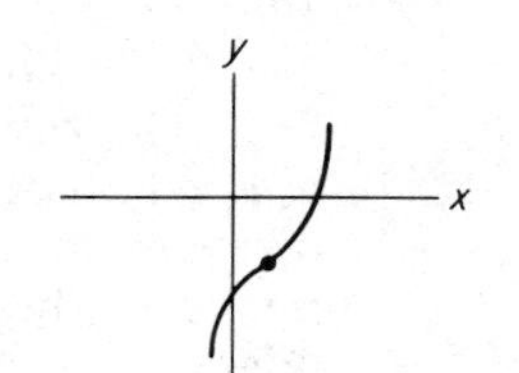

23. Int. (0, 0), (± 2, 0); inc. $(-\infty, -\sqrt{2})$, $(0, \sqrt{2})$; dec. $(-\sqrt{2}, 0)$, $(\sqrt{2}, \infty)$; rel. max. when $x = \pm\sqrt{2}$; rel. min. when $x = 0$; conc. down $(-\infty, -\sqrt{2/3})$, $(\sqrt{2/3}, \infty)$; conc. up $(-\sqrt{2/3}, \sqrt{2/3})$; inf. pt. when $x = \pm\sqrt{2/3}$; sym. to y-axis.

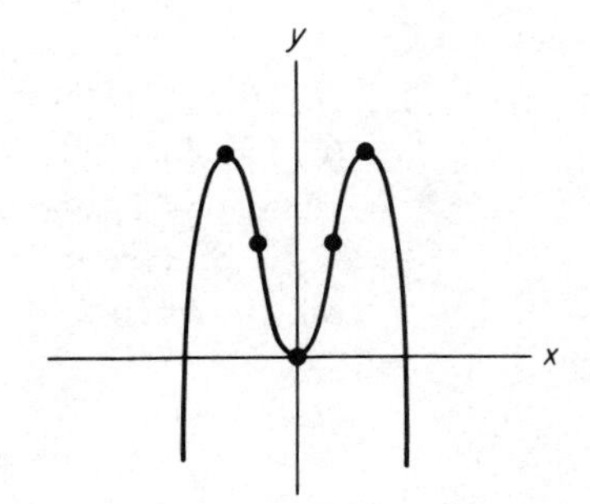

25. Int. (0, 0), (4/3, 0); inc. $(-\infty, 0)$, $(0, 1)$; dec. $(1, \infty)$; rel. max. when $x = 1$; conc. up $(0, 2/3)$; conc. down $(-\infty, 0)$, $(2/3, \infty)$; inf. pt. when $x = 0$, $x = 2/3$.

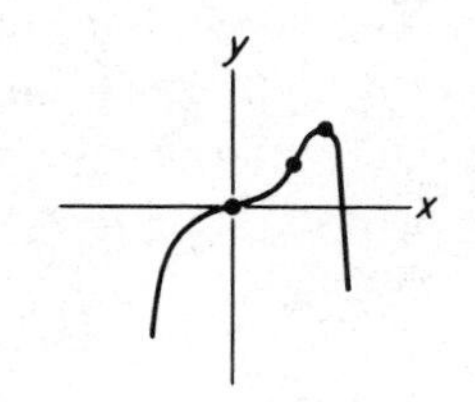

27. Int. $(0, -2)$; dec. $(-\infty, -2)$, $(2, \infty)$; inc. $(-2, 2)$; rel. min. when $x = -2$; rel. max. when $x = 2$; conc. up $(-\infty, 0)$; conc. down $(0, \infty)$; inf. pt. when $x = 0$.

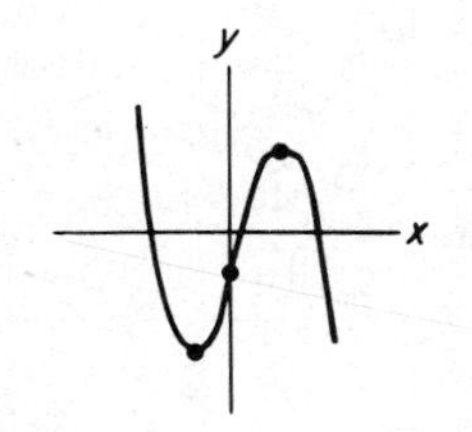

29. Int. $(0, -6)$; inc. $(-\infty, 2)$, $(2, \infty)$; conc. down $(-\infty, 2)$; conc. up $(2, \infty)$; inf. pt. when $x = 2$.

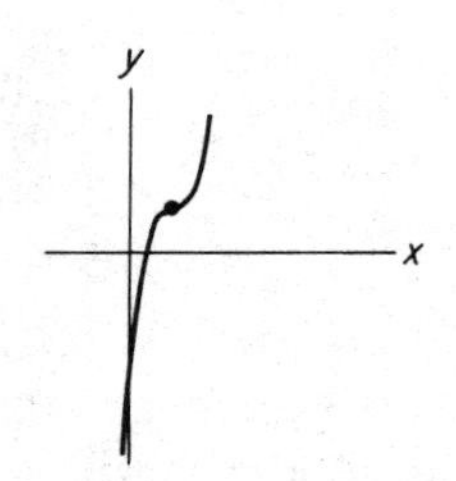

31. Int. $(0, 0)$, $(\pm\sqrt[4]{5}, 0)$; dec. $(-\infty, -1)$, $(1, \infty)$; inc. $(-1, 1)$; rel. min. when $x = -1$; rel. max. when $x = 1$; conc. up $(-\infty, 0)$; conc. down $(0, \infty)$; inf. pt. when $x = 0$; sym. to origin.

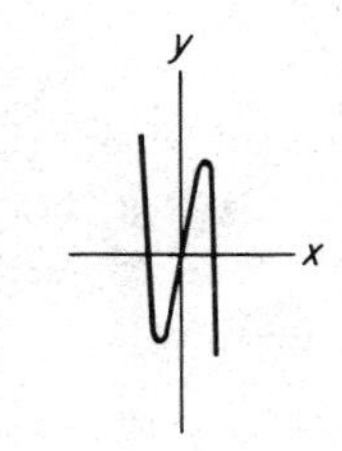

33. Int. $(0, 1)$, $(1, 0)$; dec. $(-\infty, 0)$, $(0, 1)$; inc. $(1, \infty)$; rel. min. when $x = 1$; conc. up $(-\infty, 0)$, $(2/3, \infty)$; conc. down $(0, 2/3)$; inf. pt. when $x = 0$, $x = 2/3$.

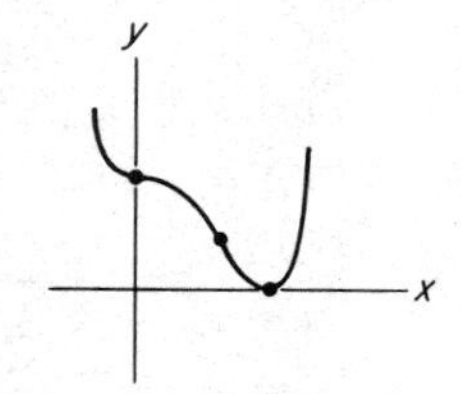

35. Dec. $(-\infty, 0)$, $(0, \infty)$; conc. down $(-\infty, 0)$; conc. up $(0, \infty)$; sym. to origin; asymptotes $x = 0, y = 0$.

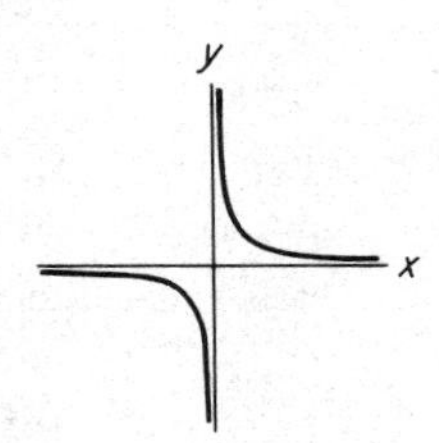

37. Int. $(0, 0)$; inc. $(-\infty, -1)$, $(-1, \infty)$; conc. up $(-\infty, -1)$; conc. down $(-1, \infty)$; asymptotes $x = -1, y = 1$.

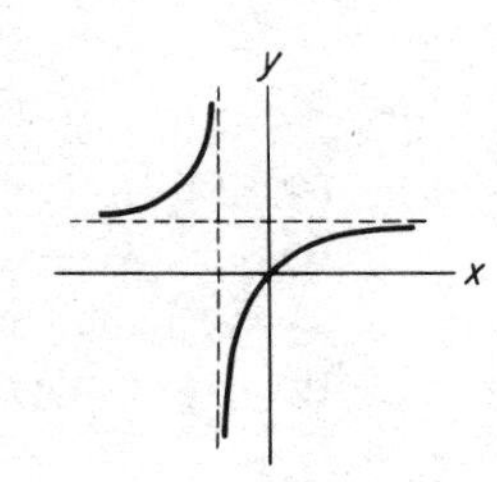

39. Dec. $(-\infty, -1)$, $(0, 1)$; inc. $(-1, 0)$, $(1, \infty)$; rel. min. when $x = \pm 1$; conc. up $(-\infty, 0)$, $(0, \infty)$; sym. to y-axis; asymptote $x = 0$.

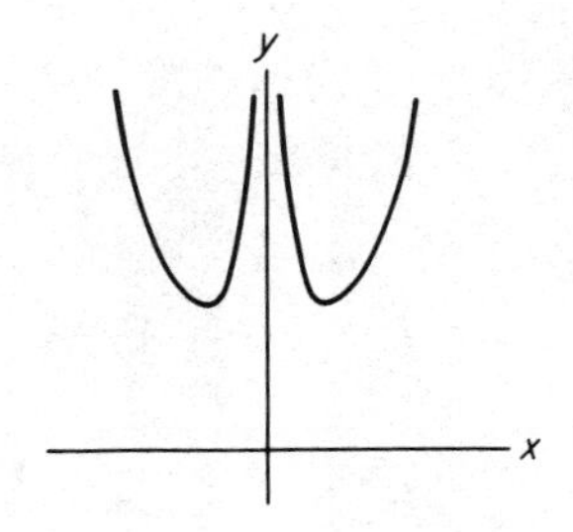

43.

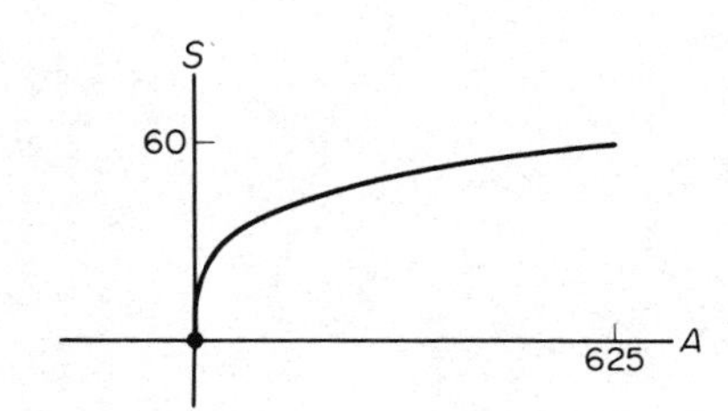

Exercise 4-5

1. Rel. min. when $x = \frac{5}{2}$; abs. min. **3.** Rel. max. when $x = \frac{1}{4}$; abs. max.

5. Rel. max. when $x = -3$, rel. min. when $x = 3$.

7. Rel. min. when $x = 0$, rel. max. when $x = 2$.

9. Test fails, when $x = 0$ there is rel. min. by first-deriv. test.

Exercise 4-6

1. 100. **3.** \$15. **5.** 525, \$51, \$10,525.

7. 625, \$4. **9.** 12, 105. **11.** (a) 110 grams. (b) $51\frac{9}{11}$ grams.

13. \$3, \$9000. **15.** 20 and 20. **17.** 300 feet by 250 feet.

19. 4 feet by 4 feet by 2 feet. **21.** 21 in. by 14 in.

25. 130, $p = 340$, $P = 36{,}980$; 125, $p = 350$, $P = 34{,}175$.

27. 250 per lot (4 lots). **29.** 35 **31.** 60 mi/hr. **33.** 10.

Exercise 4-7

1. $3\,dx$. **3.** $\dfrac{2x^3}{\sqrt{x^4+2}}\,dx$. **5.** $-\dfrac{2}{x^3}\,dx$.

7. $6x(2x+1)\,dx$. **9.** $\dfrac{2(3-x-x^2)}{(x^2+3)^2}\,dx$.

11. $-.14$. **13.** $\frac{2}{15}$. **15.** 10.05. **17.** $3\frac{47}{48}$.

19. $\frac{1}{2}$. **21.** $\dfrac{1}{6p(p^2+5)^2}$. **23.** $-p^2$.

25. $-\frac{4}{5}$. **27.** 44; 41.80. **29.** 2.04.

31. $(1.69 \times 10^{-11})\pi$. **33.** $35\frac{8}{9}$. **35.** .7.

Review Problems—Chapter 4

1. Int. $(-4, 0)$, $(6, 0)$, $(0, -24)$; inc. $(1, \infty)$; dec. $(-\infty, 1)$; rel. min. when $x = 1$; conc. up $(-\infty, \infty)$.

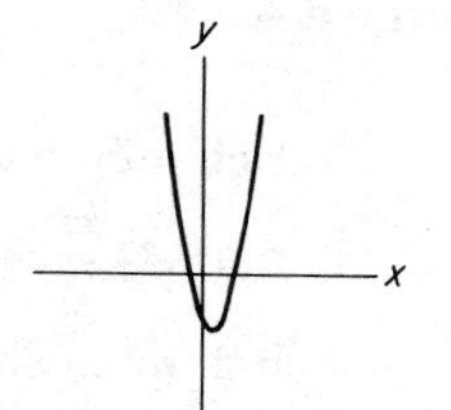

3. Int. $(0, 20)$; inc. $(-\infty, -2)$, $(2, \infty)$; dec. $(-2, 2)$; rel. max. when $x = -2$; rel. min. when $x = 2$; conc. up $(0, \infty)$; conc. down $(-\infty, 0)$; inf. pt. when $x = 0$.

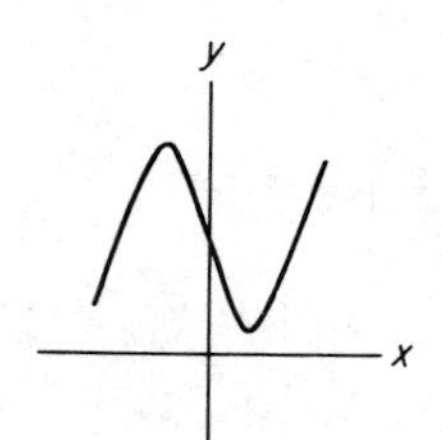

5. Int. $(0, 0)$; inc. $(-\infty, \infty)$; conc. down $(-\infty, 0)$; conc. up $(0, \infty)$; inf. pt. when $x = 0$; sym. to origin.

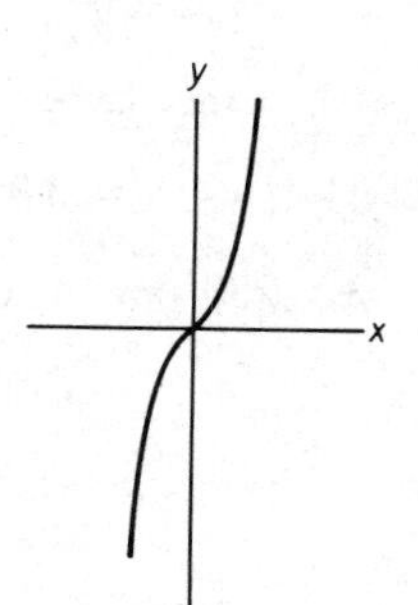

7. Int. $(-5, 0)$; inc. $(-10, 0)$; dec. $(-\infty, -10)$, $(0, \infty)$; rel. min. when $x = -10$; conc. up $(-15, 0)$, $(0, \infty)$; conc. down $(-\infty, -15)$; inf. pt. when $x = -15$; horiz. asym. $y = 0$; vert. asym. $x = 0$.

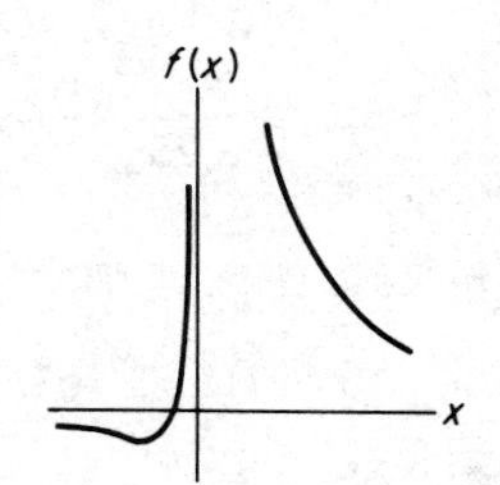

9. $(\frac{1}{3}, 1)$. **11.** 20. **13.** \$2800.

15. $\bar{c} = .01q + 5 + \dfrac{100}{q}$; $q = 100$. **17.** 20 in. by 25 in.

19. $\dfrac{x^2 - 14x - 5}{(x-7)^2}\,dx$. **21.** $11\frac{23}{24}$.

Exercise 5-1

1. $5x + C$. **3.** $\dfrac{x^9}{9} + C$. **5.** $\frac{2}{15}t^{15/2} + C$.

7. $-\dfrac{1}{6x^6} + C$. **9.** $-\dfrac{1}{9x^9} + C$. **11.** $-\dfrac{5}{6y^{6/5}} + C$.

13. $\dfrac{5\sqrt[5]{x^{11}}}{11} + C = \dfrac{5x^2\sqrt[5]{x}}{11} + C$. **15.** $8\sqrt[8]{x} + C$.

17. $11x + C$. **19.** $\dfrac{x^{\sqrt{2}+1}}{\sqrt{2}+1} + C$. **21.** $\dfrac{3x^8}{8} + C$.

23. $8u + \dfrac{u^2}{2} + C$. **25.** $\dfrac{y^6}{6} + \dfrac{5y^2}{2} + C$. **27.** $t^3 - 2t^2 - 5t + C$.

29. $\dfrac{x^2}{14} - \dfrac{3x^5}{20} + C$. **31.** $\dfrac{4x^{3/2}}{9} + C$ **33.** $-\dfrac{1}{2y^3} - \dfrac{3y^5}{10} + C$.

35. $\dfrac{p^{3.1}}{3.1} + \dfrac{1}{1.1p^{1.1}} + C$. **37.** $-\dfrac{1}{x} + \dfrac{5}{2x^2} - \dfrac{2}{3x^3} + C$.

39. $\dfrac{x^{9.3}}{9.3} - \dfrac{9x^7}{7} - \dfrac{1}{x^3} - \dfrac{1}{2x^2} + C$. **41.** $\dfrac{x^4}{12} + \dfrac{3}{2x^2} + C$.

43. $\dfrac{w^3}{2} + \dfrac{2}{3w} + C$. **45.** $\dfrac{3x^{4/3}}{4} - \dfrac{4x^{5/4}}{5} + \dfrac{5x^{6/5}}{6} + C$.

47. $\dfrac{4x^{3/2}}{3} - \dfrac{12x^{5/4}}{5} + C$. **49.** $-\dfrac{10}{x^{1/5}} + C$.

51. $-\frac{3x^{5/3}}{25} - 7x^{1/2} + 3x^2 + C.$ **53.** $y = \frac{3x^2}{2} - 4x + 1.$

55. $y = -\frac{x^4}{12} - \frac{x^3}{3} + \frac{4x}{3} + \frac{1}{12}.$ **57.** $c = 1.35q + 200.$

59. 7715. **61.** $p = .7$ **63.** $p = 275 - .5q - .1q^2.$

65. $y = -1.5x - .5x^2 + 59.3.$

Exercise 5-2

1. $\frac{x^4}{4} - x^3 + \frac{5x^2}{2} - 15x + C.$ **3.** $\frac{2x^{5/2}}{5} + 2x^{3/2} + C.$

5. $\frac{4u^3}{3} + 2u^2 + u + C.$

7. $\frac{2v^3}{3} + 3v + \frac{1}{2v^4} + C.$ **9.** $\frac{3x^2}{2} + x + \frac{1}{x} + C.$

11. $\frac{(x+4)^9}{9} + C.$ **13.** $\frac{(x^2+16)^4}{4} + C.$

15. $\frac{3}{5}(y^3 + 3y^2 + 1)^{5/3} + C.$ **17.** $-\frac{(3x-1)^{-2}}{2} + C.$

19. $\frac{2(x+10)^{3/2}}{3} + C.$ **21.** $\frac{(7x-6)^5}{35} + C.$

23. $\frac{(x^2+3)^{13}}{26} + C.$ **25.** $\frac{3}{20}(27 + x^5)^{4/3} + C.$

27. $-\frac{3}{4}(z^2 - 6)^{-4} + C.$ **29.** $\frac{2}{15}(5x)^{3/2} + C = \frac{2}{3}x\sqrt{5x} + C.$

31. $\sqrt{x^2 - 4} + C.$ **33.** $-\frac{1}{24}(3 - 3x^2 - 6x)^4 + C.$

35. $3(y + 1)^{4/3} + C.$ **37.** $-\frac{1}{25}(7 - 5x^2)^{5/2} + C.$

39. $\sqrt{2x} + C.$ **41.** $\frac{x^5}{5} + \frac{2x^3}{3} + x + C.$

43. $\frac{r^7}{7} + \frac{5r^4}{2} + 25r + C.$ **45.** $\frac{8x}{3} + \frac{2}{x} - \frac{x^2}{6} + C.$

47. $\frac{2}{9}(\sqrt{x} + 2)^3 + C.$ **49.** $-\frac{10}{21}(5 - t\sqrt{t}\,)^{1.4} + C.$

51. $y = -\frac{1}{6}(3 - 2x)^3 + \frac{11}{2}.$ **53.** $y = \frac{1}{2x} + \frac{3x}{2} - 2.$

55. $p = \dfrac{100}{q+2}$. **57.** $C = 2(\sqrt{I} + 1)$.

59. $C = \frac{3}{4}I - \frac{1}{3}\sqrt{I} + \frac{71}{12}$.

Exercise 5-3

1. 35. **3.** 0. **5.** 25. **7.** $-\frac{3}{16}$. **9.** $-\frac{7}{6}$.

11. $\sum_{k=1}^{15} k$. **13.** $\sum_{k=1}^{4} (2k-1)$. **15.** $\sum_{k=1}^{12} k^2$.

17. 101,475. **19.** 84. **21.** 273. **23.** 8, \$850.

Exercise 5-4

1. $\frac{2}{3}$ sq unit. **3.** $\frac{14}{27}$ sq unit. **5.** $\frac{1}{2}$ sq unit.

7. $\frac{1}{3}$ sq unit. **9.** $\frac{16}{3}$ sq unit. **11.** 36.

13. -18. **15.** $\frac{5}{6}$.

Exercise 5-5

1. 12. **3.** $\frac{9}{2}$. **5.** $\frac{100}{3}$. **7.** -24. **9.** $\frac{14}{3}$. **11.** $\frac{7}{3}$.

13. $\frac{15}{2}$. **15.** $-\frac{7}{6}$. **17.** 0. **19.** $\frac{5}{3}$. **21.** $\frac{32}{3}$. **23.** $-\frac{1}{6}$.

25. $\frac{65}{8}$. **27.** $\frac{3}{4}$. **29.** $\frac{38}{9}$. **31.** $\frac{15}{28}$. **33.** $\frac{13}{8}$.

35. 1. **37.** .02. **39.** $\alpha^{5/2}T$. **41.** 1,973,333.

43. \$160. **45.** \$2000. **47.** 696; 492.

Exercise 5-6

In Problems **1–29**, *answers are assumed to be expressed in square units.*

1. 8. **3.** $\frac{19}{2}$. **5.** 8. **7.** $\frac{19}{3}$. **9.** 9.

11. $\frac{50}{3}$. **13.** 36. **15.** 8. **17.** $\frac{32}{3}$.

19. 18. **21.** $\frac{26}{3}$. **23.** $\frac{3}{2}\sqrt[3]{2}$. **25.** 36.

27. 68. **29.** 2. **31.** $\dfrac{t-a}{b-a}$ sq units.

33. (a) $\frac{1}{8}$. (b) $\frac{3}{4}$. (c) $F(t) = \dfrac{t^2}{16}$.

Exercise 5-7

In Problems **1–21**, *the answers are assumed to be expressed in square units.*

1. $\frac{4}{3}$. **3.** $\frac{16}{3}$. **5.** $8\sqrt{6}$.

7. 40. **9.** $\frac{125}{6}$. **11.** $\frac{32}{81}$.

13. $\frac{125}{12}$. **15.** $\frac{9}{2}$. **17.** $\frac{44}{3}$.

19. $\frac{4}{3}(5\sqrt{5} - 2\sqrt{2})$. **21.** $\frac{1}{2}$. **23.** $\frac{20}{63}$.

Review Problems—Chapter 5

1. $\dfrac{x^4}{4} + x^2 - 7x + C$. **3.** $\frac{117}{2}$.

5. $-(x+5)^{-2} + C$. **7.** $-\dfrac{7(3x^2-8)^{-2}}{12} + C$.

9. $\dfrac{11\sqrt[3]{11}}{4} - 4$. **11.** $\frac{2}{27}(3x^3+2)^{3/2} + C$.

13. $\dfrac{y^4}{4} + \dfrac{2y^3}{3} + \dfrac{y^2}{2} + C$. **15.** $-\dfrac{1}{2x^2} - \dfrac{2}{x} + C$.

17. 11.1. **19.** $\frac{7}{3}$. **21.** $4 - 3\sqrt[3]{2}$.

23. $\dfrac{3}{t} - \dfrac{2}{\sqrt{t}} + C$. **25.** $(.5x - .1)^5 + C$.

In Problems **27–39**, *answers are assumed to be expressed in square units.*

27. $\frac{4}{3}$. **29.** $\frac{16}{3}$. **31.** $\frac{125}{6}$. **33.** $\frac{20}{3}$. **35.** $\frac{2}{3}$.

37. 36. **39.** $\frac{125}{3}$. **41.** $p = 100 - \sqrt{2q}$.

43. (a) $\frac{11}{60}$. (b) $\frac{1}{3}$. (c) $F(t) = \dfrac{11t}{60} + \dfrac{t^2}{20}$.

45. \$5000.

Exercise 6-1

1.

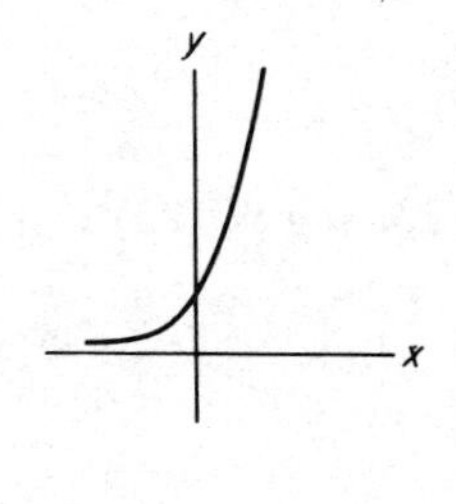

3.

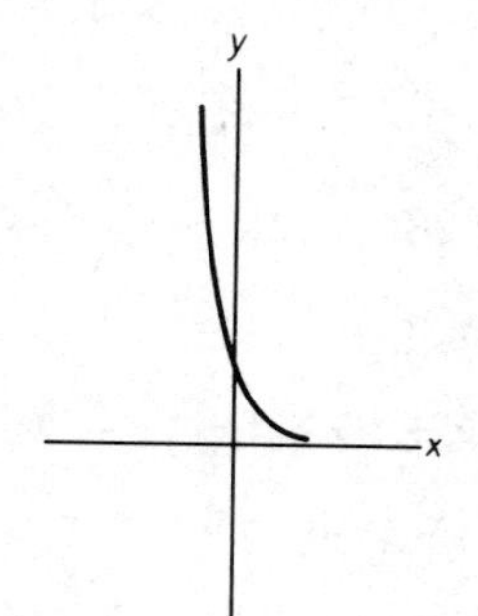

5.

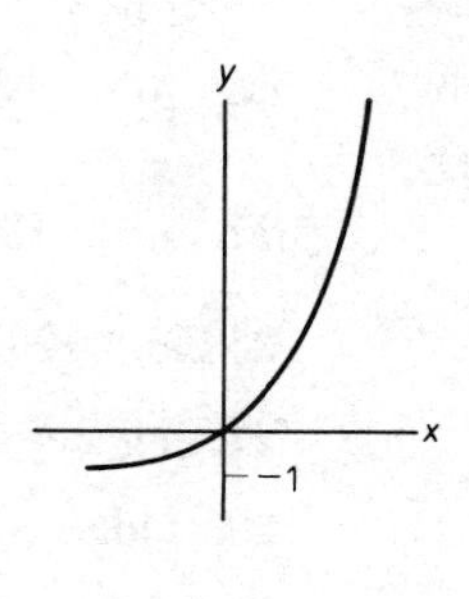

7. ∞. **9.** 4. **11.** 0. **13.** $\frac{1}{4}$. **15.** 4.4817. **17.** .67032.

19. 140,000. **21.** .2241. **23.** 50. **25.** .1466. **27.** 1.00.

29. \$1616.10. **31.** \$740,820. **33.** \$26,960. **35.** 6065.

Exercise 6-2

1. $\log_{16} 4 = \frac{1}{2}$. **3.** $\log 10{,}000 = 4$. **5.** $2^6 = 64$.

7. $2^{14} = x$. **9.** $\ln 7.3891 = 2$. **11.** $e^{1.09861} = 3$.

13.

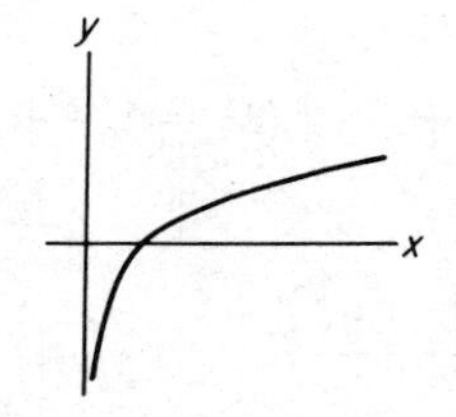

15. 16. **17.** 125. **19.** $\frac{1}{10}$. **21.** e^2.

23. 2. **25.** 6. **27.** 2. **29.** 4.

31. $\frac{1}{2}$. **33.** $\frac{1}{81}$. **35.** 2. **37.** $\log_2 5$.

39. $\dfrac{\ln 2}{3}$. **41.** $\log_3 8$. **43.** $\dfrac{5 + \ln 3}{2}$. **45.** 1.60944.

47. 2.00013. **49.** 41.50. **51.** 22 years. **53.** $E = 2.5 \times 10^{11+1.5M}$.

55. (a) 91. (b) 432. (c) 8.

Exercise 6-3

1. 1.5441. **3.** .3521. **5.** 1.3980. **7.** 3.3010.

9. 48. **11.** -4. **13.** 1.5850. **15.** $\frac{1}{3}$.

17. 4. **19.** log 28. **21.** $\log_2 \frac{2x}{x+1}$. **23.** $\log\,[7^9(23)^5]$.

25. $\log\,[100(1.05)^{10}]$.

27. $\log x + \log\,(x+2) + \log\,(x-3)$.

29. $\frac{1}{2}\log x - \log\,(x+2) - 2\log\,(x-3)$.

31. $\frac{1}{2}[2\log x + 3\log\,(x-3) - \log\,(x+2)]$.

33. 2. **35.** $\frac{1}{12}$. **37.** $\frac{5}{2}$. **39.** ± 2. **41.** 5.

43. $\frac{4}{3}$. **45.** 8. **47.** (a) 3. (b) 3.78.

49. $225 \ln \frac{9}{4}. \approx 182$. **53.** (a) 3. (b) $2 + M_1$.

55. $S = 12.4A^{.26}$. **57.** $p = \frac{\log\,(80-q)}{\log 2}$; 4.32.

Exercise 6-4

1. $\frac{3}{3x-4}$. **3.** $\frac{2}{x}$. **5.** $-\frac{2x}{1-x^2}$.

7. $\frac{6p^2+3}{2p^3+3p} = \frac{3(2p^2+1)}{p(2p^2+3)}$. **9.** $t\left(\frac{1}{t}\right) + (\ln t)(1) = 1 + \ln t$.

11. $\frac{2\log_3 e}{2x-1}$. **13.** $\frac{z\left(\frac{1}{z}\right) - (\ln z)(1)}{z^2} = \frac{1-\ln z}{z^2}$.

15. $\frac{3(2x+4)}{x^2+4x+5} = \frac{6(x+2)}{x^2+4x+5}$.

17. $\frac{x}{1+x^2}$. **19.** $\frac{2}{1-l^2}$. **21.** $\frac{x}{1-x^4}$. **23.** $\frac{4x}{x^2+2} + \frac{3x^2+1}{x^3+x-1}$.

25. $\dfrac{2(x^2+1)}{2x+1} + 2x \ln(2x+1)$. **27.** $\dfrac{3(1+\ln^2 x)}{x}$. **29.** $\dfrac{4 \ln^3(ax)}{x}$.

31. $\dfrac{x}{2(x-1)} + \ln\sqrt{x-1}$. **33.** $\dfrac{1}{2x\sqrt{4+\ln x}}$.

35. $3 + 2\ln x$. **37.** $16(2x-1)^{-3}$. **39.** $y = 4x - 12$.

41. $2(1+\ln x)\, dx$.

43. Dec. on $(0, e^{-1})$; inc. on (e^{-1}, ∞); rel. min. when $x = e^{-1}$.

45. $-.03$. **47.** $\dfrac{1}{x(2y+1)}$. **49.** $25\dfrac{(q+2)\ln(q+2) - q}{(q+2)\ln^2(q+2)}$.

51. $\dfrac{1.5E}{\log e}$. **53.** (b)

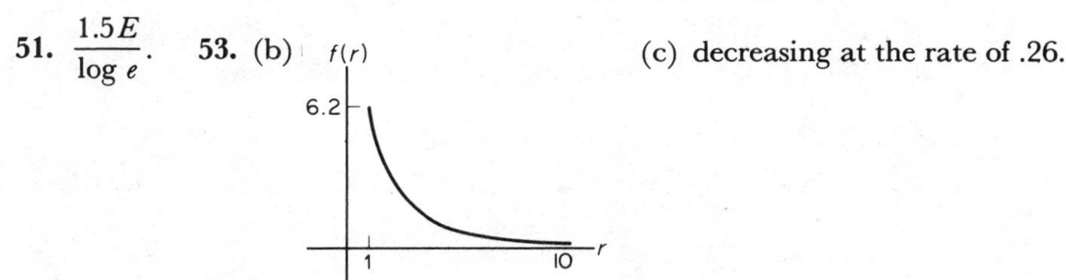

(c) decreasing at the rate of .26.

Exercise 6-5

1. $2xe^{x^2+1}$. **3.** $-5e^{3-5x}$.

5. $(6r+4)e^{3r^2+4r+4} = 2(3r+2)e^{3r^2+4r+4}$. **7.** $x(e^x) + e^x(1) = e^x(x+1)$.

9. $2xe^{-x^2}(1-x^2)$. **11.** $\dfrac{e^x - e^{-x}}{2}$. **13.** $(6x)4^{3x^2} \ln 4$.

15. $\dfrac{2e^{2w}(w-1)}{w^3}$. **17.** $\dfrac{e^{1+\sqrt{x}}}{2\sqrt{x}}$. **19.** $\dfrac{2e^x}{(e^x+1)^2}$.

21. $12e^{3x}(e^{3x}+1)^3$. **23.** $e^{e^x}e^x = e^{e^x+x}$. **25.** 1.

27. $\dfrac{e^x(x\ln x - 1)}{x\ln^2 x}$. **29.** $(\log 2)^x \ln(\log 2)$. **31.** $3x^2 - 3^x \ln 3$.

33. $y - e^2 = e^2(x-2)$ or $y = e^2x - e^2$. **35.** e^x. **37.** $e^z(z^2+4z+2)$.

39. $3x^2e^{x^3+5}\, dx$. **41.** 1.01.

43. Dec. on $(-\infty, -1)$; inc. on $(-1, \infty)$; rel. min. when $x = -1$.

45. $-\dfrac{e^y}{xe^y+1}$. **47.** $\dfrac{dp}{dq} = -.015e^{-.001q}$, $-.015e^{-.5}$.

49. 50. **51.** $\dfrac{dc}{dq} = 10e^{q/700}$; $10e^{.5}$; $10e$. **53.** $100e^{-2}$.

55. Rel. max when $x = 0$; inf. pt. when $x = \pm 1$.

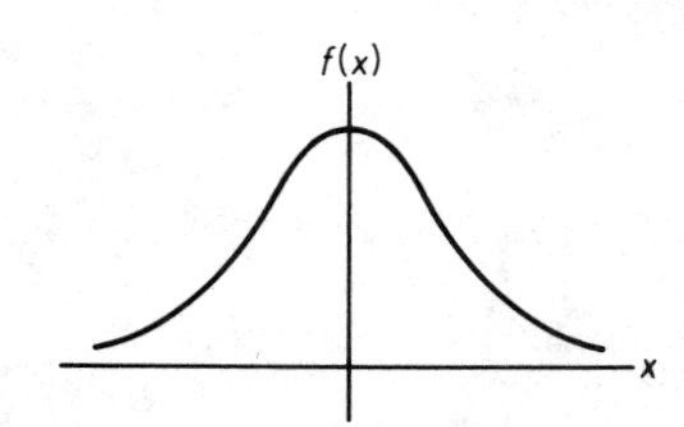

57. $-b(10^{A-bM})\ln 10$.

59. (a) $\dfrac{dp}{dt} = .8811(.85)^t \ln(.85)$; this represents the rate of change of proportion of recalls with respect to length of recall interval. (b) $-.10$.

63. (a) 0.

65. (a) 1; as time goes on, practically the entire group will be discharged. (b) .0036.

Exercise 6-6

1. $e^{3x} + C$.

3. $e^{t^2+t} + C$.

5. $\ln|x + 5| + C$.

7. $\ln|x^3 + x^4| + C$.

9. $\frac{1}{10}e^{5x^2} + C$.

11. $-3e^{-2x} + C$.

13. $-\frac{3}{4}(z^2 - 6)^{-4} + C$.

15. $4\ln|x| + C$.

17. $\frac{1}{3}\ln|s^3 + 5| + C$.

19. $-\frac{7}{3}\ln|5 - 3x| + C$.

21. 1.

23. $\frac{1}{2}e^{y^4+1} + C$.

25. $-\frac{1}{6}e^{-2v^3+1} + C$.

27. $-\frac{1}{5}e^{-5x} + 2e^x + C$.

29. $\frac{1}{3}(e^8 - 1)$.

31. $\frac{1}{3}\ln|x^3 + 6x| + C$.

33. $2\ln|3 - 2s + 4s^2| + C$.

35. $\dfrac{3}{2} - \dfrac{1}{e} + \dfrac{1}{2e^2}$.

37. $\frac{1}{2}\ln 3$.

39. $\frac{1}{2}(e^{2x} - e^{-2x}) - 2x + C$.

41. $\frac{1}{2}\ln(x^2+1) - \dfrac{1}{6(x^6+1)} + C.$

43. $\frac{1}{3}(2x+3)^{3/2} - \ln\sqrt{x^2+3} + C.$

45. $2e^{\sqrt{x}} + C.$

47. $\frac{3}{2}x^2 + x - \ln|x| + C.$

49. $\dfrac{4^{7x}}{7\ln 4} + C.$

51. $7x^2 - 4e^{x^2/4} + C.$

53. $\frac{3}{2}\ln(e^{2x}+1) + C.$

55. $\frac{1}{2}(\ln^2 x) + C.$

57. $\frac{1}{3}\ln^3(r+1) + C.$

59. $e^{(x^2+3)/2} + C.$

61. $\ln|\ln(x+3)| + C.$

63. $-\frac{1}{7}e^{7/x} + C.$

65. $x^2 - 3x + \frac{2}{3}\ln|3x-1| + C.$

67. $x + \ln|x-1| + C.$

69. $c = 20\ln|(q+5)/5| + 2000.$

71. $e^2 - 1$ sq. units.

73. $\frac{3}{2} + 2\ln 2 = \frac{3}{2} + \ln 4$ sq. units.

75. $y = \frac{1}{2}\ln\dfrac{x^2+4}{5} = \ln\sqrt{\dfrac{x^2+4}{5}}.$

77. \$8639.

79. $\dfrac{1-e^{-at}}{a} - \dfrac{1-e^{-bt}}{b}.$

81. $\dfrac{Rr^2}{4K} + B_1\ln r + B_2.$

83. (a) 1. (b) $\ln\frac{5}{3}$. (c) $-1 + \ln t$.

85. $\dfrac{i}{k}(1 - e^{-2kR}).$

Exercise 6-7

1. $y\left[\dfrac{2}{x+1} + \dfrac{1}{x-1} + \dfrac{2x}{x^2+3}\right].$

3. $y\left[\dfrac{18x^2}{3x^3-1} + \dfrac{6}{2x+5}\right].$

5. $\dfrac{y}{2}\left[\dfrac{1}{x+1} + \dfrac{2x}{x^2-2} + \dfrac{1}{x+4}\right].$

7. $y\left[\dfrac{4x}{x^2+1} - \dfrac{2}{x+1} - \dfrac{3}{3x+2}\right].$

9. $y\left[\dfrac{4}{8x+3} + \dfrac{2x}{3(x^2+2)} - \dfrac{1}{2(1+2x)}\right].$

11. $y\left[\dfrac{x}{x^2-1} + \dfrac{2}{1-2x}\right].$

13. $x^{2x+1}\left(\frac{2x+1}{x} + 2 \ln x\right)$. **15.** $\frac{x^{1/x}(1 - \ln x)}{x^2}$.

17. $xy(1 + 2 \ln x) = x^{x^2+1}(1 + 2 \ln x)$. **19.** $e^x x^{3x}(4 + 3 \ln x)$.

21. $x^x(1 + \ln x)$.

Review Problems—Chapter 6

1. $\log_3 81 = 4$. **3.** $\frac{7}{5}$. **5.** 3. **7.** $\frac{1}{100}$.

9. 3. **11.** 3. **13.** 3.3980.

15. $2y + \frac{1}{2}x$. **17.** $2x$. **19.** $y = e^{x^2+2}$.

21. $2e^x + e^{x^2}(2x) = 2(e^x + xe^{x^2})$. **23.** $\frac{1}{r^2 + 5r}(2r + 5) = \frac{2r+5}{r(r+5)}$.

25. $e^{x^2+4x+5}(2x + 4) = 2(x + 2)e^{x^2+4x+5}$.

27. $e^x(2x) + (x^2 + 2)e^x = e^x(x^2 + 2x + 2)$.

29. $\frac{y}{2}\left[\frac{1}{x-6} + \frac{1}{x+5} - \frac{1}{9-x}\right] = \frac{y}{2}\left[\frac{1}{x-6} + \frac{1}{x+5} + \frac{1}{x-9}\right]$.

31. $\frac{e^x\left(\frac{1}{x}\right) - (\ln x)(e^x)}{e^{2x}} = \frac{1 - x \ln x}{xe^x}$. **33.** $\frac{2}{q+1} + \frac{3}{q+2}$.

35. $-7(\ln 10)10^{2-7x}$ **37.** $\frac{4e^{2x+1}(2x - 1)}{x^2}$.

39. $\frac{16 \log_2 e}{8x + 5}$. **41.** $\frac{1 + 2l + 3l^2}{1 + l + l^2 + l^3}$.

43. $y[1 + \ln(x + 1)] = (x + 1)^{x+1}[1 + \ln(x + 1)]$.

45. $2\left(\frac{1}{t}\right) + \frac{1}{2}\left(\frac{1}{1-t}\right)(-1) = \frac{5t - 4}{2t(t - 1)}$.

47. $$y\left[\frac{3}{2}\left(\frac{1}{x^2+2}\right)(2x) + \frac{4}{9}\left(\frac{1}{x^2+9}\right)(2x) - \frac{4}{11}\left(\frac{1}{x^3+6x}\right)(3x^2 + 6)\right]$$
$$= y\left[\frac{3x}{x^2+2} + \frac{8x}{9(x^2+9)} - \frac{12(x^2+2)}{11(x^3+6x)}\right].$$

49. 18. **51.** 2. **53.** $y = 2x + 2(1 - \ln 2)$ or $y = 2x + 2 - \ln 4$.

55. $\frac{1}{3}$. **57.** .8.

59. Inc. $(0, \infty)$; dec. $(-\infty, 0)$; rel. min. when $x = 0$; conc. up $(-\infty, \infty)$.

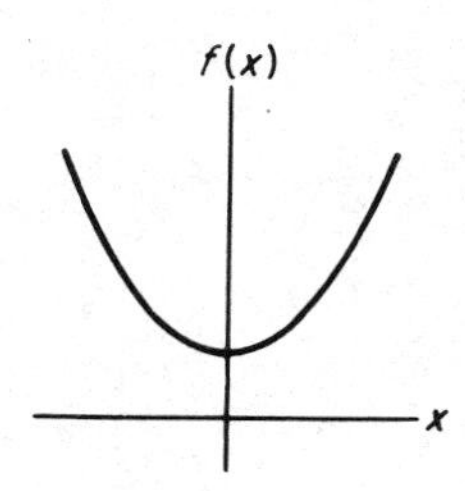

61. $\frac{4}{3}e^{3x+5} + C$. **63.** $\frac{1}{2}\ln|x + 1| + C$. **65.** $2 \ln|x^3 - 6x + 1| + C$.

67. $\frac{1}{3}\ln\frac{10}{3}$. **69.** $7 + e$. **71.** $\frac{1}{2}(e^{2y} + e^{-2y}) + C$.

73. $\ln|x| - \dfrac{2}{x} + C$. **75.** $\dfrac{5^{2x}}{2 \ln 5} + C$. **77.** $\frac{3}{2} - 5 \ln 2$.

79. $6 + \ln 3$. **83.** 134,064; 109,762. **87.** 10.

89. $.008e^{-.01t} + .00004e^{-.0002t}$.

Exercise 7-1

1. $-e^{-x}(x + 1) + C$. **3.** $\dfrac{y^4}{4}[\ln(y) - \frac{1}{4}] + C$.

5. $x[\ln(4x) - 1] + C$.

7. $\dfrac{2x}{3}(x + 1)^{3/2} - \frac{4}{15}(x + 1)^{5/2} + C = \frac{2}{15}(x + 1)^{3/2}(3x - 2) + C$.

9. $\frac{2}{3}x^{3/2} \ln(x) - \frac{4}{9}x^{3/2} + C$. **11.** $\frac{1}{4}e^2(3e^2 - 1)$.

13. $\frac{1}{2}(1 - e^{-1})$, parts not needed. **15.** $\frac{2}{3}(9\sqrt{3} - 10\sqrt{2})$.

17. $e^x(x^2 - 2x + 2) + C$.

19. $\dfrac{x^3}{3} + 2e^{-x}(x + 1) - \dfrac{e^{-2x}}{2} + C$.

21. $2e^3 + 1$ sq units.

Exercise 7-2

1. $\frac{1}{6}\ln\left|\dfrac{x}{6 + 7x}\right| + C$. **3.** $\frac{1}{3}\ln\left|\dfrac{\sqrt{x^2 + 9} - 3}{x}\right| + C$.

5. $\frac{1}{2}\left[\frac{4}{5}\ln|4 + 5x| - \frac{2}{3}\ln|2 + 3x|\right] + C.$

7. $\frac{1}{8}(2x - \ln[4 + 3e^{2x}]) + C.$

9. $2\left[\dfrac{1}{1 + x} + \ln\left|\dfrac{x}{1 + x}\right|\right] + C.$

11. $1 + \ln \frac{4}{9}.$

13. $\frac{1}{2}(x\sqrt{x^2 - 3} - 3 \ln|x + \sqrt{x^2 - 3}|) + C.$

15. $\frac{1}{144}.$

17. $e^x(x^2 - 2x + 2) + C.$

19. $2\left(-\dfrac{\sqrt{4x^2 + 1}}{2x} + \ln|2x + \sqrt{4x^2 + 1}|\right) + C.$

21. $\frac{1}{9}\left(\ln|1 + 3x| + \dfrac{1}{1 + 3x}\right) + C.$

23. $\dfrac{1}{\sqrt{5}}\left(\dfrac{1}{2\sqrt{7}}\ln\left|\dfrac{\sqrt{7} + \sqrt{5}\,x}{\sqrt{7} - \sqrt{5}\,x}\right|\right) + C.$

25. $\dfrac{1}{3^6}\left[\dfrac{(3x)^6 \ln(3x)}{6} - \dfrac{(3x)^6}{36}\right] + C = \dfrac{x^6}{36}[6 \ln(3x) - 1] + C.$

27. $\dfrac{4(9x - 2)(1 + 3x)^{3/2}}{135} + C.$

29. $\frac{1}{2}\ln|2x + \sqrt{4x^2 - 13}| + C.$

31. $\frac{1}{4}[2x^2 \ln(2x) - x^2] + C.$

33. $-\dfrac{\sqrt{9 - 4x^2}}{9x} + C.$

35. $\frac{1}{2}\ln(x^2 + 1) + C.$

37. $\frac{1}{6}(2x^2 + 1)^{3/2} + C.$

39. $\ln\left|\dfrac{x - 3}{x - 2}\right| + C.$

41. $\dfrac{x^4}{4}\left[\ln(x) - \frac{1}{4}\right] + C.$

43. $\dfrac{e^{2x}}{4}(2x - 1) + C.$

45. $x(\ln x)^2 - 2x \ln(x) + 2x + C.$

47. $\frac{2}{3}(9\sqrt{3} - 10\sqrt{2}).$

49. $2(2\sqrt{2} - \sqrt{7}).$

51. $\frac{7}{2}\ln(2) - \frac{3}{4}.$

53. $\ln\left|\dfrac{q_n(1 - q_0)}{q_0(1 - q_n)}\right|.$

55. (a) \$37,599. (b) \$4924.

57. (a) \$5481. (b) \$535.

Exercise 7-3

1. $\frac{16}{3}$. **3.** -1. **5.** 0. **7.** $\frac{13}{6}$. **9.** 12,400.

11. \$3156.

Exercise 7-4

1. .340; .333. **3.** 1.388; 1.386. **5.** .883.

7. 2,361,375. **9.** 2.967. **11.** .771.

Exercise 7-5

1. $\frac{1}{3}$. **3.** div. **5.** $\dfrac{1}{e}$. **7.** div.

9. $-\frac{1}{2}$. **11.** 0. **13.** div. **15.** (a) 800. (b) $\frac{2}{3}$.

17. $\alpha = e^{-x_c}$, $\beta = e^{-(1/8)x_c}$. **21.** 5000 increase.

Exercise 7-6

1. $y = -\dfrac{1}{x^2 + C}$. **3.** $y = Ce^x$, $C > 0$.

5. $y = Cx$, $C > 0$. **7.** $y = \frac{1}{3}(x^2 + 1)^{3/2} + C$.

9. $y = \sqrt{2x}$. **11.** $y = \ln\dfrac{x^3 + 3}{3}$.

13. $y = \dfrac{4x^2 + 3}{2(x^2 + 1)}$.

15. $N = 20{,}000e^{.018t}$; $N = 20{,}000(1.2)^{t/10}$; 28,800; **17.** $2e^{.945}$ billion.

19. .01204; 57.57 sec. **21.** 2900 years.

23. $N = N_0e^{k(t-t_0)}$, $t \geqslant t_0$. **25.** 12.6 units.

27. $q = q_0e^{-kt}$; 0; 13.5%.

29. (a) $A = 400(1 - e^{-t/2})$. (b) 157 grams. (c) 400.

Exercise 7-7

1. 430,000. **3.** 1990.

5. (a) $N = \dfrac{49.25}{1 + 195.1e^{-2.128x}}$.

(b) .2511 cm^2.

7. 1:06 A.M. **9.** \$62,500. **11.** $N = M - (M - N_0)e^{-kt}$.

Review Problems—Chapter 7

1. $\dfrac{x^2}{4}[2\ \ln(x) - 1] + C$. **3.** $5 + \frac{9}{4}\ln 3$.

5. $\frac{1}{21}(9\ \ln|3 + x| - 2\ \ln|2 + 3x|) + C$.

7. $\dfrac{1}{2(x+2)} + \frac{1}{4}\ln\left|\dfrac{x}{x+2}\right| + C$.

9. $-\dfrac{\sqrt{9 - 16x^2}}{9x} + C$. **11.** $\frac{3}{2}\ln\left|\dfrac{x-3}{x+3}\right| + C$.

13. $\dfrac{e^{7x}}{49}(7x - 1) + C$. **15.** $\frac{1}{2}\ln|\ln 2x| + C$.

17. $x - \frac{3}{2}\ln|3 + 2x| + C$.

19. 34. **21.** (a) 1.405. (b) 1.388. **23.** $\frac{1}{18}$.

25. div. **27.** $y = Ce^{x^3 + x^2}$, $C > 0$.

29. 144,000. **31.** .0005; 90%. **33.** $N = \dfrac{450}{1 + 224e^{-1.02t}}$.

35. \$1443.

Exercise 8-1

1. 3. **3.** 6. **5.** 6. **7.** 88. **9.** 3.

11. $2x_0 + 2h - 5y_0 + 4$. **13.** $y = -4$. **15.** $z = 6$.

BOOKMARK

UNIVERSITY OF PITTSBURGH AT GREENSBURG
THE BOOK CENTER

—IMPORTANT—

RETURN POLICIES
READ CAREFULLY

Refunds on textbooks will be authorized on or prior to the 17th calendar day (first day of classes and the following 16 calendar days) of each new term and only on presentation of a **"CHANGE IN COURSE FORM"** or a **"WITHDRAWAL FORM"** and must be accompanied by a **CASH REGISTER RECEIPT.**

CHECK YOUR RECEIPT — KEEP IT!

DO NOT WRITE IN ANY BOOK unt'' certain there is no necessity for returning it.

THANK YOU.

17.

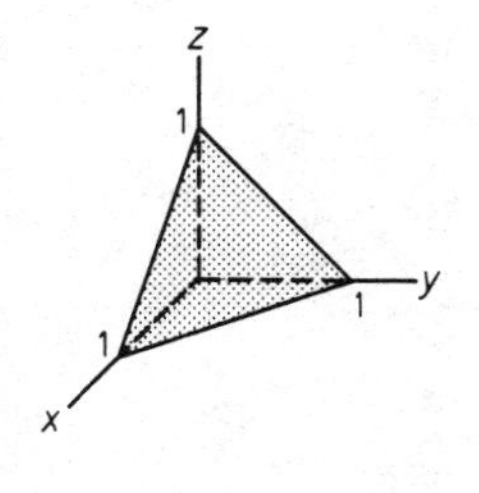

19.

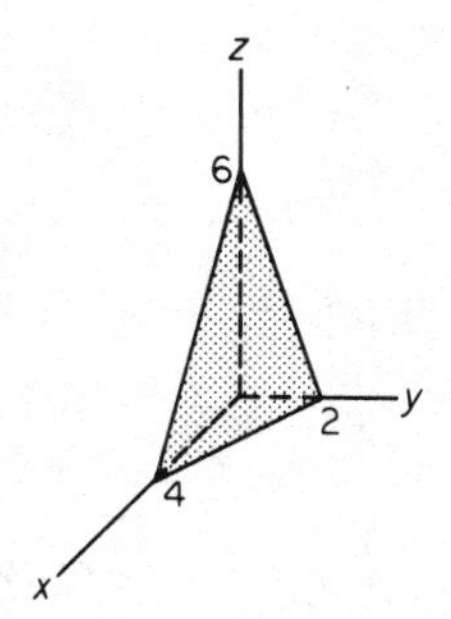

21.

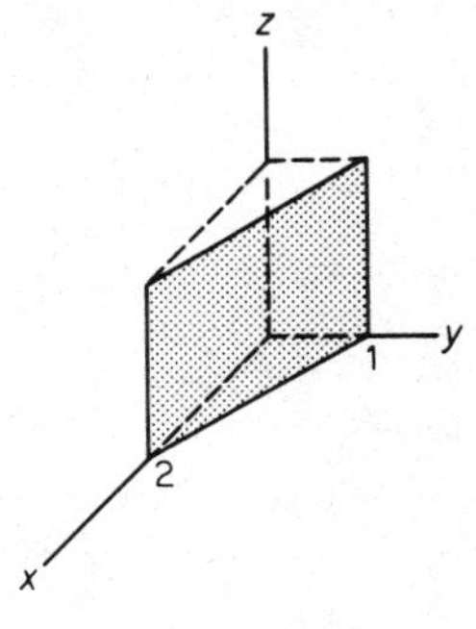

23.

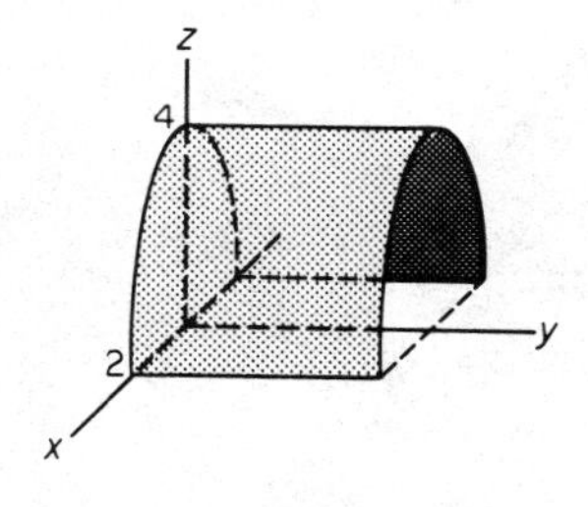

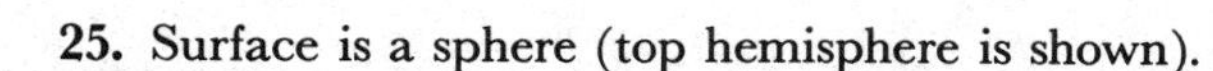

25. Surface is a sphere (top hemisphere is shown).

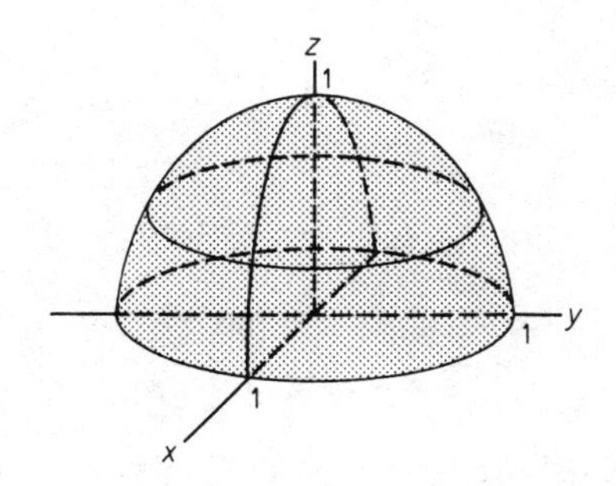

27. 2000. **29.** $P(r, 5) = \dfrac{5!\left(\frac{1}{4}\right)^r\left(\frac{3}{4}\right)^{5-r}}{r!\,(5-r)!}$.

Exercise 8-2

1. $f_x(x, y) = 1; f_y(x, y) = -5$.

3. $f_x(x, y) = 3$; $f_y(x, y) = 0$.

5. $g_x(x, y) = 5x^4y^4 - 12x^3y^3 + 21x^2 - 3y$;
$g_y(x, y) = 4x^5y^3 - 9x^4y^2 + 4y - 3x$.

7. $g_p(p, q) = \dfrac{q}{2\sqrt{pq}}$; $g_q(p, q) = \dfrac{p}{2\sqrt{pq}}$.

9. $h_s(s, t) = \dfrac{2s}{t-3}$; $h_t(s, t) = -\dfrac{s^2+4}{(t-3)^2}$.

11. $u_{q_1}(q_1, q_2) = \dfrac{3}{4q_1}; u_{q_2}(q_1, q_2) = \dfrac{1}{4q_2}$.

13. $h_x(x, y) = (x^3 + xy^2 + 3y^3)(x^2 + y^2)^{-3/2}$;

$h_y(x, y) = (3x^3 + x^2y + y^3)(x^2 + y^2)^{-3/2}$.

15. $\dfrac{\partial z}{\partial x} = 5ye^{5xy}$; $\dfrac{\partial z}{\partial y} = 5xe^{5xy}$.

17. $\dfrac{\partial z}{\partial x} = 5\left[\dfrac{2x^2}{x^2 + y} + \ln(x^2 + y)\right]$; $\dfrac{\partial z}{\partial y} = \dfrac{5x}{x^2 + y}$.

19. $f_r(r, s) = \sqrt{r + 2s}\,(3r^2 - 2s) + \dfrac{r^3 - 2rs + s^2}{2\sqrt{r + 2s}}$;

$f_s(r, s) = 2(s - r)\sqrt{r + 2s} + \dfrac{r^3 - 2rs + s^2}{\sqrt{r + 2s}}$.

21. $f_r(r, s) = -e^{3-r}\ln(7 - s)$; $f_s(r, s) = \dfrac{-e^{3-r}}{7 - s}$.

23. $g_x(x, y, z) = 6xy + 2y^2z$; $g_y(x, y, z) = 3x^2 + 4xyz$;

$g_z(x, y, z) = 2xy^2 + 9z^2$.

25. $g_r(r, s, t) = 2re^{s+t}$; $g_s(r, s, t) = (7s^3 + 21s^2 + r^2)e^{s+t}$;

$g_t(r, s, t) = e^{s+t}(r^2 + 7s^3)$.

27. 50. **29.** $\frac{1}{3}$. **31.** 0.

Exercise 8-3

1. 20. **3.** 784.5.

5. $\dfrac{\partial P}{\partial k} = 1.208648l^{.192}k^{-.236}$; $\dfrac{\partial P}{\partial l} = .303744l^{-.808}k^{.764}$.

7. $\dfrac{\partial q_A}{\partial p_A} = -50$; $\dfrac{\partial q_A}{\partial p_B} = 2$; $\dfrac{\partial q_B}{\partial p_A} = 4$; $\dfrac{\partial q_B}{\partial p_B} = -20$; competitive.

9. $\dfrac{\partial q_A}{\partial p_A} = -\dfrac{100}{p_A^2 p_B^{1/2}}$; $\dfrac{\partial q_A}{\partial p_B} = -\dfrac{50}{p_A p_B^{3/2}}$; $\dfrac{\partial q_B}{\partial p_A} = -\dfrac{500}{3p_B p_A^{4/3}}$;

$\dfrac{\partial q_B}{\partial p_B} = -\dfrac{500}{p_B^2 p_A^{1/3}}$; complementary.

11. $\frac{\partial p}{\partial B} = .01A^{.27}B^{-.99}C^{.01}D^{.23}E^{.09}F^{.27}$;

$\frac{\partial p}{\partial C} = .01A^{.27}B^{.01}C^{-.99}D^{.23}E^{.09}F^{.27}$.

13. 1120; if a staff manager with an MBA degree had an extra year of work experience before the degree, the manager would receive $1120 per year in extra compensation.

15. (a) -1.015; $-.846$. (b) one for which $w = w_0$ and $s = s_0$.

17. (a) No. (b) 70%.

Exercise 8-4

1. $-\frac{x}{z}$. **3.** $\frac{4y}{3z^2}$. **5.** $\frac{x(yz^2+1)}{z(1-x^2y)}$.

7. $-e^{y-z}$. **9.** $\frac{yz}{1+z}$. **11.** $-\frac{3x}{z}$.

13. $-\frac{9}{10}$. **15.** 1. **17.** $\frac{5}{2}$.

Exercise 8-5

1. $6xy^2$; $12xy$. **3.** $3xe^{3xy} + 4x^2$; $9xye^{3xy} + 3e^{3xy} + 8x$; $9x(3xy + 2)e^{3xy}$.

5. $(2x + y)(2x^2 + 2xy + y^2 + 1)$; $6x^2 + 8xy + 3y^2 + 1$.

7. $3x^2y + 4xy^2 + y^3$; $3xy^2 + 4x^2y + x^3$; $6xy + 4y^2$; $6xy + 4x^2$.

9. $x(x^2 + y^2)^{-1/2}$; $y^2(x^2 + y^2)^{-3/2}$. **11.** 0. **13.** 1758. **15.** $2e$.

Exercise 8-6

1. $\frac{\partial z}{\partial r} = 13$; $\frac{\partial z}{\partial s} = 9$. **3.** $\left[2t + \frac{3\sqrt{t}}{2}\right]e^{x+y}$.

5. $5(2xz^2 + yz) + 2(xz + z^2) - (2x^2z + xy + 2yz)$.

7. $3(x^2 + xy^2)^2(2x + y^2 + 2xy)$.

9. $-2s(2x + yz) + r(xz + 3y^2z^2) - 5(xy + 2y^3z)$.

11. $15s(2x - 7)$. **13.** 324. **15.** -1.

Exercise 8-7

1. $(\frac{14}{3}, -\frac{13}{3})$.

3. $(2, 5), (2, -6), (-1, 5), (-1, -6)$.

5. $(50, 150, 350)$.

7. $(-2, \frac{3}{2})$, rel. min.

9. $(-\frac{1}{4}, \frac{1}{2})$, rel. max.

11. $(1, 1)$, rel. min.; $(\frac{1}{2}, \frac{1}{4})$, neither.

13. $(0, 0)$, rel. max.; $(4, \frac{1}{2})$, rel. min.; $(0, \frac{1}{2})$, $(4, 0)$, neither.

15. $(122, 127)$, rel. max.

17. $(-1, -1)$, rel. min.

19. $l = 24$, $k = 14$.

21. $p_A = 80$, $p_B = 85$.

23. $q_A = 48$, $q_B = 40$, $p_A = 52$, $p_B = 44$, profit $= 3304$.

25. $q_A = 3$, $q_B = 2$.

27. 1 ft by 2 ft by 3 ft.

29. $(\frac{105}{37}, \frac{28}{37})$.

Exercise 8-8

1. $(2, -2)$.

3. $(3, \frac{3}{2}, -\frac{3}{2})$.

5. $(\frac{4}{3}, -\frac{4}{3}, -\frac{8}{3})$.

7. $(6, 3, 2)$.

9. $(\frac{2}{3}, \frac{4}{3}, -\frac{4}{3})$.

11. $(3, 3, 6)$

13. Plant 1, 40 units; Plant 2, 60 units.

15. 74 units (when $l = 8$, $k = 7$).

17. $x = 12, y = 8$. **19.** $x = 10, y = 20, z = 5$.

Exercise 8-9

1. $\hat{y} = .98 + .61x$; 3.12.

3. $\hat{y} = .057 + 1.67x$; 5.90.

5. $\hat{q} = 82.6 - .641p$.

7. $\hat{y} = 100 + .13x$; 105.2.

9. $\hat{y} = 8.5 + 2.5x$.

11. (a) $\hat{y} = 35.9 - 2.5x$.
(b) $\hat{y} = 28.4 - 2.5x$.

Exercise 8-10

1. 18. **3.** $\frac{1}{4}$. **5.** $\frac{2}{3}$. **7.** 3. **9.** 324.

11. $-\frac{58}{35}$. **13.** $\frac{8}{3}$. **15.** $-\frac{1}{3}$. **17.** $\dfrac{e^2}{2} - e + \frac{1}{2}$. **19.** $-\frac{27}{4}$. **21.** $\frac{1}{24}$.

23. $e^{-4} - e^{-2} - e^{-3} + e^{-1}$. **25.** $\frac{3}{8}$.

Review Problems—Chapter 8

1. $4x + 3y$; $3x + 2y$.

3. $\dfrac{y}{(x+y)^2}$; $-\dfrac{x}{(x+y)^2}$.

5. $2xze^{x^2yz}(1 + x^2yz)$.

7. $\dfrac{y}{x^2 + y^2}$.

9. $2(x + y)$.

11. $xze^{yz} \ln z$; $\dfrac{e^{yz}}{z} + ye^{yz} \ln z = e^{yz}\left(\dfrac{1}{z} + y \ln z\right)$.

13. $2(x + y)e^r + 2\left(\dfrac{x + 3y}{r + s}\right)$; $2\left(\dfrac{x + 3y}{r + s}\right)$.

15. $\dfrac{2x + 2y + z}{4z - x}$.

17. (2, 2), rel. min.

19. (3, 2, 1).

21. $\hat{p} = 85.15 - .43t$.

23. $\dfrac{\partial P}{\partial l} = 14l^{-.3}k^{.3}$; $\dfrac{\partial P}{\partial k} = 6l^{.7}k^{-.7}$.

25. .530; $-.027$.

27.

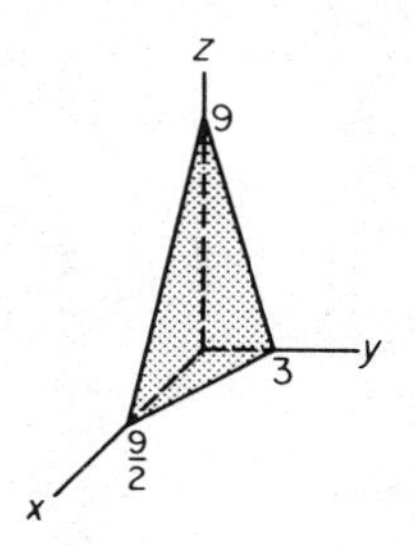

29.

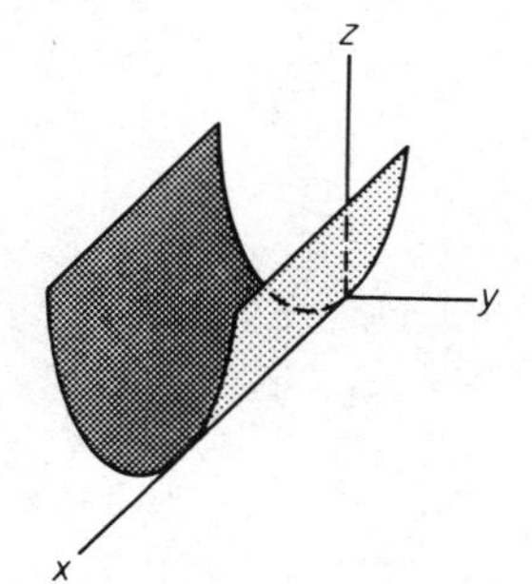

31. $\frac{7}{2}$. **33.** $\frac{81}{5}$. **35.** 4 ft by 4 ft by 2 ft.

Exercise A-1

1. -6. **3.** 2. **5.** 11. **7.** -2. **9.** -63.

11. -6. **13.** $6 - x$. **15.** $-12x + 12y$ (or $12y - 12x$).

17. $-\frac{1}{3}$. **19.** -2. **21.** 18. **23.** 25. **25.** $3x - 12$.

27. $-x + 2$. **29.** $\frac{8}{11}$. **31.** $-\dfrac{5x}{7y}$. **33.** $\dfrac{2}{3x}$. **35.** 3.

37. $\dfrac{7}{xy}$. **39.** $\frac{5}{6}$. **41.** $-\frac{1}{6}$. **43.** $\dfrac{x - y}{9}$. **45.** $\frac{1}{24}$.

47. $\dfrac{x}{6y}$. **49.** Not defined. **51.** Not defined.

Exercise A-2

1. 5. **3.** -2. **5.** $\frac{1}{2}$. **7.** 10. **9.** 8.

11. $\frac{1}{4}$. **13.** $\frac{1}{32}$. **15.** $4\sqrt{2}$. **17.** $x\sqrt[3]{2}$. **19.** $4x^2$.

21. $3z^2$. **23.** $\dfrac{9t^2}{4}$.

25. $\dfrac{x^3}{y^2z^2}$. **27.** $\dfrac{2}{x^4}$. **29.** $\dfrac{1}{9t^2}$.

31. $7^{1/3}s^{2/3}$. **33.** $x^{1/2} - y^{1/2}$. **35.** $\dfrac{x^{9/4}z^{3/4}}{y^{1/2}}$.

37. $\sqrt[5]{(8x - y)^4}$. **39.** $\dfrac{1}{\sqrt[5]{x^4}}$.

41. $\dfrac{2}{\sqrt[5]{x^2}} - \dfrac{1}{\sqrt[5]{4x^2}}$. **43.** $\dfrac{2x^6}{y^3}$. **45.** $t^{2/3}$.

47. $\dfrac{64y^6x^{1/2}}{x^2}$. **49.** xyz. **51.** $\frac{1}{9}$.

53. $\dfrac{4y^4}{x^2}$. **55.** $x^2y^{5/2}$. **57.** $\dfrac{y^{10}}{z^2}$.

59. $\dfrac{1}{x^4}$. **61.** $-\dfrac{4}{s^5}$. **63.** $\dfrac{4x^4z^4}{9y^4}$.

Exercise A-3

1. $x^2 + 9x + 20$. **3.** $x^2 + x - 6$. **5.** $10x^2 + 19x + 6$.

7. $x^2 + 6x + 9$.

9. $x^2 - 10x + 25$. **11.** $2y + 6\sqrt{2y} + 9$. **13.** $4s^2 - 1$.

15. $x^3 + 4x^2 - 3x - 12$. **17.** $2x^4 + 2x^3 - 5x^2 - 2x + 3$.

19. $5x^3 + 5x^2 + 6x$. **21.** $3x^2 + 2y^2 + 5xy + 2x - 8$.

23. $z - 4$. **25.** $3x^3 + 2x - \dfrac{1}{2x^2}$. **27.** $x + \dfrac{-1}{x+3}$.

29. $3x^2 - 8x + 17 + \dfrac{-37}{x+2}$. **31.** $t + 8 + \dfrac{64}{t-8}$.

33. $x - 2 + \dfrac{7}{3x+2}$.

Exercise A-4

1. $2(3x + 2)$. **3.** $5x(2y + z)$.

5. $4bc(2a^3 - 3ab^2d + b^3cd^2)$. **7.** $(x - 5)(x + 5)$.

9. $(p + 3)(p + 1)$. **11.** $(4x - 3)(4x + 3)$.

13. $(z + 4)(z + 2)$. **15.** $(x + 3)^2$.

17. $2(x + 4)(x + 2)$. **19.** $3(x - 1)(x + 1)$.

21. $(6y + 1)(y + 2)$. **23.** $2s(3s + 4)(2s - 1)$.

25. $x^{2/3}y(1 - 2xy)(1 + 2xy)$. **27.** $2x(x + 3)(x - 2)$.

29. $4(2x + 1)^2$. **31.** $2(x + 3)^2(x + 1)(x - 1)$.

33. $2(x + 4)(x + 1)$. **35.** $(x^2 + 4)(x + 2)(x - 2)$.

37. $(y^4 + 1)(y^2 + 1)(y + 1)(y - 1)$. **39.** $(x^2 + 2)(x + 1)(x - 1)$.

41. $x(x + 1)^2(x - 1)^2$. **43.** $x(xy - 5)^2$.

45. $(x - 2)^2(x + 2)$. **47.** $(y - 1)(y + 1)(y^4 + 4)^2$.

Exercise A-5

1. $x + 2$. **3.** $\dfrac{5}{3t}$. **5.** $-\dfrac{1}{p^2 - 1}$.

7. $\frac{2x^2 + 3x + 12}{(2x - 1)(x + 3)}$. **9.** $\frac{2x - 3}{(x - 2)(x + 1)(x - 1)}$.

11. $\frac{35 - 8x}{(x - 1)(x + 5)}$. **13.** $-\frac{y^2}{(y - 3)(y + 2)}$.

15. $\frac{3 - 2x}{3 + 2x}$. **17.** $\frac{x}{2}$. **19.** $\frac{2}{3}$. **21.** $\frac{2(x + 4)}{(x - 4)(x + 2)}$.

23. $-\frac{(2x + 3)(1 + x)}{x + 4}$. **25.** $\frac{x + 1}{3x}$.

27. $\frac{(x + 2)(6x - 1)}{2x^2(x + 3)}$. **29.** $\frac{x - \sqrt{5}}{x^2 - 5}$.

Exercise A-6

1. $\frac{15}{2}$. **3.** $\frac{12}{5}$. **5.** 2. **7.** 90. **9.** $-\frac{26}{9}$.

11. $-\frac{37}{18}$. **13.** 3. **15.** $\frac{1}{4}$. **17.** $\frac{3}{2}$. **19.** $\frac{5}{3}$. **21.** 3. **23.** $P = \frac{I}{rt}$.

25. $x = \frac{p + 1}{6}$. **27.** $r = \frac{S - P}{Pt}$. **29.** $-2, -1$. **31.** 4, 3.

33. ± 2. **35.** 0, 8. **37.** $\frac{1}{2}$. **39.** $0, \pm 8$.

41. $0, 1, -2$. **43.** $2, -1$. **45.** $0, 2, -3, 4$.

47. $0, \pm 1$. **49.** $-5, 3$. **51.** $\frac{3}{2}$.

53. $\frac{5 \pm \sqrt{13}}{2}$. **55.** No real roots. **57.** $\frac{1}{2}, -\frac{5}{3}$. **59.** $4, -\frac{5}{2}$.

61. $\frac{-2 \pm \sqrt{14}}{2}$.

Exercise A-7

1. $x > 2$.

3. $x \leqslant 2$.

5. $y > 0$.

7. $x \leqslant -\frac{1}{2}$.

9. $s < -\frac{2}{5}$.

11. $x \geqslant -\frac{7}{5}$.

13. $x > -\frac{2}{7}$.

15. No solution.

17. $x < \dfrac{\sqrt{3} - 2}{2}$.

19. $y \leqslant -5$.

21. $-\infty < x < \infty$.

23. $t > \frac{17}{9}$.

25. $x \geqslant -\frac{14}{3}$.

27. $r > 0$.

29. $x \leqslant -2$.

31. $y < 0$.

Index